Frontispiece.—A Petroleum Refinery (the Nobel Factories at Baku). From a photograph communicated by MM. André (Vol. i.).

TECHNOLOGY OF PETROLEUM

THE OIL FIELDS OF THE WORLD

THEIR HISTORY, GEOGRAPHY AND GEOLOGY
ANNUAL PRODUCTION
PROSPECTION AND DEVELOPMENT

OIL-WELL DRILLING

EXPLOSIVES AND THE USE OF THE "TORPEDO"

TRANSPORT OF PETROLEUM
BY SEA AND BY LAND

THE STORAGE OF PETROLEUM

TECHNICAL DATA—TABLES, FORMULÆ, PROBLEMS, ETC., ETC.

BY

HENRY NEUBURGER AND HENRI NOALHAT

WITH 153 ILLUSTRATIONS AND 26 PLATES

TRANSLATED FROM THE FRENCH
BY
JOHN GEDDES McINTOSH

LATE LECTURER ON GEOLOGY AND MINERALOGY, PEOPLE'S PALACE, LONDON, AND ON
ILLUMINATING AND LUBRICATING OILS, THE POLYTECHNIC, REGENT STREET,
AND THE BOROUGH POLYTECHNIC, ETC., ETC.

LONDON

SCOTT, GREENWOOD AND CO.

Publishers of "The Oil and Colourman's Journal"

19 LUDGATE HILL, E.C.

1901

[The sole right of translation of this work into English rests with the above firm]

TN 870
.N4713

THE ABERDEEN UNIVERSITY PRESS LIMITED.

𝕿𝖍𝖎𝖘 work is dedicated by the Authors

TO THE

MINISTER OF COMMERCE AND INDUSTRY

AND TO

THEIR COLLEAGUES OF
THE SOCIETY OF COMMERCIAL GEOGRAPHY

IN TESTIMONY OF SINCERE CONFRATERNITY
AND OF DEVOTION TO THE COMMON CAUSE.

HENRY NEUBURGER.
HENRI NOALHAT.

AUTHORS' PREFACE TO FRENCH EDITION.

The authors make no pretension of having accomplished an unimpeachable work. But they have got the conviction that they have written a useful book. This book is not intended solely for the man of science, but rather and more especially, for the business man, engaged in the industry.

It points out to him, at one and the same time, the end to be accomplished, the obstacles to be surmounted, the stumbling blocks to be avoided, on which pioneers have too often fallen; it tells him exactly and prosaically, in plain language, what are the risks to be run, and the illusions to be evaded.

In their career as engineers, in their adventurous life as " *oilmen*," the authors have been able to recognise the riches which the ignorance, or rather the indifference, of their compatriots has abandoned to foreign enterprise, even in such fertile and unexplored regions as those of our Colonial domains, which, nevertheless, have cost France such great sacrifices.

In their opinion, the Frenchman has not been to blame for this *inertia*. No one has ever spoken to him of the riches to which these lines allude, no one has ever indicated to him, in a methodical manner (with illustrative examples in support thereof), the means of discovering and profiting thereby. Even in the French scientific world, in default of French documents, the same ignorance prevailed.

A revival of our industrial activity would bring prosperity and fortune into regions abandoned to foreign initiative; it would bring back millions to the country.

Should the authors, by means of this book, contribute to this revival, and guide their fellow countrymen, along an unknown road, towards an industry of which, in default of French documents, they knew nothing, they will be highly recompensed for their trouble, and will be proud to have been the initiators thereof.

The authors specially thank their *confrères*, engineers, and others, engaged in the industry, who have been good enough, by furnishing valuable information and communicating graphic documents, to take an active part in the elaboration of this book, and thus have, by their disinterested concurrence, earned the title of *collaborateurs*.

The following have contributed in this way to this volume :—

1. The Society of Commercial Geography, Paris.
2. The Government of Algeria.
3. L'Union Coloniale Française.
4. M. Lippman, Paris.
5. M. André, Paris.
6. M. Zipperlen, Paris.
7. MM. Fenaille et Despeaux, Paris.
8. M. Desmarais, Paris.
9. MM. Nobel, Baku.
10. M. Franzl, Vienna.
11. M. Seeger, Ottynia.

The authors, in equal measure, acknowledge a debt of gratitude to their predecessors, who have smoothed the inherent difficulties of their task by their remarkable works, to which quotations here and there in this treatise render justice ; to MM. Boverton Redwood, Arrault, F. Hue, W. de Fonvielle, the tribute of their admiration and thanks.

N.B.—The authors will be grateful to their *confrères*, engineers, and others engaged in the industry, who will be good enough to forward to their publisher, J. Fritsch, 30 Rue Jacob, à Paris, any communications of a nature to contribute to the elaboration of their second volume of the *Technology of Petroleum: Industrial Treatment, Products and Bye-Products*.

An appendix to the first volume will also be published, :and all communications relating thereto will be equally acceptable.

The authors, in anticipation, express their thanks to .their future *collaborateurs*.

H. N. AND H. N.

TRANSLATOR'S PREFACE.

THE authors' competency to deal with the subject may be judged from the fact that, with the view of profiting by his experience in Galicia, etc., one of them, M. Neuburger, was recently commissioned by the French Government to survey and report upon the Algerian Oil Fields. In their preface, they give but a cursory allusion to the subjects dealt with in this treatise. For instance, prominent features of the work are the descriptions of the Physical Geography and Geology of the different Oil Fields. The technical education which the translator underwent, many years ago, during his attendance on the sessional courses of the Royal School of Mines, in order to qualify as a trained Government teacher of Geology and Mineralogy, together with his subsequent experience, both on the field, and as a lecturer on these subjects, in various London Polytechnics, has been of service to him, in his endeavours to render these descriptions into idiomatic and, at the same time, he hopes, scientific English.

Moreover, the whole subject is one with which, from the nature of his calling as an oil expert, and lecturer *inter alia* on the Technology of Oils and Fats in connection with the City and Guilds of London Institute, he is of necessity familiar and in sympathy.

He thinks it but right to mention these facts, in order that his readers may have some guarantee that such a highly technical work has not suffered to any very appreciable extent, in the translation. He does not, however, imagine, for one moment, that his translation is without flaws. He can, conscientiously, say that he has tried to minimise them.

It may be pointed out that the section on oil-well drilling is presented to the reader in a more logical sequence, and in a more systematic and rational manner, and with greater detail than has hitherto been done.

To aid the reader to grasp in his mind's eye the approximate quantities, measurements, and cost, in the great majority of instances, if not in all, the extensive numerical data are given in both English and French equivalents. But, as far as measurements are concerned, the intelligent engineer or prospector, who takes this book as a guide, may easily get over the difficulty by the use of a metre steel tape line, which may be had with both English and French equivalents.

The translator has taken the opportunity of correcting several typographical and other errors in the original French edition of the work ; many of these were due to the misplacement of the decimal point. We have reason to be thankful that our much deprecated system of vulgar fractions does not lead us to commit such palpable mistakes—mistakes which have rendered the inherent difficulties of translating such a work far more onerous than they had need be. Moreover, in the description of natural phenomena, etc., the Frenchman soars to the highest flights of imagination, and is difficult to follow in our more prosaic language. The translator hopes that he has not fallen from the sublime to the ridiculous in his efforts to retain some of the verve and vivacity of the original. Therein lies the main difficulty of French translation.

To relieve monotony and facilitate reference numerous headings have been given to the different paragraphs. The table of contents has been enlarged, and a compendious index added.

The innuendos, as to our ubiquity on the oil fields of the world, and as to our scenting rich oil fields in China, might well have given place to a more impartial account of Britain's share in the history and development of the petroleum trade, together with some information as to the amount of indigenous petroleum produced annually in

Britain. It certainly is small, but then the authors have chronicled much smaller deposits. The otherwise excellent account of the German oil fields is marred by political prejudice, which the translator is afraid has something to do with the authors having somewhat ignored Britain's *rôle* in the petroleum world.

Nevertheless, as the English translator of a French technical work remarked so far back as 1789 :—

"Whether our good neighbours the French be our natural enemies or not, certainly they are our most powerful rivals in commerce and manufactures. In this sense they are our enemies ; let us not, therefore, from pride or ill-humour, spurn their instructions—*fas est et ab hoste doceri.*"

As a matter of fact, however, we, in any case, can afford to be magnanimous. We have got a long way ahead of the French in most departments of industry, and with perhaps the single exception of light automobiles, in no department do they at all approach us except in industries of forced growth, *viz.*, those which are mouth-fed and hand-nursed by the unhealthy bounty system. Moreover, it is only but a year or two ago that the French Government had to become a foster-mother to the crude shale oil and petroleum refining industry. In view, therefore, of these facts and of the numerous wails of the authors at the general inertia of French industry, the translator can only regard Mr. T. P. O'Connor's optimistic report on the recent immense strides made by France as being biassed by a perfervid Celtic love of Gaul. It would certainly not be to the credit of French intelligence and their elaborate system of technical training if they, amongst all the nations of the world, were to remain stationary.

But, in order that we may hold our own with the French, in the matter of petroleum as well as in other articles of commerce, it may be useful to point out that several years ago a prolonged discussion took place on the question, "What to do with our boys," and, in this connection, the translator may be excused if he suggests that the professions of petroleum prospector and oil well drilling engineer or even that of oil well drilling contractor have not received from British parents and guardians the attention which, from their lucrative nature, they deserve. It goes without saying, however, that to succeed in these professions, all preliminary investigations, and subsequent developments, have to be conducted with skill, ability, system, discretion and energy, backed by the necessary capital and experience.

What preliminary training or technical education, therefore, should a young man undergo in order to qualify himself for a situation of this nature?

The translator would advise him to take the course for the degree of the Royal College of Science, in Mining, with simultaneous attendance on the evening course on oils and fats, including candle manufacture, at some one or other of the London Polytechnics, gaining practical experience in rock-boring, during his academical vacations, with some firm of well-sinking contractors or mining engineers, or on some prospecting works, such as those now in progress on the Kent Coal Fields. The distillation and refining processes, adopted in the case of crude petroleum oil, do not differ so very materially from those in use for crude shale oil. Practical experience, therefore, in the laboratories and works of one or more of the Scotch Shale Oil Companies would be of undoubted value, many of the important posts in the Baku and other refineries being filled by men trained in the works referred to. Similarly, experience in the crude petroleum oil refineries of this country, like that at Thames Haven, would be equally if not more valuable.

As far as Scotsmen are concerned, perhaps the best plan would be to attend Professor Mill's three years' course of technical chemistry in connection with the Young Chair, and also the mining and mine surveying classes of the Glasgow Technical College, together with the University geology course, following up the whole course of study by one or two years in the laboratories and works of the Shale Oil Companies, availing himself of every opportunity of instruction in well and shaft sinking, mine surveying, geological surveying, etc.

These are some of the methods by which young men of the present day might qualify themselves for posts of this nature. But the importance of a thoroughly systematic training, for such an important career, and the necesssity for opportunities to specialise, are so great that the translator recommends to the favourable consideration of the Technical Department of the Board of Education for England, and the Secretary for Scotland, the question whether or not some such systematic specialised course of training could be included in the mining curriculum of the Royal Colleges of Science of London and Dublin and the Technical Colleges of Edinburgh and Glasgow.

The frequent drilling of numerous deep wells in the neighbourhood of London and other large towns in connection with breweries, chemical works, etc., could be taken advantage of for practical instruction, as well as frequent visits paid to such prospecting work and trial borings as those now in operation in the Kent Coal Fields.

Be this as it may, it is not to the credit of the technical education system of the present day that entrance to this highly lucrative profession should be by circuitous byepaths. That the French engineers are highly trained even in the minutest details of their profession, this treatise affords abundant evidence. The translator has no doubt but that the Englishman or Scotsman is the most practical man of the two, but, to say the least, his calibre would not be deteriorated had he the scientific training of the Frenchman. The authors twit us with being to the fore wherever petroleum is to be found. Let us show them our right to be in the foreground, not only from our practical skill and ability, but by our scientific knowledge and systematic technical training. The headquarters of many of the European oil companies, including Russia, are in London. These should not have to seek their oil well drilling engineers abroad. The question is one the consideration of which might well commend itself to the City and Guilds of London Institute for the Advancement of

Technical Education, as well as the Technical Education Committee of the London County Council and the London Chamber of Commerce.

The more recent petroleum statistics will be found in the *Oil and Colourman's Journal Diary* for 1901, and, in the monthly issues of the journal in question, particulars of our imports of petroleum are also given, together with the petroleum news of the world in general, for instance, full particulars are given in recent issues of such recent developments as the prodigious Texas gushers which are bursting heavenwards from day to day, and whose enormous contingent goes to swell the world's already immense production.

THE TRANSLATOR.

LONDON, *May*, 1901.

CONTENTS.

LIST OF ILLUSTRATIONS xxxi
LIST OF PLATES xxxv

PART I.

STUDY OF THE PETROLIFEROUS STRATA.

CHAPTER I.

PETROLEUM.—DEFINITION.

Forms and uses—Appearance and properties—Composition, density, origin of petroleum—History--Petroleum as a liquid fuel and illuminating agent—The first
shale oil works—Price of land in Pennsylvania and Canada after discovery of
petroleum—European contrast—European oil fields—Short history of oil fields.
of France, Alsace, Hanover, Italy, Galicia, Roumania, Russia—Present condition of European oil fields—Their geological horizon and association—Main
petroliferous line of strike—The Mediterranean secondary line of strike—Map
of the oil fields of the world—Map of the European oil fields . . *pages* 1-9·

CHAPTER II.

THE GENESIS OR ORIGIN OF PETROLEUM.

Leopold de Buch's hypothesis—Turner and Reichenbach's theory—Rozet and
Millet's opinion. I.—Organic Origin of Petroleum: (1) Sterry-Hunt's hypo-
thesis ; (2) Lesley and Lesquereux's theory ; (3) Hitchcock's opinion ; (4) Ges-
ner's hypothesis ; (5) Canadian geological survey, Surveyor's hypothesis ; (6)
Credner compares formation to that of guano ; (7) Rupert Jones and Watson
Smith's opinion ; (8) Von Stromberg's hypothesis ; (9) Le Bel's theory ; (10)
Orton's classification of petroleum-producing agents ; (11) Engler's experiments
and conclusions ; (12) Phillip's hypothesis ; (18) Hofer's contentions. II.—
Formation of Petroleum by Volcanic Action : (1) Virlet d'Aoust's opinion ; (2)
Daubrée's review ; (3) Lapparent's views. III.—Formation of Petroleum by
Chemical Action : (1) Berthelot's hypothesis ; (2) Pelouze and Cahours' decision ; (8) Coquand's opinion ; (4) Benoit's conclusion ; (5) Cloez's experiments ;
(6) Mendéléef's hypothesis ; (7) Abich's conclusion ; (8) Ross's experiments and
deductions ; (9) Sokoloff and Grabowski's decision ; (10) Peckham's conclusion :
(11) Maquenne's experiments and deductions ; (12) Moissan's experiments and
deductions ; (18) Stanislas Meunier's decision. IV.—Authors' summary and
conclusion . . . . *pages* 10-23

CHAPTER III.

THE OIL FIELDS OF GALICIA, THEIR HISTORY.

Early struggles at Boryslaw—Final success—Rush and subsequent failure of shallow
wells — Discovery of Ozokerite saves the situation — Annual production of
Galicia—Inaccuracy of Austrian Government Statistics . *pages* 24-27

CHAPTER IV.

PHYSICAL GEOGRAPHY AND GEOLOGY OF THE GALICIAN OIL FIELDS.

Map of the Oil fields of Galicia—The Carpathians: Their mineral riches—The
general trend and nature of the petroliferous strata—Conformation and arrangement of strata—Highly inclined strata difficult to work but prolific in yield—
Signs of Petroleum—Depth to be bored before striking oil—Diagram to illustrate
variation in depth—Duration of yield of Galician oil wells—Geological systems
to which the Galician oil fields belong—The bearing of the geology of the
Galician oil fields on the origin of petroleum—The Eastern and Western
Galician oil fields—Comparative geology of both divisions—Geological section
of the West Galician oil fields (illustrated)—The exceptionally productive
petroliferous strata—Vertical section of strata encountered in boring No. 2
Well of the French Co. (La Société Française) at Holowecko, Galicia—The
Western Galician oil fields—The different centres—The amount and nature of
their products—View of Ropa oil field (illustrated)—Section of strata at Wietrzno
showing (in A) productive and (in B) barren wells—Krosno pumping station (illustrated—Bobrka: The first petroleum installation in Galicia—Facts which led
to boring at Bobrka—The Central Galician oil fields—The railway station of
Uherce Galicia (illustrated)—Section of the Carpathians between the San and
the Stryja (illustrated)—The French Company's works at Holowecko (illustrated)—Boryslaw—Ozokerit—Statistics of its annual production, showing its
fluctuations from year to year—An old works in the Boryslaw district (illustrated)
—Dangers run in working ozokerit—The Eastern Galician oil fields—Sloboda—
Producing centres of Stanislawow, Nadworna, Rungury, Sloboda Rungurska—
View of (illustrated)—Sloboda Kopalnia—M'Intosh the American and his famous
flowing well—Cal—Table showing variation in density of Sloboda petroleum—
Table giving the geological formation, density and colour of the crude oil from
different Galician producing centres .   .   .   .   .   .   pages 28-49

CHAPTER V.

PRACTICAL NOTES ON GALICIAN LAND LAW—ECONOMIC HINTS ON
WORKING, ETC.

I. NOTES ON GALICIAN LAW, ETC.—(a) Precautions to be taken in purchasing land ;
(b) Price of land ; (c) Leasing land ; (d) Cost of concessions ; (e) Legal formalities
to be observed in renting land ; (f) Restrictions on minors—Women's rights ;
(g) Formalities to be observed before starting work : (1) Application to Council
of Mines for permission to work ; (2) Registration of manager or overseer ; (3)
Registration by Prefect of District ; (h) Distance from other wells and from
habitations and fires generally—An oil well on fire at Mrzaniecz (illustrated).
II. GALICIAN LAWS AS TO MASTER AND SERVANT.—(a) Notice of dismissal and intention to leave. Workmen : Failure to give notice entails penalties or forfeits
on either side ; (b) Professional staff : Failure to give notice entails penalties or
forfeits on either side ; (c) Compensation when incapacitated by accident—
Master contributes to men's accident fund ; (d) State officials must be notified
of all accidents.
III. HINTS ON ECONOMICAL WORKING IN GALICIA.—(a) Prices : High qualit and
home consumption keep up price ; (b) Causes which tend to lower pric s ; (c)
Prices ruling in different localities ; (d) Cause of variation of prices : (1) Want
of communication, (2) Good or bad quality, (3) Influence of season on prices ;
(e) The selling of petroleum—Contracts : (1) Selling per fixed number of barrels ;
(2) Selling whole production for fixed period ; (f) Filling the barrels—Effects of
warm and frosty weather on barrels filled too full ; (g) Contents of American
and Galician barrels—Table of the properties of different Galician petroleums
with the geological strata in which they occur .   .   .   .   pages 50-57

CHAPTER VI.

ROUMANIA—HISTORY, GEOGRAPHY, GEOLOGY.

Geographical position of the oil fields—Lines of strike of the petroliferous strata—
Geological classification and association of petroliferous strata with common
salt—Map of the Roumanian oil fields—Why Roumania seems doomed to perpetual bankruptcy—Causes : (1) Geographical situation ; (2) Bad roads and want

of railway communication; (3) Protective tariffs (of other countries); (4) Land laws in regard to aliens—Futile attempts at working by natives—"Pockets" and their illusions—Division of Roumania into two oil fields: (1) The Western Roumanian oil field; (2) The Eastern Roumanian oil field—Producing centres —An old oil works in the Bacau district (*illustrated*)—Annual production— Roumanian mining laws . . . . *pages* 58-65

CHAPTER VII.

PETROLEUM IN RUSSIA—HISTORY.

Trend of the petroliferous strata—Continuation of Galician oil field—History of Russian petroleum—The fire-worshippers—Marcus Polo and Russian petroleum —Paul Duval on Baku petroleum and its geology—Kœchlin Schwartz on physical geology of Baku—Jonas Hanway and petroleum—Summary—Recent developments—Petroleum production of Baku, 1863-1897—Ludwig Nobel— Annual production of other Caucasian centres . . . *pages* 66-73

CHAPTER VIII.

RUSSIAN PETROLEUM.

THE CAUCASIAN OIL FIELDS—GEOGRAPHY AND GEOLOGY.

Map of the Caucasian oil fields (p. 76). I. The Crimea. II. The North Caucasian oil fields: (1) Tamansk, (2) Kouban, (3) Terek, (4) Grosnaïs, (5) Daghestan. III. The South Caucasian oil fields: (1) Anaclie, (2) Tiflis, (3) Elizabethpole, (4) Baku— Map of the Apshéron peninsula—Baku's magnificent natural harbour—Trans-Caucasian Railway—Fernand Hue's contrast between the geological structure of the American and Baku oil fields—Geological horizon of Caucasian petroliferous strata—Theory of intermittent action of Baku spouting wells—Mud volcanoes and underground phenomena—Oil level situated at very varying depths—The torpedo impracticable—Duration of yield of an oil well—Spouting wells or fountains cause loss and devastation—A Balakhany spouting well (*illustrated*)— Precautions against spouting—Spouting phenomena—Remedial measures—No. 3 spouting well of the Schibaeff Company—The Droojba spouting well—Other spouting wells—Table showing the depth, period of spouting, total production and density of oil of the spouting oil fountains of the Apshéron peninsula— Different deposits in Baku oil field—Sourakhany—Group of wells at Saboun-tchany (*illustrated*)—The temple of everlasting fire described—Group of wells at Balakhany (*illustrated*)—Geological section of strata encountered in boring No. 8 Schibaeff Well—Bibi Aïbad (Bibi Eibat)—Binagadine—Vertical section of Binagadine oil field—Present condition of Baku petroleum industry—Graphic representation of the petroliferous production of Baku and of the United States (1863-97)—Much ground still available—Recent free grant of concessions
pages 74-103

CHAPTER IX.

RUSSIAN PETROLEUM (*continued*).

PRACTICAL TRADE HINTS—RUSSIAN PETROLEUM LAWS: (1) Wages: Comparison with Galician wages; (2) Plant and raw material; (3) Contractors: Comparative cost of boring a well 300 metres deep in Galicia and in Baku; (4) Price of land; (5) Capital; (6) Schedule of expenses incidental to an oil well installation at Baku; (7) Royalty; (8) The State the proprietors of the greater part of the oil fields; (9) History of the early Russian petroleum monopoly; Its abolition; (10) Civil proprietorship and its effects; (11) Indiscriminate sale of land stopped; (12) Classification of land; (13) Exportation of crude oil—Depôts—Table showing the general properties of the different Caucasian petroleums . *pages* 104-110

CHAPTER X.

THE SECONDARY OIL FIELDS OF EUROPE—NORTHERN GERMANY, ALSACE, ITALY, ETC.

GERMANY.

The North German oil field—The Alsace oil fields (formerly French territory)— Geology of North German oil field—Remote history—The "fettlöcher"—Recent history—Oelheim—Properties of the North German oil—Principal deposits— Oelheim and the "petroleum swindle"—Total production—General properties of North German petroleum—The Alsace oil field—Pechelbronn asphaltum quarries—Petroleum bored for in asphaltum quarries—Petroleum oil works in Pechelbronn oil field (*illustrated*)—Walsbronn—Statistics—General properties of Alsatian petroleum . . . . *pages* 111-118

ITALY.

The ancients and petroleum—Modern history of Italian petroleum—The Abruzzi—
Geology—Properties of the oil—Classification of Italian oil fields—The Sicilian
oil field—The Southern Italian oil field—The Northern Italian oil field—French
oil works at Villeja—Photograph of (Zipperlen)—Statistics—Photograph of
petroleum oil works in Italy (Zipperlen) .   .   .   .   .   *pages* 118-123

MINOR EUROPEAN OIL FIELDS.

Zante—Hungary—Spain—Russia—Sweden—Britain—Holland   .   *pages* 123-125·

CHAPTER XI.

PETROLEUM IN FRANCE.

French capital diverted to Italian oil fields—Gabian oil—Comparative geology of
the oil fields of France, Galicia, Roumania, and the Caucasus—An oil
field classification or correlation impracticable in France—The Autun shale oil
industry—Gabian—History of the healing fountain of Gabian—Dr. Rivière's
account—Praised by the Bishop of Beziérs—Recent developments at Gabian—
Vertical section of strata met with in boring for petroleum at Gabian oil
fountain (1884)—Summary—Vertical section of strata encountered in second
boring for petroleum 900 metres north of first boring—Cause of failure—Diagrammatic section to illustrate reason of failure—Vaucluse—The Roque-Salière
bituminous shales—Limagne—Geology of the district—Bitumen—Small bitumen mines—Recent borings in Limagne—Le Credo (Ain)—Le Gua (Isere)—The
" burning " fountain—Geology of its neighbourhood—Its gas pressure—Analysis
of its gas and Krafft's deductions—Piret's attempts to capture and utilise it
gas—Châtillon (Upper Savoy)—Geological features of the district—Natural gas
utilised as fuel and illuminating agent—The " petroleum stone "   *pages* 126-146·

CHAPTER XII.

PETROLEUM IN ASIA.

TRANS-CASPIAN AND TURKESTAN TERRITORY.

Isle of Tchéléken—" Naphtonia, Naphtnia Gora "—Geological structure of the island
—Development retarded by situation—Tentative borings—Average yield—Properties of the oil—Mikhailowsk—Backward condition of the industry—A store
for future use—General properties of different Trans-Caspian petroleums
                                                            pages 147-148

TURKESTAN.

Fergane oil field—Altitude of wells—Small yield—Geological formation—Probable
existence of undiscovered springs   .   .   .   .   .   *pages* 148-150·

PERSIA.

Extent of oil field—Northern section—Central section—Southern section—General
properties of the different Persian petroleums .   .   .   .   *pages* 150-151

BRITISH INDIA.

Beluchistan—Punjaub—Assam —Early attempts at developing—Geological structure
—Properties of the oil—Number of borings and total yield   .   *pages* 151-153

BRITISH BURMAH.

Divisions of oil field—Coast district (Arakan Isles and mainland)—Interior or Upper
Burmah district—Coast district (Arakan Isles)—Geological structure and phenomena—Mud volcanoes—Native methods—The different Arakan oil fields—Ramri
—Cheduba—Flat—Barongah—Coast oil field—(mainland)—Production—Properties of the oil—The Irrawaddy or interior oil fields ("Rangoon Oil")—
Warren de la Rue's process—History—Twingoung oil field—Native methods of
excavating wells—Thickness of oil-bearing stratum—Yield—Present yield of
Twingoung oil field—Properties of the oil—Yenangyaung oil field—General properties of Yenangyaung petroleum—Number and depth of wells—Yield—Chaotic
structure of the ground—Geological horizon—Great variation in properties of
oil—Annual production—General properties of the different petroleums of the·
British possessions in Asia .   .   .   .   *pages* 153-160·

CHINA.

Chinese petroleum an unknown factor—Huart's memoir—Native hostility and superstition—Taï-li-Chen oil field—Yield—Boring tools—Production does not meet home consumption—The English scent rich oil fields—French concessions and treaties—Properties of Chinese petroleums . . . *pages 160-162.*

CHINESE THIBET.

Supposed extension towards the east of the Turkestan oil field—Natural gas—Tse-liou-tsu (fire wells)—Brine springs—Use of natural gas to evaporate brine—Se Tchouen oil field—Geological formation—The song of the earth—The song of the gas of a fire well . . . . *pages 162-163.*

JAPAN, FORMOSA AND SAGHALIEN.

Japan—Ancient knowledge of petroleum (discovered 7th century A.D.)—Producing centres—Production most minimum—Primitive methods of excavating wells—Description—Air machine—Oil raised in buckets—Progress impeded by want of capital and cost of drilling machinery—Kozodzu: the oldest wrought district—Best well—Summary—Yield—Property of oil—Schinano—Production—Properties of oil—District of Ugo—Present progress in Japan—Production, 1882 to 1897—The Formosa and Saghalien oil fields—Properties of Japanese petroleum *pages 163-168.*

CHAPTER XIII.

PETROLEUM IN OCEANIA.

SUMATRA, JAVA, BORNEO.

General remarks—Geological structure—Volcanic origin—Map of the Asiatic Archipelago . . . . *pages 166-170*

SUMATRA.

Langkat oil field—Geological structure and horizon—Development of oil field—Properties of Langkat petroleum—Palembang—Magnificent profits of English and Dutch Companies . . . . *pages 170-172.*

JAVA.

Geological structure—Division into two oil fields—South coast oil fields—Northern coast oil field—Mud volcanoes—Properties of Java petroleum . *pages 172-175*

BORNEO.

Working less prosperous formerly than in Sumatra—Now no reason to envy its neighbours—Geological structure and horizon—North coast oil fields (English)—The southern coast oil fields (Dutch) . . . . . *pages 175 176*

ISLE OF TIMOR.

Its petroliferous deposits—Geological horizon—Yield . . . *page 176.*

APPENDIX TO SUMATRA, JAVA AND BORNEO.

Rapid growth of Dutch-Australasian oil fields—Annual production—Exports—Anticipated future progress—General character of the petroleum of the Dutch-Australasian possessions . . . . . . *pages 176-177*

PHILIPPINE ISLES.

(1) Isle of Cébu: Surface wells; (2) Panay Archipelago—Inertia of French industry—European population—Monopolies . . . . *pages 177-179*

NEW ZEALAND.

Mud volcanoes—Geological structure—Sugar Loaves—Poverty Bay—Manutahi—General properties of New Zealand petroleum . . . . *pages 179-181*

CHAPTER XIV.

THE UNITED STATES OF AMERICA.

HISTORY.

The mound builders and petroleum—The Red Indians and petroleum—The early
French settlers and petroleum—Delaroche, Father Joseph, and the Marquis
de Montcalm—The trappers and petroleum—The salt-makers and petroleum—
Dr. Hildreth's first use of petroleum as an illuminant—Early theories as to
origin of petroleum—Crude petroleum, sale of, pushed as a medicine as
a medicine but succeeds as an illuminant—The shale oil industry paves the
way—The Pennsylvanian Rock Oil Company—Its vicissitudes—Bissel's introduction to petroleum—Early difficulties of the "Oil Kings"—Colonel Drake—
First use of iron piping to "case" the bore hole—Uncle Billy Smith—The first
oil well—The rush to the oil fields—Bissell and the "oil kings"—Colonel Drake
and speculation—Retires ruined—Drake pensioned by a grateful country—The
fate of American petroleum decided by Drake's success—The first spouting well
—The "Fountain": The "Oil Creek Humbug"—Other spouting wells—Mushroom towns—Pithole, its rise and fall—Cherry grove—Oil city—Clarion and
Bradford districts   .   .   .   .   *pages* 182-201

CHAPTER XV.

PHYSICAL GEOLOGY AND GEOGRAPHY OF THE UNITED STATES OIL FIELDS.

Map of the North American oil fields—The oil fields of the Eastern Basin (course of
the Ohio and Mississippi)—Northern and Central Basin (course of the Missouri and
Upper Mississippi)—Southern Basin (banks of the rivers of the Mexican Gulf)—
Western Basin (Rocky Mountains)—General trend and dip of the petroleum-
bearing strata—Geological formations—Silurian petroliferous strata—Devonian
petroliferous strata—Carboniferous petroliferous strata—Geological section of
Canadian and Pennsylvanian oil fields—Western oil field—Cretaceous and tertiary petroliferous strata—New York and Pennsylvanian oil fields—Subdivisions:
(1) Richburg district (New York State), (2) Bradford district (on the north-east),
(3) Middle Field district, (4) Lower Field district, (5) Washington district (on the
south-west)—Ideal section of New York and Pennsylvanian petroliferous strata
—Strata encountered in actual boring near Pittsburg—Description of individual
oil fields—(1) The oil fields of New York State—Richburg district—Properties of
the oil—Pennsylvania: (2) Bradford district, Geology: The MacKean sands
(Silurian); History of Bradford district and the Barnsdall spouting well; Annual
production. (3) The Middle Field district. Geology: The Warren group; The
Venango group; Warren County, oil producing centres; The history of Cherry
Grove; Forest County, localities. (4) The Lower Field district: Crawford
County; Venango County: Clarion County; Butler County; Beaver County.
(5) Washington district: The Alleghany Pool; Washington County; Green
County; Elysée Réclus on Pittsburg; Elk County—General remarks on
Pennsylvanian petroleum—Density—Pipe lines—Depth at which oil is struck—
Ratio of barren wells to productive wells (dry holes)—Annual production of
Pennsylvania from 1859 to 1895—Decline of Pennsylvanian production.
WESTERN VIRGINIA.—Recent rapid development—The great oil belt—Its geological
structure—Properties of Virginian petroleum—Total annual production from
1878 to 1894. OHIO.—Its important production—Resemblance of its petroleum
to that of Canada—Geological structure of the Ohio oil fields: (1) The Macksburg
oil field to the east; (2) The Lima oil field to the north-west—Progressive annual
production, 1875 to 1895—In 1895 superior to Pennsylvania by 1,313,791 barrels.
INDIANA.—Of recent development—Montpelier district rich in natural gas—
Terre Haute and its fluctuating production—Progressive nature of Indiana production, 1890 to 1895. KENTUCKY.—*Niles's Register*, 1829, and the American
oil well—One of the oldest wrought oil fields of the Union—Minimum production. TENNESSEE.—Its deep wells in Dickson district—Unremunerative yield of
both oil and gas. WISCONSIN.—"But little petroleum, much gas." MINNESOTA.
—Insignificant yield. DAKOTA. Negative results in spite of gas and petroleum
signs. ALABAMA AND LOUISIANA.—Alabama surface traces—Louisiana, a minute
quantity of viscous oil. WYOMING.—Rich in hope, but present production may
be neglected. UTAH.—Mineral wax of Salt Lake City—Little petroleum.
COLORADO.—The United Florence Rocky Mountain Co. and the Triumph.
MISSOURI. — Signs only. TEXAS. — Recent searches require many borings.
KANSAS.—Recent productive discoveries. CALIFORNIA, LOS ANGELES.—Sub-
marine oil wells.
Table showing the total annual production of the United States and the different
States of the Union, 1859 to 1895, both inclusive   .   .   .   *pages* 202-231

CHAPTER XVI.

CANADIAN AND OTHER NORTH AMERICAN OIL FIELDS.

Extent—History—Spouting wells--The famous legendary Shaw well—Shaw's sad fate—The seed-bag: its discovery—Yield of Canadian spouting wells—Gaspé oil field—Its geology—The Gaspi Upper Silurian sandstones and Devonian limestones—The "Petroleum Oil Trust"—Densities of different Gaspé crude oils—Sir W. Logan's opinion of Gaspe oil field, and Obalski's report on mineral riches. ONTARIO.—Enniskillen—Oil Spring—Petrolia, etc. WESTERN CANADA.—Alberta and British Columbia—Petroliferous strata—Total Canadian production, 1875- to 1894, both inclusive—Newfoundland, Nova Scotia, Alaska: its alleged lake of almost pure petroleum   .   .   .   .   *pages* 232-241

CHAPTER XVII.

ECONOMIC DATA OF WORK IN NORTH AMERICA.

UNITED STATES.

Wages—The rig builder—Oil well drilling in Russia and America—Comparative monthly wages bill—Comparative cost of oil well drilling in America, Baku and Galicia—Depth of wells—Yield and duration of flow—Minimum remunerative yield—Ratio of barren to productive wells—Land purchase.

CANADA.

Cost of rig—Cost of boring a well 800 metres deep—Mining laws—American export trade   .   .   .   .   *pages* 242-246

CHAPTER XVIII.

PETROLEUM IN THE WEST INDIES AND SOUTH AMERICA.

I. WEST INDIES.

(1) Cuba—Cuban asphaltum—Products from the destructive distillation of chapapote; (2) Barbadoes—Barbadoes asphaltum or manjak; (3) Trinidad—Configuration of the island—Products of the destructive distillation of Trinidad asphaltum.
pages 247-250

II. SOUTH AMERICA.

(1) Columbia; (2) Venezuela; (3) Ecuador; (4) Peru—Extent and boundaries of the Peruvian oil fields - Réclus' delineation of the Peruvian oil fields—Their history according to W. de Fonvielle—The Incas and petroleum—The subdivision of the oil fields: (1) Tumbez oil field—History—The Faustino Piaggio Company—Its refinery at Zorritos—Properties of the oil; (2) The Paita oil fields—Négritos—"The London and Pacific Petroleum Company," "La Mina Bréa y Pariñas"—H. W. Tweddle's concession sold to the "London and Pacific"—Distillation products of London and Pacific crude petroleum; (3) Puerto Grau oil fields—Compagnie Française des Pétroles de l'Amérique du sud —Extent of their concession—Compressed air as a motive power—Their tank steamer -State of the wells—Production—Richness of the Puerto Grau oil field— (4) Piura oil field—Total production of Peru—Price of Peruvian crude petroleum —Brazil, Bolivia, Chili, and the Argentine Republic   .   .   *pages* 250-258

CHAPTER XIX.

PETROLEUM IN THE FRENCH COLONIES.

ALGERIA.

The Black Spring—Its mythological history—Algerian pirates come to it for tar— The Spaniards and the Black Spring—Geological horizon of the Algerian oil fields—The depth at which oil is struck—The Oran oil field—Map—(1) Deposits on first line of strike: Port-aux-Poules—Sidi-Brahim—Properties of Sidi-Brahim petroleum—Beni Zenthis—Ain Zeft—Geological section in north to south direction of Aïn Zeft oil field—Properties of Aïn Zeft petroleum—Viscous charged with paraffin wax—Distillation products—Tarria—Evrard's report—(2) Deposits on second line of strike—Boring near Relizane in the Oran oil field, Algeria (*illustrated*)—Departments of Alger and Constantine—Future hopes in regard to Algerian petroleum production .   .   .   .   .   .   *pages* 259-272

TUNIS.

Difficulty of obtaining authentic information—Researches too recent to even discount their results   .   .   .   .   *page* 273*

MADAGASCAR.

Explorers' accounts—Gautier's and Grosclaude's accounts—Mining laws—Taxes—
Annual rent charges .   .   .   .   *pages* 273-275

NEW CALEDONIA.

Surface oil of Island of Wagap—Its properties .   .   .   .   .   *page* 275

APPENDIX TO PART I.

GEOLOGICAL CLASSIFICATION OF THE DIFFERENT OIL FIELDS.

, Table showing the geological classification of the different oil fields of the world
with the characteristic fossils of the petroliferous strata .   .   *pages* 276-279

PART II.

EXCAVATIONS.

CHAPTER XX.

HAND EXCAVATION OR HAND DIGGING OF OIL WELLS.

Hand digging—The first European method of oil well sinking—The slowest and most
costly method—Plant and tools—Method of procedure—Section of an oil well excavated to the point where surface infiltration ceases (*illustrated*)—Oil well and
surface water outflow tunnel or drain (*illustrated*)—Aëration of excavated oil
wells—Section and plan of an oil well being excavated by hand (*illustrated*)—
Air propeller—Lowering a man into an insufficiently aërated well for urgent
purposes—Precautions to adopt—Remedy for asphyxiation—Signalling and
alarum bell—Davy lamp explosions—Smithy tools—Carpentry and joinery
work—Arrangement of battens intended to consolidate the wooden casing (*illustrated*)—Use of dynamite—Staff—Allocation of work—Spouting wells rare owing
to wide diameter of bore hole—Seed bag to be used to cap the well—More substantial cap—Estimate for plant, etc., required in the sinking of a well 150
metres (492 feet) by manual labour: (1) Cost of construction of the shed covering
the well ; (2) Cost of tools and plant for sinking the well ; (3) Cost of building
and furnishing the smithy—Freight of building materials and plant variable
and not included—Wages—Progress and cost per running foot of well excavated
—Daywork *versus* piecework   .   .   .   .   *pages* 280-291

PART III.

METHODS OF BORING.

CHAPTER XXI.

METHODS OF OIL-WELL DRILLING OR BORING.

General ideas and history of drilling—Differentiation from coal and ozokerit mining
—Classification of different systems of boring—Shock systems of boring: (1)
Boring with the rope by hand ; (2) Free-fall boring by hand and by steam power;
(3) Canadian system (steam power) ; (4) Combined system (steam power) ; (5)
American system (steam power) ; (6) Hydraulic system with the bit (hand and
steam power). Rotation systems of boring: (1) Auger system ; (2) Diamond
system—Boring or drilling involves greater initial outlay, but is the most rapid
and finally the cheapest and most efficient system of oil-well sinking—The false
economy of excavation—History—Boring an ancient process—The Jesuits and
the Chinese system of boring—Introduction of Canadian methods into Europe
by MacGarwey—Enhanced results obtained by Canadian system   *pages* 292-295

CHAPTER XXII.

BORING OIL WELLS WITH THE ROPE.

History—Impracticable in European strata—Advantages and disadvantages—The drilling tackle—Bit, description of—Method of drilling—Diagram to illustrate first method of rope boring—Sand-pumping—Sand-pump (*illustrated*)—Examination of the detritus—Advantages of filling bore hole partially with water—Land-slips—Lining well with timber—Colonel Drake's discovery—Casing the well—Casing with cast-iron pipes—Lowering the casing—Snapping the fastening —Plant easily procured and simple — System impracticable in majority of cases and uneconomical—Improved rope-boring systems—Jobard's system (*illustrated*)—Description of his drills—Goullet-Collet system (*illustrated*)—Staff and progress—Cost per foot—Lippmann on rope boring—Chronological nomenclature of the different systems of rope boring   .   .   .   .   *pages* 296-308

CHAPTER XXIII.

DRILLING WITH RIGID RODS AND A FREE-FALL. FABIAN SYSTEM.

Advantages and disadvantages—First introduction—Difference between free fall and rope system—Greater initial outlay and staff—The work of the foreman borer—Walking beam independent of the derrick (*illustrated*)—Sand pumping—Derricks—Derivation of the word—Precautions to be observed in constructing —Minimum height—Long section rods save time in unscrewing—Types of derricks—Tripod derricks used for sand pumping purposes at Galicia (*illustrated*) —Four legged derrick (*illustrated*)—Cost of building such a derrick (detailed estimate)—American derrick (*illustrated*)—Description—Cost of erection with detailed estimate—Canadian derrick (*illustrated*)—Advantages—Description—Cost of erection—Detailed estimate—Special kind of derrick—Instrumentation derrick (*illustrated*)—Transportable derricks—Installations (rigs) without derricks (*illustrated*)—Boring machines—The walking beam or " bascule "—Description and function—How wrought—Walking beam fixed to woodwork of derrick (*illustrated*)—Excavation in which foreman borer works to escape blows from " walking beam " (*illustrated*)—Point of support—Counterpoise—Duplex " walking beam " —Rods, temper screw, rotating handle, rope or rod turner (*illustrated*)—Rods (*illustrated*)—Lengthening the drilling tackle or system of rods—Function of retention key (*illustrated*)—Temper screw (*illustrated*)—Function of, description—Working—Another method of lengthening rods—Disadvantages—Working bar—Elongation winch (*illustrated*)—Free-fall instruments—Various systems—Fabian free-fall system—Merits and function—Description—Upper part—Lower part (*illustrated*)—Section of Fabian's instrument at the shoulder (*illustrated*)—Fabian's instrument modified by Fauck (*illustrated*)—Clutching—Moment to unclutch—Plan of the unclutching of Fabian's instrument—Drill and auger stem—Description and function—Auger stem guide (*illustrated*)—Its function—The drill—Transformations and modifications (*illustrated*)—Present shape (*illustrated*)—Free-fall drill, etc. (*illustrated*)—Function of small blades—Length of neck—Necessity for long neck (*illustrated*)—Screwed or collared connections—Tempering of the blades—Machines for raising rods—Windlasses—Ladder wheel (tread mill) (*illustrated*)—Hoisting winch (*illustrated*)—Safety hook (*illustrated*) —Mechanical advantage of ladder wheel—Faults of ladder wheel—Sand-pumps —Function—Description—Sand-pump cable—Sand-pump windlass—American piston sand-pump (*illustrated*)—Sand-pumping by drilling tackle (rods)—Casing the wells—Object—Prevention of landslips—Use of water in preventing landslips —Method of casing the wells—Thickness of tubes—Rolling and riveting (*illustrated*)—Holes or eyes in pipes for raising—Lowering the casing—Precautions—Lowering by sand-pump cable or by the rods—Weight of column of casing to be used—Casing a last resource—How casing pipes are riveted and joined together (*illustrated*)—Provisional point—Shape of lower end of casing (*illustrated*)—Evil effect of too short casing—Method of suspending casing when being lowered (*illustrated*)—Disadvantage of lost tubes or " tubes perdus "—Complete casing and partial casing (*en colonne perdue, illustrated*)—Lippmann's opinion on both systems—Surface well—Casing surface well (*illustrated*).

FABIAN SYSTEM—HAND POWER—PRACTICAL HINTS AS TO ITS EMPLOYMENT.

Staff—Rests or spells from working—Signs of necessity of lengthening rods—Signs of necessity for sand-pumping—Duration of and intervals between the various operations (*a*) in drilling ; (*b*) In sand-pumping ; (*c*) In raising the rods ; (*d*) Lowering of the rods—Scrutinising the drill—Scrutinising the debris and drill simultaneously—Foreman borer drilling, free-fall system (*illustrated*)—Surface

installation of boring on Fabian's hand-power system (*illustrated*)—Detailed estimates and inventory for construction of derrick—Cost of equipment for sinking well and the cost of equipment of smithy—Wages—Total cost of boring a well 300 metres or 984 feet—Cost per running foot—Method of commencing to drill without a surface well (*illustrated*)—Provisional guiding casing
pages 304-338

CHAPTER XXIV.

FREE-FALL DRILLING BY STEAM POWER.

Steam engines—Uns iitability of locomobiles—Their use ruinous—Their uses interdicted both at Baku and Galicia—The engine used in drilling and its parts : (1) the boiler ; (2) the engine —Kind of engine required— Wood used as fuel —Consequently furnace large—Crude oil used as fuel—Calorific intensity of crude petroleum oil —Selection of specially designed drilling engine—Various uses which it has to serve—Steam engine manufacturing countries—Merits and defects of engines from different countries--Engines of French construction—Suitable French boiler—Pumping engines - Engines of American construction—Engines of British construction—Engines of German and Austrian construction—Steam-power drilling machines, windlasses and hoisting windlasses—Old steam power surface installation (elevation and plan)--Details of rod windlass in first steam power drilling plant (*illustrated*)—The " walking beam "—Driving rod—Play—Hoisting Crane—Combination of " walking beam " and hoisting winch—Plassy boring made under Kind's directions (*illustrated*)—Diagram of first rig or machine combining in a single apparatus all the necessary parts of both drilling and hoisting machinery—" Walking beam " may if necessary be put out of gear—Horse power—Objection—Drills effectively—Fauck's oil well drilling machine (*illustrated*)—Fauck the great improver of the system—Number of his machines or rigs at work—Description—Modifications of " walking beam " —Substitute for temper screw—The bouncing post (*illustrated*)—Practical ideas on the working of Fauck's steam power system—The workmen's posts—Technical data—Time occupied and cost of drilling a well 300 metres—Surface installation —Fauck's system—Elevated " walking beam " (*illustrated*)—Fauck's system of boring—Economical data—Detailed estimate and inventory for construction of derrick and smithy and boring equipment of well, etc.--The staff and their wages—Cost of repairs and renewals—Cost of housing of workmen
pages 339-352

CHAPTER XXV.

OIL-WELL DRILLING BY THE CANADIAN WOODEN ROD OR POLE SYSTEM.

Its peculiarities—The lightness of all its parts—Its rapidity—A revelation to contractors and owners of wells—Engine—Its horse power—The rig—Wood the dominating material of its construction—Wooden " walking beam " (described and *illustrated*)—Elongation windlass (described and *illustrated*)—Adjustment of auger band wheel and bull wheel—Diagram to illustrate—Transmission of power for hoisting—Oscillating cylinder (*illustrated*)—Its function—System of rods and drill—Top point of the Canadian " walking beam " (*illustrated*)—Canadian rig (*illustrated*)—View of oil field wrought on Canadian system—Rods lose almost all their weight in water—Weight of wooden rods per square centimetre of section in air and water—The " jars " (described and *illustrated*)—Set of Canadian drilling tools (described and figured)—The auger stem, the bit, the sand-pump—Practical working hints—Foreman borer guiding the work (*illustrated*)—Interior view of boring installation (*illustrated*)—Manipulating the rods—Rôle of the workmen (*illustrated* and described)—Cost, estimate and inventory for construction of derrick and equipment of well—Time occupied in and cost of drilling a well of 300 metres (984 feet)—Cost per running foot
pages 353-863

CHAPTER XXVI.

DRILLING OIL WELLS ON THE COMBINED SYSTEM.

Definition of the system—A combination of the Canadian and Fabian—Adopted for pseudo-economical reasons—Objections to the system : (1) Deficient unclutching ; (2) Reduced speed ; (3) Less effective blow ; (4) Bending of rods - Speed, the main advantage of the Canadian system, lost—Fauck's light steam power rig (described and *illustrated*)—Practical hints– Observations made at Ropa—Time occupied in and cost of drilling a well 300 metres (984 feet) deep—Cost per running foot
pages 864-867

CHAPTER XXVII.

COMPARISON BETWEEN THE COMBINED FAUCK SYSTEM AND THE CANADIAN.

Lippmann's objections to the Canadian system—Legendary origin of wooden drilling rods or poles—Modifications suffered by wood under heavy water pressure—Shrinkage of wood from iron joints loosens the drilling tackle—Van Dijk's wood-enveloped iron rods—Fauck's objections to the combined system—Jurski's points in favour of Canadian system: (1) Speed ; (2) Repairs less frequent and rods raised more easily than in other systems ; (3) Wide drills may be used—Léon Syroczinski cites Wiertzno's results—Canadian system emboldens contractors to undertake large contracts to be executed within stated time—Authors' remarks on the discussion—Arguments in favour of Canadian system— Testimony quoted from Galicia, Italy, Germany and the United States . *pages* 368-373

CHAPTER XXVIII.

THE AMERICAN SYSTEM OF DRILLING WITH THE ROPE.

The principle of the system—Liability to deviation—Only practicable in easy ground—The rope and drilling tools—Rope socket (described and *illustrated*)—The bit—The auger stem —The jars—Temper screw—Combined rope and free-fall systems—American bits, narrow section (*illustrated*)—Surface arrangement of American drilling machinery ((*illustrated*) —American rig (photo of)—"Walking beam"—How wrought—The band wheel—The bull wheel—The sand reel—Practical hints on the American system—Staff—Slow progress in hard ground—Impracticable with wide heavy drills—Galician rope-bored wells—Data in regard thereto—System impossible in bad ground—Economical data—Detailed cost estimate of derrick and buildings, engines and windlasses—Rope drilling tools and other expenses—Cost and speed of drilling . . . *pages* 374-378

CHAPTER XXIX.

HYDRAULIC BORING WITH THE DRILL BY HAND AND STEAM POWER.

Principle —Simultaneous drilling and hydraulic sand-pumping—Method—The inventor of the system, the famous Perpignan boring—Degoussée's criticism—Lippmann supports Degoussée—Bits on the Fauvelle system (*illustrated*)—Rod and auger stem on the Fauvelle system (*illustrated*)—Gas piping tried and failed—Fauck's arrangement for hydraulic drilling (*illustrated*)—Portable installation (*illustrated*)—Special pipes—Dimensions of the hollow rods used in drilling—Fauvelle's system —Hydraulic sand-pumping equipment—By natural fall of water—By pressure pumps—Sand-pumping by reversed current of water —Arrangement of a boring on Fauvelle's hand system (*illustrated*)—Fauck's combined oil-well drilling machine, fitted up for hydraulic boring (*illustrated*)—Speed of current varies with nature of disintegrated matter brought up—Table of speeds with different detritus—Practical hints on the Fauvelle system: (1) Hand system : handy in trial borings ; its use in the colonies in seeking for water; hand power applications to petroleum boring ; "walking beam"; the foreman borer's post ; (2) Steam power ; staff; division of labour; progress; method of adjusting the water conduits (*illustrated*)—Fauvelle drill for soft, diamond drill for hard ground—Hydraulic oil well drilling machine, Fauck's (*illustrated*)—Average progress—Steam power —Time occupied and cost of sinking a well 300 metres (984 feet)—Cost per running foot—Detailed estimate . . *pages* 379-389

CHAPTER XXX.

ROTARY DRILLING OF OIL WELLS—BITS, STEEL-CROWNED TOOLS, DIAMOND TOOLS—HAND POWER AND STEAM POWER—HYDRAULIC SAND-PUMPING.

Principle of rotary drilling—Advantages of this system on hard ground—Extensively adopted in America but not in Europe—Used in trial borings to ascertain structure and nature of rocks—Deductions from sand-pump *débris* unsafe—The core or circular section of ground cut out and brought to surface intact by diamond drill—General principles of rotary boring—The process is continuous—Different kinds of augers (*illustrated*)—The rods and the

crown—System primarily used in soft beds—Its metamorphoses—Corkscrew
motion of auger—Equalising the sides of bore hole—Reamer (*illustrated*)—
Cylinder with conical auger (*illustrated*) - The crown—Crown tools steel teeth
(*illustrated*)—Fixed crowns (*illustrated*)—Its metamorphoses—The diamond replaces steel—Diamond crown (*illustrated*)—Artificial substitutes for diamond :
iridium crystals, pure alumina—The "carottier" or core barrel—Neuburger's
improvements on rods—Knuil (swivel)—Rotary oil-well drilling machinery—
Motive power—Circulation of the water—Casing of rotatory-drilled oil wells
(*illustrated*)—Eccentric crown diagram, to illustrate—Circular drilling with the
diamond—Economical data—Working staff—Jerking—Raising and emptying
the carottier or core barrel—Examination of crown—Speed obtained—Continental
Diamond Rock-Boring Company's results—Average and maximum speeds (progress)—The Villefranche d'Allier boring—Duration of the crown

                                                     pages 390-401

CHAPTER XXXI.

IMPROVEMENTS IN AND DIFFERENT SYSTEMS OF DRILLING OIL WELLS.

Peculiar features in the boring systems of different countries—Peculiarities of
French boring machinery — Complicated, costly, highly adjusted — Simple
Colonial machinery—American machinery—Characteristic detail—Comparative
cost between boring by French and Austrian machinery—Britain—Diamond
boring a feature—Germany—Soundness and simplicity—Austria—The scene of
the struggle between the free-fall and the Canadian system—Austrians constructing on American models—Principle of Austrian machines—Rigid rod system—
Different models of installation—Artesian wells, their chronology — Mulot's
successors and their machinery—Degoussée's machinery—Laurent's improvements on Degoussée's system — Lave's simultaneous sand-pumping and
drilling plant—Boring installation on Lippmann's hand system (*illustrated*)—
Machines on Mulot's and free-fall principle—Machines on Degoussée's
system—Modified Laurent system adopted in the Caspian—Modified American
system at Baku—Inside view of derrick at Baku showing boring equipment
—Different models of rods—Different models of free-fall instruments—Fabian's
the model and typical instrument—Origin of the free-fall—Kind's first
free-fall instrument (*illustrated*)—Boring installation at Baku (*illustrated*)—
Manner in which it acts—Acts automatically—The free-fall systems of Kind,
Rost, Fabian, Mulot, Gault, Werner, Wlach, Zobel, Seckendorff, Degoussée,
Esche, Greifenhagen, Ermeling, Wilke, Maldener—Merits and defects—Léon
Dru's water pressure free-fall instrument (*illustrated*)—Description, function and
working of Dru's instrument—Laurent's free-fall bayonet system (*illustrated*)—
Free-fall instruments at the Paris Exhibition—Lippmann on free-fall instruments
—Principle of Kind's instrument—Kind's idea improved by Dru—Dehulster's
application of Kind's idea—Van Dijk's free-fall instrument—Sigmondi's free-fall
instrument—François' free-fall instrument—Lippmann's free-fall instrument—
Fauck's free-fall instruments (bayonet) (*illustrated*)—Different models of "jars"
—Origin—Invented in 1884 by Oenhausen—Rope drilling—Various modifications
—Round ropes *versus* flat ropes—Rope rods—Rope substitutes—Free-fall applicable to boring with the rope—First attempt by Gaiski—Other inventions on
the same principle—Sontag, Straka, Sparre, Noth, Sisperle, Benda—Drilling by
circulation of water—Different modifications—Neuburger's improvements—
Hollow augur stem and bit (*illustrated*)—Free-fall in hydraulic boring introduced by Köbrich—Various other attempts—Fauck's method—More simple
apparatus—Total substitution of sheath for hollow rods (*illustrated*)—Fauck's
hydraulic "jars" (*illustrated*)—Hydraulic boring by hand—Prolongation of crane
substituted for derrick—Diagrammatic representation of Fauck's hydraulic
system—Crank motion—Manner in which it works—Raising the rods—Speed of
boring—Cost—The Raky rapid system of hydraulic boring—Number of blows of
Raky drill per minute—Rapidity of boring—Rotary boring—Different modifications—Augers—Diamonds—Borts, etc.—Artificial diamonds—The combined
diamond and Fauck system of boring (*illustrated*)—Borts, artificial diamonds—
Carbons—Carborundum—Combined systems—The free-fall system with the
auger system—The Fauvelle system (hollow rods) with the free-fall system and
circulation of water—The Fauvelle system with diamond drilling—The diamond
system, combined with the Fauck system, using hollow rods for both—The record
speed in drilling obtained by the latter   .   .   .   .   .   *pages* 402-425

PART IV.

ACCIDENTS.

CHAPTER XXXII.

BORING ACCIDENTS—METHODS OF PREVENTING THEM—METHODS OF REMEDYING THEM.

Risk involved in boring—Different causes of accidents—Deviation of drill—Landslips —Fractures of rods and drill and deformation of casing—Fall of foreign objects into bore hole—Flooding—Unforeseen spouting : (1) Deviation of drill—Partial and total deviation—Partial deviation—Causes—Hard highly-inclined strata—How partial deviation may be avoided—Total deviation—Remedies ; (2) Landslips : General prevalence—The cause of various accidents—Precautions, use of water —Strong casing pipes—Danger of " lost " pipes—Forewarnings—Measures to adopt—Landslips may render daily progress *nil* or negative—Exasperation aggravates disaster—Necessity for prompt action—Wrenching—Use of temper screw in freeing the drill, a safe operation—Wrenching proper—Tools (*illustrated*) —Raising an embedded sand pump—Precaution—(3) Damaged rods : 1. Deformation and fracture—" Pear " pipe expander (*illustrated*)—Tap grab (*illustrated*)— Tools for cutting casing pipes—Fauck's (*illustrated*)—Dru's (*illustrated*) ; 2. Fracture—Broken drills—Fractured rods—Unscrewed tackle—Unscrewed drill— Unscrewed rods—(4) Fall of foreign objects into the bore hole—The " fishing " tools numerous and varied—Principal types—Pike-mouthed pincers and tapped piece (*illustrated*) – " Ram's horns " (*illustrated*)—(5) Flooding—Preventive measures— Hermetically sealed casing—Toothed spring and spring pincers—Dru's expanding drill (*illustrated*)—Dimensions of hermetically sealed pipes and cost per running foot—Ratio of diameter of bore hole to that of casing—Expanding drills— Fauck's and Dru's (*illustrated*)—Eccentric drill—Its use now casual—Use of freezing machines to prevent flooding—(6) Unforeseen spouting—Regarded as accident—A spouting well (*illustrated*)—Previous signs and subsequent manifestations—Intervals between signs and actual spouting—The damage done by the Potok Wielki spouting well of Galicia—Precautions and procedure to stop unforeseen spouting—Capping the well—Method of procedure—Bredt's patent oil-well cap (*illustrated*)—Telescopic method of capping .   .   *pages* 426-443

CHAPTER XXXIII.

EXPLOSIVES AND THE USE OF THE "TORPEDO"—LEVIGATION.

I. The " torpedo "—When used—Causes which tend to stop the yield of a well— Chemical properties of dynamite—Nitro-glycerine unsafe and its use discontinued in Europe—Same effects produced by more safe dynamite—Varieties of dynamite—Dynamite with an inert base—Dynamite with an active base— Properties of good quality dynamite—Appearance—Behaviour on ignition— Comparative sensibility to shock of frozen dynamite and soft dynamite—Dynamite cartridges—Blasting gelatine—Properties of blasting gelatine—Inferiority of dynamite to nitro-glycerine greater than that deduced from ratio of nitroglycerine in the dynamite to that of the inert body therein—Blasting gelatine— Theoretical explosive force—Pictet's fulgurite—Properties of an ideal explosive —(1) Safety in manufacture ; (2) Bear handling in transit; (3) Unaffected by weather ; (4) Non-volatile and non-pulverent on storage ; (5) Easily made from abundantly occurring raw materials—Comparative explosive properties of Pictet's fulgurites and the principal explosives in use—Other explosives—Ammonium nitrate—Explosive force, volume and weight of gases from different explosives —Firing by electricity—Firing outfit—The principal conductor—The return wire—Wire must never be " run to earth "—Insulators—Priming—Fulminating capsule—Method of using explosives in petroleum borings—Precautions—Operation done by contract—The " torpedo " monopoly in America—Colonel Roberts' patent—First use—The phenomenal results at Thorn Creek—" Torpedoing " an oil well in America before delegation of the French Society of Civil Engineers— The president's description of the process; II. Levigation .   *pages* 444-457

CHAPTER XXXIV.

STORING AND TRANSPORT OF PETROLEUM.

Storing oil from spouting or flowing wells—Oil from non-flowing wells: pumping-wells, draw-wells—Pumping-wells—Capacity of pumps—Necessity for thorough and exhaustive pumping—Beneficial effects of extreme pumping—Balakhany oil-pumps (*illustrated*)—Hand pumps—Steam pumps, their dimensions and output —Wooden, metallic and natural reservoirs—Natural reservoirs—Wooden reservoirs—Wrought-iron reservoirs—Construction—Site—Transport—Bulk transport—Barrels—Filling barrels at Baku (*illustrated*)—The American barrel—The English barrel—The Galician barrel—Tank waggons—Lepage's system—French tank waggon, State railway model sections to scale (*illustrated*)—American tank cars—Tank-charging stations—Railway Station, U.S.A. oil field (*illustrated*) —Tank steamers—Tank steamer moored at Baku (*illustrated*)—Nobel Brothers' innovations—The end in view—The method adopted—Tank barges—Conveyance in bulk from Baku to St. Petersburg—*Per mare et per terram*—Effect: Wide-spread use of Russian petroleum—Petroleum reduced in price Illustration—Initial difficulties—Railway and steamship opposition—Finally determine to carry out proposals personally—Ultimate universal adoption—Railways add tank waggons to rolling stock—Present strength of Nobel's tank waggons, rolling stock and fleet of tank steamers and lighters—El Gallo tank steamer *photograph of*—Routes—Caspian Sea and Volga route—The Siberian route—German and Scandinavian route—Trans-Caspian route *via* Batoum—Batoum reservoirs (*illustrated*)—Docks and workshops—Tank waggons *versus* barrels—Rapid transport—Leakage reduced to a minimum—Charging and discharging tank steamers—Fuel—Mazout—Tzarytzin mechanical cooperage—Other depôts on the Volga—Domnino depôt (*illustrated*)—Storage capacity of depôts—Capital of company—Total transport and storage capacity—Tank steamers—Fireproof compartments—Typical observation—Pipe lines—American origin—Circumstances which led to their adoption—French Consul's gloomy picture of results of want of communication in early days of petroleum—Cost of delivering a barrel of petroleum (f.o.b.), Port Sarnia—Lack of practical methods of conveyance—Conflagrations in transit—First pipe line—The laughing-stock of the oilmen—Final success—Glut of pipe line companies with insufficient capital—The United Pipe Lines Company—The Standard Oil Company—Extent of pipe lines and dimensions of piping—Dimensions, manufacture and testing of pipes for mains—Laying down a pipe line—Intervals between pumping stations—American pumping station (*illustrated*)—Nobel's pumping station at Baku (*illustrated*)—Receiving stations—Nobel's at Baku (*illustrated*)—Stocks of Pennsylvanian and Ohio petroleum in depôts, 1885-1896—The Standard Oil Company's monopoly—The work of Samuel Andrew, a workman, and Rockefeller, a commercial employé—The Russian and American fight for European trade—Tide Water Pipe Company—Pipe line companies' method of transacting business—European ground unsuited for pipe lines—Pipe lines in Galicia—The first pipe line there—Java and Peru—Specification for pipes for pipe lines—Petroleum depôts and forwarding stations: Galician—Roumanian—Russian—German—French—Spanish—American   .   .   .   .   *pages* 458-481

CHAPTER XXXV.

GENERAL ADVICE—PROSPECTING, MANAGEMENT AND CARRYING ON OF PETROLEUM BORING OPERATIONS.

Prospecting—Necessity for rational and methodical work—Necessity for technical knowledge—Capitalists deceived by speculators—Preliminary prospecting work a necessity—Too much importance not to be attached to surface signs—The prospectors' examination should bear on four principal points: (1) The presence of the oil; (2) Its approximate depth; (3) The geological arrangement of the strata; (4) The nature of the strata—The presence of oil determined (*a*) by deductive reasoning; (*b*) surface signs—Trial borings —Collection of samples of strata being bored—Nature of boring journals or diaries—Different entries to be made in journal—Registration of accidents and dimensions of drilling tools—Tools to be repaired after use and put back into stock in good condition—Inspection and lubrication of machinery —Engine driver subordinate to foreman—Relations with other foremen on the establishment—Information from wells outside establishment to be regarded with distrust—Model boring journal of an actual oil well—Additional documents—Routine of day and night shifts—A few of the customs connected with drilling—General boring journal of the different wells of the establishment—Table of results of comparison of samples .   .   .   .   .   *pages* 482-492

PART V.

GENERAL DATA.—CUSTOMARY FORMULÆ.

MEMORANDA. PRACTICAL PART.

GENERAL DATA BEARING ON PETROLEUM.

Table of the specific gravity of liquids heavier than water (Baumé's "rational scale") —Table of specific gravities of liquids lighter than water (petroleum) (comparison of Baumé's scale with gravities)—Table of corrections for bringing the densities of Russian petroleum, taken at various temperatures, to 15° C. (according to Riche and Halphen)—Table showing the general properties of the crude mineral oil from the producing centres of the principal oil fields of the world—Weights and measures used in petroleum countries and their metric equivalents—Unit of measure adopted in boring—Table of inches and their equivalents in millimetres—Table for the reduction of foreign money into francs—Preliminary Work—Earthworks—The weight of a cubic metre of soil, sand, clay, etc.—Time occupied in excavating a cubic metre of sand, gravel, potter's earth, etc.—Time occupied in wheeling away a cubic metre to the distance d—Data for the construction of derricks and buildings—Resistance of materials, oak, pine, iron, etc.: (a) to traction parallel to fibre; (b) to crushing—Ratio of load to height and thickness of a post of timber—Resistance of derrick to the wind—Resistance of ropes and cables—Round hemp, iron or steel wire cables—Their weight and ratio of their load to their dimensions—Calculations used in estimating the output flow of a conduit—Table of the viscosity of petroleum and heavy oils at different temperatures—Weight of rods—Weight per current metre of square section and round section iron rods both in the air and in water—Formulæ for converting these indications into English measures, i.e., lbs. per running foot—Tables for gauging barrels from their interior length, height at the bung, and width of bottom—Continental regulations under which mineral oils are accepted for transport—Memoranda—Theoretical part—Useful formulæ—Factors commonly used in calculations—Various factors of π—Various factors of g Square roots—Cube roots—Algebra—Equations of the second degree—Progressions: arithmetical, geometrical—Depth of a well—Geometry and mensuration of surfaces—Volumes—Timber measuring—Gauging of barrels—Centre of gravity of lines—Centre of gravity of surfaces—Centre of gravities of solids—Mechanical data and formulæ—Fall of bodies—Simple pendulum—Conical pendulum—Kilogrammetre—Tonnemetre—Mass of a solid body—Motive force—Work done by a force—Momentum or quantity of motion—Vis viva—Velocity—Different methods of cartage—Daily work of a horse—Traction of vehicles—Ratio of the traction to the load drawn, vehicle included, on different kinds of roads—Ordinary ground, causeway or highway, tramways, etc.—The number of horses required to draw a given load on horizontal ground—Resistance of solids—Horizontal piece morticed at one of its extremities—Horizontal piece resting on two points of support . . . . *pages* 493-532

GLOSSARY OF TECHNICAL TERMS USED IN THE PETROLEUM INDUSTRY *pages* 533-535

INDEX . . . . *page* 537

LIST OF ILLUSTRATIONS.

FIG.   PAGE

1. Depth to be Bored before Striking Oil 31
2. Geological Section of the West Galician Oil Field 34
3. View of the Works of Ropa (Galicia) 36
4. Section at Wiertzno showing Productive and Barren Wells . . . 38
5. Pumping Station in the neighbourhood of Krosno (Galicia) . . . 40
6. The Station of Uherce (Galicia) 42
7. Geological Section of the Carpathians between the San and the Stryja . 43
8. Works of the French Company of Holowecko (Galicia) 44
9. An Old Works in the District of Boryslaw 45
10. View of Sloboda Rungurska 47
11. An Oil Well on Fire at Mrzaniecz (Galicia) 53
12. Map of the Roumanian Oil Fields 59
13. An Ancient Work in the Bacau District 63
14. Map of the Oil Fields of the Caucasian Isthmus 76
15. Group of Wells in the Grosnaia District 78
16. Map of the Apshéron Peninsula 82
17. Cause of Intermittent action of Baku Spouting Wells . . . 86
18. A Petroleum Fountain at Balakhany 89
19. Group of Wells at Sabountchany 96
20. Group of Wells at Balakhany 98
21. Petroleum Oil Works in Pechelbronn Oil Field 117
22. French Oil Works at Villeja (Italy) 121
23. Petroleum Oil Works in Italy 123
24. Showing how Nos. 1 and 2 Wells failed to strike the Petroliferous Tertiary
    Strata at Gabian 135
25. The Oil Fields of the Asiatic Archipelago (Oceania) 170
26. Geological Section of the Canadian and Pennsylvanian Oil Fields . 204
27. Oran Oil Field (Algeria) 261
28. Geological Section in North to South Direction of Aïn Zeft Oil Field,
    Algeria 269
29. Boring near Relizane, in the Oran Oil Fields (Algeria) . . . 271
30. Section of a Well excavated to the point where Surface Infiltration ceases 281
31. Well and Outflow Tunnel 281
32. Section and Plan of a Well being excavated by Hand 283
33. Arrangement of Battens intended to consolidate the Boarding of the
    Wooden Casing 286
34. First Method of Rope Boring 297
35. Sand Pump 298
36. Jobard's System of Rope Drilling, Cylindrical Ram and Ram Sand Pump 301
37. Goullet-Collet Drill 302
38. "Walking Beam" independent of the Derrick 305
39. Tripod Derricks used for Pumping Purposes at Zagorz, Galicia . 308
40. Four-legged Derrick (to scale) 308
41. American Derrick (to scale) 309
42. Canadian Derrick (to scale) 311
43. "Instrumentation" Derrick 312
44. Fauck's Plant without Derrick 313
45. "Walking Beam" fixed to Woodwork of Derrick 314
46. Excavation in which Foreman Borer Works to Escape Blows of "Walking
    Beam" 314
47. Double "Walking Beam," its Support forming part of Woodwork of Derrick 315
48. Rod of Drilling Tackle 316

FIG.																			PAGE
49. Fork or Rod Retention Key 316
50. Function of Retention Key 316
51. Temper Screw 317
52. Elongation Winch 318
53. Working Bar 318
54. Upper and Lower Parts of Fabian's Free-fall Instrument . . . 319
55. Section of Fabian's Instrument at the Shoulder 319
56. Fabian's Instrument modified by Fauck 319
57. Plan of the Unclutching of Fabian's Instrument 320
58. Auger Stem Guide 320
59. Free-fall Instrument, Auger Stem and Drill 321
60. Successive Modifications in Shape of Drills 321
61. Blades of Bit with Lugs 322
62. Bit 322
63. Difficulty of Raising Short-necked Drills when accidentally detached . 322
64. Free-fall System of Drill 323
65. Hoisting Winch 324
66. Safety Hook 324
67. Ladder-wheel Treadmill Windlass 324
68. Section and Plan of Ladder-wheel Hoisting Installation . . . 325
69. Piston Sand Pump Section and Plan 326
70. How Casing Pipes are Riveted and Joined together 327
71. Provisional Point ; Shape of Lower End of Casing 327
72. Deviation of too Short Casing 327
73. Method of Suspending Casing when being Lowered 328
74. A, Partial Casing (en colonne perdue) ; and B, Complete Casing . . 329
75. Supports of Casing Pipes in Surface Wells 331
76. Foreman Borer Drilling. Free-fall System 333
77. Surface Installation of Boring on Fabian's Hand-power System . . 334
78. Commencing to Drill without a Surface Well 338
79. Old Steam-power Surface Installation. Elevation and Plan . . 342
80. Details of Rod Windlass in the first Steam-power Drilling Plant . . 343
81. Plassy Boring made under Kind's Directions 344
82. First Windlass or Machine combining in a Single Apparatus all the
	Necessary Parts of both Drilling and Hoisting Machinery . . 345
83. Fauck's Drilling Machine 346
84. Oscillating Head-piece of " Walking Beam " 347
85. Diagrammatic Representation of Fauck's " Walking Beam " . . 347
86. Surface Installation. Fauck's System. Elevated " Walking Beam " . 350
87. Elevation and Plan of Woodwork. American Drilling Machine . . 353
88. Canadian " Walking Beam ". Profile and Plan 354
89. Canadian Rod Elongation Windlass or Reel 354
90. Adjustment of Auger Band Wheel and Bull Wheel 354
91. Transmission of Power for Hoisting—Canadian System . . . 355
92. Oscillating Cylinder 356
93. Canadian Rig 357
94. Top of Canadian " Walking Beam " 356
95. " Jars " in Sheath 359
96. Canadian Bit 359
97. Canadian Drilling Tools 359
98. Foreman Borer guiding the Work—Canadian System . . . 360
99. Interior View of a Canadian Installation 361
100. Manipulating the Rods. Rôle of the Workmen 362
101. Fauck's Combined Drilling Machine 365
102. Rope Socket 374
103. American Bits; Narrow Section 375
104. Surface Arrangement of American Drilling Machinery . . . 375
105. Bits on the Fauvelle System 381
106. Rod and Auger Stem on the Fauvelle System 381
107. Fauck's Hydraulic Oil-well Drilling Rig 382
108. Portable Oil-well Drilling Rig 384
109. Arrangement of a Boring on Fauvelle's Hand System . . . 385
110. Fauck's Combined Oil-well Drilling Machine fitted up for Hydraulic Boring 385
111. Method of Adjusting the Water Conduits 387
112. Hydraulic Oil-well Drilling Rig. Fauck's Machine, Capable of being
	driven indifferently by either Hand or Steam Power . . . 388
113. Different Kinds of Augers 392
114. Cylinder with Conical Auger 393
115. Reamer 393

FIG.   PAGE
116. Crown Tools with Steel Teeth   394
117. Fixed Crowns   394
118. Diamond Crown   395
119. Working Levers   395
120. Rotary Oil-well Drilling Rig for Shallow Depths   396
121. Rotary Oil-well Drilling Machinery   397
122. Function of Eccentric Shape of Crown   399
123. Boring Installation on Lippmann's Hand System   408
124. Boring Installation at Baku after a Model deposited at the *Conservatoire des Arts et Métiers*   409
125. Kind's Free-fall Instrument   411
126. Léon Dru's Water-pressure Free-fall Instrument   414
127. Laurent's (Bayonet System) Free-fall Instrument   415
128. Free-fall Instrument on Dru's Principle   416
129. Fauck's Free-fall Instrument (Bayonet Principle)   418
130. Hollow Auger Stem and Bit   419
131. Completely Sheathed System of Rods   420
132. Fauck's Hydraulic "Jars"   420
133. Diagrammatic Representation of Fauck's Hydraulic System   .   .   421
134. The Combined Diamond and Fauck System of Boring   .   .   .   424
135. "Fishing" Tools (Wrenches)   430
136. "Pear" Pipe Expander   432
137. Tap Grab   432
138. Fauck's Instrument for Cutting Casing Pipes   433
139. Dru's Instrument for Cutting Casing Pipes   433
140. Pike-mouthed Pincers and Tapped Piece   435
141. Worm Grabs ("Ram's Horns")   435
142 and 143. Cylinders   436
144. Toothed Spring and Spring Pincers   436
145. Dru's Expanding Drill   436
146. Fauck's Expanding Drill   438
147. A Spouting Well   441
148. Bredt's Patent Oil-well Cap   442
149. Oil Pumps at Balakhany   459
150. French Tank Waggon ; State Railway Model (to scale)   .   .   .   463
151. On the Pipe Line   477
152. Area of a Quadrilateral   527
153. Resistance of Solids   532

LIST OF PLATES.

PLATE
  I. A Petroleum Refinery (the Nobel Factories at Baku) . . *Frontispiece*
  II. Map showing Distribution of Petroleum over the Globe . *Facing page* 6
  III. Map of Petroliferous Europe . . . . ,, 8
  IV. Map of the Oil Fields of Galicia . . . . . ,, 28
  V. The Quay at Baku . . . . ,, 81
  VI. The Working of Petroleum at Baku . . . . ,, 88
  VII. View of the Oil Wells of Balakhany . . . . ,, 95
  VIII. Comparative Graphic Representation of the Petroliferous
    Production of Baku and of the United States, 1863-1897 . ,, 102
  IX. Oil Creek in 1867 . . . . ,, 198
  X. North American Oil Fields . . . . . . ,, 202
  XI. General View of Oil City . . . . ,, 214
  XII. Group of Oil Wells in Butler County . . . . ,, 216
  XIII. Crude Oil Reservoirs near New York . . . . ,, 246
  XIV. View of an Oil Field wrought according to the Canadian
    Method . . . . ,, 357
  XV. An American Rig . . . . ,, 375
  XVI. Boring Installation at Baku. Interior of Derrick showing
    the Drilling Machinery . . . . . . ,, 409
  XVII. Filling Petroleum Barrels at Baku . . . . . ,, 462
  XVIII. Loading Petroleum Tank Waggons at a Railway Station in
    United States Oil Fields . . . . . . ,, 463
  XIX. Crude Oil Receiving Tanks—Nobels' Works at Baku . ,, 464
  XX. *El Gallo*, Tank Steamer . . . . ,, 466
  XXI. Group of Petroleum Reservoirs at Batoum . . . ,, 467
  XXII. Batoum Petroleum Depôt . . . . ,, 468
  XXIII. Russian Petroleum Depôt at Domnino . . . . ,, 469
  XXIV. Tank Steamer moored before the Offices of the Nobel Ad-
    ministration at Baku . . . . ,, 470
  XXV. Pumping Station of the "United Pipe Line" at Clarion . ,, 474
  XXVI. Nobel Brothers' General Pumping Station at Baku . . ,, 474

PART I.

STUDY OF THE PETROLEUM-BEARING STRATA.

CHAPTER I.

PETROLEUM.—DEFINITION, PROPERTIES AND USES.—HISTORY, GEO-
GRAPHY AND GEOLOGY OF EUROPEAN OIL FIELDS.

DEFINITION.

WHAT is petroleum ? [1]

To this question, the dictionary replies that it is a mineral oil proceeding from wells, situated in Asia and in America.

This definition, however exact, is very incomplete. What is petroleum then ? Ah, well ! Petroleum is a thousand different things, according to the modifications which it has been made to undergo—a Proteus, under a hundred different aspects.

FORMS AND USES.

It is, in the first place, the illuminating oil, known to every one. Then it is the benzine used by the dyers and cleaners ; a lubricating machinery oil : solid briquettes (block fuel) for heating purposes ; or colours of brilliant tints. So much for the industrial point of view.

You will meet it also on the toilet table in the shape of a pomade pot of perfumed vaseline. Moreover, if you fight shy of these philodermic preparations, but, none the less on that account, are fond of good cheer, you have often, perhaps, partaken of petroleum under the guise of an excellent butter with a nutty flavour. Oh, adulteration ! Have no fear, however, that petroleum will injure your health ; on the contrary, although

---

[1] The chemical and physical properties of petroleum will be dealt with more completely in the second volume of the *Technology of Petroleum* devoted to the distillation and treatment of mineral oils.

1

it was sold formerly as a universal panacea under the name of
Seneca oil, or *Gabian oil,* it is an undisputed fact at the present
time that petroleum is an excellent remedy for diphtheria.

If we desired to exhibit petroleum, in all its forms, we should
further find it as an agent of destruction in melinite, and even
then we should not have done with it.

Suffice it to say that, at the present day, the tendency of
industry is more and more to substitute petroleum for coal, that
the railways of Russia, in large part, heat their locomotives with
it, and that, in fact, the Italian Navy, imitating in this respect
the example shown by a large number of merchant steamers, feed
the furnaces of their ships with the new fuel. Although, normally,
the gaseous carbides which escape from petroliferous soil may be
associated with petroleum, as well as the solid asphaltums which
may be worked in quarries, liquid naphtha [1] will alone occupy our
attention.

APPEARANCE AND PROPERTIES.

If the hand of man can transform petroleum in a hundred
ways it is but rarely met with in nature with a uniform appearance and composition.

It is, at one time, a bright, almost transparent, liquid of
exceeding great mobility and volatility (petroleum of Sourakhany, Russia), whilst at another time, it may be brown, viscous,
exhibiting a greenish aspect in daylight (Galician petroleum).
Again, it is often met with in an almost pasty condition, with
an appearance and consistency, recalling that of black soap; in
such a case its density is sensibly increased and may, in exceptional cases, exceed that of water (*Cal* of Sloboda, Galicia).

COMPOSITION.

In any case, the composition of naphtha remains always the
same, so far as the number of the principal elements is concerned;
only the proportion of these elements differ.   These are $C^n H^{2n+2}$.

> Carbon,
> Hydrogen,
> Oxygen (in small proportion),

[1] Crude oil.—*Tr.*

the aggregate of which constitutes the following industrial factors :—

1. Gasoline.
2. Benzine.
3. Illuminating oil.
4. Lubricating oil.
5. Paraffin wax.
6. Heavy oil.
7. Tar (Mazout).
8. Residue (combustible.)

DENSITY.

The predominance of one of these factors, whether heavier or lighter, modifies the average density of petroleum :—

The density of the petroleum of Galicia is	0·870—0·885
,, ,, Roumania	0·878
,, ,, Caucasus	0·882
,, ,, Baku	0·838—0·924
,, ,. Germany	0·892
,, ,, Northern Italy . . .	0·787—0·823
,, ,, Southern Italy . . .	0·942
,, ,, Gabian	0·894
,, ,, Pennsylvania . . .	0·784—0·886
,, ,, Virginia	0·824—0·897
,, ,, Canada	0·844
,, ,, China . . . · .	0·860
,, ,, Burmah	0·875
,, ,, Java	0·823—0·923

ORIGIN OF PETROLEUM.

It will be seen from these examples that naphtha is a light liquid of a density less than that of water; in a word, it is an oil. What then is the origin of this oil?

Chemists are far from being agreed upon this point.

Some assert that coal and petroleum have a common origin; that the continued action of the heat of the globe, or according to others, the momentary heat produced by the volcanic upheavals of relatively recent geological epochs, have produced in the carboniferous strata a kind of distillation, the distillate of which constitutes petroleum oil. As a matter of fact the oil produced by the artificial distillation of coal has great analogy with petroleum; but, on the other hand, it must be admitted that not one of the petroliferous regions of Europe are found in the carboniferous strata, or even in proximity thereto.

Others believe that they see in naphtha the product of a slow decomposition of marine organic matter which has accumulated for centuries. This opinion is that which has received most general acceptance. The presence of petroleum, in sheets of fresh water, may be explained by the phenomena of subterranean infiltration. Whilst fully holding this opinion, certain geologists have attributed the transformation of marine debris into petroleum, not to a slow decomposition, but to a sort of distillation, produced during volcanic upheavals. We shall consider, further on, these different theories in regard to the origin of petroleum.

HISTORY.—PETROLEUM AS A LIQUID FUEL AND ILLUMINATING AGENT.

It was about 1859 that the first petroleum wells, prolific enough to give rise to commercial transactions, were found in Pennsylvania.

This discovery was to revolutionise all the industries depending on heating, and above all, with illumination. For a long time previous the probable and relatively near exhaustion of the coal beds had been discussed. Even the critical moment when this substance, the very life of commerce, would be sure to be wanting, had been foreseen, and the question only was how to cope with the situation which would result from this eventuality.

The question was thus far advanced, when petroleum made its appearance. Nature had taken upon itself to reply to the alarmists.

The new material, a real liquid coal, was as good as the best coal of England or Belgium as a fuel for furnaces and foundries, its calorific intensity, weight for weight, was three times as great; moreover, it responded to a pressing need at that epoch of industrial progress, viz., that of a cheap illuminating oil.

In fact, the illuminating power of petroleum almost equals that of gas, and petroleum gas much surpasses, even economically, coal gas.

THE FIRST SHALE OIL WORKS.

1. IN FRANCE.

From 1832, attempts had been made in France by Selligues, the chemist, to extract by distillation a strong nauseous-smelling illuminating oil from the bituminous shales of Autun. This industry was crowned with a certain amount of success, and shale oil was sold in France before petroleum was officially known.

2. IN BRITAIN.

Britain was not long in following suit. From 1850 onwards, she strove to extract from the coal which her territory furnishes in abundance, a fœtid, dangerous illuminating oil. These attempts, however, more meritorious than practical, were crowned with success. One year the sale of this kerosené oil, as it was called, reached the figure of 300,000 hectolitres, averaging in price 1 franc per litre (say 660,000 gallons at 3s. 7d. the gallon).

SHALE OIL PAVES WAY FOR AMERICAN PETROLEUM.

This explains the sale of the first tons of Oil Creek petroleum, at 3 francs the litre, and the fabulous fortune which Colonel Drake, the joint proprietor of the first "flowing well," made in so short a time—a fortune, moreover, which has also descended to those who continued his work: the "oil kings".

PRICE OF LAND IN AMERICAN OIL FIELDS.

The price of land in Pennsylvania and in the lake district of Canada rose. People became infatuated. They bought land, no matter at what price. In 1862, in Canada, land was sold at £3,000 sterling per acre, which brings the hectare to more than 180,000 francs. This is an exception, but the fact is perfectly well known in that country.

The average price in the United States was $2,000, or about 10,000 francs per hectare (say $800 or £167 per acre).

EUROPEAN CONTRAST.

At the same period of time, viz., about 1850, Europe, which at the present day produces 59,653,000 barrels of naphtha per annum (when the rest of the world only produces 58,151,000) did not work her oil fields in any way whatever. Sixty francs the hectare (say 20s. the acre) was the ridiculous average price paid at that time for land in the oil field regions of Galicia. Whilst at Baku not a single transaction had been made in the sale of petroliferous ground.

EUROPEAN OIL FIELDS—HISTORY.

It is interesting to pass in review the history of the present European oil fields. Undoubtedly petroleum was known to exist on the European Continent, since without the help of man it revealed its presence on the surface of the ground in many places.

1. FRANCE.

In France, at Gabian, the natural springs of oil were wrought with a purely medicinal end in view. The oil of Gabian was collected at the spring for centuries, as a panacea for all maladies. This marvellous well was worked in 1717 by Dr. Rivière, in 1735 by a Bishop of Béziers, in 1752 by Bovillet, in 1834 by A. Hugo. But it was not until 1884 that its industrial working was organised under the direction of M. Zipperlen.

2. ALSACE.

In Alsace, about 1735, Antoine le Bel, the naturalist, wrote a report on the petroleum of that country, the working of which commenced in 1883.

3. HANOVER.

In Hanover, in Germany, from time immemorial, the country people used, as an illuminant, the crude oil which they collected by plunging brooms, into the small pools, on which floated a layer of petroleum, and which on account of this phenomena were called *fettlöcher* or fat-holes. Serious attempts at working only commenced in 1865.

4. ITALY.

In Italy, petroleum was not wrought until 1880.

5. GALICIA.

At the time of the oil fever in America, the rich strata of Galicia had not been the object of any research. The country people only collected the crude petroleum, on the surface of the ground, and used it masked as boot-blacking for their boots, and in the guise of pomade for their hair. The annual production of Galicia now amounts to 1,500,000 barrels.[1]

[1] These are presumably the Austrian Government figures. See page 27 where authors give production as 2,500,000 barrels.—*Tr.*

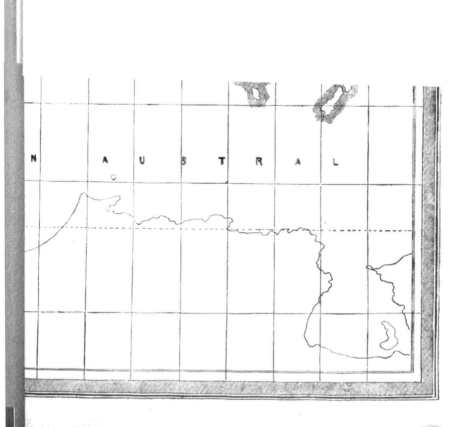

6. ROUMANIA.

In Roumania, a country immensely rich in petroleum, working is still in a rudimentary stage.

7. RUSSIA.

In the Caucasus at Baku, the Russian oil fields, which have, at the present day, far exceeded the American production, were not the occasion of any regular working. Petroleum, however, had centuries before been shown to exist there by Marcus Polo,[1] Father Duval, etc. It was used for a purely platonic purpose, worshipped by the Guèbres, as representing the divinity of Ormuz, as the source of eternal fire. It also cured camels of the scab.

PRESENT CONDITION OF EUROPEAN OIL FIELDS.

Now, however, at the present day, our little Europe would seem to be a vast petroliferous reservoir. Released by the drill of the miner, the fountain of oil bursts forth, the jet of which breaks everything in its passage, and the earth, in its haste to prodigally spend its treasures, inundates, with the precious liquid, those who have never had any doubts of its presence—the oil prospectors.

THEIR GEOLOGICAL HORIZON AND ASSOCIATION.

Petroleum is most generally met with in beds of sandstone. The shales of the same formation may also yield it on distillation, but in much smaller quantity.

Moreover the petroleum stone (*pierre à pétrole*) generally consists of a friable coarse grained conglomerate of a brown colour and characteristic smell. Although in America petroliferous strata occur in almost all the formations of the geological series, but, above all, in the older stratified rocks, it is not so in Europe. As previously mentioned, the strata of the tertiary system, almost alone, present any certain sign of the presence of petroleum, in the latter continent. Petroleum is found :—

At Gabian in France in the tertiary, miocene.

---

[1] A celebrated Venetian traveller born 1250, died 1823, who traversed Asia and published an account of his travels.—*Tr.*

In Limagne in the base of the tertiary, miocene and upper eocene strata (paludina beds).

At Gau (Isère, France), in the tertiary, eocene.

At Châtillon (Haute Savoie) France, in the tertiary.

In Italy, in the tertiary, miocene.

In Alsace, in the tertiary.

In Hanover, in the Upper Jurassie (tithonic).

In Galicia, in the tertiary, eocene (nummulite beds).

In Roumania, in the tertiary, eocene (congerian beds).

In the Crimea, in the tertiary, eocene (nummulite beds).

In the Caucasus, in the tertiary, eocene (nummulite beds).

At Baku, in the tertiary, eocene (nummulite beds).

MAIN EUROPEAN PETROLIFEROUS LINE OF STRIKE.

According to this classification, it is easy to see that all the riches of the Caucasus and Carpathian deposits are found in the eocene formation. These are moreover the best determined or outlined deposits; they follow with the single interruption of the Black Sea an almost straight line, which, commencing in the Carpathians at the foot of the volcanic upheaval of Tatra, stretches from the north-west to the south-east and crossing the Caspian Sea, after having pierced the whole of oriental Europe, loses itself in Asia.

This line, which we will call the great or main boundary line of the old continent, constitutes an eocene formation of the first importance. We have said that it commences in Galicia. In that country, it follows the foot of the northern projecting flanks of the Carpathians, presenting its most rich oil fields at that spot where the plain of Poland commences to rise in slight undulations or uplands at the approaches of the mountain. In the same way it continues into Roumania, and its course is sharply delineated even to the shores of the Black Sea at Dudeschi. We find it further on, crossing the Crimea, always in the same direction, then, dividing into two branches, it follows, in the same way as it did in the Carpathians, the north and south watersheds of the Caucasus; flanking Baku on both sides, it loses itself in the Caspian Sea which it impregnates with the naphtha which it distils. It continues on in Asia, passing through Persia,

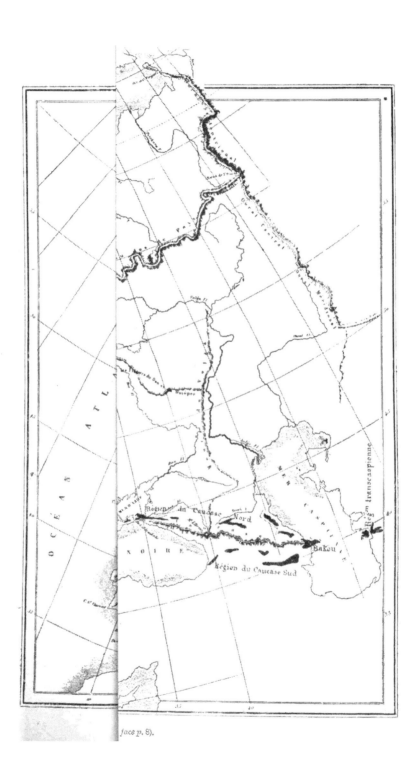

face p. 8).

Beluchistan, Western India and Burmah, terminating in a rectilinear prolongation across the islands of Oceania.

THE MEDITERRANEAN SECONDARY LINE OF STRIKE.

Another line, of very secondary production, which we shall call the Mediterranean, appears to follow a direction a little parallel to the main line, also in the direction of from north-east to south-west. It is far better indicated by signs of petroleum than by important results. Commencing in the French department of the Pyrénées Occidentales (Western Pyrenees), it follows the northern watershed[1] of these mountains, passes on to Gabian, is staked out by the petroleum traces of the department of Vaucluse, and finally, passing through Italy, runs parallel with the right bank of the Po towards the deposits in the environs of Plaisance. The principal petroleum indications of this line originate in the miocene system and nearly always manifest themselves in the neighbourhood of important volcanic upheavals ; its direction ollows in a great measure that of the nummulite beds of the Mediterranean.

The study of the geology of each petroleum-bearing region will be deferred until later.[2] That is the reason why we terminate here this short sketch, and conclude this chapter by asserting that, if much has been done of late years in regard to petroleum, there still remains, however, much to be accomplished, consequently, much to be gained.

[1] American " Divide ".—Tr.   [2] See chap. iii., p. 24, et seq.—Tr.

CHAPTER II.

THE GENESIS OR ORIGIN OF PETROLEUM.

THE first man of science, who attempted to find the origin of petroleum, was Leopold de Buch.

LEOPOLD DE BUCH'S HYPOTHESIS.

In 1801, this *savant*, according to the commands of his sovereign, the King of Prussia, studied the geology of the Jura. "There is not, in proximity to this bitumen, any imprints of vegetable petrifications, there are no leaves, no reeds, and it is more than probable that these masses owe their origin, more to the animal than to the vegetable kingdom. The quantity of shells in the neighbourhood gives rise to this assumption, even if no attention be paid to their nature and to the volatile alkali which they contain " (*Gesammelte Schriften*, Vol. i., Berlin, 1867).

TURNER AND REICHENBACH'S THEORY.

Turner and Reichenbach were equally conclusive in ascribing an organic origin to petroleum, similar to that of coal.

ROZET AND MILLET'S OPINION.

Rozet (*Bull. Soc. Géol.*, 1836, Vol. vii., p. 138), and Millet (*Bull. Soc. Géol.*, 1840, Vol. xi, p. 343), ascribed a purely organic origin to the hydrocarbides.

At the present time, the opinion has met with many detractors; many hypotheses have been put forth, amongst which three are surrounded by a great number of adepts.

CLASSIFICATION OF THEORIES.

1. The first of these opinions still remains that of Leopold de Buch, which pre-supposes the organic origin.

2. The second involves a volcanic origin.

3. The third concludes that petroleum is the result of chemical reaction.

Reserving in the meantime our own opinion in the matter, which will be expressed farther on, we will proceed to pass in review the arguments which the *savants*, who have taken part in this debate, have brought forth in support of their conclusions.

I.—ORGANIC ORIGIN OF PETROLEUM.

1. STERRY-HUNT'S HYPOTHESIS.

According to Mr. Sterry-Hunt,[1] petroleum is indigenous to the different strata where it is met with. In his opinion the oil is the result of a peculiar transformation of vegetable organic matter which takes place at the bottom of the waters where sediment is deposited; this transformation being of the same nature as that which produces bituminous coal; for chemical analyses show a series of products passing, by insensible gradations from anthracite to petroleum (*Bull. Soc. Géol.*, xxiv., p. 570).

2. LESLEY AND LESQUEREUX'S THEORY.

Lesley and Lesquereux have originated an almost identical opinion. This latter is to the effect, however, that, if coal has been formed by the decomposition of woody fibre, petroleum proceeds from that of non-fibrous vegetables, such as sea-weeds.

3. HITCHCOCK'S OPINION.

C. H. Hitchcock considers the origin of petroleum as essentially organic—the greater part being from a vegetable source—but animals might furnish a little in the form of fish oil.

He does not think that mineral oil can be the product of the distillation of coal, because it has existed prior to this formation. Perhaps, according to him, the salt water of the primitive saline lagoons, by hindering the disengagement of gaseous hydrocarbides, evolved from the vegetable matter underneath, forced these gases to condense into the liquid form.

---

[1] An eminent Canadian professor of geology, who, had he lived in the days of Hutton, would have been a staunch supporter of Werner in the war between the Huttonians and Wernerians, with which the present controversy as to the origin of petroleum has so much in common.—*Tr.*

4. Gesner's Hypothesis.

Abraham Gesner says (*A Practical Treatise on Coal Petroleum*, New York, 1861), that the sources of oil are met with, in general, in the immediate neighbourhood of the beds of coal and are disseminated in the same basin, and that there is no reason to doubt but that petroleum and coal have the same origin. " The idea has been put forth," he adds, " that petroleum is produced by the distillation of the anthracite of Pennsylvania and surrounding regions, because anthracite contains no bitumen. But the anthracite beds are separated from the bituminous coals, and the petroleum wells by the chain of the Alleghanies, and it is not at all reasonable to suppose that the distilled bitumen could traverse the enormous mass of these high mountains consisting of primitive and metamorphic rocks."

5. Canadian Geological Survey, Surveyor's Hypothesis.

An article in the *Geological Survey of Canada*, 1862, says : " The limestone, which contains the mineral oil, extends over a surface of 1·125 square kilometres in Western Canada ; this limestone is of marine origin and does not contain any other organic remains but those of marine animals, which leads us to infer petroleum hydrocarbides are produced by a decomposition of those tissues. We know that the tissues in question differ but little from those plants which in more recent formations have given rise to bituminous coal."

" We may suppose that many gelatinous animals and perhaps even plants, the remains of which have disappeared, may have contributed to the formation of petroleum."

6. Credner Compares Formation to that of Guano.

M. Credner almost assimilates or classifies the petroleums with the guanos on account of their similitude of organic origin.

7. Rupert Jones and Watson Smith's Opinion.

Rupert Jones and Watson Smith are equally in favour of an exclusively organic origin.

8. VON STROMBERG'S HYPOTHESIS.

According to M. von Stromberg (*Deutsche Geologische Gessel-schaft*, tom. xxxiii., p. 277), in the petroleum oil fields of Germany the oil is produced by the decomposition of the lignites of the strata in proximity thereto.

9. LE BEL'S THEORY.

Le Bel (1885) acknowledges an organic origin, but which cannot come into play except under the influence of a reaction originating in other strata than the petroliferous strata.

10. ORTON'S CLASSIFICATION OF PETROLEUM-PRODUCING AGENTS.

Orton (1888), who studied the Ohio deposits (*Annuaire Géol. Univ.*, 1889, p. 890), comes to the same decision. According to him :—

1. Petroleum is derived from organic matter.

2. The typical petroleum of Pennsylvania is derived from organic matters, yielded by bituminous shale, and probably of vegetable origin.

3. Canadian petroleum is produced from limestone, and is probably of animal origin.

4. Petroleum has been produced by a normal heat, and not under the influence of high temperatures capable of producing the destructive distillation of bituminous shale.

5. The quantity of petroleum contained in the rock cannot be augmented.

11. ENGLER'S EXPERIMENTS AND CONCLUSIONS.

Engler, in 1888, commenced researches on the origin of petroleum, which he attributed to the combined action of heat and pressure on the soft portions of the organisms of the fauna of the paleozoic seas.

In support of his assertions he made the following experiment :—

In an autoclave connected with a condenser he submitted 472 kilogrammes of cod-liver oil, of density 0·930, to destructive distillation.

At the commencement of the experiment, the pressure was

10 atmospheres, but during the course of distillation, the pressure reduced to 4 atmospheres from 320° C. of initial temperature, the autoclave was raised to 400° C.

Combustible gases were soon given off from the condenser, then an oil which, containing acids or fatty bodies, was in part saponifiable. The first distillate was redistilled under pressure and under the same conditions as previously, and 299 kilogrammes of oil of a density of 0·810 were collected. This oil gave off inflammable vapour at 44° C., its smell was not disagreeable, it had no smell of acrolein, its colour was brownish with a pronounced green fluorescence, of 100 cc. of oil; there were :—

		Per Cent.
Soluble in water	. . . .	0·4
,, potash lye	. . . .	4·8
,, sulphuric acid, 66·9° Baumé	. . . .	20·8
,, Fuming sulphuric acid mixture at 66° Baumé	.	6·0
		32·0

The remainder was distilled under the same conditions, but fractionally. The following results were obtained :—

Below 150° C. . . . 29·5 per cent. by volume of density 0·712
Between 150° C. and 300° C., 57·5 ,, of density . . . 0·817
Above 300° C. . . . 13·0 ,,

There were now found in the distillates fatty acids, nonsaturated hydrocarbides and saturated hydrocarbides. These were separated and studied, by chemical methods, and the most exact rectification, especially

	Density.	Boiling point.
Normal Pentane	622	37
,, Hexane	664	67
,, Octane	688	98

Engler repeated his experiment upon 492 kilogrammes of fish. He obtained an almost identical result.

It would follow from these experiments that petroleum is the result of distillation under pressure of animal organic matter, a distillation carried out in Nature's own laboratory.

12. PHILLIPS' HYPOTHESIS.

Phillips (*Am. Chem. Journal*, p. 409) embraces Engler's theory. He has no doubt whatever that petroleum is formed by the decomposition of vegetable organic matter, this decomposition having taken place in a space deprived of air, under water for example.

13. HOFER'S CONTENTIONS.

During the last fourteen years, M. Hofer has defended the organic animal origin of mineral oils (*Allgem. Oesterr. Chem. und Tech. Zeit.*, 1891, p. 230). One of the arguments used against M. Hofer by his opponents was the absence of nitrogen, or nitrogenous compounds, in petroleum. To this objection the Austrian *savant* replies in the following manner : It is important to remark that in the decomposition of animal matter gases are formed, with the result that the nitrogen would disappear in the gases accompanying the petroleum.

Nevertheless, all mineral oils are not destitute of nitrogen ; according to Engler the percentage of nitrogen in Pechelbronn gas oscillates between 8·9 and 17 per cent. Moreover, the presence of nitrogen in petroleum is not necessary, even according to the theory of animal origin. Animals are constituted in their soft parts by a nitrogenous fleshy portion and by a non-nitrogenous portion composed of fat.

During the putrefaction of a corpse or carcass, the nitrogenous matter decomposes rapidly, whilst the fat is characterised by its remarkable stability.

In support of this fact, Engler quotes the presence of adipocere, or corpse wax, in all tombs.

II.—FORMATION OF PETROLEUM BY VOLCANIC ACTION.

Alongside this hypothesis, of a purely organic origin, several geologists have contended for an organic formation under the influence of a reaction due to volcanic phenomena, or these phenomena themselves as an initial cause.

1. VIRLET D'AOUST'S OPINION.

Virlet d'Aoust was one of the first to decide in favour of a volcanic origin ; in his dictionary of the natural sciences the French geologist declaims against the theories of the two German *savants*, Turner and Reichenbach, who ascribed a common origin to both coal and bitumen.

2. Daubrée's Review.

In 1867 M. Daubrée, in his report on mineral substances, expressed himself in these terms : " The origin of this substance has not yet been determined with certainty. It is generally said to have been derived from organised substances, vegetable or animal, by a transformation analogous to that which has produced coal.

"Nevertheless, two lines of thought have led to the belief that petroleum may have a strictly mineral origin : (1) on the one hand, the association of bituminous deposits with eruptive phenomena as in Auvergne, or at least with the dislocations arising from internal phenomena, as well as the thermal springs which often accompany them ; (2) on the other hand, the possibility of reproducing liquid hydrocarbides by direct synthesis and without the help of bodies which had passed through a living stage of existence."

3. Lapparent's Views.

M. de Lapparent (*Traité de Géologie*, 1881) would also appear to range himself along with those in favour of this theory.

III.—Formation of Petroleum by Chemical Action—
1. Berthelot's Hypothesis.

Thirdly, we have the opinion first given expression to by Berthelot that petroleum results from a chemical reaction, *viz.*; from the direct combination of carbon with hydrogen. This theory presupposes the existence of an intensely heated metallic kernel in the centre of the earth. Carbonic acid, in its passage through the mass of the earth, produces in contact with this kernel containing alkaline metals, alkaline acetylides such as the monosodic acetylide, C^2HNa.

On the other hand, the diffusion of water through the mass of the globe has been proved by volcanic eruptions ; this water, arriving at the central kernel, is decomposed with liberation of free hydrogen.

Under the influence of heat and the free hydrogen, the alkaline acetylide is decomposed, and acetylene becomes the centre of a series of reactions ; it yields condensation and addition products, which constitute precisely the hydrocarbides of which mineral oils are composed.

2. PELOUZE AND CAHOURS' DECISION.

Pelouze and Cahours (*Comptes Rendus de l'Académie des Sciences*, tom. lvii., p. 69, 1863) range themselves on the side of Berthelot. "Amongst the numerous samples which we have received from different sources," they say, "we have never met with benzine or any of its homologues"; which would seem to indicate that these hydrocarbides could not have been derived from coal, or, if they did so originate, that this substance had undergone a different decomposition to that which it undergoes either when submitted to a slow or to a rapid distillation at a low temperature or at a high temperature.

3. COQUAND'S OPINION.

According to M. Coquand, who, in 1867, made a special study of the petroliferous strata of the Carpathians, the petroleum of that country is contemporaneous with the beds which contain it, and its arrival dates from the very moment when these sandstones and clays were deposited at the bottom of the eocene and miocene waters. The conclusion of the French geologist is that the birth of petroleum is due to a chemical reaction. This reaction, in his opinion, must be sought for in the depths of the incandescent furnace in the centre of the earth (*Bull. Soc. Géol.*, tom. xxiv., p. 562).

4. BENOIT'S CONCLUSION.

Benoit (*Bull. Soc. Géol.*, tom. iii., p. 440, 1875) holds that the origin of petroleum is the result of chemical combinations formed under the powerful and still unknown influence of subterranean heat and pressure.

5. CLOEZ'S EXPERIMENTS.

In 1877, M. Cloez succeeded in preparing a carbide of manganese which, with boiling water, gave off carbides of hydrogen analogous with the carbides of natural petroleum.

6. MENDÉLÉEF'S HYPOTHESIS.

M. Mendéléef, who, in 1877, made important researches into the matter, also decided in favour of a formation by chemical reaction (*Bull. Soc. Chim. de Paris*, tom. i., p. 501). In his opinion it is difficult to imagine that petroleum results from the decomposition of animal and vegetable organisms, for it would then be impossible to picture to oneself the origin of petroleum without the formation of a corresponding quantity of carbon. On the other hand, it is impossible to imagine the existence of a grand number of organisms in the epochs preceding the Silurian and Devonian. These considerations have led M. Mendéléef to suppose that petroleum has not in any sense an organic origin. Starting from the hypothesis of Laplace on the origin of the earth, by applying the law of Dalton to the gaseous state, in which the elements constituting the terrestrial globe were bound to exist, and taking into consideration their relative densities, he concludes that there was a condensation of different metals towards the centre of the earth. Amongst the number of these metals, iron predominates. Supposing also the existence of metallic carbides, he finds a solution not only for the origin of petroleum, but also for the manner of its distribution in the districts where the terrestrial strata, at the time of the upheaval of the mountain chains, were filled with crevasses in their lower part, these crevasses having in the first place opened an access for water to the metallic carbides ; the action of the water thereon at an elevated temperature and great pressure, gave rise to metallic oxides and to saturated hydrocarbides, which, being impelled upwards with the vapours, reached the strata where they might condense easily and impregnate the highly permeable beds of sandstone.

7. ABICH'S CONCLUSION.

M. Daubrée made the following report upon a communication of M. Abich (*Comptes Rendus de l'Académie des Sciences*, tom. lxxxviii., p. 891, 1879) : According to observations which he has made on the ground, as well as following a chemical work of M. Mendéléef, M. Abich rallies himself completely on the side of those whose opinion is that petroleum is not of organic origin, but is the product of internal action. He also finds an argument in favour of this conclusion, in the presence of carbon combined with the iron of certain meteorites, which gives an idea of the nature of the profound parts of the terrestrial globe.

8. Ross's Experiments and Deductions.

In 1891, D. C. D. Ross attributed the origin of petroleum to the action of the volcanic gases upon limestone. He supports his theory by laboratory experiments, founded on the theory of Bischoff, that sulphur could be obtained by the action of volcanic gases upon lime.

9. Sokoloff and Grabowski's Decision.

Sokoloff and Grabowski equally maintain the theory of formation by chemical reaction.

10. Peckham's Conclusion.

Peckham looks upon liquid hydrocarbides as simply condensation products of marsh gas.

11. Maquenne's Experiments and Deductions.

In 1892, a French chemist, M. Maquenne, presented to the Academy of Sciences the following consecutive observations relating to an experiment upon a definite carbide of barium :—

" This direct and remarkably easy formation of a metallic acetylide supplies a new method for the synthesis of acetylene, which is not, perhaps, without interest, from the point of view of the formation of the natural carbides. It is in fact quite likely that this property of barium is not confined to itself, and that other metals are capable of directly yielding stable acetylides at high temperatures. Calcium, according to Wöhler, would appear to be one of these. If we recall that, according to the beautiful researches of M. Berthelot, acetylene is the point of departure of the synthesis of all the cyclic carbides, we are thus enabled, by the simple decomposition of natural acetylides, as this *savant* has stated (*Ann. de Ch. et de Ph.*, 4ᵉ série, tom. ix., p. 481), to explain in a satisfactory enough manner the origin of the Russian petroleum, which would appear to consist of aromatic hydrocarbides, which the industry of to-day has been able to convert into true benzine hydrocarbides.

12. M. MOISSAN'S EXPERIMENTS AND DEDUCTIONS.

In 1896, M. Moissan gave to the world a new theory of the formation of petroleum. The numerous definite and crystallised compounds of carbon with metals, which this man of science has obtained by the aid of the electric furnace, have led him to form interesting geological conclusions.

Certain metallic carbides yield, by simple decomposition with water, a large quantity of carbides of hydrogen. The liquid thus obtained contains both saturated and non-saturated hydrocarbides. However far it may be established that, under the action of heat alone, a non-saturated carbide can fix hydrogen, and finally yield compounds analogous to petroleum, M. Moissan is careful not to generalise too much upon this action of water upon metallic carbides. Certain bituminous shales have every appearance of being of organic origin, but in certain districts in Limagne [a district of France (Auvergne), included in the department of Puy-de-Dôme], for example, the carbides of hydrogen may be ascribed to the action of water upon metallic carbides. The same explanation holds good in regard to certain volcanic phenomena, accompanied by evolution of carbides of a very varied nature, going from asphalt and petroleum down to the ultimate term of all oxidation, carbonic acid.

M. Moissan has already made experiments having a direct bearing upon this very point. He succeeded, with the aid of the electric furnace, in preparing a whole series of metallic carbides, decomposable by water, by treating the oxides of carbon. Some of the carbides thus obtained (calcium, strontium, barium, lithium carbides) yielded with water pure acetylene, the others (aluminium, glucinum, and uranium carbides) yielded methane, the first term of the family of petroleum carbides. These liquid carbides were of the same series as those of petroleum, and commenced to boil between 70° C. and 200° C.; solid bituminous products were obtained at the same time.

13. STANISLAS MEUNIER'S DECISION.

M. Stanislas Meunier (*Comptes Rendus de l'Académie des Sciences*, tom. cxxiii., 1897, p. 1329), inspired by these researches, says : " We are not justified in looking upon asphaltum, as has so often been done, as a product of the subterranean decomposition of organic bodies, animal and vegetable . . . bitumen is the result of purely chemical actions, of which the type is the double decomposition which takes place between metallic carbides and water ".

IV.—AUTHORS' SUMMARY AND CONCLUSIONS.

We have passed in review without any comment whatever the arguments brought forward by the leaders of the scientific world in support of such and such a theory with the simple end in view of demonstrating to the reader how divergent are the opinions held in regard to the point at issue.

Nevertheless, as a matter of fact, we are only in presence of two hypotheses—that of the chemists (Berthelot's theory), who suppose an instantaneous reaction, a direct combination of carbon with hydrogen; and that of the geologists, who regard petroleum as the product of the slow decomposition of organic substances.

The chemists, as the result of very beautiful laboratory experiments, have been able to make artificial petroleum, as they could, if need be, no matter by what reaction, make almost all substances by means of the rational combination of their chemical elements.

Geologists have based their assertions upon observation and by the comparison of the different petroliferous strata which they have had the opportunity of studying. They have not been able to see that an origin can be given to a liquid combustible, that is, petroleum, so altogether different to that which is given to other mineral combustibles to which it presents more than one point of resemblance, having regard, above all, to the liquid hydrocarbides which may be extracted from the latter. They consider that combination of carbon and hydrogen may very well be brought about, and that without breaking any of the principles of chemistry, by the decomposition of marine organisms resulting in the natural formation of hydrogencarbides.

Engler has amply demonstrated that these suppositions stand the test of experiment.

This hypothesis attacks or wounds nothing in chemistry; and, on the other hand, the theory of the chemists, as we shall demonstrate to them, appears to be in very slight accord with the geological conditions which overrule the establishment of the majority, if not of all the petroliferous basins of the world. As a matter of fact, petroleum would appear to have no affinity whatever for those regions of the earth which consist of plutonic rocks ; important petroliferous deposits have never been found in those regions, and rarely in their neighbourhood.

Nevertheless, igneous rocks abound in fissures, in which, according to the theories of Mendéléef, petroleum should accumulate, just as it does in the fissures of the sedimentary strata. Now, the fissures of plutonic rocks never contain hydrocarbides in appreciable quantity. The mining engineer, however inexperienced he may be, knows it so well that he stops boring as soon as he reaches rocks of this nature. Whence comes, then, this preference of petroleum for the sedimentary rocks ? In our opinion it can only proceed from one cause, and that is that *petroleum is indigenous to the strata where it is met with in large quantity.*

In order to corroborate our opinion—the opinion of practical men—in order to fortify it against the absolutely theoretical speculations of chemists, we only ask of these an explanation of the following fact : If petroleum be the result of a chemical reaction, which has produced, by the sudden combination of well-defined bodies and by a unique action, a single body—mineral oil—how comes it to pass that the petroleum produced by this unique action differs in physical properties and chemical composition from one deposit to another ? Whence come these differences ? They arise, perhaps, someone will say, from the nature of the strata traversed ; from the materials which they have drenched and assimilated in their passage through the fissures of the globe.

This is not a valid reason in our opinion. Petroleum differs, as regards chemical composition, not in the nature of its elements, which are almost always the same, but in the percentage of those elements hydrogen and carbon of which the liquid is composed. It will be asserted without doubt that certain strata, certain phenomena, met with during the long journey of the fluid matter through the subterranean conduits, have been of such a nature as to assimilate a certain portion of the constituents of petroleum. In such a case it will be necessary to acknowledge that each vein has been submitted to particular conditions. Now, this would almost be admissible if in the same district neighbouring wells did not produce different oils. These oils may even present notable differences in the same petroleum-bearing bed.

This fact alone will amply show that petroleum is indigenous to the strata in which it is met with, and that it originates from the decomposition of organic matters, in which the quantity or the quality, almost alike in certain points, may however vary.

Salt springs, the deposits of salt in the neighbourhood of petroliferous basins, are not, in our opinion, proof of volcanic action as asserted by some, but rather of the presence of sheets of salt water, the organic matter of which by a progressive decomposition would, in the first instance, form a kind of peat almost analogous to that which is formed in marshes (peat mosses), and from which, moreover, hydrocarbides may be extracted. Petroleum would flow from this peat eventually, either by pressure or by a kind of natural distillation.

We, therefore, rank ourselves, and do so with confidence, on the side of the theory of the geologist—on the side of the hypothesis based upon the purely organic and progressive genesis of mineral oils.

CHAPTER III.

THE OIL FIELDS OF GALICIA, THEIR HISTORY.

It is about forty years (1859) since petroleum oil was first struck in Pennsylvania in marketable quantity.

At that time the Galician peasants greased their boots and their waggon wheels with petroleum from the natural oil wells so abundant in that country, whilst land in the petroleum oil fields sold at sixty francs the hectare on the average.

In 1859, an English engineer, was travelling through Galicia, and, at the spot where the Derricks of Wietrzno now rise, he exclaimed : " There is a fortune in your mountains ! "

EARLY STRUGGLES AT BORYSLAW.

So far back as 1817, two men of genius, Joseph Hecker and Johann Mitis, had already foreseen this, and made trial of a distillery at Boryslaw.

Later on, the Jews of the country, eventually saw that the petroleum, which was sent to them from America, could be collected at their feet ; which they did. Some—and amongst the first of these was Abraham Schreiner, who died in the deepest misery at Boryslaw, his native town—excavated the ground in this country. For the greater part they were poor men, their only possession being their intelligence.

During the time that they in America wrought with the aid of steam and a relatively perfect equipment, they at Boryslaw were obliged to scrape the rock with the pick and shovel. Moreover, the Austrian Government had interdicted the use of lamps in the wells, and the unfortunate miners had therefore to dig in darkness in an almost unbreathable atmosphere. The small amount of pure air necessary was obtained by the aid of a ventilator of the most rudimentary construction.

FINAL SUCCESS.

Nevertheless, after many years of patient research, their efforts were crowned with success, mineral oil made its appearance in large quantity, almost all the wells yielded, daily, several hundred hectolitres (hectolitre = 22 gallons).[1]

RUSH AND SUBSEQUENT FAILURES.

Then came, as in America, an infatuation, but a local infatuation. They thought they had found a new California. Every one wanted to have his well, from the trader in easy circumstances to the poor peasant whose only possession was his field. It was sheer madness. It is an avowed fact that no result of any importance can be obtained until a depth of 130 to 150 metres has been reached. These unfortunate persons had over-estimated their own strength and above all their fortune. In spite of their desire to reach the vein of oil, their patience began to give way, and they abandoned their wells after reaching a depth of from five to eighty metres. These wells yielded but little oil, the deepest only about five hectolitres a week, but the latter was only obtained at the cost of great sacrifices ; it was, in fact, necessary to maintain the wells, and keep them in good working order, etc., etc. From this cause the large output of the Boryslaw wells gradually ceased.

DISCOVERY OF OZOKERIT SAVES THE SITUATION.

At Boryslaw, in spite of these failures, some still had hope in the future. Though looked upon as madmen, they were not discouraged, hoping that, by dint of patience and labour, fortune would one day smile upon them. They bought, at ridiculous prices, the wells bored by their neighbours. The earthy matter excavated out of these wells consisted of a soft brown substance which they threw away, although what they sought was to be found in it—not petroleum but fortune.

In fact, the substance which they threw away so nonchalantly was no other than mineral wax ozokerit.

This wax, which, on account of its cheapness can replace all

---

[1] Where the British equivalents of the French weights and measures are not given simultaneously in the context, they may easily be ascertained by consulting the tables in the appendix.

other waxes, sells at the present on the spot at about 35 florins or 70 francs the 100 kilogrammes. Millions of quintaux of it are extracted annually.

At the present day Boryslaw belongs in great part to the firms of Wagmann, Liebermann, Gartenberg, and to a French company (*La Société Française*). Besides, as we have already seen, foreigners, more especially Frenchmen, are at work in prospecting for petroleum oils and mineral wax in Galicia. Already, in different places, one or other of these substances have been found. Every day new deposits are being discovered. Whole towns have been created by this industry—Boryslaw, Wietrzno, Sloboda-Rungurska.

Thanks to improvements in the method of working introduced in part by M'Garwey, who brought them from Canada in 1883, and to the more and more extended use of steam power, Wietrzno, Polona, Sloboda, Siari, etc., an output has been attained of more than 1,000 hectolitres per day.

It is to be borne in mind that two factors contribute to form the value of the annual production of crude petroleum—the quality and the quantity. Now, whilst Galician petroleum yields on an average 60 per cent. of burning oil, the crude oils of Baku only yield but 18 per cent. The latter almost always contain 30 to 40 per cent. of sand.

Let us interpolate here a few statistics :—

				Quintaux.[1]	Value per Quintaux in francs.	
Annual production of petroleum in Galicia—1872				209,000	14·00	
,,	,,	,,	1877	120,976	9·93	According to Gintl.
,,	,,	,,	1878	245,000	9·41	
,,	,,	,,	1879	300,000	7·11	
,,	,,	,,	1880	320,000	7·47	
,,	,,	,,	1881	400,000	6·52	
,,	,,	,,	1882	461,000	6·23	
,,	,,	,,	1883	510,000	5·36	
,,	,,	,,	1884	600,000	5·08	
,,	,,	,,	1885	700,000	4·67	According to Hans Urban.
,,	,,	,,	1886	1,000,000	4·18	
,,	,,	,,	1887	1,200,000	4·06	
,,	,,	,,	1888	1,700,000	4·05	
,,	,,	,,	1889	2,100,000	4·00	
,,	,,	,,	1890	2,400,000	4·00	
,,	,,	,,	1891	3,800,000	4·00	
,,	,,	,,	1892	5,100,000	4·00	
,,	,,	,,	1893	6,400,000	4·00	

[1] Quintal = 220·548 lb.

Alongside these figures given above, the source of which is
indicated, let us place the official statistics of the State of
Austria :—

				Barrels.
The annual production of petroleum in Galicia was—1883				166.500
,,	,,	,,	1884	233,000
,,	,,	,,	1885	333,000
,,	,,	,,	1886	433,000
,,	,,	,,	1887	532,000
,,	,,	,,	1888	665,000
,,	,,	,,	1889	740,000
,,	,,	,,	1890	816,000
,,	,,	,,	1891	1,088,168
,,	,,	,,	1892	1,096,242
,,	,,	,,	1893	1,192,016
,,	,,	,,	1894	1,200,129

We take the figures of Gintl as almost correct, those of Urban
appear to us slightly exaggerated. We estimate the annual production of petroleum in Galicia in round figures as 2,500,000
barrels.

We are perfectly aware of the fact that this figure exceeds
those given by the official statistics; we further know that a
certain German author, and not one of the least, has reprimanded
one of our French predecessors for a similar addition to that
which we have given to the figures of the annual production of
petroleum in Galicia.

We would recall to this German author that the statistics of
the Austrian custom houses, so far as the valuation of the production is concerned, cannot be regarded as articles of faith, seeing
the extensive scale in which fraud predominates in Galicia, and
that almost all the declarations of production are steeped in false-
hood. The means of verification being *nil*, the fiscal authorities
are satisfied with what they are told. They believe it, or pretend
to believe it, and thereupon base their statistics at the ridiculous
figures which the German author has given us as exact.

CHAPTER IV.

PHYSICAL GEOGRAPHY AND GEOLOGY OF THE GALICIAN OIL FIELDS.

THE CARPATHIANS.

THE chain of the Carpathians which, as everybody knows, forms the whole of the orographical (mountainous) system of Eastern Austria, extends itself like an immense arc, embracing Hungary, with the Danube as its chord. These mountains are particularly rich in minerals of all kinds, from gold and the emerald to common salt and petroleum. Here is what M. Jules-Radu said, amongst other things, of these mountains so far back as 1851 :—

"The Carpathian system is the richest in Europe; it yields all the metals; gold abounds there. Galicia possesses the richest salt mines known."

It is a matter for regret that, at the time when Radu wrote his remarkable work, petroleum was hardly known, for he could have supplemented his description with the fact that the Carpathians possessed riches which certainly were no more worthy of contempt than salt, nor even than gold itself.

As a matter of fact, as we have previously indicated, the petroleum trade in this country has acquired a very considerable importance indeed.

THE GENERAL TREND AND NATURE OF THE PETROLIFEROUS STRATA.

Line of Strike.—A great number of borings have been made upon an almost straight line, somewhat parallel to the mountain chain, making with the latter an angle of 9°. This line is called *Streichungs-Linie* (*Ligne de Direction, Line of Direction. Cf.* English geological term *strike*, the direction at right angles to the *dip*; and

outcrop, the direction of the edges of the strata as they appear at the surface). This line of strike is inclined from north-west to south-east, forming an angle of almost 70° with the meridian.

The railway line, Neu-Sandec-Zagorz-Chyrow-Stry-Kolomea, almost follows this line of strike. The petroliferous strata commence in the west of Galicia, somewhat near to the town of Neu Sandec, and from there more or less important petroliferous centres follow each other in almost unbroken succession until Sloboda-Rungurska is reached.

Conformation and Arrangement of Strata.

The conformation of the ground differs essentially from that met with in America. Here we are more often confronted with disturbed and reversed strata, faults, sudden plunges, in a word, with a difficult underground.

Highly Inclined Strata, Difficult to Work but Prolific in Yield.

The strata are often highly inclined, we have frequently brought to the surface layers making an oblique angle of 77° to 81° with the horizon.

That is, however, rather a good sign if we are to believe what M. Gauldrée-Boileau, the Canadian French Consul wrote some years ago in a learned and interesting discussion on the origin of mineral oils. "*Petroleum,*" he says, "*has been met with in Virginia, in Pennsylvania, and in Ohio, in almost vertical fissures. The richness of the ground in mineral oil would appear to be in proportion to the number of fissures which it contains. The occurrence of petroleum in horizontal strata is not an isolated fact, it is found, on the contrary, in abundance in more or less inclined strata. The principal deposits brought to light proceed from ground exhibiting sure traces of dislocation.*" But the boring of such highly inclined strata becomes very difficult, the drill is very apt to jar on a hard rock, and, instead of following a strictly vertical course, to deviate in the almost vertical direction of a soft bed.

We shall indicate in its proper place what must be done in such a case. But this is the reason why well sinking with the rope, as done in America, cannot be taken advantage of in

Galicia. Rigid rods are required, and even these are not always sufficient to warn the miner against deviation.

Moreover, the schistose or argillaceous strata, when they are highly inclined, are apt to fall in with their own weight, and fill up the well as fast as it is being bored.

Nevertheless, all these wells are far from having the same beds underground, and good strata are often met with ; that is to say, few moving beds ready to fall in ; or if they are met with, they have a certain amount of stability due to their horizontal position. This is frequently the case at Iwonicz and at Polana. It will thus be seen that, at any rate, the presence of an engineer is not altogether useless when the choice of ground is being made.

Signs of Petroleum.

There is often met with, in the petroliferous districts of Galicia, a thin layer of oil, floating on the waves of the streams, and imparting an iridescence thereto.

In those countries where no oil wells exist, this phenomena indicates, in a certain fashion, the presence of naphtha; but where ?

"These traces," says M. Ducasse (*Rapport à la Société de l'Industrie Minérale*) " are not, moreover, always met with, even above the deposits; they simply indicate their existence, at a more or less remote distance, without pointing out more precisely the exact spot where they are to be found." In fact, the same engineer adds : " The strata are completely reversed in certain points ; the beds which form the saddle-backs (anticlines) and depressions are folded parallel to the line of the Carpathians, and form undulations which do not at all correspond with the mountains and valleys on the surface. The line of strike of the petroliferous strata follows the summit of the anticlines (Fig. 2), or approximately so. As a matter of fact, it hardly ever follows the synclines.

Depth to be Bored Before Striking Oil.

Moreover, in consequence of this inclination of the strata, petroleum is met with, in Galicia, at very variable depths. What depth of strata must be bored through before the petroliferous strata are tapped? *God alone knows!* The whole thing depends on striking on the slope of the strata on the saddle (*sattel*), according to the trade term. This is precisely where the difficulty comes in.

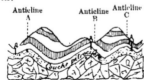

FIG. 1.—Depth to be Bored before Striking Oil.

We have ourselves seen two wells sunk in ground difficult to work. No. 1, at the depth of 94 metres, yielded 100 hectolitres daily. No. 2 has already reached a depth of 280 metres, and is far from yielding the same result.

The same thing took place in connection with two wells under our management. The first yielded, at a depth of 15 metres, one barrel per day, then at the end of a month the flow ceased; at 32 metres the same result and the same duration of flow; at 40 metres the well gave two barrels per day, and this yield has been kept up uninterruptedly until the present time.

Its immediate neighbour, at a distance of 20 metres, is 362 metres in depth, and up to now, it has not given us the slightest hope.

But, although located at very variable depths, the petroliferous strata are rarely met with before a depth of 150 metres has been bored. Oil may certainly be struck before this depth has been reached, but never in large quantity. This surface petroleum comes, as in the case of the wells cited above, from fissures where it has accumulated, or from which it emanates pressed out by the gases.

In such cases, the yield will go on diminishing, and finally gradually disappear. It may also proceed from petroliferous veins of the slightest importance, which are almost always met with before striking the petroliferous sandstone.

DURATION OF YIELD OF THE GALICIAN OIL WELLS.

Here comes the question: What is the lifetime or duration of the yield of a petroleum oil well in Galicia? To this we will reply that the duration of yield is very variable, but in general longer than in America or Russia, and in this connection we shall quote the remarks of M. Julius Noth to the Vienna Mines Congress of 1888 :—

" The yield of the well at Bobrka has lasted ten years, at Ropianka ten years, at Schodnica, Pohar, Harklowa, more than sixteen years ; finally, in a well of the French Company (La Société Française) more than twenty years. At Wietrzno, Well No. 2 has for two and a half years yielded twenty quintaux (say twenty cwt.) of oil daily, and that as regularly as clockwork.

" We may say that the yield of petroleum by a well may lower after ten years, but that it is never exhausted before the end of five to ten years and upwards."

GEOLOGICAL SYSTEMS TO WHICH THE GALICIAN OIL FIELDS BELONG.

The line of strike of the petroleum, in the west of Galicia, can be included in the cretaceous strata as far as Siari, then it is met with, almost without exception, in the eocene strata. The soft eocene sandstones (*Ropianka Schiefern*) are the richest in petroleum ; a cubic metre of these arenaceous rocks may contain more than a hectolitre of oil (*i.e.*, more than 10 per cent. by volume).

The schists and clay shales, rich in fossils. may likewise yield petroleum, but in small quantity.

The chalk sands (upper cretaceous and neocomien) of the Ropianka formation are likewise fairly rich in oil. They generally form the southern portion of the petroliferous line. The northern portion consists almost exclusively of miocene strata, rich in salt deposits : Wielycka, Starasol, Drohobycz. This peculiar conformation would appear to us to prove the marine origin of petroleum deposits.[1]

[1] *Comparative Geology of Galician and American Oil Fields.*—It seems to us that it would be interesting for the reader to make a parallel comparison between the Galician strata and the still more varied American strata.

In Kentucky and Tennessee, petroleum occurs in the lower Silurian ; in Canada, in the lower Silurian, and often, in the south-east of that country, in the upper Silurian.

In Virginia, the oil is very often met with in the carboniferous rocks. In Connecticut and North Carolina, the oil proceeds from the trias. Finally, in Colorado and Utah, petroleum is extracted from the cretaceous, and also often from the tertiary strata.

THE BEARING OF THE GEOLOGY OF THE GALICIAN OIL FIELDS
ON THE ORIGIN OF PETROLEUM.

It may be added that the petroleum oils of this formation are dense and bituminous, or are strongly charged with paraffin.

On the other hand, plates of lignite are often met with in the black shales above the petroliferous sandstone. In a well under our management we found in these same beds, at a depth of 101 metres, lumps of carbon of the size of a man's fist. According to Gesner, this fact proves the purely vegetable origin of petroleum. "*The oil springs*," he says, " are met with generally in the immediate neighbourhood of coal disseminated in the same basin. There can be no doubt but that coal and petroleum have a common origin."

For our own part, we repeat that we have often met with petroleum and lignite in the same well. However, the opinion of M. Gesner is disputed by that of our eminent chemist, M. Berthelot, who looks upon petroleum as the result of a chemical reaction, *viz.*, the direct combination of carbon with hydrogen.

THE EASTERN AND WESTERN GALICIAN OIL FIELDS.

Between Schodnica and Stanislawow we meet with a sinking of the petroliferous vein which divides the Galician oil fields into two regions—the Western and the Eastern.

COMPARATIVE GEOLOGY OF BOTH DIVISIONS.

The geology of this latter region is far more difficult to study than that of the Western on account of the rarity of fossils. The eocene would appear to be mixed with pliocene beds and important cretaceous deposits.

H. Walter, our esteemed master, in his report entitled *A Section in the Middle Carpathians from Chyrow to the Hungarian Frontier*, extracted from the *Annual of the Austrian Imperial and Royal Geological Institute* (vol. xxx., 1880), describes the geological conditions of the Galician oil fields.

The author there remarks that in Galicia the eocene sandstone is always mixed with a yellow argillaceous earth, which, in the Carpathians, is a certain index of eocene strata. In regard to those beds of the eocene formation, composed of hard gravelly sandstone—in a word, almost porphyries—these may already be included in this formation. We sometimes encounter shales forming part of the eocene; they are soft and flat, and most generally of a

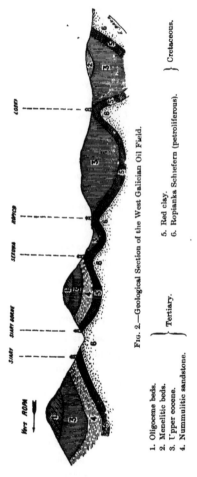

Fig. 2.—Geological Section of the West Galician Oil Field.

1. Oligocene beds.
2. Menelitic beds.
3. Upper eocene.
4. Nummulitic sandstone.
5. Red clay.
6. Ropianka Schiefern (petroliferous).

Tertiary.

Cretaceous.

deep grey colour, but lanceolate gypsum is never present; the hieroglyphic markings are thin and in straight lines; finally these beds are intermingled with narrow bands of a friable sandstone.

THE EXCEPTIONALLY PRODUCTIVE PETROLIFEROUS STRATA.

Amongst the petroliferous strata of Galicia the "Oilmen" have differentiated four principal beds as exceptionally productive of petroleum. To these they have given special names based upon a classification altogether local. These are :—

1. *The Magura Sandstone* (*grès de la* Magura), very micaceous and quartzose, almost always mixed with white clays containing fossils (remains of plants).

2. *The Belowosa Shales* (*Belowosa Schiefern*), sandstone mixed with red micaceous clay. Presence of fucoïds.

3. *The Smilno Shales* (*Smilno Schiefern*), undulated and contorted black laminated shales.

4. *The Ropianka Shales* (*Ropianka Schiefern*), micaceous and quartzose clays of a bluish grey colour; numerous veins of soft granular sandstone. The *Ropianka Schiefern* are *par excellence* the productive petroliferous beds of Galicia.

The following are the beds met with in the boring of No. 2 Well of the French Company (*la Société Française*) at Holowecko :—

VERTICAL SECTION OF STRATA ENCOUNTERED IN BORING No. 2 WELL OF THE FRENCH COMPANY (LA SOCIÉTÉ FRANÇAISE) AT HOLOWECKO.

	Metres of 3·28 feet.
Marl, with sands .	12·54
Grey clay and sand	4·22
Sandstone, mixed with solidified petroleum (*Ropianka Schiefer*)	1·03
Shaly limestone .	5·18
Grey clay and sand	12·44
Green shales	41·29
Loam .	4·11
Grey petroliferous clay	2·58
Quartzose conglomerate (*Ropianka Schiefer*)	1·33
Shaly limestone .	0·80
Green shales	29·62
Sand and clay (gas)	13·08
Grey petroliferous clay	9·21
Green shales	33·18
Grey clay .	12·24
Black shales	2·45
Grey clay .	3·22
Brown clay	9·16
Loam .	9·16
Brown clay .	3·12
Black shales (*Ropianka Schiefer*)	1·24
Shaly limestone (gas) .	2·31
Grey clay and petroliferous sandstone	11·17
	224·63

The petroliferous sandstone goes down as far as 229·59 metres.

It will be seen from the above vertical section that the Galician strata differ essentially in composition from those of America. In spite of the presence of shales, they are more nearly related to the Caucasian strata.

THE WESTERN OIL FIELDS—THE DIFFERENT CENTRES—THE AMOUNT AND NATURE OF THEIR PRODUCTS.

As we have already said, these strata almost follow the chain of the Carpathians, and the working of the petroliferous deposits commence on the western side in the neighbourhood of Neu

FIG. 3.—View of the Works of Ropa (Galicia).

Sandec. We shall enumerate the most important of these works.

Commencing then with the Western oil fields, we find

KLECZANY, with more than 100 wells of a depth of from 200 to 400 metres (all bored by Fauck's system). Reddish-yellow petroleum of density 0·779.

ROPA, which yields a red petroleum.

In these two first localities a very light oil has been found (about 0·780 in density) mixed with a large proportion of paraffin. This petroleum appears, from that cause alone, absolutely yellow. It is highly esteemed on the market, and a high price is got for it because there is no tax on paraffin.

· To the east of these two places, and ranged along the flanks of a mountain, we have :—

ROPICA POLSKA.

SIARI, with 24 machines working and more than 300 wells which yield a dark brown oil of density 0·885.

SEKOWA, with 8 machines working and about 80 derricks (petroleum greenish black, of density 0·837) ; further along are found :—

KOBILACKA (13 machines working) and more than 150 derricks.

KRYG (6 machines, Fauck's system of working) 100 derricks. Petroleum, dark brown ; density, 0·898.

ROPICA RUSKA (3 machines working) about 30 wells, a reddish brown fluorescent petroleum of density 0·808.

Coming south, we meet MÉÇINA, one of the oldest oil fields in Galicia which has no wells but those excavated by hand, the product of which is a black greenish petroleum of density 0·853.

GORZCE.

ROPIANKA.

BARWINECK.

KROCIENKO, a rich deposit yielding 900 tons of petroleum per month.

WOJTKOWA, a deep green petroleum of density 0·820.

LIBUSZA, a greenish black petroleum of density 0·842.

STARUNIA is situated on ground of a saline formation. The petroleum furnished by this district is blackish and of a density of 0·845.

PAGORZIN, density of petroleum 0·847.

LIPINKI, brownish petroleum of a density of 0·849. This oil is much valued by distillers, and yields the following results on distillation :—

		Per cent.
Light Oil	.	9·8
Illuminating Oil	.	45·5
Heavy Oil	.	40·6
Residue and Loss	.	4·2
		100·1

Then coming back to the railway line we find once more :—
TLOKI.

ROWNA, which possesses 10 machines at work, is situated in
the immediate neighbourhood of Wiertzno (25 machines at work
and more than 150 wells), is remarkable for the difficulties
experienced in boring and also for the yield of its wells, one
of which gave more than 8,000 hectolitres per day. This well,
famed through all Galicia, belongs to Messrs. Bergheim & Mac-
Garwey ; it is 240 metres deep, and played during ten months from
August, 1887, until a part of its yield was diverted by a neighbouring well. The gross profit on this well alone during the first
year of its yield was 1,500,000 florins or about 3,300,000 francs
(say £132,000).

WIERTZNO.

A fact worthy of notice is that Wiertzno possesses borings
dating from 1865, and that different contractors have abandoned
these wells, either owing to the difficulty of boring or want of
capital. However, in 1885, M. Klobassa, proprietor of the

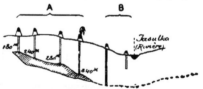

FIG. 4.—Section of the Ground at Wiertzno, showing (in A) the Productive Wells,
and (in B) the Barren Wells.

Bobrka Wells, acquired, less a royalty—averaging 5 to 10 per
cent. on the production—to the different proprietors of the ground,
the right to sink wells at Wiertzno. This right he ceded to
Messrs. Bergheim & MacGarwey for a royalty of 50 per cent. on
the production. The first wells sunk by the latter at once yielded
encouraging results (ten barrels per day at a depth of 220 metres),
and the work was already flourishing when the grand Wiertzno
well spouted.

Messrs. Bergheim & MacGarwey have built, after American
principles, a pipe line 13 kilometres in length which connects
their works with the depôt at Krosno.

Wiertzno petroleum is brown, and becomes greenish when
viewed through light. Its density is 0·872.

This petroleum, so to speak, constitutes the average of Galician petroleum.   It yields on distillation :—

	Per cent.	
Benzine . .	5·0	
Illuminating Oil	55·0	
Heavy Oils .	20·0	
Tar . . .	10·0	
Loss . .	10.0	(Bergheim & MacGarwey's refinery)
	100·0	

The average quality of petroleum extracted at Baku yields us the following figures :—

	Per cent.	
Benzine . .	4·0	
Illuminating Oil	27·0	
Heavy Oils .	45·0	
Tar . . .	14·0	
Loss . .	10·0	(Nobel Brothers' refinery)
	100·0	

We may add further that Russian crude oil is mixed with a quantity of sand which is often present to the extent of 40 per cent. of the weight of the crude oil.

BOBRKA.—THE FIRST PETROLEUM INSTALLATION IN GALICIA.

Quite near to Wiertzno is a village which has a right—to more than a mere heading—to a special notice.   It is BOBRKA.   It was, in fact, at this place, where there are already more than 200 wells sunk, and 4 machines at work sinking others, that the first boring was made in Galicia.   Even at that time, the inhabitants of Boryslaw had brought to market from a remote period the petroleum which they collected on the surface of the ground and used it in the preparation of leather.

FACTS WHICH LED TO BORING.

But here are the facts and circumstances which led to the first installation of a methodical petroleum works at Bobrka.

There existed a well at Bobrka, a well which was always bubbling.   The country people of the neighbourhood attributed to this water—and above all to the grease which floated in the well—a supernatural power against the diseases of cattle.

In 1863, a farmer of the neighbourhood, named Tytus Trzecieski, required to use this oil for his cattle, and caused some bottles of it to be sent to him.

One day, having to go to Lemberg,[1] he took a sample of it with him, and asked one of his friends, the Secretary of the Chamber of Commerce of that town, M. de Lance, to be good enough to analyse it for him. The latter sent the farmer to a pharmacist of Gorlice, M. Ignatius Lukasiewicz, who recognised in the so-called curative liquid all the characters of naphtha, which, even at that time, was much spoken about. M. Lukasiewicz was a man of action, he soon set out to find the proprietor of the ground on which this famous well was sunk (he was M. Klobassa, who afterwards made a colossal fortune and was the first "oil king" of Galicia), and asked his authority to sink wells. The first borings did not exceed 20 metres in depth, but these wells only gave a scanty remunerative yield.

FIG. 5.—Pumping Station in the neighbourhood of Krosno (Galicia).

It was then decided to go farther down, and a well of 150 metres was dug with the pick and shovel and deepened further by the Fabian system installed in 1865 by M. Henri Walter, at the present day Councillor of Mines at Cracow.

Finally, by the aid of steam power, and under the management of M. Fauck, a depth of 400 metres was reached. This was the first works worthy of the name. This well gave 1,200 barrels per day.

M. Lukasiewicz also established the first petroleum refinery in Galicia at Chorkowka.

In 1878, he received from the hand of the Emperor of Austria a gold medal as a souvenir of the riches which he had imparted to his country.

[1] The capital of Galicia, pop. 127,943.—*Tr.*

At the present day the monthly production of the Bobrka oil-field may be valued at 1,500 barrels.

Quite near to Bobrka are the works of IWONICZ, the ground of which is easy to bore, but where it is always necessary to go beyond 400 metres.   Ten machines belonging to a French company are at work in this place.   A small distance from Iwonicz, there is situated a most picturesque place, Bad-Iwonicz, which, as its name indicates, possesses a bathing establishment fed by iodide springs, and also petroleum wells ; there are 3 machines at work.

In the neighbourhood are mounted the derricks of RYMANOW, the property of the Austro-Belgian Company.

We again meet along the line of railway with :—

TARGOWISKA.

KLEMKOWKA.

NOWOSIELCE-GNIEWOSZ with 2 machines and finally Zagorz with 2 machines working, and a dozen of derricks where the oil obtained is pumped.

THE CENTRAL GALICIAN OIL FIELDS.

At this place the San separates the western division from the central division of the Galician oil fields.   We must, therefore, pass to these regions where, first of all, we find :—

OLSANICA.

POTOK, noted for its sources of natural gas.

UHERCE.

ROPIANKA, 7 machines at work and 50 derricks.

WANKOWA.

USTIANOWA.

LESCHZOWATE.   To the north of these works rise the derricks of STARZVWA, a place newly wrought by the Austro-Belgian Company.

LODYNA, an excellent place.   Then starting from the station of Ustrzyki, the railway line quits the line of strike to go in a northerly direction.   We must therefore penetrate into the moun-tains, the outlines of which begin to accentuate themselves.   We find upon this route :—

HOSCHOW.

GALOWKA.   In this place, says M. H. Walter, " the presence

of oil is certain, the position of the beds is regular, and I regard this place as a hope for the neighbourhood ". Further on, and to the south of Galowka, we find a remarkable place :

POLANA. Certain wells have yielded daily more than 7,000 hectolitres of oil at a depth of 150 metres.

STOPUCIANI.

The density of the petroleums of this district is from 0·835 to 0·844.

Finally, we enter the valley of the Dniester, and the first village which we come across at the foot of the Magura in a deep, narrow valley is :—

FIG. 6.—The Station of Uherce (Galicia).

HOLOWECKO, which possesses 13 machines working, 2 borings in hand, and 20 wells already bored. The ground of this place is one of the most difficult, but, in certain wells, the depth of which is not beyond 100 metres, the daily yield exceeds 100 hectolitres of excellent quality oil, having a density of 0·861. No deep wells are to be found in this village, seeing that the working is barely six months old.

Here is the opinion of M. H. Walter on this place : "At Lomna only, at the foot of the Magura, is there any energetic relief, and near there, at Holowecko, close to the mill, we have noticed

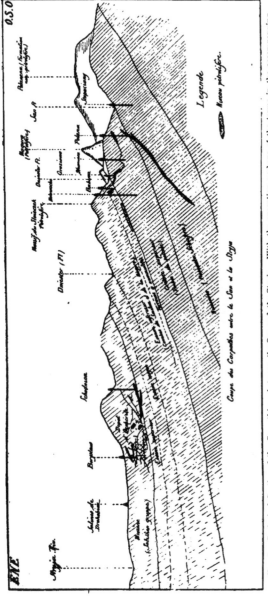

FIG. 7.—Geological Section of the Carpathians between the San and the Stryja. With the exception perhaps of the terms in the accompanying table, the geological and geographical terminology descriptive of the above section will be readily understood.

French.	English.
Legende.	Explanation.
Salines de Drohobycz.	Drohobycz salt works.
Niveau petrolifere.	Petroleum bearing level.
Poches de petrole.	"Pockets" of petroleum.

beds of the same kind as those of Ropianka; we walk here on sure signs of oil, and there is already established in this district a well-located works, with great expectations ".

A French and a German company divide Holowecko between them.

To the north of Holowecko, and in the neighbourhood of the salt works of STARASOL, we find, in the middle of the woods :—

STRZELBYCZ, which, at the present moment, possesses 8 machines working. The petroleum extracted in this district is black, heavy, and bituminous, and commands a very good price in the market; it is specially rich in by-products; its specific gravity is 0·898.

FIG. 8.—Works of the French Company of Holowecko (Galicia).

Finally, to the south-west of this last works, at the entrance to the Galician plains, we find

OZOKERIT (MINERAL WAX).

BORYSLAW. But very little petroleum is extracted at Boryslaw, but, as a set-off against this, it is by thousands of quintaux that the ozokerit (mineral wax) production must be reckoned ; it is by thousands (there are more than 15,000) that the workmen engaged in extracting this substance must be counted ; finally, the number of wells at work in this town exceed 1,200. It is remarkable that ozokerit is nowhere found in notable quantity except at Boryslaw ; its price is very high.

				Quintaux of about 220 lb.
Annual production of Ozokerit at Boryslaw in year			1877	89,610
,,	,,	,,	1878	103,420
,,	,,	,,	1879	90,666
,,	,,	,,	1880	105,270
,,	,,	,,	1881	106,491
,,	,,	,,	1882	99,300
,,	,,	,,	1883	106,299
,,	,,	,,	1884	119,699
,,	,,	,,	1885	130,258
,,	,,	,,	1886	139,254
,,	,,	,,	1887	80,500
,,	,,	,,	1888	87,800
,,	,,	,,	1889	75,600
,,	,,	,,	1890	61,699
,,	,,	,,	1891	61,586
,,	,,	,,	1892	56,376
,,	,,	,,	1893	56,248

Fig. 9.—An Old Works in the District of Boryslaw.

The working of ozokerit at Boryslaw is very primitive, in consequence of the ownership of the wells being spread over a great number of small proprietors, who do not possess large capital. Only two large companies have utilised the progress of modern science ; these are *la Societe de la Banque galicienne de Credit*, with an annual production of 1,500 tons, and the French Company (*la Société française*), 1,000 tons.

DANGERS RUN IN WORKING OZOKERIT.

Besides, the miner is at all times exposed to great dangers : to fire-damp, sudden eruptions of wax, petroleum and sand (Matka's), which often rise to a height of 100 metres, in the well.

Here is what a French engineer [1] said of Boryslaw in 1887 : " I do not believe there is a mine in the world where pressure is more to be feared than at Boryslaw ; pieces of oak, several decimetres square, are often broken like matches ".

The proportion of deaths by accident is therefore enormous ; it is 13 per 1,000 workmen, whilst in ordinary mines the proportion is 1·88 per 1,000.

Farther along than Boryslaw, once more penetrating into the mountains, we find the wells of SCHODNICA (14 machines at work). One of the first wells in Galicia was dug in this place by a Jew named Lindenbaum ; at the present day it is wrought in an energetic manner by the Austro-Belgian Company.

THE EASTERN GALICIAN OIL FIELDS.

Leaving Schodnica, we do not come across any more petroleum works within a distance of 150 kilometres. At last, in the neighbourhood of STANISLAWOW, we find a machine at work and several hand borings.

At NADWORNA there are 3 hand borings. From this point to Sloboda, on the banks of the Pruth, many oil fields are met with. The forests, covering almost the whole of this tract of country, rise up from the imperial treasure, and no working operations are to be found there.

SLOBODA.

As regards the villages of RUNGURY, SLOBODA RUNGURSKA, SLOBODA KOPALNIA and its neighbourhood, over 1,500 derricks may be counted there at least.

The first who wrought on the large scale was a Frenchman, M. Rice, who, after a loss of several hundreds of thousands of francs, was forced to abandon the works. These same works, fallen into the hands of McIntosh, the American, yielded a short time afterwards more than 10,000 barrels a month.

It is at Sloboda, at the works of this same McIntosh, that the deepest well in Galicia is to be found ; it is 508 metres deep, and is unproductive. At the present day, Sloboda is one of the most important places in Galicia.

The petroliferous zone of Sloboda produces a total of about 1,700 barrels per day. It is worthy of notice that here the yield is

[1] M. A. Rateau.

more uniform than elsewhere, and may be estimated at about 25 barrels per well, per day.

It often happens at Sloboda that a reddish-coloured petroleum, which has been baptised with the name of *Cal*, is obtained. This *Cal* is a cause of terror to the well-sinkers ; because this oil, intimately mixed with sand, is of very mediocre quality ; it is almost unsaleable, and is a true misfortune for those engaged in the industry. In spite of this inconvenience, Sloboda produces in greater part good quality petroleum, the density of which, as

Fig. 10.—View of Sloboda Rungurska.

shown by the following table, borrowed from Boverton Redwood, is very variable :—

DENSITY OF SLOBODA PETROLEUM.

Number of the Well.	Date of the Boring.	Depth in metres of 3·28 feet.	Density of the Oil.
1	1865-84	213	0·842
2	1887	194	0·868
3	1881	189	0·835
4	1883	164	0·850
5	1884	225	0·838
6	1885	275	0·845
7	1885	282	0·844
8	1886	274	0·833
9	1886	202	0·863
10	1886	280	0·837
11	1886	305	0·839
12	1886	280	0·837
13	1886	282	0·864
14	1886	250	0·830
15	1886	311	0·839

GEOLOGICAL ORIGIN AND GENERAL PROPERTIES OF GALICIAN PETROLEUM.

Place of Origin.	Geological Formation.	Density of Oil.	Colour.
Kleczany	Cretaceous (middle Neocomian)	0.779	Reddish, yellow
Ropa	Cretaceous (upper Neocomian) *Ropianka Schiefern*	0·800	Brownish red
Siari	" "	0·885	Deep brown
Kryg	Eocene	0·898	"
Ropica ruska	Cretaceous	0·808	Reddish brown, green fluorescence
Mécina	Eocene	0·853	Greenish black
Ropianka	Cretaceous (upper Neocomian) *Ropianka Schiefern*	"	"
Vostkowa	Eocene	0·820	Dark green
Libusza	Eocene middle (menilitic limestone)	0·842	Greenish black
Starunia	Miocene (saline clays)	0·845	Blackish
Pagorzin	Eocene	0·847	"
Lipinki	Eocene	0·849	"
Wietrzno	Eocene	0·872	Greenish brown
Bobrka	Eocene middle (menilitic limestone)	"	"
Iwonicz	Eocene	"	"
Wankowa	Eocene lower	"	Blackish brown
Ustianowa	"	"	"
Lodyna	"	"	"
Polana	"	"	Greenish brown
Holowecko	Cretaceous (upper Neocomian) *Ropianka Schiefern*	0·961	Deep green
Strzelbycz	Miocene (saline clays)	0·898	Black
Boryslaw	Eocene (Magura sandstone)	"	Ozokerit
Schodnica	" "	"	Reddish brown
Stanislawow (oil field)	Cretaceous (Neocomian) nummulitic sandstone		
Sloboda	Lower eocene	0·880-0·897	Greenish brown

Leaving Sloboda, the signs of petroleum are lost in the abrupt valleys of Bukovina. Oil shows itself at KIMPOLUNG, but no works worthy of the name are to be met with there. Further on in Roumania we again find the petroliferous line of strike.

A short tabular description of the principal Galician deposits and the nature of their productions (p. 48) will finish this chapter.

CHAPTER V.

I. NOTES ON GALICIAN LAW, ETC.

(a) *Precautions to be taken in Purchasing Land.*

THE greatest circumspection must be used in the purchase of land in Galicia, for certain laws of the country are so vague that they may be interpreted in several ways. An advice of the first importance is to conclude all purchases (however insignificant they may be) before a notary. A second precaution is to take actual possession, no matter by what means—placard or sign-board, boundary ditch or some sort of building—of each piece of ground acquired. If need be, possession may be taken by deed drawn up before the judge of the district (*juge de la commune*) and signed by witnesses. None of these precautions should be neglected, as such neglect would be repented of later on.

According to the adage that "possession is as good as a title" ("*possession vaut titre*"; *anglice* "possession is nine-tenths of the law"), should there be a flaw in the deeds, and the purchaser has neglected to take actual possession, he may very well be dispossessed, and have to engage in lawsuits which often cost more in expenses than the value of the property in litigation.

The reflection made by Heurteau, the French engineer, gives food for thought : "*It seems impossible to carry out effectually a serious business transaction with this hypocritical, thievish population of a country, where property is badly defined and its mode of transfer often uncertain*".

(b) *Price of Land.*

The price of land in Galicia, as in new countries in general, is very variable. We have seen ground sold at prices varying from

60 to 100 francs the hectare.   We shall give but a single instance. We bought ground in 1884 at Wiertzno at the rate of 25 florins (about 50 francs the hectare), and in the interval of 15 years, the price has multiplied more than a hundredfold.

Petroliferous land, excellently situated, may, at the present day, be still bought at the lowest prices.   But, on the other hand, we know owners of property who have paid more than 1,000 francs the hectare, and that in situations absolutely unfit for cultivation.

(c) Leasing Land.

In those districts, where land has acquired such a high value, it is better to lease the ground for 25 years from the landlord, the lease—be it well understood—being renewable on the same conditions, and ceding the right to sink wells at a rate averaging so much per well.

(d) Cost of Concessions.

As in buying land, the price of these concessions is very variable.

Here is a transaction of this nature : In 1890, we bought at Holowecko the right to sink wells in a certain place, at the rate of 170 francs per well.   In 1891, after a fortunate stroke of boring, we were desirous of taking a lease in a neighbouring spot. Naturally, competitors, attracted by the news, had made purchases in the same village.   We could only remain masters of the situation by paying to the ancient proprietor a slump sum down beforehand of 2,000 francs for each well opened, and by afterwards paying him a royalty of 10 per cent. on the gross revenue of the wells.

And these prices may still further increase at a great rate. A sudden spouting may increase the value of the ground surrounding that of the fortunate proprietor one hundredfold.

(e) Legal Formalities to be Observed in Renting Land.

In the case of renting, just in the same way as that of purchase, it is necessary to cause the deeds to be drawn up with the greatest of care, and to take actual possession of the land, because failing this latter precaution the lessee may very well forfeit his titles at the end of five years.

(f) Restriction on Minors. Women's Rights.

Before any one has the legal right to buy outright or lease
land, he must either have attained his majority, or be free from
tutelage; and a woman may buy, lease, or sell underground
goods (*des biens fonds*) even without the authority of her husband,
unless the laws of her natal country prohibit it.

In Austria, woman has the same civil rights as man.

(g) Formalities to be Observed before Starting Work.

In regard to the purchasing or leasing of a location before
commencing work, the following formalities should be complied
with by the purchaser or lessee :—

1. *Application to Council of Mines for Permission to Work.*—
Make a declaration before the Mining Supervision Council at
Cracow or at Drohobycz of his intention to open up petroleum
wells, and the mode of working he intends to follow (whether by
machine or by hand); also the place where the works will be
situated. The right of working is conceded, and 4 florins,
about 8 francs, have to be paid annually to preserve this
concession.

2. *Registration of Manager or Overseer.*—It is understood that
the Commissary of Mines be informed of the person responsible
for safe working and the observation of the laws.

3. *Registration by Prefect of District.*—When permission to
work is granted, it has to be registered at the office of the prefect
of the district in which the works are situated.

These formalities accomplished, boring may be commenced.

*(h) Distance from other Wells and from Habitations and Fires
Generally.*

1. *Distance from Habitations.*—The wells are to be situated at
least twenty metres from any dwelling-house, forge, or steam-
engine—from, in fact, every place where an open fire is burnt.

2. *Distance from other Wells.*—Their distance from other wells
must not be less than sixteen metres. Finally, it is well understood that if the well be hand wrought, and consequently
ventilated, or even only uncovered, no fire can be made within
twenty metres of the mouth of the well.

II. Galician Laws as to Master and Servant.

We now give, believing it may prove useful to our readers, a sketch of the Austrian laws regulating the relations of master and servant.

(a) *Notice of Dismissal and of Intention to Leave.*

1. *Workmen.*—The Austrian mining laws prohibit the engagement of a workman unless he be known, without his having a pass-book. This pass-book is kept by the proprietor of the works during the whole of the time that the workman remains in his service. It must be shown whenever requested to the authorities.

Fig. 11.—An Oil Well on Fire at Mrzaniecz (Galicia).

When the workman leaves, he has the right of requiring the inscription, in his pass-book, of the date at which he commenced work, and the date at which he left; a certificate may be appended thereto.

2. *Penalties.*—No workman can leave (at least without an agreement between the two parties) without having beforehand given fourteen days' clear notice. The same holds good in regard to the master who dismisses a workman; the latter should receive notice of his impending dismissal fourteen days beforehand. In cases where such notice has not been given, the employer is obliged to pay the workman—the same as though he had been at work—fifteen days' wages.

(b) Professional Staff.

As far as regards employés and engineers responsible to the Supervising Council of Mines, they ought to have three months' notice beforehand or be paid for that time; it goes without saying that in a contrary case the employé or engineer ought to give three months' notice of his intention to leave, or in case of his abrupt departure to forfeit his last three months' salary.

(c) Compensation when Incapacitated by Accident. Master contributes to Men's Accident Fund.

The custom is that the doctors' bills and medicine of those injured whilst at work are defrayed by the employer, and that half their wages be paid to them so long as they are incapacitated from work. These disbursements are paid, one-half by the employer and the other half by the body of workmen. That is to say, that these latter ought to pay to the accident fund of the district, 2 per cent. of their wages, and that the employer should contribute to this fund a sum equal to that contributed by the whole of his workmen.

(d) State Officials must be Notified of all Accidents.

In case of accident, the Supervising Council of Mines must be immediately informed.

Many works do not conform to these latter rules, but on that account they are scarcely ever free from squabbles afterwards. Their example is not one to follow, the more so as these rules protect both employer and workman.

III. Hints on Economical Working in Galicia.

(a) High Quality and Home Consumption keep up Price.

The petroleum of Galicia is of superior quality. It far surpasses in value that of Baku. If it has never reached the high prices at which American petroleum was sold in the earlier days of its discovery, it has never on the other hand sunk to such a low price as that of Baku. This is due to one essential reason: Galician mineral oil is wholly consumed by the Austrians themselves; they have the advantage over all comers of being in close proximity to the great manufacturing and consuming centres, to the important distilleries of Kolomea, Drohobycz, Fiume, Trieste, Buda-Pesth, Florisdorf near Vienna.

This proximity necessarily causes a very great reduction in the cost of freight to these centres.

(b) Causes which tend to Lower Prices.

The price of petroleum has however lowered more than 50 per cent. in Galicia during the last ten years. The reason of this is : (1) the discovery of important new deposits ; and (2) competition, which grows stronger every day.

(c) Prices Ruling in Different Localities.

In the Eastern and Central Oil Fields petroleum is sold on an average at the rate of $4\frac{1}{8}$ florins the 100 kilogrammes, the average weight of a barrel being 150 kilogrammes ; this brings the cost of the latter to $6\frac{1}{3}$ florins, or about 13·75 francs per barrel.

In the Western Oil Fields the prices of petroleum, as well as the manner in which it is measured, vary slightly.

The hectolitre of petroleum is worth on an average $3\frac{1}{2}$ florins, and the barrel 7 florins—about $15\frac{1}{4}$ francs.

(d) Causes of Variation of Prices.

1. *Want of Communication.*—The difference in the price obtained for the product of those two fields is due to the fact that in the one case communication is more easy and direct than in the other.

2. *Good or Bad Quality.*—Be it well understood that we are now only speaking of average prices ; it goes as a matter of course that certain oils, on account of their bad quality, command but very low prices. The petroleum of Strzelbycz, in the district of Staremiasto for example, fetches no higher price than $4\frac{1}{2}$ florins— about 9·8 francs (say 7s. 6d. the barrel). On the other hand, other oils fetch much higher prices than our average.

3. *Influence of Season on Prices.*—Seasons also influence the prices of mineral oil. In winter, petroleum is dearer than in summer; in fact the days are short and a great deal of light is required, and the freight trains having started, the freight of oil is cheaper. It is better, therefore, where the production is but

little, to try and store it during the summer, and to dispose of it in the winter.

(e) The Selling of Petroleum—Contracts.

1. *Selling per Fixed Number of Barrels.*—The Galician refineries buy petroleum and generally pay for it in cash. The custom is for the owner of the wells to obtain a bond of security to an amount commensurate with the number of barrels, which he undertakes in writing to deliver, and which the distiller is bound to take delivery of, at the price agreed upon in his deed of purchase, no matter however far prices may have fallen.

2. *Selling whole Production for Fixed Period.*—It sometimes happens that instead of making a contract for a fixed number of barrels, the owner of the oil well binds himself to deliver, during a specified period of time, as agreed upon (say three months, six months or one year), all the oil which he extracts. It is equally advisable in this case for the well owner to assure himself, in regard to the validity of the terms of his contract, by obtaining a bond of security.

Moreover, we recommend oil works' directors to provide their own barrels, and not to use those sent them by the distillers, who, in such cases, deduct 25 per cent. from the gross weight for the tare.

It will be easily understood that it is a fool's bargain as far as the vendor is concerned (an empty barrel weighing on an average 21 kilogrammes and holding 150 kilogrammes of petroleum). We cannot, therefore, insist too much on the recommendation given. It is of course well understood that the tare of these barrels has to be marked in such a manner as will facilitate the calculation of the weight of the quantities delivered.

(f) Filling the Barrels.

Effects of Warm and Frosty Weather on Barrels filled too Full.— A last advice is never to fill the barrels completely. In summer the petroleum expands considerably, and loss may often occur in this way during great heat to the extent of one-fifth of the original weight of the oil. In winter the frost may cause the barrels to burst, and the loss of the contents is added to that of the containing vessel.

(g). *Contents of American and Galician Barrels.*

With regard to barrels, it may be useful to state that generally the American barrel, or small Galician barrel, contains about two hectolitres (say 44 *English* gallons), and that the large barrel (only used in countries with good roads) contains double. In the present case we must be understood as speaking of the American barrel.

PROPERTIES OF THE DIFFERENT GALICIAN PETROLEUMS.

Place of Origin.	Density.	Colour.	Remarks.
Kleczany	0·779	Reddish yellow	Rich in paraffin
Ropa	0·800	Brown red	,,
Ropa (2)	0·780	Yellow	,,
Siari	0·885	Deep brown	,,
Sekowa	0·837	Greenish black	,,
Kryg	0·898	Brown	,,
Ropica ruska . . .	0·808	Brown red	,,
Méçina	0·853	Greenish black	,,
Wojtkowa	0·820	Deep green	,,
Libusza	0·842	,,	,,
Pagorzin	0·847	,,	,,
Lipinki	0·849	Brown	,,
Wietrzno	0·872	Greenish brown	,,
Wankowa	0·835-0·844	,,	,,
Holowecko	0·861	Deep green	,,
Strelbycz	0·898	Black	Bituminous
Starunia	0·845	,,	,,
Sloboda	0·833-0·868	Deep green	,,

CHAPTER VI.

GEOGRAPHICAL POSITION OF THE OIL FIELDS.

THE Roumanian oil region, as far as its geographical position is concerned, presents more than one analogy with the Galician oil fields which we have just described.

We may even say that, geologically, the oil fields of Galicia and Roumania, are one and the same. Only, political conventions have, by their acts, established a distinction indispensable to the understanding of this treatise.

Like Galicia, Roumania extends along the arc of a circle formed by the Carpathians, and occupies the eastern portion of that arc. It is in the flanks of these mountains, at a point almost equi-distant between the Roumanian *Thalweg* (channel of the river) and the chalk of the Carpathians that the line of strike of the petroliferous strata is to be found. This line divides into two parts.

LINES OF STRIKE OF THE PETROLIFEROUS STRATA.

1. The first extends in a direction somewhat parallel to the line of railway which runs between Lemberg in Galicia, and Bucharest, or Bukharest (Bucuresci), the capital of Roumania.

2. The second line is somewhat parallel with the course of the Danube, which it follows to its embouchure.

GEOLOGY.

Association with Common Salt.—In the same way as in Galicia, petroleum is found in Roumania in association with common salt (*Roumania produces more than* 100,000 *tons of common salt annually*).

But little progress has been made in the study of the geology of Roumania. The reports of MM. Cobalescu and Porumbaru contain precious information upon this subject.

[1] Sometimes written Rumania; at other times Romania.—*Tr.*

The petroliferous strata of the Western oil fields are generally
miocene; but, according to the nature and configuration of these
strata, they may be divided into three distinct groups.

1. Congerian beds.
2. The Paludina beds.
3. The Argilo-saline formation.

Fig. 12.—Map of the Roumanian Oil Fields. (The Deposits are indicated by the
sign **X**).

The first (the congerian beds) are distinguished by large banks
of sandstone and sand, with an almost entire absence of fossils,
and containing isolated blocks of a harder sandstone; these sandstones have a yellowish-white appearance, are coarse-grained and
exceedingly friable. It is not rare to find, in these same beds, in
certain points, developments—more or less important—of a coarse-

grained conglomerate equally friable and analogous with the Galician conglomerates. It contains, like them, chlorite, mica, gneiss, and quartzose particles.

We meet, for instance, with a very well-developed example of these conglomerates at DRAGONÈSE, near to the petroleum mine of M. Cantaguzène. Above this bed we find brown and plastic argillaceous shales with prints of plants and yellow deposits of sulphur (*Terrains de Truckawice près de Boryslaw, Galicia*).

M. Pelisse has also studied very completely the strata which descend from the Carpathians towards the Danube. According to the latter, the salt-bearing strata are intimately mixed with the petroleum-bearing strata. Nevertheless, in the Neogene beds, M. Pelisse distinguishes two different kinds of strata :—

    1. The Sarmatic strata, rich in common salt.

    2. The Congerian strata, essentially petroliferous.

As regards ease of boring, the nature of the Roumanian strata renders it equally difficult with that of Galicia ; their arrangement and composition are identical. The signs of petroleum are met with equally frequently, and are equally powerful arguments in favour of the working and development of the Roumanian oil fields.

Given such conditions, therefore, why then does not the petroleum industry develop in Roumania ? How is it, seeing that Galicia prospers, Roumania seems doomed to perpetual bankruptcy ?

This is due to four principal causes :—

1. GEOGRAPHICAL SITUATION.

The Carpathians may be compared to a fortress, Galicia and Roumania occupying the outside glacis. The Galician glacis fronts the north and west. It looks towards the countries of industrial activity and commercial enterprise. It is thus by its position in the neighbourhood of the district, whose needs it has to satisfy, an advantage accentuated by the fact that the countries of which we speak, Germany, France, and the remainder of Austria, have up to now a production of petroleum which may be regarded as *nil*. Roumania, on the contrary, located on the south-eastern glacis, would appear to offer its products to Turkey, which, not distilling, receives its products all in the *refined state* from Austria or Russia, and, on the other hand, to Russia, which does not know what to do with its own enormous production.

2. BAD ROADS AND WANT OF RAILWAY COMMUNICATION.

Roumania is a mountainous country, especially in the oil fields district. The roads are bad, railways are awanting; civilisation, and, consequently, industry, can only be introduced with difficulty.

3. PROTECTIVE TARIFFS.

The third obstacle to the extension of the petroleum industry falls quite within the domain of political economy. Roumania possesses few refineries; it exports its crude oil, and as all the producers have to sell to foreign countries, principally Austria, they are obliged, so as to cope with Austrian competition, to accept lower prices caused by the cost of freight, and the burden of a protectionist tariff imported upon them by the Austrians.

4. LAND LAWS IN REGARD TO ALIENS.

The Roumanian law interdict aliens from becoming proprietors of land in any district of Roumanian territory. It follows that the aliens who have made the fortunes of Galicia and Baku but very rarely wend their way towards Roumania.

These facts explain why the processes of working in Roumania are almost everywhere of a primitive nature. The petroleum industry there is in its infancy. It is the golden age of sinking wells by hand digging.

FUTILE ATTEMPTS AT WORKING BY NATIVES.

Just as it was in the first years of the discovery of petroleum in Galicia, all the proprietors whose ground may be foreshortened to a small extent by the line of strike of the petroliferous strata are oil prospectors; they have their well, or, rather, their hole, to be more accurate. In this manner, whilst having on hand a mass of wells excavated by the pick and the shovel, instead of by studied boring, they burst up and ruin themselves without any result, and in this way gradually bring discredit on the Roumanian oil fields.

Wells are rarely pushed below a depth of 100 metres. It will be understood that with strata analogous to that of Galicia this depth is not sufficient. Looking, however, upon these embryonic workings as serious undertakings, several of our learned engineers have peremptorily declared " there is no petroleum in Roumania ". To this we would reply that oil is not to be found by scratching the ground, but that by working seriously and in normal conditions success may be attained. We could cite to them the example of the grand spouting well, the " Suspiro " of DRAGONÈSE (District of Prahova).

" POCKETS " AND THEIR ILLUSIONS.

Another objection that has been formulated against the Roumanian oil fields is as follows : " The yield of the Roumanian wells does not last ". Ah, well, the oil wells of Galicia have been accused of the same fault. The statistics which we have already given are, as far as Galicia is concerned, a conclusive reply to this accusation.

In Roumania, cavities are often met with at a shallow depth, thin layers of porous sandstone from which the petroleum sweats in greater or less abundance. These beds or cavities we call *pockets*. The total duration of their flow is subordinate to their volume and also the rapidity of the flow.

As far as we are concerned, and every other serious engineer as well, a *pocket* is at the most but an incident—a mere detail. The worthy country squire—the owner of a Roumanian oil well—does not, however, look at it from that point of view; the first day of the sweating he thinks he has found all the gold of Peru. He would not part with his hole for millions, not he! But at the end of a fortnight the *pocket* is drained dry. Distressed, but not discouraged, our worthy gentleman abandons his well (30 metres in depth) as exhausted (!) and, in hope of better results, digs another.

That is the way they work in Roumania.

We, naturally, in making these remarks, exempt such intelligent proprietors as Hardy, Lang, etc.

We have said that the petroliferous strata of Roumania may be divided into two oil fields :—

1. The Western oil field—an immediate continuation of the Galician oil field.

2. The Eastern oil field, extending as far as the Black Sea.

1. THE WESTERN OIL FIELD.

The first of these oil fields—the Western—forms a line which extends between the river Sereth (affluent of the Danube) and the Carpathians—a line which bends strongly to the south following the direction of those mountains.   There are numerous petroliferous deposits in this oil field.   Commencing in the north, we find few works worthy of the name.   But, on the other hand, we must say that unequivocal signs of petroleum succeed each other without break (unless it be during some miles in a sinking of the strata in Bukovina) from Sloboda in Galicia as far as Bacau in Roumania, where we find for the first time the deposits wrought.

To the north of Bacau we come on MAINESCI with 100 wells, almost all hand dug, which yield about 20 barrels per day.

FIG. 13.—An Ancient Work in the Bacau District.

ITANESCU.

TIRGU OKNA, terminus of the railway which connects with the line from Lemberg to Bucharest, with Adjud.

TETZARU.

COMANESKU, which is the petroliferous centre of the district on account of its situation on an important route.

In the district of ROMNICULU SARATU we find numerous natural oil wells.   The petroleum-bearing lands there are unfortunately neglected on account of the want of ways of communication.   Rudimentary works may be pointed out as existing here, producing some tons of petroleum of superior quality in the neighbourhood of DIMITRESCI, DDOBESTI and SAVEJA.

The petroliferous centre of the Western oil field of Roumania is in the district of BUZAU (Buzcu or Buzco).   The most earnest of those engaged in the industry are located there.   There the most flourishing establishments are to be found. It is but right to add that three lines of railway intersect this district, *viz.* :—

1. Those going towards Bucharest.

2. Those coming from Lemberg and Kronstadt in Austria (Transylvania).

3. The small railway with a single line of rails from Filipesci to the petroliferous centre of ILANICA.

The principal places at work in this district are :—

FENDUL SARATU, the 300 hand-bored wells of which produce 30 barrels daily.

BREZA, on the line of railway coming from Hungary.

Starting from this spot, *the Western Roumanian Oil Field* bends towards the *west*, occupying the district situated to the north of Bucharest. We there find :—

MATITZA.

DRAGONEZA where a well was sunk according to the accepted rules of boring—the "Suspiro"—which has yielded very beautiful results and has spouted for a long time.

COLIBAZI, the property of the Government, which possesses more than 200 wells, one of which has yielded more than 50 barrels per day.

2. THE EASTERN ROUMANIAN OIL FIELD.

The Eastern Roumanian Oil Field forms a line almost parallel with the Danube, which runs from Braïla to the embouchure of this river passing through BATOY JALOMITEI, DUDESCHI.

This oil field is badly wrought. Appearances lead us to prophesy rich deposits. The poor success which oil prospectors have had in this district is due to the depth of the petroliferous strata, the working of which would require plant able to pierce beyond the alluvial beds of the Danube. Sure signs of petroleum are manifest at GARVAN, TATAR, KIOI, HIRSOVA, DOIAN, and BABADAG.

We must remark here that the Eastern Roumanian oil field would appear to be in direct connection across the Black Sea with the petroliferous strata of the Crimea.

Roumanian petroleum varies in density between 0·839 and 0·896.

The production of petroleum in Roumania in 1885 was estimated by Fernand Hue at 125,000 barrels.

In 1896, the Custom House statistics gave 133,200 barrels, but this total would appear to us to be below the truth. In fact, according to the report of the Austrian vice-consul at Ploesti, it would appear that the Roumanian production is as follows:—

District of Glodeni	.	.	.	.	.	9,000	Tons.
,, Campina	.	.	.	.	.	8,000	,,
,, Doftaneti	.	.	.	.	.	8,000	,,
,, Matitza	.	.	.	.	.	800	,,
,, Buzau	.	.	.	.	.	6,000	,,
,, Tega	.	.	.	.	.	1,000	,,
Other Districts	.	.	.	.	.	2,000	,,
						34,800	Tons.

or about 275,000 barrels.[1]

ROUMANIAN MINING LAWS.

Some notes on the laws of Roumania may well terminate this study of the Roumanian oil fields.

The mining administration of the country divides underground working into two kinds, *viz.,* *mines* and *quarries*. Mines are conceded directly by the State in conformity with the organic law. The quarries (including the working of petroleum) are the exclusive property of the proprietor of the ground. Working is free, without concession, subject only to simple supervision exercised by the mining administration. The proprietor pays a tax to the Government of 1 per cent. on the net production of his works.

[1] The European Oil Co. now occupy under lease, varying from fifteen to twenty-one years, about 2,000 acres of land in the Roumanian oil fields. The first deep boring sunk on the property spouted at a depth of 222 metres (72·8 feet), on 22nd February, 1900. After a heavy upheaval of sand the eruption settled down, leaving the casings full of oil up to 30 metres (98·4 feet) of the surface. A recent report gives the production of the well as 10,000 kilogrammes (say 10 tons) of oil per day. A pipe line twelve miles long to the railway, with the necessary storage tanks, has been completed.—*Tr.*

CHAPTER VII.

PETROLEUM IN RUSSIA—HISTORY.

TREND OF THE PETROLIFEROUS STRATA—CONTINUATION OF GALICIAN OIL FIELD.

THE Roumanian oil field, as we have just seen, is a prolongation of the Galician oil field, which it continues on in a line which wends its way towards the east. This line is staked off throughout its entire length by more or less important oil workings, the last of which is found at Dudeschi, not far from the Black Sea.

It is quite clear to us that under the waters of the old *Pont Euxin* (*Pontus Euxinus* = *the Black Sea*) are to be found the petroliferous strata which we meet with further on in the Crimea.

Opposite this peninsula, on the other side of the Straits of Ienikalé,[1] another peninsula is met with, that of Tamansk, consisting also in great measure of strata from which the oil flows naturally to the surface of the soil, and which is the first station of the line which, traversing the Caucasian isthmus from sea to sea, comes to a halt in the Caspian at the *Russian* "Petrolia" of Baku.

We may add that the grand petroliferous vein of the old continent does not stop at Baku; it continues across Asia in the relatively straight line which it has traced in Europe, losing itself first in the Caspian Sea, then mapped off by the Russian Trans-Caspian oil field of Tcheleken, on to Merv, to India, through the Punjaub to Burmah, to Bhœma, to Pathar, to Barongah, to Java, to Australia.

HISTORY OF RUSSIAN PETROLEUM.

But to return to Europe, the knowledge of petroleum in Russia is very ancient. The natural springs of naphtha served there

[1] Kertch.

according to the epoch—firstly, as objects of worship, then as medicinal remedies, and, finally, as fuel.

The Fire-Worshippers.—Baku was always the holy city of the fire-worshippers. Zoroaster, the founder of the *Parsee* religion, of the third or sixth century B.C., whose doctrines are to be found in the *Zendavesta*, had indicated fire, but especially eternal fire (that which burnt without apparent fuel by a phenomena then incomprehensible, and the cause of which seemed to proceed from an occult divine will), as the emblem of all knowledge, as the incarnation of the beneficent deity.

These fires, proceeding from the natural petroleum gas, were numerous in the Caucasian isthmus. The Caspian Sea, with its naphtha-covered waters, which an accidental cause might very well inflame and cause it to look like an infernal ocean, with waves fringed with flame, should suffice to explain how the worship of fire, fostered by such extraordinary phenomena, first originated and then maintained itself in the neighbourhood of Baku even to the present time.

Dupuis, Ch. Fr. (author of *The Origin of Religions*, 1742-1809), speaking of this religion, said in 1795: "The ethereal fire which circulates throughout the whole universe, and of which the sun is the most apparent hearth, was represented in the *Pyrées* by the holy and everlasting fire kept up by the Parsee priests.

"Each planet, which contains a portion of it, had its *Pyrée* or its particular temple, where incense was burned in its honour. The people went into the chapel of the Sun to render homage to this star and to celebrate its festal day; into that of Mars and Jupiter to honour Mars and Jupiter, and so on with the other planets.

"Before going to battle with Alexander, Darius, the King of Persia, invoked the Sun, Mars and Eternal Fire. On the top of his tent was an image of this star, enclosed in a crystal, and which reflected its rays to afar. Amongst the ruins of Persepolis (the capital of Ancient Persia, captured by Alexander, B.C. 350) there may be distinguished the figure of a king kneeling before the image of the sun; quite near is the holy fire preserved by the priests, and which Perseus they say, had formerly caused to descend upon the earth.

"The Parsees, or the descendants of the ancient disciples of

Zoroaster, still address their prayers to the sun, to the moon, to the stars, and principally to fire as being the most subtle and the most pure of all elements. This fire was specially preserved in the Aderbighian, where was the grand *Pyrée* of the Persians, and at Asàse in the country of the Parthians.

"To the north of Asia, the Turks established *near the Caucasus* had a great respect for this fire."

Let us be permitted here to make a diversion from religion in favour of the fable of Prometheus, who was enchained on the Caucasus for having carried away the fire of Heaven. From that very day an eagle unceasingly devoured his liver, and this eagle vomited (according to Appolonias of Rhodes, the Greek poet, B.C. 276-186, who wrote the *Argonauts*, in which the fable is narrated) a blackish liquor, which the Greeks called *Naphté*, which was to render Jason, the chief of the Argonauts, who was to capture the golden fleece, guarded by the dragon, invulnerable. This fable may be regarded as a mythological narration of the origin of the Caucasian petroleum.

Fire-worship was the cause of many persecutions—first of the Christians, then of the Mahomedans. The Guèbres gradually dispersed themselves. They fled to the coast of India and established a *Pyrée* at Seate, where they re-formed themselves under the name of Parsees.

But the city of everlasting fire, the holy city, always attracted them, and, so far back as the twelfth century A.D., they re-established at Baku a temple, still in existence, where the sacred fire has never ceased to burn. About this time petroleum began to be used as a medicine.

Marcus Polo and Russian Petroleum.[1]

In the thirteenth century A.D., Marcus Polo described the properties of the petroleum of Georgia (a province in Russian-Asia to the south of the Caucasus). He undeniably establishes its abundance, and celebrates its therapeutic virtues. "They make use of it," says he, "to coat the camels with, which have the scab." He also makes the remark, "If the oil was not fit for eating, it could very well be burnt".

[1] See page 7.

PAUL DUVAL ON BAKU PETROLEUM AND ITS GEOLOGY.

So far back as the seventeenth century, Caucasian petroleum was used for illuminating purposes. We, in fact, read in the *Universal Geography* of Paul Duval (1688) : " The Sultan of Turkey holds there (in Georgia) two strong places, Tiflis and Cori-Derbent, often disputed by the Turks and Persians, and in the passage known as the iron gate. These are the remainder of the Caspian ports which one sees from Mount Barmach, with some medicinal oil springs. And, further on, Bakouë (Baku) gives its name to the Caspian Sea, and has in its neighbourhood *a source (well) of oil which is burnt throughout all Persia.*"

We cannot resist the desire to quote one more passage from this same author. This is what Duval remarks with great discrimination : " All this is found between the Black Sea and the Caspian Sea, which are believed to communicate with each other, because they both have the same species of fish, and because the ground between the two resounds and seems to be hollow when one rides across it on horseback."

KŒCHLIN SCHWARTZ ON PHYSICAL GEOLOGY OF BAKU.

Compare this remark of Duval with those made in 1879 by a manufacturer of Mulhouse, M. Kœchlin Schwartz, in his book, *Un touriste au Caucase* : " . . . . These hills, these inflations of the crust of the earth, may very well, people say, disappear and efface themselves some fine morning. One feels under his feet, as it were, a mass in perpetual fusion. We comprehend that subterranean work has gone on there, and that it is still going on ; but work of quite a special nature, quite peculiar, which surpasses all our comprehension. Evidently this ground has nothing in common with what we folk, dwellers of ancient Europe, call the old *terra firma.*[1] We are in presence of the unknown."

JONAS HANWAY AND PETROLEUM.

In 1754 an Englishman (Englishmen are to be found wherever there is petroleum), Jonas Hanway, entrusted by his Government

---

[1] This clause does not admit of a literal translation, the suggestion intended to be conveyed is far more forcible in the original, *viz.* : " Avec ce que nous autres habitants de l'antique Europe, nous appelons *le vieux plancher des vaches !* "—*Tr.*

with a commercial mission on the shores of the Caspian Sea, visited Baku. He has left us his impressions in his book, *An Historical Account of the British Trade over the Caspian Sea.*

Certain passages from this work have been translated by M. Fernand Hue in his excellent work *Le Pétrole.* As these quotations are important at this point, we borrow a few lines of his translation : " What the Parsees or fire-worshippers call eternal fire is a very curious phenomenon. The object of their worship is situated about an English mile to the north-east of the village of Baku on an abrupt and barren rock. A few feet from the temple is a recess in the rock, in which is an underground opening six feet long by three feet wide, from which there issues a perpetual flame very similar to that produced by alcohol in burning, but even more pure. When the wind blows it rises sometimes to the height of eight feet. It is not so ardent in calm weather. The ground in the neighbourhood possesses a surprising property. By digging two or three inches and applying a glowing coal to the denuded ground it takes fire immediately. The flame heats the ground without consuming it, and without communicating its heat to neighbouring objects. Near to this locality sulphur is extracted and springs of naphtha. are found. Baku supplies Ghilan-Mazanderan and the other neighbouring countries with naphtha. The principal point of extraction of the brown or black oil is the isle of Wetoy (Swiatoï, or Holy Isle) situated at the extremity of the Aspsheron peninsula.

" The Persians load the naphtha in bulk in their miserable ships in such a way that at times it is spread on the waves for many miles. When the weather is dull and misty, the springs of the Holy Isle spout more energetically ; the naphtha often takes fire on the surface of the ground and rushes into the sea setting it on fire for miles. During bright weather the spring hardly rises two feet above the ground ; as it spreads out, the oil gradually coagulates to such an extent as to completely choke up the orifice ; then the naphtha finding no further vent pierces the soil a little further on and recommences to run. The natives transport the naphtha in buckets into pits or reservoirs, the one connecting with the other, leaving the heavy water with which it is mixed as it comes from the spring at the bottom of the first. It gives off a strong nauseous smell, and is but little used except by the Persians and the neighbouring poor people for burning in their lamps or in the cooking of food, to which it communicates a disagreeable taste. It is stored away from houses in earthenware vessels, buried in the ground so as to avoid all contact with fire, for it is highly inflammable."

Further on Jonas Hanway mentions a lighter oil of a brighter colour. It is that which is still found at Sourakhany of which he speaks [1] : "The Russians drink it as a cordial and as a medicine, but it does not intoxicate. Taken inwardly, it is, they say, an excellent remedy for gravel, chest affections, headaches. Outwardly it is applied for scurvy, gout, etc., but it ought not to be applied to an open wound because it penetrates instantaneously into the blood and may occasion great pain. It possesses also the properties of spirits of wine in removing grease spots from silk or woollen goods, but the cure is worse than the disease, because it imparts an abominable smell. It is exported, they say, to the Indies, where it is employed as a varnish, which is the most beautiful and the most durable known."

SUMMARY.

We thus see that so far back as this time a commercial use was made of naphtha. Some years afterwards, in 1800, the local commerce, represented by Armenian caravans or Persian barques, began to assume a total of 120,000 barrels per annum; and it is to be remarked that at this time not a single well had been bored, and that the whole of the oil thus exported was the product of natural springs like that of the Holy Isle of which (Hanway) the English traveller speaks.

RECENT DEVELOPMENTS.

In 1801, the Russian Government, seeing the profit that could be made out of the sale of naphtha produced in its own territory, monopolised the sale and the working of the wells. This monopoly was a great misfortune, for it stopped the scope of this new industry; but, as the Government drew good profits from it, they

[1] It will be understood that these are translations of Hue's rendering of Hanway's rare work, and not exact quotations from the original.—*Tr.*

made it last a long time. It was only in 1872, that the petroleum industry was declared free. There will be no need of literary periods in writing the history of Russian petroleum during the latter part of this century; we need no longer recall to mind either Darius or the Argonauts—a simpler table of statistics suffices. The figures are quite eloquent enough. Let them speak for themselves :—

Petroleum Production of Vicinity of Baku from 1863 to 1897.			
Remarks.	Year.	Baku Production. Tons of 8 Barrels.	American Production. Tons of 8 Barrels.
Monopoly period	1863	5,000	326,000
	1864	8,700	
	1865	8,900	340,000
	1866	11,100	
	1867	16,100	
	1868	11,900	
	1869	27,100	
	1870	22,800	671,000
	1871	22,700	
	1872	24,800	
Impost period	1873	65,000	
	1874	80,000	
	1875	95,000	1,100,000
	1876	195,000	
	1877	250,000	
	1878	333,000	
	1879	380,000	
	1880	400,000	3,250,000
	1881	660,000	
	1882	833,000	
	1883	900,000	
	1884	1,480,000	
Inauguration of the Trans-Caucasian Railway	1885	1,916,000	2,600,000
	1886	1,980,000	
	1887	2,750,000	
	1888	3,200,000	
	1889	3,200,000	
	1890	3,282,638	3,638,000
	1891	4,728,760	
	1892	5,122,600	
	1893	5,520,000	
	1894	5,066,800	
	1895	6,409,000	6,500,000
	1896	6,298,450	
	1897	6,875,400	

These millions which we align upon paper, these figures which perhaps people will not take the trouble to read, represent all the vitality, all the industry of a country. We must add that not only the Caucasus, but Baku particularly, still contains underground almost inexhaustible stores of oil which would seem to render the Caspian underground the largest petroliferous basin of the world.

LUDWIG NOBEL.

Whilst enumerating these riches, it would be unjust if we did not recall one name, that of Ludwig Nobel, the promoter—we should say the fortunate promoter—of the industrial movement in the Caucasus.

The initiative of Nobel at Baku has been crowned with success. In the remainder of the Caucasus the difficulties of communication have stopped even the most audacious spirits. Nevertheless, we may say that if Baku produces 6,800,000 tons annually—

The annual production of the Government of Tamansk			is	40	Tons.
,,	,,	Kouban	,,	24,732	,,
,,	,,	Daghestan	,,	4,595	,,
,,	,,	Tiflis	,,	8,921	,,
,,	,,	Elizabethpole	,,	50	,,
		Grosnaia District,,	155,000		,,

Although these statistics (1895) are the most recent, they are already much below the actual production at the present time, especially as far as Kouban is concerned. Moreover, only an approximative estimate of the actual production of all the Caucasian districts can be made. A lucky bit of boring may alter in a single day the most carefully compiled statistics.

It will be sufficient to note that in the space of two years from 1891 (the date of the report of M. A. Leproux) to 1893, the production of Tiflis increased from 921 tons to more than 7,000 tons.

CHAPTER VIII.

THE CAUCASIAN OIL FIELDS—GEOGRAPHY AND GEOLOGY.

I.—THE CRIMEA.

THE petroliferous strata of the Roumanian oil fields lose themselves towards the embouchure of the Danube under the waters of the Black Sea.

They again crop up in the Crimea where they are well developed.

At KERTCH, an important establishment has commenced work, under the management of M. Birkel; the system of boring employed is that of Fauvelle.

The hopes of succeeding appear as if they were to be realised in the very near future. About ten kilometres to the south of Kertch we come upon CHINGALEK.

One of the wells bored in this locality yields a product of about 30 barrels per day. This output was attained at a depth of nearly 300 metres (984 feet).[1]

Farther on, at KOP-KUT-CHIGAN, an important oil field is met with, comprising more than 100 wells. These works, stopped for the moment, are shortly to be resumed. The other points wrought in the Crimea are TEMSEH, ZAMOSKAYA, KELECHI, and TESCHEWLI.

A peculiarity of these Crimean deposits is that in this country, as in the Carpathians, the nummulitic bed reappears; the same as that which M. Walter points out as being essentially petroliferous. We notice here under the petroliferous beds, the characteristic fossil of which is the *Nummulites Leymeriei*, important beds of chalk with *Belemnites mucromatus*, a very curious circumstance of two orders of things usually separated.

---

[1] The actual depth would appear to have been 941 feet, and the production given only lasted for a short time. Its total production was 3,500 barrels.—*Tr.*

II.—THE NORTH CAUCASIAN OIL FIELDS.

1. TAMANSK.

The Straits of Kertch separates the Crimea from the Caucasian isthmus. On the other side of this branch of the sea we encounter the peninsula of TAMANSK.

There we find, in bluish clay belonging to the eocene and covered with hard sandstones, clay and green shales of the same formation, a clear, light petroleum of a greenish hue, poor in by-products, but almost fit for illuminating purposes.

The strata of Tamansk would appear to differ from the remainder of the Caucasian strata in the fact that the sands so frequent at Baku are rare there, and that, as in America, hard rock predominates. Tamansk oil is but little abundant; it is produced by a slow sweating, the result of a subterranean distillation. We are rather inclined to think that the oil found at 'Tamansk proceeds from distant small beds, and that this petroleum, so different in appearance from ordinary Russian petroleum, has undergone, during a long passage through porous strata, a kind of filtration—a sort of natural distillation which gives it a different appearance from that of the petroleum wrought in the other beds of the region.

As in Galicia, it is met with at the bottom of narrow valleys, near to salt or iodide springs. This oil weighs, on an average, 0·765, and gives 80 per cent. of burning oil.

2. KOUBAN.[1]

The Government of KOUBAN is one of the most ancient oil fields; we there meet with the general characteristics of the Caucasian deposits.

The principal establishments are located in the environs of NOWOROSSIRSK in the valley of Ilskaïa. The French were, so far back as 1873, the first pioneers of the petroleum industry in that district, but, crushed by expenses of every kind, they did not succeed. At the present time, the Standard Oil Company of Russia works the deposits of the valley of Ilskaïa, and possesses there about 100 wells.

[1] Also written Kuban when the name is taken from a German source. So also Kudako, etc.

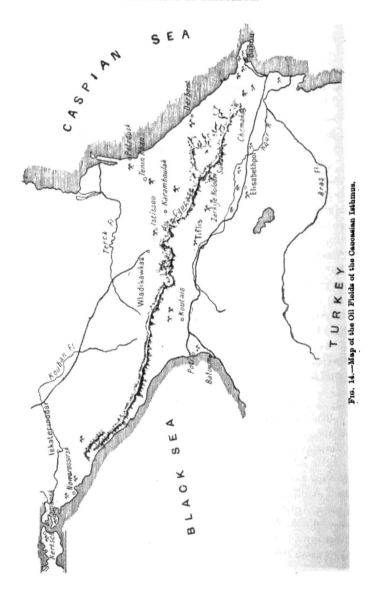

Fig. 14.—Map of the Oil Fields of the Caucasian Isthmus.

The petroleum obtained at Noworossirsk is viscous, of density 0·980.

To the north-east of Noworossirsk is GLINOÏ BALKA. As a recent establishment, this place has already yielded proven results —one of its wells of a depth of 119 metres has given nearly 800 barrels per day. The density of the oil of Glinoï Balka is 0·970.

More to the east extends the KOUDAKO oil field, of which several, amongst the large number of wells, have exceeded an annual daily production of 150 barrels. The sandstones, which form the underground strata of Koudako, are particularly favourable for boring. The oil obtained has a density of 0·815.

Numerous other petroliferous deposits are found in Kouban, amongst which we may mention PSIPH, NEPITEL, CHEKONSPCE, etc.

The sources of petroleum in the Kouban are very often directly associated with reservoirs of brine or alkaline water. At the present day the Kouban produces 237,840 barrels a year of a petroleum which weighs on. an average 0·960.

3 TEREK.

The line which passes Noworossirsk extends to the north parallel to the chain of the Caucasus; the petroliferous strata occur without interruption until towards Wladikawkas[1] in the Government of TEREK, but, unfortunately, they are but little wrought in consequence of the want of communication. The oil field of Terek extends along the line of railway from Ekaterinograd to Wladikawkas, a length of about forty kilometres.

To the east of Wladikawkas, and not far from that town, are the rich deposits of ISTISSOU and KARAMBULAK.

The Government of TEREK produces annually 36,750 barrels of petroleum.

4. GROSNAÏA.[2]

Near to GROSNAÏA extend rich deposits which produce a large quantity of a heavy blackish oil. The oil field of Grosnaïa has been wrought for a long time, but it was not until 1893 that rational methods of working were adopted. The result was not long in being forthcoming. In fact, on the 6th October, 1893, when the first well had reached the depth of 145 metres, an

---

[1] Also written Vladikavkas, etc.   [2] Also written Groznii.—*Tr.*

important spouting occurred. The first day the output of the well reached 10,000 barrels in 24 hours; this production was maintained for 15 days, but the daily yield of the well then fell to about 2,000 barrels. The oil obtained has a density of 0·892.

We said that the trial made at Grosnaïa was lucky. This was not all. A month after, this fortunate bit of boring (to be exact, on the 18th of November, 1893), a neighbouring well, the depth of which was 70 metres, spouted in its turn. This spouting was superb; the jet of oil was so powerful that it rose more that 60 metres in the air; 100,000 barrels were thus projected during the

Fig. 15.—Group of Wells in the Grosnaïa District.

first day. For a long time the yield of the well kept up to more than 3,000 barrels per day. The oil had a density of 0·8973.

The annual production of the Grosnaïa district may be estimated at 1,300,000 barrels.

5. DAGHESTAN.

In the DAGHESTAN, petroleum is met with over a wide stretch of ground; the oil field widens out towards the shores of the Caspian. The principal deposits are met with not far from the sea, concurrently with sulphur, in a land of bubbling sands. The neighbourhoods surrounding TEMIR-KHAN, PETROWSK, and DERBENT are especially rich in oil.

"In 1898, operators on a large scale began to occupy these regions. At the present day, quite a crowd of oil prospectors wishing to take land may be seen along the shore of the sea from Baku to Petrowsk, in the territory of Daghestan. Amongst the crowd are to be found not only the largest firms, but also employes, clerks, etc., all seeking to get a plot of ground where petroleum deposits or other mineral riches are supposed to exist. Cases are quoted of persons who, having acquired ground at the right moment, have gained enormous sums by selling them to foreign capitalists. For the lots belonging to the State, it is necessary to make a declaration of intention to search, and on one lot near Derbent over 140 declarations have been made.

"At Derbent some lots have been already resold at 1,500 to 2,000 roubles, but latterly prices have risen higher.

"Forty verstes[1] from Derbent, in the district of Kaitavotabassar, on the ground of the bey Abdoul Djalil Outchieff, naphtha had been drawn from the wells for a long time, and, naturally, attracted the attention of enterprisers. It is difficult to say what gave rise to this great movement, but the fever for searching for naphthiferous land shows no signs of abating.

"At eighteen verstes to the north of Derbent, an inhabitant of this town, M. Kosliakovsky, has some borings on hand ; the work is carried on by mining engineers. He has also made declarations in the district of Kiurinsk and in the forest of ' Oulouchema,' which belongs to the State. These lots are all passed by the line of railway—Petrowsk-Derbent-Baku—now in course of construction.

"Eighteen verstes to the south of Petrowsk, several borings have been commenced, of which five are in the hands of MM. Linbimoff, Kokoreff and Akhverdoff. In this locality a piece of ground of 400 déciatines has been taken on lease by an English company with a view to its immediate working."[2]

Daghestan produces annually 184,256 barrels of petroleum.

Here ends the line of petroliferous strata which extends along the Caucasus from Tamansk to Derbent.

III.—THE SOUTH CAUCASIAN OIL FIELDS.

The Southern oil field commences, like the Northern, on the shores of the Black Sea, between Poti and Batoum, ending on the shores of the Caspian at Baku, after having followed a line parallel to the Caucasus.

[1] Verste = 1,166 yards.—*Tr.*　　　　[2] *Moniteur des Pétroles.*

The western portion of this oil field is constituted by some undertakings between POTI and BATOUM, then farther on, in the mountains, to the north of KOUTAIS.

The first division of this field is mapped out by SUPSA, NAROUDJA, JACOBI, SAMKTO, MICHEL GABRIEL, GOÜRIAMTI, GOU-LIANI, and MAGHÉLÉ. The whole of this oil field is situated within a few kilometres of the Black Sea. Up to now, in spite of the abundant signs of petroleum existing there, it cannot be regarded as productive. Workings are rare. The ground, which consists of sandstone and bluish sands, lying upon beds of shales and limestone, may be considered as forming part of the miocene; it is absolutely impregnated with *kir* (*solidified petroleum*), which is treated on the spot for a low-quality illuminating oil.

1. ANACLIE.

An important geological discovery has recently been made in the Western Caucasus in the neighbourhood of Anaclie, a town situated near to the embouchure of the Ingour in the Black Sea. Naphtha springs have been found there. The land containing this mineral oil belongs to the Prince of Mingrélie, and the right of working to a capitalist of Moscow, M. Mindowsky. The geological researches which led to this discovery were conducted by MM. Young & Tzouloukidzé, the first of whom is the representative of a French banking firm, and the other a Russian mining engineer.

These explorers wrought from the 2nd to the 14th May, 1898, in excavating the ground, and soon came upon beds of sand, impregnated with naphtha. The surface of the water liberated during the working was soon covered with liquid naphtha.

According to the report of these two explorers, it has been decided to commence immediately the boring of the ground and the working of the petroleum as at Baku, of which Anaclie will be a fortunate competitor, because it is a port of the Black Sea, and not of the Caspian, isolated from Europe by the whole width of the Caucasus (900 kilometres).[1]

2. TIFLIS.

The department of TIFLIS is rich in petroleum; the most of

[1] *Moniteur des Pétroles*, July, 1868.

Pl. V.—The Quay at Baku. From a Photograph communicated by M. André (Vol., i., *to face p.* 81).

the works are recent undertakings, and are most numerous to the north-east of this town, in the direction of SAKATALI.

The simple village of KOLODZI produces more than 50,000 barrels per annum, whilst the total production of Tiflis may be estimated perhaps at 164,000 barrels.

3. ELIZABETHPOLE.[1]

In the administrative district of ELIZABETHPOLE, workings have multiplied so rapidly in a few years that it is impossible to control the production.

The town of Elizabethpole is surrounded on all sides by deposits. The strike of the petroliferous strata here divides into two parts, the most vertical of which follows first the river Kour, and then running away from the other for about forty kilometres, again converges with it in the neighbourhood of BAKU. The Elizabethpole oil field produces annually 1,400 barrels.

Near to this town, the deposits become more and more numerous. In the same way as the Northern, the Southern oil field widens, in the neighbourhood of the Caspian. Workings become crowded there, and seem to form a guard of honour to the petroleum metropolis, Baku.[2]

4. BAKU.

The earth would appear to be desirous of spending, with more liberality at Baku than elsewhere, the inexhaustible riches which it carries in its bosom.

Baku, in an almost barbarous country, is an advanced post of our old European civilisation. A live, policed, manufacturing town, the innumerable factory chimneys of which shoot up to heaven thick clouds of smoke, which from afar mark its position in the midst of the solitudes of oriental Armenia.

Baku, which in 1870 scarcely numbered 12,000 souls, has, in the interval to now, increased its population tenfold, and the antique Persian city, the seat of cruel Sultans, the straggling village hidden in the sands of an arid country, to-day possesses streets aligned like those of our capitals, rich shops in imitation of those of Paris, a magnificent quay several kilometres long,

---

[1] Also written Elizavetpol.—*Tr.*   [2] Also written Bakou.—*Tr.*

where ride at anchor or in shelter quite a mercantile fleet.
The quay is lined with grandiose buildings, hotels of the first
rank, and in the evening sparkles with the glare of the electric
lamp. And yet this conquest over barbarism was altogether

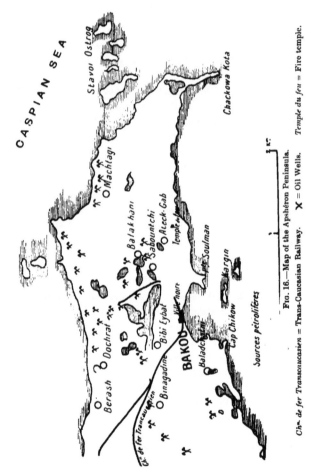

FIG. 16.—Map of the Apshéron Peninsula.   **X** = Oil Wells.   *Temple du feu* = Fire temple.

Ch⁰ de fer Transcaucasien = Trans-Caucasian Railway.

pacific; the agent of this metamorphosis was industry, and the
factor of that industry was petroleum.

Petroleum is the substance which flows naturally from the

fissures of this ground, the material which is so abundant that it is allowed to run away, which is cheaper than water (very rare it is true at Baku), and which is used to "water" the streets. Here one lives in an atmosphere saturated with petroleum. The walls are stained with this substance, and the stream which flows in the gutter rolls black waves of this liquid.

Down there, in the north-east, beyond the rich quarters of the modern town, spread the dark, always petroleum-imbued, roofs of the *black town*, folded in smoke. There are the innumerable refineries where they treat, almost on the spot, the naphtha which this ground produces. We cross men on the way with black skins and oily clothes. We splash about in a viscous mud; everything here sweats petroleum. We ourselves are soon impregnated with it.

And then one would believe himself almost in a town of dreams in the middle of these dark lanes which cross each other, intersected with pipes which we must stride over, lined with tall chimneys, which unceasingly, night and day, darken the heavens with clouds and smoke. We are seized with a sort of respect for this immense industry, and with pride in reflecting on the prodigious rapidity with which it has been created.

A Magnificent Natural Harbour.

Although Baku owes its prosperity to its extraordinary riches in petroleum, it owes it also, in a certain measure, to its admirable situation. In fact the petroleum metropolis is situated on the least expensive line of communication—the sea. The port of Baku is situated in the heart of an admirable bay, closed and sheltered from the wind by an island, and which, consequently, forms a magnificent natural harbour.

This harbour, thanks to the quay of which we have spoken, extends the whole length of the town, thus presenting sufficient berthing places for the ships of all nations, which may come alongside and berth at ease.

Trans-Caucasian Railway.

Moreover, since 1884, Baku is in direct communication with the Black Sea, thanks to the Trans-Caucasian railway which, starting from this town, joins the valley of the Kour, ascends to

Elizabethpole, bends towards the north to connect Tiflis, passes
to the south of Koutaïs, then dividing into two branches, terminates on the Black Sea at Poti and Batoum, having thus followed,
as near as may be, the strike of the petroliferous strata of the
Southern Caucasus.

It is to be noticed that the Trans-Caucasian railway increased
the production of petroleum by more than 4,000,000 barrels
in one year, 1884, by facilitating its exportation.

Since that time the production has continued to increase
still further, for, in 1891, the increase had reached 12,000,000
barrels, that is it had tripled itself, bringing the production of
1890, which was 25,685,714 barrels to 37,830,080 barrels. The
following years were almost stationary; but in 1895 the production of Baku reached 42,299,400 barrels, and 55,003,200 barrels
in 1897.

The oil field of Baku is situated entirely on the Apshéron
peninsula.

FERNAND HUE'S CONTRAST BETWEEN THE GEOLOGICAL STRUC-
TURE OF THE AMERICAN AND BAKU OIL FIELDS.

Let us be permitted here to quote a few lines from the excellent treatise of M. Fernand Hue. "The stratum consists of
tertiary beds covered by miocene strata, situated at very variable
depths. There, as in America, *savants* have formulated numerous
theories on the shape and arrangement of the beds; they have
thought themselves capable of determining in a sure manner
the position of the stratum, basing their assertions upon the
hypothesis of the existence of a lake, of a vast sheet of petroleum, spreading out horizontally, and feeding all the wells.
Experience has proved their error. It follows, in fact, from
numerous borings which have been made, that the wells, even
the nearest to one another, are fed by reservoirs, by species of
cells, more or less abundantly filled, independent the one of
the other, and situated at different stages of depth. Near to
the village of Strikoff, for example, there existed 4 wells in
full swing, bored a few metres the one from the other; the first
was 259 feet deep, the second 560, the third 286, and the fourth
350. Not far from that, an old well only 70 feet deep has yielded

from time immemorial a certain quantity of oil. Hoping to strike oil at a similar level to this, a boring was made quite close to this latter well, but 420 feet was reached before oil was struck.

"The theory of the common reservoir is refuted by the fact of the spouting wells which, in spite of the abundance of their output, do not diminish, in any way, the yield of the neighbouring wells. During all the time that the fountain of Droojba, of which we will speak farther on, shot into the air a jet of 300 feet in height, the nearest wells continued to yield their customary amount.

"We must therefore suppose that instead of being, as in Pennsylvania, in presence of a bed of sandstone and sands saturated with petroleum, or of cavities of relatively small dimensions enclosing gas and liquid, the oil would appear to be contained in vast reservoirs, near to one another, and shaped like the alveoles of a honeycomb. In regard to the difference in the depths to be bored, Nobel, proprietor of numerous wells and important refineries, attributes it to a volcanic convulsion which, by displacing the cells, give to them a diagonal position with a very irregular slope from above downwards."

Geological Horizon.

M. Coquand (*Bull. Soc. Géol.*) has classified the Caucasian petroliferous strata as belonging for the greater part to the upper oligocene corresponding to the level of the Aix gypsum.

The clays and sands of the tertiary formation which contain the petroleum, are impregnated with chloride of sodium, to which M. Coquand attributes the saline lakes of the region.

Theory of Intermittent Action of Baku Spouting Wells.

The spouting of the wells at Baku may be attributed to the fact that the petroleum existing, as our learned colleague has just said, in cells of greater or less capacity, the gases, which are extremely strong and numerous at Baku (the pressure of the gases during a spouting has been estimated at twenty atmospheres), press with great force all the mass of oil through the aperture made by the borehole. The frequent intermittance of these spoutings may be explained in the following manner : The ground underneath Baku, as well as that of all the other European deposits, belongs to the category of soft strata (*terrains mous*). When the petroleum, having spouted for a certain time, descends to a level lower than the orifice of the bore hole (Fig. 17 A), the gases, thus finding a free vent, rush outwards—the pressure thus gradually diminishing in the petroliferous cell and no longer sustaining its sides ; these, composed in greater part of clay and shifting sands, slip from the top of the cell and overwhelm the lower part.

Consequently the level of the petroleum rises in the cavity and stops the escape of gas (Fig. 17 B). Often the naphtha sweats afresh by the landslip thus produced, and brings its contingent to the spouting : the gases thus accumulate, and the well commences to spout afresh until the level of the oil has again

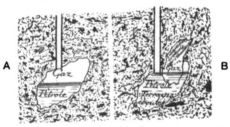

FIG. 17.—Cause of Intermittent Action of Baku Spouting Wells.

Terrains eboutes = Ground fallen from top to bottom. *Excavation produit per l'eboulement* = Gap left behind.

descended below the orifice of the bore hole. This state of affairs tends to continue for years until the source is exhausted.

MUD VOLCANOES AND UNDERGROUND PHENOMENA.

The abundant presence of gas at Baku, as well as the internal modifications of the situation of the beds in soft ground, explains the frequent occurrences there of numerous *salses*—mud volcanoes : small natural spouts of petroleum, gas and mud.

"These eruptions," says Leymerie in his *Géologie*, "are always accompanied by a disengagement of carburetted hydrogen, and often of bituminous matter, and are formed by openings of the ground with formations all round of those kinds of craters, which, consisting of small cones of mud, eventually dry up, assuming a certain degree of consistency. The name of salses is derived from the fact that these muds are generally salt. They also bring in their train a certain quantity of gypsum. This curious phenomenon is very well developed in Sicily, in the environs of Modena, in the Crimea, on the borders of the Caspian. It is especially to be noticed in America at the foot of the Rocky Mountains, where aqueous eruptions have been already shown to exist, as well as concretionary calcareous deposits. It is also necessary to quote Carthagena in New Granada."

"Springs of naphtha, petroleum and the deposits of maltha or asphaltum, which these bitumens sometimes form, are frequently associated with salses. They, however, show themselves in many localities, independent of these small pseudo-volcanic cones. In any case they share with them the property of being accompanied by saline water and a disengagement of proto-carburetted hydrogen. Bituminous springs are most abundant in North America, in Canada, in Kentucky and in Pennsylvania. In the western part of this latter State, petroleum not only sweats from the surface of the ground and collects together in unlimited quantities at the bottom of wells sunk with this object in view, but it may also be obtained, by boring more or less deep, in the form of showers, which spurt up in the air to the height of 30 metres. These substances abound at Baku, on the shores of the Caspian."

The sum of all this is that the Apshéron peninsula, in constant petroliferous eruption, can only be compared to an almost fluid mass, blown up with bubbles full of gas and naphtha.

Under the action of weight this mass varies unceasingly: it is always at work; the bubbles disappear, others form; the oil, seeking an issue, circulates continuously through the sands which the pressure of the gases, in default of their own weight, causes to flow and shift about so as to make room for their expansion. Over all this movement, over this decomposition agitated by antediluvian life, spreads a vast bank of limestone, the thickness of which reaches 600 metres at Balakhany, and increases as it wends its way southwards. The soft nature of the Baku underground involves frequent casing, which is equivalent to saying that if screwed pipes be not used the boring must be started with a diameter wide enough to reach the petroliferous bed.

OIL LEVEL SITUATED AT VERY VARYING DEPTHS.

Here more than anywhere else does the depth of the petroliferous strata vary. In support of our assertion, we give further on the section of two borings; oil was struck in the first at a depth of 140 metres, and in the second at 635. It is to be noted that in these two cases the strata perforated were almost similar.

THE TORPEDO IMPRACTICABLE.

The soft ground of the Baku oil field has another effect. Underground blasting, *torpillage*, is impracticable, and besides it would be ineffectual.

DURATION OF YIELD OF AN OIL WELL.

Here comes the question. How long does the yield of a Baku oil well last? We acknowledge that it is difficult to give a categorical answer. Their lifetime is very variable. It is, in fact, subordinate to the mobility of the petroliferous beds of a shifting ground. By using tubes of wider diameter a higher yield but of less duration will evidently be obtained. The flow may perhaps be stopped abruptly by landslips, due to the gas pressure or other causes difficult to determine.

These same causes may also block up the subterranean conduits which convey the oil to the bottom of the bore hole. What we can affirm—and this reality has no need of proof—is that in normal conditions, and in a general manner, the duration of the total yield is in inverse proportion to the amount of the daily yield.

We find, for instance, at Sourakhany, wells which have yielded a small quantity of petroleum daily from time immemorial.

On the other hand, we have seen fountains produce a quantity of oil, totalling millions of barrels for a few days, and then stop abruptly.

SPOUTING CAUSES LOSS AND DEVASTATION.

At Baku, the quantity of telluric gases tends to increase the spouting which, in a very short time, yields an enormous quantity of petroleum, then exhausts itself in a few days. The spouting of these fountains is a very frequent spectacle in the Apshéron peninsula.

Pl. VI.—The Working of Petroleum at Baku (Vol. i., *to face p.* 88).

But the appearance of these fountains of petroleum, instead of being a subject of rejoicing on the part of the owner of the wells, is rather an object of which they have the greatest fear. In fact, hardly has a fortunate bit of boring discovered an important reservoir of petroleum than, as quick as may be, this fortune, which the oil prospector had been trusting to for so long, gushes out with an extraordinary violence, crushing everything, gasping with a sort of rage, to dry up very shortly. And often millions of

FIG. 18.—A Petroleum Fountain at Balakhany.

hectolitres of this oil, which ought to have recompensed the owner for his sacrifices, are lost, without his having been able to collect a drop.

PRECAUTIONS AGAINST SPOUTING.

Therefore, when such a *dénouement* is foreseen, when a well begins to dissolve gas with the noise of a boiler, when the water in the bore hole is in a kind of effervescence, the " kalpak " must be placed on the well with all possible speed. This is an iron cover made fast to the orifice of the well and furnished with a tubulure closed by a tap, by means of which the output of petroleum may be regulated at will.

Should this precaution not have been taken beforehand, everything is lost. When the petroleum has "packed up," it is very difficult, almost impossible, even to place the "kalpak" on a well.

Spouting Phenomena.

Along with the petroleum, sand and rocky fragments, often very large, are hurled into the air, thus rendering it dangerous to approach the *fountain*.

It is, in fact, a veritable eruption, analogous to a volcanic eruption. "The sight of one of these fountains produces a stupefying impression never to be forgotten. The naphtha projected by the pressure of the subterranean gases brings in its train sand and rocks, breaking everything in its passage, falling down again upon everything surrounding, carried afar by the wind, burying, under miniature mountains of sand and the debris of rocks, the other undertakings in the neighbourhood, including pumps, engines and the whole plant. The naphtha spouts in this manner for long weeks, forming streams which excavate a bed across the sand as they flow to lower levels where they form real lakes."—Paul Vibert.

Remedial Measures.

What then should be done in the face of one of those sudden spoutings? If there be not time to cap the well, the best plan will be to collect the petroleum. It would therefore be well to provide beforehand a supply of wooden or metallic reservoirs and natural reservoirs. These reservoirs are made by forming near to the wells or below them, if they be on a slope, deep ponds on the natural route of flow of the streams of oil produced by the spouting. They can be easily established very cheaply, either by banks of earth stopping the flow on a slope, or by excavating channels from the well which lead the oil to natural or semi-natural reservoirs, such as dammed-up valleys or ravines.

In this way, one may still hope to store a little of the petroleum which, without this precaution, would be completely lost. It is right to say that during one of these eruptions a new natural reservoir always absorbs a certain quantity of petroleum. But gradually there forms on its sides a solidified mixture of sand and petroleum (*Kir*), which renders it impermeable.

No. 3 Spouting Well of the Schibaeff Company.

Owing to neglect in capping the wells, spoutings are frequent at Baku. It may even be said to be an ordinary occurrence which does not cause the least astonishment.

Two of these spoutings are remembered by every one. These are that of the No. 3 well of the Schibaeff Company and that of the Droojba well of the Armenian Company. In 1872, the well No. 3 commenced to spout, but with a force unknown up to then. It ripped up its derrick and smashed the boring instruments. It was a veritable deluge; the reservoirs were filled in an instant, and the petroleum, in the shape of pressing waves, formed real rivers flowing towards the sea. The shower of oil flew to a prodigious height and the sand, which it carried in its train, made it shine in the sun like real living gold.

For more than a year this well vomited 8,000 hectolitres of oil daily, but this enormous quantity of naphtha was lost because it was found impossible to "cap" the well. Under these circumstances, the Grand Duke Michael, Governer of the Caucasus, happening to pass through Baku whilst on one of his travels, desired to satisfy himself in regard to this gigantic phenomenon.

We have mentioned how frequent are the spoutings of petroleum. We have often spoken of the caprices of veins of oil favouring one and ruining his neighbour. What then was the caprice of the vein which at this moment fed No. 3. Suddenly, without any apparent reason, the evening of the arrival of the august visitor, the well, No. 3, stopped "spouting". At the end of three days the Grand Duke leaves without seeing anything, and— fresh caprice—the fountain starts spouting again with fresh passion, and strange to say in the interval they had not profited by this lull to put on its "kalpak". But by this time the manner of the spouting was changed; it had become intermittent. It flowed during six minutes and then stopped for half an hour, still giving however, in spite of that, 2,840 hectolitres daily.

The yield continued for a year further, when the well became choked up with sand. It was repaired but its flow had become reduced almost to *nil*. At the present time it is quite exhausted.

THE DROOJBA SPOUTING WELL.

The spouting of the Droojba was still more gigantic. The eyes of the Armenian Company to whom this well belonged were, to use a vulgar phrase, bigger than their belly. They bored the well with a final diameter of 22 centimetres. Imagine then the bulk of the column of liquid projected from the well! The eruption was sudden; without warning, the column of petroleum flew to a height of 100 metres smashing the derrick.

Then came the magical spectacle of this new geyser. A column of water, 20 centimetres in diameter, rose to more than 100 metres, then, with the graceful curve of a cyclopean fountain, widened into a cloud of spray of greater or less extent, according to the force of the prevailing wind, and which darkened the whole of the horizon.

Several kilometres from the neighbourhood of the well dull rumbling noises were heard mixed with the frequent detonations which it produced.

Often, when the tubes became choked up with the sand which this column of naphtha charioteered, the height of the fountain diminished, but very soon the enormous pressure of the column of oil overcame this obstacle and recommenced shooting in the air its torrent of petroleum, sand and stones. One of these in falling had sufficient force to shatter a thick copper roof.

To give an idea of the quantity of naphtha vomited by the Droojba, it may be mentioned that a certain refiner had no difficulty in filling a reservoir with it of a capacity of 75,000 barrels. This quantity of petroleum only cost him 300 roubles or as near as may be at the rate of one centime the barrel.

The quantity of sand vomited by the Droojba was such that the neighbouring works within a radius of 300 metres were engulfed in sand. In the immediate neighbourhood of the well, the deposit of sand rose to a height of 5 or 6 metres, and on this sand rolled green waves of petroleum—an inundation, in very truth, producing rivers navigable by boats. These rivers rushed into all the excavations, bringing with them materials belonging to other undertakings. Then when all the excavations were full, the naphtha, through a grand channel, rushed into the sea, choking the works in the neighbourhood with its waves of sand, running the risk everywhere of causing fearful conflagrations, for the petroleum had penetrated even to the engine rooms which, had it not been for the promptitude of the mechanics in extinguishing all lights and fires, would have been the cause of terrible disasters. The Droojba soon became a public danger, a plague on this industrial country. The unfortunate Armenian Company to whom the Droojba belonged, and who had expended its last fraction in the boring of this well, was overwhelmed with lawsuits and actions for damages and interests.

And the well continued to vomit 40,000 barrels daily and this oil could not be collected; for want of sufficient reservoirs it ran to the sea. A commission was despatched from St. Petersburg to advise as to the means to be adopted to stop the flow when suddenly the well ceased spouting. The friction of the sand, carried in the train of the petroleum, had gradually worn away the tube, which was crushed.

The rejoicings at this turn of affairs at the termination of the reign of terror had hardly begun when, a few hours afterwards, the Droojba began to spout again with fresh vigour. Ten days afterwards the well was put under control.. The well then yielded about 4,500 barrels per day for some time further, when gradually the production diminished, and fell to the ridiculous figure of 1 barrel per day.

It is estimated that during the spouting of the Droojba this well lost more than 4,000,000 barrels.

OTHER SPOUTING WELLS.

Spouting wells are still to be met with, the flow of which amounts to considerable figures.

The ancient land of everlasting fire would appear to be inexhaustible. In 1882, the "Company of Petroleum Participators" struck oil in its No. 9 well, at a depth of 476 feet; the diameter of the pit was $\frac{1}{4}$ metre. The spouting was formidable; in a month the well had vomited 8,000,000 gallons of petroleum, averaging 6,660 barrels per day.

A short time afterwards Nobels' No. 9 well, 624 feet deep, spouted in turn, and the column of oil rose in a shower which iridised the sun at the respectable height of 80 metres. A rain of petroleum mixed with sand inundated the derricks within the radius of 100 metres. As was the case with the Droojba, rivers of naphtha, in very truth, rolled along the slopes of the plateau of Balakhany. The yield attained in a month the colossal figure of 300,000,000 gallons, which gives the respectable production of 25,000 barrels per day.

The owners despaired of ever mastering this fountain, when the sands, aggregating at the mouth of the casing, regulated naturally the flow of the oil which was then easily put under control. During this second period the well continued to give the handsome yield of 600 barrels an hour, or 14,400 barrels per day.

Recently (1898), a boring at Balakhany, belonging to M. Vichau, spouted to the extent of 50,000 barrels daily.

Other fountains play without ceasing at Baku. Thanks to the data of Han and Kobolow, we are able to tabulate (p. 95) the statistics in regard thereto.

Total production of spouting wells in eighteen months 1,235,365 tons.

A fair conception may be made from the above table of the abundance of oil at Baku; but this oil, by its very abundance and also by its poor percentage of lamp oil, is sold very cheaply.

ANALYSES OF BAKU CRUDE OIL.

The following results, obtained at the Topolnica refinery, show its analysis from a commercial point of view :—

	Per cent.
Benzine	6
Kerosene (1)	30
„ (2)	8
Mazout	45
Loss and gas	11
	100

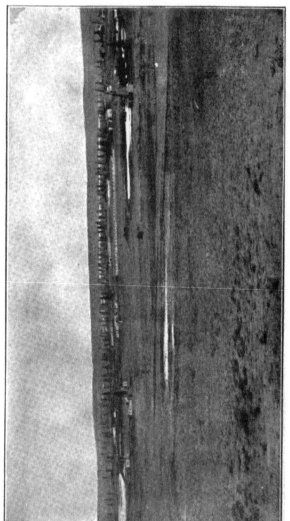

Pl. VII.—View of the Oil Wells of Balakhany.   Photograph communicated by M. André (Vol. i., *to face p.* 95).

Spouting Oil Fountains of the Apshéron Peninsula
in 1889 and 1890.

Name of Well.	Depth in metres. of 3·28 feet.	Spouting Period in Months.	Total Production in Tons of 8 Barrels.	Density of Oil.	Remarks.
Abibekow and Lianosow, No. 1 . .	236	4	20,636	0·878	
Armawir, No. 2	266	5	36,283	0·884	
Baku Co., No. 2	260	2	44,250	0·887	
De Bur Bros., No. 6	217	3	7,583	0·865	
Kalantarow, No. 1	307	4	70,660	0·878	
Caspian Co., No. 14	261	4	80,290	0·865	
„ No. 17	266	4	85,801	0·865	
„ No. 37	266	3	18,320	0·868	
Caspian-Black Sea, No. 14 . . .	243	3	17,780	0·870	Year 1889.
„ „ No. 19 . . .	238	4	4,330	0·872	
Kastscheef, No. 3	230	2	14,430	0·868	
Lianosow, No. 23	247	2	8,217	0·873	
Massis Co., No. 3	251	6	44,170	0·886	
Nobel Bros., No. 24	298	9	11,600	0·854	
„ No. 35	217	4	7,083	0·899	
„ No. 50	266	5	50,812	0·876	
„ No. 70	304	1	7,760	0·882	
„ No. 80	277	7	47,380	0·870	
Tagiew, No. 15	217	12	126,460	„	
Ararat, No. 4	217	4	29,961	„	
Astchik, No. 2	228	3	84,710	„	
Baku Co., No. 2	272	3	60,120	„	
De Bur Bros., No. 6	217	1	550	„	
Caspian-Black Sea, No. 17 . . .	302	5	96,298	„	
„ „ No. 23 . . .	217	1	2,000	„	First six months of 1890.
Kalantarow, No. 1	307	2	18,850	„	
Nobel Bros., No. 21	299	6	5,141	„	
„ No. 51	266	6	174,158	„	
„ No. 77	304	6	23,063	„	
„ No. 81	278	3	5,098	„	
Massis Co., No. 6	243	3	47,516	„	
Mirzoeff Bros., No. 24	228	1	13,300	„	
Tagiew, No. 15	217	6	18,041	„	
„ No. 21	219	6	59,200	„	

The results obtained at Nobel's refinery are:—

		Per cent.
Benzine	. . . .	4
Kerosene (1)	. . . .	27
„ (2)	. . . .	14
Mazout	. . . .	45
Loss and gas	. . . .	10
		100

Different Deposits in Baku Oil Field.

The deposits in the neighbourhood of Baku occupy all the Apshéron peninsula from Sourakhany on the east to Binagadine on the west. The summit of the plateau on which Balakhany is situated would appear to be particularly rich. We shall now pass in review the principal deposits.

SOURAKHANY.

At the extreme end of the Apshéron peninsula, ten miles
to the north-east of Baku, and connected therewith by a railway
completed about ten years ago, is Sourakhany.

Sourakhany is the oldest wrought locality in the Baku oil
fields. From a very remote time there existed at Sourakhany
two wells which produced a different kind of petroleum from
that produced in this region.

This petroleum, produced without doubt by similar pheno-

FIG. 19.—Group of Wells at Sabountchany.

mena to that which we have explained as coming into play
at Tamansk, by a sort of natural distillation, is white, very
light, rich in kerosene, of density 0·760. Not long since they
did not refine it, but used it in its crude condition. The two
wells which produce it together yield the modest amount of
3 to 4 barrels daily. But we must add by way of amendment
that for many long years this yield has never varied. There
then, are 2 wells which differ greatly from their Balakhany
confrères !

At the present time, on the Sourakhany plateau, few wells are
being worked ; the remainder have been abandoned as exhausted.

The district is used up, says the one faction. It is the "pearl of Apshéron," says the other; you have only to seek for it.

Our opinion is that Sourakhany is beginning to become exhausted, and that prospectors should direct their steps towards the unbroken virgin land on the north-west.

THE TEMPLE OF EVERLASTING FIRE.

It is at Sourakhany the temple of everlasting fire was built, for the description of which we pass the pen to one more eloquent than we are—M. A. Kœchlin Schwartz.

"But, let us, for one moment, leave industry and its discoveries to visit a temple, or, more correctly speaking, a monastery. Formerly, when the temple was isolated in the plain, and especially when nightfall came on there, at the moment when all its fires were lit up, it must have presented a highly imposing spectacle in the midst of this naked immensity. At the present day, the impression is that of a kind of deception. The door of the surrounding wall is closed; we knock. A Parsee comes to admit us; he is the only one left. Ten years ago the monastery still contained three; they are dead, the one after the other. The last was assassinated, it would appear, so as to rob him of his hoard. For some time back the monastery has not been occupied. During this time the naphtha distillery has been installed.

"The monastery itself has been respected, but the factory has diverted the greater part of the gas for its own use.

"We find ourselves in the presence of a square temple, constructed right in the middle of this court. A sort of arched gallery, looking upon the court through large arched openings, circles round the temple, which is gained by mounting five or six steps of a wide perron, which encompasses it everywhere. The interior is only lit up by a door, pierced in the side, facing the east, where there is a sort of sanctuary, under which burn a certain number of jets of gas, all deprived of ornament. The walls are festooned with calico prints, and, in the middle, is a dome forming an exit for a multitude of small chimneys.

"From the top of this dome and all the chimneys with which

7

it bristles, and from the points of all the festoons, issue jets of flame, thin meagre jets I ought to call them. But, it will be easily understood that the effect would be altogether different, during the time of the splendours of the monastery, when the temple absorbed all the gas of the plateau.

"Then the small flames of to-day were strong and well fed; immense sheets of fire rose towards heaven, which the slightest breeze wafted towards the dome of the chapel.

"The spacious court is surrounded by a high wall decorated with printed calico, that is to say, embattled with festoons above and arches underneath. One part of these arches is surmounted

FIG. 20.—Group of Wells at Balakhany.

with stones bearing very ancient inscriptions. Our guide told us that a Frenchman who visited the temple some years ago had offered a thousand roubles for one of these inscriptions.

"On the right and on the left, against the surrounding wall, project small buildings—small lodges—some perched aloft, others, on the contrary, very low down. They were used as cells by the ancient monks.

"Everywhere from these battlements, from these domes, from all the walls, from all the roofs, jets of flame issued forth formerly.

"These flames are extinguished to-day for want of gas."

BALAKHANY.

To the north of Baku, and eight kilometres from that town are erected on a denuded plateau, the derricks of Balakhany.

Balakhany is in the centre of the Apsheron oil field, with about 1,000 wells. A railway cuts through the works, passing the end of the salt lake Ozera ; it stops at the SABOUNTCHANY station, which is the key to the principal works.

At Balakhany, are crowded the works of the principal "Oilmen" of Europe. In front of us, arriving by the route which, across the hill of Balakhany, leads to Mochtaghi, we see the wells of the Bnito Company. To the left are enormous reservoirs of sheet iron, containing sufficient oil to light up every town in the world. Farther along, in the same direction, is the pumping station of the Baku Company, behind which rise the derricks of the works of Schibaeff, Franke de Tiflis, etc., whilst, on the horizon, is profiled the gigantic installation of Nobel Brothers. To our right, between us and the railway, are sheets of naphtha, real lakes reflecting the derricks of the French Company. A "Pipe Line," eight kilometres long, conveys all this petroleum to the refineries of the Black Town. The route which we follow is horrible, hardly kept in repair, broken with ruts full of naphtha, crossed by petroleum pipes, barely covered with earth.

The transmission levers of the pumps grate above our heads, whilst the whistling of the engines, the concussion of the parts of the boring plant, make a sad accompaniment to this far from melodious symphony.

Finally, and this will pass over many details, each well, wrought at Balakhany, has an average production of 400 barrels daily which makes this petroliferous station the richest in the whole world. Here is a section of the strata, taken in the centre of Balakhany. It is that of well No. 8 belonging to Schibaeff.

GEOLOGICAL SECTION OF STRATA ENCOUNTERED IN BORING No. 8
SCHIBAEFF WELL.

	Metres of 3·28 feet.
Alluvial sand	2·74
Shifting sand	3·13
Grey clay and sand	4·24
Sand mixed with solidified oil (Kir)	1·83
Grey clay	0·91
Sand (gas)	0·80
Sandstone	1·52
Grey clay and sands	8·84
Petroliferous sands	11·59
Grey clay	27·74
Petroliferous clays (minimum yield)	3·05
Grey clays and sands	28·95
Sands (gas)	8·59
Petroliferous clay (minimum yield)	4·88
Grey clay	4·27
Brown clay	10·87
Petroliferous clay (minimum yield)	2·13
Grey clay	4·87
Brown clay	2·13
Grey clay and petroliferous sands	9·44
Petroliferous sands reach a depth of	140·79

BIBI AÏBAD.[1]

This geological section might almost serve as a type of the general conformation of the whole of the oil field. Moreover exactly the same strata are met with at BIBI AÏBAD, a petroliferous station situated to the south of Balakhany; nevertheless, according to the law which we have previously enunciated, that the solid stratum and not the petroliferous would appear to thicken in going towards the south, the oil there would appear to be deposited at a greater depth. In spite of that fact, workings are numerous and prosperous. The latest improvements are adopted. Bibi Aïbad is not so old an oil station as Balakhany.

At the present time boring by the rope is being attempted. *Audaces fortuna juvat*, says the old adage. Let us hope so.

BINAGADINE.

To the north-west of Baku is found a new station which has lately begun to be wrought. We refer to BINAGADINE. The deposits of Binagadine would appear to be immensely rich. Their depth, according to the law enunciated above, would appear to be greater than those of the Balakhany deposits. Here is, more-over, a geological section of this ground which forms the western

[1] Also written Bibi Eibat.

extremity of the Apshéron peninsula. This section was made in 1891, by a French engineer, M. A. Leproux.

VERTICAL SECTION OF BINAGADINE OIL FIELD.

	Metres of 3·28 feet.
Marl with argillaceous shifting sands	115·80
Friable sands with clays	70·20
Shifting sands with some beds of marl	26·30
,, with clays and sandstone . . .	80·00
,, ,, . . .	30·50
Coarse-grained quartzose sands	86·00
Marl	33·00
Quartzose sands with sandstone	54·50
,, with clays	87·00
Analogous bed	15·04
Argillaceous marl and petroliferous sand . . .	40·00

Petroleum was met with at the depth of 656·00 metres at Binagadine. It is not, however, in every case necessary to carry the boring to such a great depth. There is to be seen in this locality a spouting well which has reached the petroliferous strata at a depth of only 75 metres.

In the same direction we find the recently established works (1897) of Schibaeff and M. Von Veklé at KHIDIRZIND.

PRESENT CONDITION OF BAKU PETROLEUM INDUSTRY.

We extract, from a report of the Austrian Consul at Tiflis, some data regarding the present condition of the petroleum industry of Baku.

" Baku borings have been continued with feverish activity in the Caucasus. Fabulous sums have been paid for petroliferous ground. This is easily explained by the enormous yield of some of the wells. One well, for example, which started in November, produced 100,000 barrels daily. Wells producing 10,000 to 50,000 barrels daily are not exceptional.

" Last summer, when an English syndicate bought the domain of Tagieff for 5,000,000 roubles, a large number of proprietors of the Baku wells were inclined to sell, and numerous negotiations were entered into with this end in view with foreign capitalists. There was even a certain apprehension that the whole of the oil fields would pass into the hands of foreigners. But following up the fact that, on the domain of Tagieff, a well has been opened up which yields 50,000 barrels daily—that is to say a daily revenue of 40,000 roubles—the inclination to sell has gone.

"During the last quarter of 1897, a great conflagration had to be booked and to be added to the numerous fires which the Baku oil field has witnessed. From an unknown cause, four fountains of oil, producing 38,000 barrels daily, took fire, and set fire to 300,000 barrels of crude oil stored in pits in the neighbourhood and to 29 boring installations, which were entirely consumed."

EXPORTS.

The following table gives a comparison between the exports for the first quarter of 1897 and the corresponding periods of 1896 and 1895 :—

First Quarter.	To Europe.	To other Markets.
1897	28·3 per cent.	71·7 per cent.
1896	44·0 ,,	56·0 ,,
1895	47·1 ,,	52·9 ,,

The exports were distributed between the different countries as follows :—

	Ponds of 36 lb.
Great Britain	4,682,448
Austria-Hungary	1,813,894
France	1,679,568
Belgium	1,424,594
Germany	1,226,481
Bulgaria	176,100
Holland	24,000
Italy	507,249
Spain	232,599
Turkey.	1,997,521
Egypt	7,713,636
East Indies	2,505,165
China	3,342,513

NUMBER OF PRODUCTIVE WELLS AND PERCENTAGE PRODUCTION PER LOCALITY.

Name of District.	Number of Productive Wells.	Percentage of Total Production of Baku Oil fields.
Balakhany	201	19
Sabountchany . . .	294	46
Romany	59	18
Bibi Aïbad	33	10
Binagadine	19	8
Various	76	4

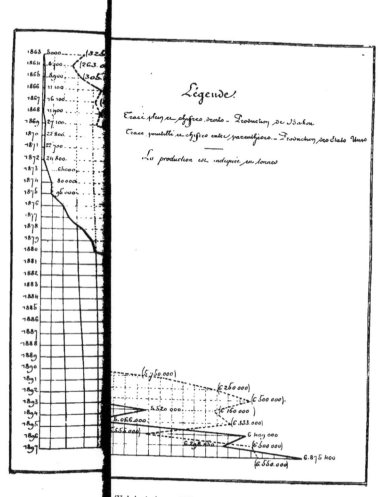

Classification of wells according to their production :—

Annual Production.	Percentage of Wells.
Up to 50,000 pouds (of 36 lb.)	71
From 500,000 to 1,000,000 pouds . . .	13
From 1,000,000 to 2,000,000 pouds . .	10
More than 2,000,000 pouds	6

Kind of Wells.	Contribution to Total Yield.
Flowing wells	26·5 per cent.
Pumping wells	73·5 ,,

Much Rich Ground Still Available.

It may be seen from this brief enumeration of the principal
Apshéron deposits that, if the territory be not wide (1,900 square
kilometres, in which only 15 are wrought) the production is little
in accord with the small surface. When we recognise that these
territories still contain wide tracts of non-ceded ground,[1] to all
appearance equally as rich as those already wrought, we need have
no fear in regard to the near exhaustion—so much spoken of—of
the stock of coal concealed in the bowels of the earth.

[1] The Russian Minister of Finance has just decided to grant boring rights on no
less than 100 new naphtha bearing lands free of charge. — *Oil and Colourman's
Journal*, 1st Oct., 1900, p. 1,895.

CHAPTER IX.

PRACTICAL TRADE HINTS—RUSSIAN PETROLEUM LAWS.

1.—WAGES.

AT Baku and in all the Caucasian oil fields, the price of manual labour varies but little from that current in Galicia. However, the wages of the foreman borer has a tendency to be rather higher, owing to the fact that many foreigners are employed in this capacity. On the other hand, the subordinate workmen, who are for the most part natives, are paid at very modest rates. The following table, which gives the rates paid for manual labour at Baku, will convey a good idea of what we state :—

	Monthly Wages.
Foreman borer	100 to 150 roubles.[1]
Assistant ,, ...	25 ,,
Labourer	15 ,,

COMPARISON WITH GALICIAN WAGES.

If we compare these wages with those of Galicia we get the following comparative monthly wages bills for the staff on an ordinary boring :—

	Galicia.	Russia.
1 Mechanic	60 florins.	50 roubles.
1 Stoker	30 ,,	25 ,,
1 Foreman borer . . .	85 ,,	100 ,,
2 Assistant borers . .	48 ,,	50 ,,
4 Labourers	84 ,,	60 ,,
1 Smith	50 ,,	25 ,,
	337 ,,	310 ,,
Or in francs . .	674 francs.[2]	806 francs.[3]

The difference of 132 francs is compensated, in the environs of Baku, by cheap transport. It may therefore be said that the working expenses are practically the same in both countries.

[1] 100 roubles = £14 according to table in "Addenda," p. 506.—*Tr.*

[2] The florin here is calculated evidently as 2 francs.—*Tr.*

[3] The rouble is calculated at 2·6 francs. The value in appendix being given as 3½ francs.—*Tr.*

2. Plant and Raw Material.

The price of plant and material is very variable in the Caucasian oil field, owing to the fact that the working is in the hands of many big companies, each of which adopts a different system, or brings improvements to bear more or less upon the methods which they have adopted.   Moreover, the greater number of these firms being foreigners, can easily procure, thanks to the facilities of transport brought to Baku by the Trans-Caucasian Railway, tools, implements, and plant, varying with the country from which they come.

This point will be dealt with later on.

3. Contractors.

Firms of contractors are not awanting at Baku.   Moreover, they have a perfectly uniform scale of charges.   Their prices again perfectly agree with contractors' charges in Galicia.   The following table of Galician mixed contract charges (that is to say, the proprietor of the ground finding the steam power), affords a comparison between the cost of the operation in the two countries :—

Comparative Cost of Boring a Well 300 Metres Deep (say 984 Feet), in Galicia and in Baku.

Galicia.	Russia.
Per metre to 300 metres : 50 florins.	Per sagène (2·13 metres) up to 10 sagènes (21·3 metres), 72 roubles. 10 roubles additional per sagène for each fraction of 10 metres above 100 sagènes.
Which for 300 metres gives a total of 15,000 florins.	Which gives a total cost for 141 sagènes to the 300·33 metres of 11,202 roubles
Or 29,125 fr. 20 (say £1,165).	or 28,350 francs (say £1,134).

The cost is much the same, with a difference of 775 francs (say £31), in favour of Baku, which an alteration in the rate of exchange of the rouble or the florin may one day cause to disappear.

4. Price of Land.

As in all petroleum countries, the cost of buying land is very variable.   In the zone of the oil fields which is being wrought, the average price in the Caucasus is about ten roubles per deciatine

per annum. Moreover, a royalty of 1 kopeck per poud of petroleum extracted is payable to the proprietor of the ground. In the
neighbourhood of Baku, the price is about 20 roubles (52 francs)
per square sagène (a little more than five square metres, which
gives a price of 100,000 francs per hectare, say £10,000 per
acre).

5. CAPITAL.

The installation of a single boring plant requires at least a
space of 400 square metres ; 4,000 francs (£160), must therefore be
allowed as the cost of the superfices of this minimum site for
boring. Further, on this space with a well at work, there would
be no room for a forge, workshops or reservoirs. Three times
as much space is therefore required for a complete installation,
costing 10,000 to 12,000 francs (say £400 to £480), at least.

These prices, be it well understood, are not at all fixed and
invariable. Land may be bought at Balakhany, even in the up to
now little wrought portion of the plateau, at the price of 5 roubles
the sagène, say about 13 francs. There the ground for a boring
including a complete installation would be 4,160 francs (say
£166 8s.).

As far as we are concerned, we would never advise a beginner
in the petroleum industry to buy such small plots of land. A
single boring is in fact only an experiment. If it succeeds,
neighbouring ground cannot be got but at exorbitant prices,
unless he means to limit himself to his single success, and
content himself with having discovered a rich oil field for the
adjacent proprietors. In that case, he has wrought too hard
for them. If it be unproductive, he will not have the chance
of being able to try again to recoup himself. Moreover the
general expenses of boring and workshops are the same for
one boring as for ten, the profits, if there be any, will be
devoured by them. It is therefore necessary for an oil operator
in order to instal himself without risk at Baku to disburse the
following sums :—

6. Schedule of Expenses Incidental to an Oil Well Installation at Baku.

	Francs.	£.
For the purchase of deciatine (about 1 hectare of land)	40,000	1,600
For acquiring the right of purchase of 2 deciatines of neighbouring land (in case of non-fulfilment of contract in three years, this sum may be wiped off as forfeited to the then proprietors) .	8,000	320
Installation of forges, workshops for two borings by steam power reservoirs, etc.	42,000	1,680
Cost of two borings each 300 metres (984 ft.) deep .	30,000	1,200
Reserve fund	20,000	800
Royalty payable to State for two years' right of working	100	4
Or a sum total of	140,100	5,604

7. Royalty.

The royalty, feu duty or ground rent, payable to the State for the right of concession, is 10 roubles (26 francs) every six months, payable on the 2nd January and the 2nd July of each year. This concession lasts twenty-four years.

In those cases where boring operations have not been successful, or the deposit has become exhausted, the deed of concession may be annulled.

In this case, the contracting party returns to the State the deeds of concession which he had obtained from it, demolishes the works and the buildings which he has erected and is no longer held liable to any tax by the fiscal authorities (law of the 1st February, 1872).

8. The State the Proprietors of the Greater Part of the Oil Fields.

The same holds good in regard to the purchaser on the Apshéron peninsula; it is principally with the State that he has to deal. In fact the Russian State is the proprietor of the greater part of the petroliferous land of the region.

Let us examine the cause of this state of things.

9. History of the Russian Petroleum Monopoly.

In 1801, the State assumed the monopoly of petroleum. It was then sole proprietor of the Baku deposits. It granted licenses to work these deposits to certain great personages, but without alienating the ground of which it desired to remain the sole proprietor. The *régime* of this monopoly, which moreover gave the right of working exclusively to the Czar's subjects, retarded for a long time the development of the petroleum industry. Accordingly, in 1872, it was decided that it had reigned long enough.

10. Civil Proprietorship and its Effects.

Then the State to encourage those engaged in the industry, put up for sale at comparatively very moderate prices parcels of half a deciatine. It even gave with these lots full proprietary rights with indemnity title deed.

That was the origin of civil proprietorship at Baku.

Very soon the results of the abolition of the monopoly began to make themselves felt. In a few years, the petroleum production of the Apshéron peninsula had multiplied a hundredfold. Petroleum, too abundant, fell to such low prices that the State, with good reason, feared a glut of the market.

11. Indiscriminate Sale of Land Stopped.

Then it stopped its sale of land, and afterwards, as a measure of forethought and also as a means of checking the decline in price of Russian petroleum, it divided the ground which still remained to it (nine-tenths) into two parts ; on one of which it granted concessions, on the other it would allow no operation until the first had become exhausted.

This wise determination has resulted in the fact that at the present day 50 per cent. of the petroleum-bearing land is kept in reserve for the future.

12. Classification of Land.

The land in the Apshéron peninsula may just now be divided into two categories :—

1. Lands bought or received as an indemnity from the State by certain persons who have afterwards kept, leased or sold them.

2. Land belonging to the State, of which it concedes a very small portion on the condition that the *cessionnaire* commences work in the two years following the deed of concession.

The lands of the first category are the most costly, because they are situated on the borders of land formerly wrought, just because they were of this same category which, as we have seen, were the first to be conceded.

It does not follow that they are the richest. To avoid all confusion we shall simply say that they are the dearest. They are also the most easily obtained, for in the transactions of buying and selling these lands the parties thereto are business men who deal with the matter in a commercial spirit. Whilst on the other hand, the obtaining of a concession on the other lands is hampered by official slowness, and the oil prospector has to submit to interminable formalities. This may, perhaps, also be one of the causes of the affluence of the purchasers of land of the first category.

These are privileged as regards means of communication. Besides railway the networks of the pipe line intersect and converge towards the collecting conduits, which join the Balakhany works with the Black Town of Baku. The oil flows naturally into these conduits without being pumped. In fact, the slope which separates Balakhany from Baku, is 8 metres per kilometre (1 in 125), which is very efficient.

13. EXPORTATION OF CRUDE OIL.

All the oil which the Apshéron oil field produces cannot be refined by the 300 refineries of the Black Sea; a large quantity, therefore, of the precious liquid is despatched in *tank* either in gigantic tankships which cut across the Caspian Sea, and wooden tugs which ascend the Volga to supply the Russian consumption, or in the tank waggons of 70 barrels of the Trans-Caucasian Railway, which despatches it all over Western Europe to the refineries of Fiume, Marseilles, etc., or by the Russian railways to Tzaritzin. Here are the Russian railway stations which have petroleum depôts.

Depôts.

Stations possessing reservoirs for more than 1,000,000 barrels :—

Baku.
Tzaritzin.
Orel.

Stations possessing reservoirs for 500,000 to 1,000,000 barrels :—

Moscow.
St. Petersburg.

Stations possessing reservoirs for 100,000 to 500,000 barrels :—

Astrakhan.	Kiev.
Saratov.	Minsk.
Nijni-Novgorod.	Warsaw.
Kharkov.	

Stations possessing reservoirs of more than 100,000 barrels :—

Kazan.	Krementschong.
Voronège.	Tchernigov.
Kozlow.	Koronew.
Riazan.	Polotsk,
Smolensk.	Dunaborg.
Toula.	Riga.
Ekaterinoslaw.	

At the present day Russian petroleum would of itself suffice to supply the whole of the European consumption. With this eloquent affirmation we finish our study of Russian petroleum.

General Properties of the Different Caucasian Petroleums.

Place of Extraction.	Density.	Colour.	Remarks.
Tamansk	0·765	Clear bright, almost translucid	
Noworossirsk . . . :	0·985	Black	Viscous
Glinoï Balka	0·970	,,	,,
Koudako	0·815	Greenish black	
Grosnaïa	0·892	,,	
,, (2)	0·873	,,	
Baku	0·886	Black	
,, (2)	0·884	,,	
,, (3)	0·938	,,	
Bibi-Eibat	0·859	,,	
Sabountchany	0·810	,,	
Sourakhany	0·770	Yellowish white	Volatile
,, (2)	0·760	,,	,,
Balakhany	0·905	Black	Rich in heavy oil
,, (2)	0·910	,,	,,

CHAPTER X.

Up to now Western Europe would not appear to be very rich in petroleum. The eastern countries, deprived of coal, would appear, on the other hand, to have the monopoly of the liquid combustible which is engaging our attention.

Nevertheless, the search, which for some years back has been made for petroleum, has yielded, in some countries, convincing results, whilst in others, at least the presence of petroleum has been more or less demonstrated.

GERMANY.

Germany possesses two oil fields :—

1. THE NORTH GERMAN OIL FIELD.
2. THE ALSACE OIL FIELD (FORMERLY FRENCH TERRITORY).

1. NORTH GERMAN OIL FIELD.

The North German oil field stretches across the north-western portion of the empire in a sharply defined direction, almost perpendicular to that of the great European oil fields, *viz.*, south-west to north-east. Entirely within the space enclosed by the lower courses of the Weser and the Elbe, the line of strike of the petroliferous strata of Northern Germany passes near to Peine, Hanover, Celle and Werden, where it bends slightly towards the north and, crossing the plain of Luneburg, terminates at Holstein.

The southern extremity of the oil field commences in the Hartz Mountains. The geological conformation of the German oil field differs essentially from that of the oil fields which we have as yet studied.

GEOLOGY OF NORTH GERMAN OIL FIELD.

We find in these petroliferous strata three different formations :—

1. Diluvial sands.
2. Neocomian limestones.
3. Jurassic limestones and sandstones.

It is in the last named strata that oil is met with in large quantity.

The petroliferous strata of the North German oil field belong, as a matter of fact, to the upper jurassic, but being, in the great majority of instances, in immediate contact with the neocomian beds, they have undergone lithological modifications which lead us, along with Oppel the geologist, to regard them as constituting a sort of transition between the jurassic and the lower cretaceous, thus forming a line of contact or junction between the two formations. The Germans call this stratum the *tithonic* bed.

Immediately above the petroliferous strata, except in the frequent cases of denudation, there is a bed of diluvial sand, which is itself covered by important alluvial beds of a country consisting in great measure of marshes.

The whole of these sand beds are easily permeated by the stagnant waters, as well as by the petroleum in the beds beneath. It follows that in consequence of the difference in density of the two liquids, and the effort to establish equilibrium, and also by the pressure of gas, there is produced between them a sort of circulation from below upwards of the petroleum and from above downwards of the water. This state of affairs, which is not altogether favourable to the quality of the oil, has the advantage, however, of showing in a decided manner its presence above ground.

REMOTE HISTORY.

It will be readily understood that, under these geological conditions, naphtha has been known to exist in the German oil fields since very remote times.

In fact, centuries ago the surface deposits of petroleum were wrought in this region. The operation is conducted in a curious way, which it may be of interest to recapitulate.

" Fettlöcher."

The inhabitants who used petroleum for illuminating purposes obtained it in the following manner. They dug deep pits, which they called *Fettlöcher*—" grease-holes "—and gradually the oil rose to the surface of the marshy waters, which filled these primitive wells. When, at the end of a few weeks, a sufficient quantity of petroleum had collected on the surface, the peasants removed it for use. Here was their method of pumping. They gathered on the borders of the marshes long herbaceous plants (reeds) which they bound into a faggot or bundle and fixed on to a pole, thus forming a sort of broom, then with this funny tool they extracted the petroleum by plunging the " broom " into the liquid, and then withdrawing it charged with oil and also with water with which it was always mixed. The water was allowed to drop off, and the broom was then twisted to separate the more adherent oil.

Recent History—Oelheim.

Later on, in 1860, when the oil fever was at its height in America, the Government caused borings to be made throughout the region between Werden and Peine in the south of Hanover.

The borings were sunk more especially in the countries north of Peine, in the neighbourhood of Oedesse, where the *Fettlöcher* were particularly numerous and productive. Wells were excavated to a certain extent everywhere; several borings even were undertaken, but with decidedly negative results. The naphtha on its way to stop short at the *Fettlöcher* followed roads only known to itself. No matter, the village of Oedesse, which is situated in the centre of the attempts to strike oil, assumed the name, quite as pompous as it is ridiculous, of *Oelheim* (Oil Country).

The Hanoverian government soon abandoned a position which seemed to greatly compromise it. Then individuals, the business men of the locality, were anxious to risk their stake. They lost it.

The money that was swallowed up by this oil country—the hopes that were deceived! Nevertheless, the Germans are a tenacious people; they continued their searches without a break, in the firm hope that the *German Petrolia*, as they have baptised the North German oil field, will produce by-and-by as much

8

petroleum as will make it an important rival to Pennsylvania and Baku. There is nothing impossible (above all for Germans); "Good luck!" then. Whilst waiting, the North German oil field, though one of the best wrought in the world, yields annually about 60,000 barrels.

PROPERTIES OF THE NORTH GERMAN OIL.

The North German product is an oil of medium density (it weighs 0·843), very much resembling Galician petroleum, as far as distillation products are concerned. It is generally associated with water, strongly charged with sodium and calcium chlorides.

The presence of this latter product may be easily explained by the calcareous nature of the strata, in which the petroleum occurs (cretaceous, lower neocomian).

We now proceed to pass in review the principal deposits of the North German oil field.

PRINCIPAL DEPOSITS.

To the south-west of Hanover, in the neighbourhood even of the town, is WENIGSEN. There are also at NEUSTADT, in the north-west of Hanover, several works with but very moderate yields.

At WEETZEN, a town in the neighbourhood of Wenigsen, we see some works conducted after the old-fashioned system of the *Fettlöchers*. There also are some borings made about 1860. Weetzen petroleum is the heaviest of the Hanoverian petroleums. One well in this locality produces about 700 barrels a year.

STEINFORDE possesses a score of wells, one of which, 500 metres deep, belongs to an English company, and is unproductive. The production, which is declining, is about 4,000 barrels annually.

OBERG, to the south of Piene, is a place newly wrought, on the large scale. The results obtained are rather encouraging.

OELHEIM AND THE "PETROLEUM SWINDLE".

OELHEIM, the metropolis of the North German oil field, is situated six kilometres north of Piene, near to the line of railway from Stendal to Hanover.

It was about 1881 that the development of the petroleum of Oelheim suddenly made a fantastic spurt under the impulsion of several German financiers. During the year 1881, 100 wells were bored here. The infatuation of the German public was overwhelming in face of the remunerative results which the dexterously got up reports of these financiers showed.

That was the times when the German oil field was baptised *deutsch Pensylvanien*, the German Pennsylvania.

At this time the newspaper press was gorged with emphatic articles announcing news for the benefit of extensive shareholders in the oil companies. Each day brought a new spouting well.

At one time, such and such a well yielded 300 barrels a day.

Again, another had not its equal in America.

We must, for the sake of truth, state that the average daily production of a good well at Oelheim is 8 barrels. We may add that the proportion of productive wells to dry wells (dry holes) is one in four. Be that as it may, so much dust was thrown in the eyes of the honest shareholders, that the promoters had to pack up, and the affair soon became known as the "petroleum swindle" (*la blague du pétrole; Petroleumsschwindel*). At the present time there is still one genuine undertaking. It is the *Petroleum Industrie Gesellschaft*. It possesses some 15 wells of an average depth of 200 metres ; its yield is about 25,000 barrels per annum.

Some individuals are also engaged in the business—some working one, some two wells. The product of these wells may be valued at 10,000 barrels per annum.

The petroleum swindle left palpable traces behind it at Oelheim in the shape of some 100 abandoned wells.

As we ascend towards the north, we find to the south of the little town of Celle, KOENIGSEN, which although only recently wrought on the large scale, would appear to us to be a rather rich locality.

Still farther along, going northwards, in the middle of LUNE-BURGER HEIDE, is SOLTAU, with a minimum production. In the same country we encounter SEHNDE, where the undertakings at least yield a result sufficient to cover working expenses.

In Brunswick, where the production is almost *nil*, working is still confined to the *Fettlöcher*.

The few borings which have been made, have yielded no appreciable results.

On the other side of the Elba, in Holstein, in the vicinity of Heide, are some works, the prospects of which, believe us, are not very brilliant.

TOTAL PRODUCTION.

The total production of petroleum in Hanover was in—

1881	22,968 Barrels.
1882 (epoch of the petroleum swindle) . .	48,114 ,,
1883	19,980 ,,
1884	29,064 ,,
1885	21,580 ,,

The present production may be estimated at about 45,000 barrels, which joined to that of the other petroleum wells of the North German oil field, estimated at 15,000 barrels, gives a total production of 60,000 barrels annually for the whole oil field.

GENERAL PROPERTIES OF NORTH GERMAN PETROLEUM.

Locality of Origin.	Density.	Colour.	Remarks.
Oedesse	0·849	Deep brown	Viscous
Wenigsen	0·850	,,	,,
Sehnde	0·865	Black	Bituminous
Oelheim	0·908	Very deep brown	Viscous
Hanover (St. Claire Deville)	0·892	,,	,,
Weetzen ,, (2)	0·955	,,	,,
,, ,, (3)	0·944	,,	,,

2. THE ALSACE OIL FIELD.

The oil field of Alsace is situated in the vicinity of Hagenau (Strasburg). The principal deposits are met with at PECHELBRONN, SCHWABWILLER, LOBSANN, OLHUNGEN and WŒRTH.

PECHELBRONN ASPHALTUM QUARRIES.

These localities, from a very remote time, possessed quarries from which asphalt was dug in very large quantities. So far back as 1735, Antoine le Bel the naturalist, wrote a report on the greasy and oleaginous bodies which sweated from the stones of Alsace. He even took in hand some researches. Some forty years ago François Degoussée had pointed out the existence of petroleum in a well excavated at Schabwiller; but the interesting communication of this engineer passed unnoticed by commercial men.

PETROLEUM BORED FOR IN ASPHALTUM QUARRIES.

It was not until 1883 that boring was commenced at Pechelbronn in the asphaltum mine itself. They had hardly struck the sandy beds of the tertiary formation, when the oil commenced to sweat abundantly. The sandstone, from which the petroleum sweated, was harder and finer grained than that of Baku. In less than a week a quantity of oil was collected in two wells estimated at 1,000 barrels.

It will be understood, in view of this result, that borings were hastily multiplied. The results, we are told, have up to now been rather encouraging. The best deposit would appear to be that of Pechelbronn.

FIG. 21.—Petroleum Oil Works in Pechelbronn Oil Field.

WALSBRONN.

A little to the east of the Pechelbronn oil field is WALSBRONN, a Lorraine village in the vicinity of Bitche. Although signs of petroleum are abundant, there are no works.

M. A. Hugo wrote as follows so far back as 1835 : " Walsbronn, village and castle, in a valley near to Bitche, has a singular fountain which ran there formerly in the midst of the forests (*Walsbronn* = well of the woods).

" This fountain, like several Persian springs, brings up petroleum. It was very famous in the Middle Ages. Inscriptions, antique medals and the remains of a Roman road lead to the belief that the Romans themselves knew its virtues. In the sixteenth century there was a basin and baths for the sick. We are ignorant of the course of events which led to the ruin of this establishment. A doctor who, in the last century, endeavoured to find the Walsbronn fountain, discovered it almost hidden under the ruins. It resumed its flow to him, but the petroleum only reappeared in small quantity. No one at the present day, it would appear, takes the trouble to collect it."

STATISTICS.

As the petroleum industry is of very recent date in Alsace, no authenticated statistics are in existence. At the present, according to our information, the annual production may be estimated at 35,000 barrels.

GENERAL PROPERTIES OF ALSATIAN PETROLEUM.

Locality of Origin.	Density.	Colour.	Remarks.
Pechelbronn (wells 146) .	0·906	Black	
„ (wells 213) .	0·885	„	} According to Engler
„ (1)	0·912	„	
„ (2)	0·968	„	} According to St. Claire Deville
„ (3)	0·892	„	
Olhungen	0·878	Deep brown	
Wœrth	0·885	„	
Schwabwiller (1) . . .	0·861	„	} According to St. Claire Deville
„ (2) . . .	0·829	„	

ITALY.

THE ANCIENTS AND PETROLEUM.

The existence of petroleum was known to the ancients; the fact is proven. Pliny, Diodorous of Sicily, and Dioscorides entertain us with the wells of Agrigentum (the modern Girgenti). The oil which they drew from it was burned in lamps under the name of *Sicilian oil*.

In the Middle Ages, the oil extracted at Miano was used for lighting purposes by the inhabitants of the vicinity.

MODERN HISTORY OF ITALIAN PETROLEUM—THE ABRUZZI.

The modern period of petroleum history in Italy commences in 1865 in the travels of an English engineer in this country—M. E. Fairman. It was at his instance that some proprietors consented to bore the ground in the neighbourhood of Bologna, Modena, and Reggio, without, however, any great success.   In 1880, a French company—*La Société des Pétroles d'Italie*—was founded in Paris ; enormous capital was placed at its disposal ; an American engineer was engaged to prospect the country.

In spite of the optimistic report which he published on his return, when we were asked our opinion on these strata, which we had already studied some years previously, we decidedly reserved it, having no faith in the great riches of the Abruzzi basin, a district of central Italy washed by the shores of the Adriatic, which it was desired to work.

The company decided otherwise, and the American engineer who conducted the works at the head of the workmen—*who had been brought* from America at· great expense—commenced borings at Tocco.   That lasted two years ; when, as the result was almost negative, the company, crushed with the expenses, ceased working.

Since that time several genuine undertakings have been installed, mostly foreign.   They have brought enormous capitals. Will they take their capitals back with them?   *Chi lo sa?*

GEOLOGY.

The strata in which petroleum is found in Italy belong to the miocene.

The Italian deposits are, in our opinion, very difficult to reach, more especially in the south of Italy.   We believe that the great cause of this is the numerous volcanic dislocations which upheave the underground strata of this district, breaking the stratifications, causing downthrusts, levelling and sudden faults.

Notwithstanding the persistent attempts of oil prospectors, the fact remains that the grand petroliferous vein has not yet been reached.

PROPERTIES OF THE OIL.

The North Italian oil is generally fairly bright and of a rather reddish tint.   Its density varies between 0·787 and 0·828.   In the Abruzzi and in Sicily it is generally thicker, with a more penetrating odour and of a deeper tint.   Its average density is 0·942.

CLASSIFICATION OF ITALIAN OIL FIELDS.

According to their geographical position we divide the Italian petroliferous strata into three oil fields :—

1. *The Sicilian Oil Field*, comprising some deposits not wrought, including, amongst others, those of GIRGENTI, which, however, may be regarded as the *doyen* of the oil fields of Europe. Pliny was its godfather.

2. *The Southern Italian Oil Field*, which includes the deposits of TOCCO, formerly wrought by *La Société des Pétroles d'Italie*, those of MANOPELLO, GARDAGRELLI, CHIETI, RIONERO DI MOLISE, TIRIOLO and GERACE ; and the lignite mines, situated to the north of Aquila in the flanks of the chain of San Lasso, an offshoot of the Apennines.

This fuel, taken from the flanks of the mountain, contains petroleum in large quantity. According to M. Giorgi it yields on distillation 66 per cent. of liquid hydrocarbides.

3. *The Northern Italian Oil Field.*—This is the oil field, the existence of which was pointed out by M. Fairman so far back as 1863. It is the largest, the best developed and, at the present day, the most extensively wrought of the Italian oil fields.

It stretches along the line of railway from Milan to Bologna, in a direction almost parallel to the chain of the Apennines, that is to say, from the north-west to the south-east. It is wholly situated on the right bank of the Po, on the level of the first undulations of the mountains, and passes the environs of Plaisance, Parma and Modena.

The principal deposits of this oil field are SALSOMAGGIORE, belonging to *La Société Française des Pétroles*, under the management of M. Zipperlen. The works here are but in an experimental stage, and although tangible results are rather slow in being forthcoming, good hopes are entertained. Seven borings are in operation. The wells of Salsomaggiore yield large quantities of natural gas, which is used to light up the village and the works.

SALSOMINORE, which numbers 5 borings.

MIANO and MEDESANO, known from the Middle Ages. So far back as 1869 a company started boring at Miano. Two

wells of 124 and 203 metres were bored, and gave but a
meagre enough yield.

At the present time 5 wells, either dug or bored by the
French-Deutsch Company, exist in this locality.   The petroleum extracted is greenish yellow; its density is 0·908.

FORNOVA DI TARO, 3 borings.

NEVIANO, first wrought in 1877 by a French company. At
the present day 3 borings are at work and produce a small
quantity of a dark yellow oil.

FIG. 22.—French Oil Works at Villeja.   Photograph communicated by
M. Zipperlen.

OZZANO DE SAN ANDREA DEL TARO has three unproductive
borings made by Messrs. Deutsch of Paris, under the management of the French contractor Lippmann.

MONTECHINO, farther to the west, where are scattered the
German works of M. Huber, under the management of M.
Schmidt. About 25 borings have been made ; results do not
exceed 4 barrels daily. The German company have given up

working at Montechino. In 1898, a French commercial man, M. Marchand, resumed working. The wells yield about 600 kilogrammes daily.

VILLEJA, a little farther to the north, is one of the most important of the oil undertakings of Italy. A French company, under the management of M. Zipperlen, has made 20 borings, which produce about 80 barrels daily.

PIETRA MALA. Messrs. Klobassa & MacGarwey, whom we have previously mentioned as having been so successful in Galicia, commenced working here in 1898.

MODENA. There are some deposits in the vicinity which have not been as yet wrought very energetically. We may mention SASSUOLO, with 4 hand-excavated wells.

STATISTICS.

The sum total of the Italian petroleum production is not great, seeing that it does not exceed 24,000 barrels annually.

Moreover, the statistics in the following table are very interesting as showing that if the petroleum industry is progressing in Italy, it is doing so very slowly indeed :—

ITALY—PETROLEUM STATISTICS.

Year.	Number of Works.	Production.		Number of Workmen Employed.
		Tons.	Value.	
1881	2	173	76,540	24
1882	2	183	86,844	121
1883	5	225	58,317	92
1884	6	397	135,452	110
1885	6	270	110,066	136
1886	7	219	91,130	145
1887	7	208	76,720	135
1888	5	174	55,630	75
1889	7	177	57,000	70
1890	9	417	120,608	177
1891	10	1,155	848,500	255
1892	7	2,541	754,500	267
1893		2,652	775,050	130
1894		2,854	847,260	194
1895		2,358		
1896		2,497		
1897		1,868		

In 1890, the year that the Italian production more than tripled itself, two important installations were established in the vicinity of Plaisance. This increase is due to the efforts of these two companies—the French Zipperlen Company at Salsomaggiore and the German Huber Company at Montechino.   These two companies—the most important in Italy—brought enormous capital with them, but their expenses figured into millions.   At the present time the French company alone survives.   It is the only work registered by us since 1894.

Fig. 23.—A Petroleum Oil Works in Italy.   Photograph communicated by M. Zipperlen.

GENERAL PROPERTIES OF ITALIAN PETROLEUM.

Locality of Origin.	Density.	Colour.	Remarks.
Salo	0·787	Bright brown	Fluid
Parma oil field	0·828	Red brown	,,
Miano	0·908	,,	,,
Tocco	0·942	Black	Bituminous

MINOR EUROPEAN OIL FIELDS.

ZANTE.

The isle of Zante—one of the Ionian Islands—possesses petroleum deposits which were mentioned in ancient times by Herodotus.

KIERI. In this valley there is a well, or rather a natural hole, where naphtha is continually bubbling. It is a sort of oval excavation of about a dozen metres wide and the same in depth. The yield of the fountain of Kieri is not large enough to admit of profitable working.

Few genuine attempts at boring have been made in this locality. Three shallow borings were made in 1871, without any great success; one of them struck a pocket of oil which yielded 30 barrels daily. This yield only lasted long enough to determine the statement just made. It stopped in a few hours.

The petroleum extracted at Zante is black, bituminous, rather heavier than water, with which it is mixed. Its density, 1·017.

HUNGARY.

Traces of petroleum, which have been the cause of fruitless attempts to reach their source, occur at SMILNO, SZINA, JOD, KONIA, SACZAL, ZIBO, UDWARHELY, SOOSMEZÖ.

They may be regarded as infiltrations of minimum importance across the Carpathian chain from the Galician oil fields.

SPAIN.

An English company—the South of Europe Exploration Company—has undertaken the working of a certain petroliferous strata at CONIL, in the vicinity of Cadiz. Up to now only 1 boring has been made, with a minimum yield of a reddish-brown petroleum of density 0·837. Two borings are in course of being sunk at Conil, where the mining engineer, D. Faustine Caro of Linares has, in his opinion, discovered an oil field 12,894 hectares in extent.

The South of Europe Exploration Company are also working experimentally near to Burgos, to the north of that town, the petroliferous strata of HUIDOBRO. At the time of writing the work has not passed the experimental stage. A horizontal tunnel has been excavated in the surface sandstone. The oil distilled from this sandstone is reddish-brown and of density 0·921.

RUSSIA.

The permian beds of Central Russia exhibit in several places traces of bitumen analogous to petroleum.

Let us cite for reference: SUKKOWO (Government of Kazan),

MICHAÏLOWKA, KAMISCHKI, CHUGOROWA, SARABILKOWA, JAKUSH-
KINO, NOWO-SEMECKINO (Government of Samara).

No regular working.

SWEDEN.

The existence in Sweden of bituminous shale for a long time
led to the hope of finding petroleum in that country.

In 1867, a boring was made with this end in view on one of
the flanks of MOUNT OSMUND. Pushed to the depth of 300 metres,
this boring did, in fact, yield petroleum, but in such small quantity
that the search was abandoned.

Quite recently further search was made at NULLABERG with
identical results.

BRITAIN.

At ASWICK COURT, Somersetshire, a liquid bitumen has been
discovered presenting all the characteristics of petroleum. This
oil, which flows from shale in small quantity, has the density of
0·816.

HOLLAND.

The majority of the Dutch trade journals describe, in an interested manner, the discovery of petroleum which has occurred in
the brickfields in the vicinity of Maestricht. A brickmaker had
purchased a piece of ground in the neighbourhood of his kilns, and
whilst sinking a well for water supply struck a pocket of oil in
the rock which constitutes the adjacent hill.[1]

[1] *Moniteur des Pétroles*, 15th July, 1898.

CHAPTER XI.

FRENCH CAPITAL DIVERTED TO ITALIAN OIL FIELDS.

FRENCH capital has, hitherto, been employed almost exclusively
in the development of the Italian oil fields. The barren nature
of the results obtained there has, to a greater or less extent,
touched French savings. Be this as it may, France, the petroleum resources of which have just begun to be developed, is
much richer in petroleum than Italy.

GABIAN OIL.

For many centuries there has been sold in France a medicinal
oil which is taken inwardly by our fellow countrymen (French)
under the name of Gabian oil. Ah, well! this oil, which flows
from a spring in the village from which it takes its name, is
neither more nor less than petroleum, and moreover petroleum
of the very best quality too. How comes it to pass then, as
we shall see in the sequel, that boring operations have but
just made a start at Gabian? The reason is that Gabian is
situated in the Hérault, on the side of a line of railway, under
our own (French) control. Ah! if it had been in Texas, or even
in the Abruzzi!

Nevertheless, there is no reason why one or more French
oil fields should not exist. The geological constitution of France
would affirm *a priori* the possibility of rich oil fields. Sure signs
clearly show the presence of oil.

But no one wants to see them, and "there are none so
blind as the wilfully blind".

COMPARATIVE GEOLOGY OF THE OIL FIELDS OF FRANCE, GALICIA, ROUMANIA, AND THE CAUCASUS.

The tertiary strata are, relatively, more extensively developed in France than in any other country of Europe. We are enabled, by experience, to say that it is in these strata that the greater part of European petroleum is found. In many localities, eocene beds, similar to those of Galicia, Roumania and the Caucasus, crop out. The nummulitic beds occupy, in the south of France, from the Atlantic to the Italian frontier, a wide belt which extends in a direction almost parallel with the Pyrenees, and identical with that of the similar beds of the Caucasus, Crimea and Carpathians, which they seem to join after passing through the Alps.

Other beds of the same nature are met with in Auvergne, in Upper Savoy; throughout which the presence of petroleum is demonstrated by unequivocal signs.

It can therefore be said that the French deposits belong to the same geological system as that in which the flowing wells of Baku, the Caucasus and Galicia occur. This is an excellent augury of their riches.

AN OIL FIELD CLASSIFICATION OR CORRELATION IMPRACTICABLE IN FRANCE.

The search for petroleum, which in France has been limited to a few isolated and scattered borings over the whole extent of the country, does not permit of any classification of the productive districts by oil fields, nor the foundation of any petroliferous system of geography.

We shall therefore confine our remarks to the statement that some localities bear undeniable signs of petroleum without trying to find whether these different deposits may or may not be correlated.

We may, however, remark in parenthesis, so as not to be under the necessity of referring to it again, that in those districts where the plutonic upheaval of the central plateau is directly confined to tertiary strata (whether in the south, in Hérault, or in the north, in Auvergne), this tertiary strata presents in our opinion undeniable signs of the presence of petroleum.

Moreover, we have tangible proofs of it at Gabian and in the plains of Limagne, in the borings that have been made in those two districts which have engendered the most favourable expectations.

Let us close our parenthesis and continue.

THE AUTUN SHALE OIL INDUSTRY.

In 1832, the mineral oil industry of France was started as a result of the efforts of Selligues the chemist, who distilled the Autun shale to obtain an illuminating oil—*shale oil*.

From the bituminous shales of Autun which he distilled in a revolving retort, he obtained 10 per cent. of crude oil for refining. This oil yielded equal proportions of illuminating oil and solid residue.

This industry attained certain proportions. Works were established, but they were obliged to close their doors in consequence of the introduction of foreign petroleum with which the French mineral oils could not compete.[1]

It was from abroad that the French people obtained their oil when all of a sudden the idea came to them that they might find petroleum at home. And it has been found in small quantity, it is true, but in time, more abundant deposits may be brought to the light of day.

Petroleum has been discovered in France at GABIAN (Hérault), at GUA (Isère), at CHATILLON (Haute-Savoie), and in LIMAGNE.

GABIAN (HÉRAULT)

Gabian is a rather important market town in the canton of Roujan. It is situated on the right bank of the river Tougue which flows into the river Hérault. In spite of manifest traces of petroleum, neither of two trial borings have yet struck oil. It is true that in one of these trial borings, which has only attained a depth of 202 metres, the petroliferous sandstone has not yet been struck; but then, as we have seen, this is not reached until a depth of 300 metres has been bored. The other boring was made quite near to the exterior manifestations of the Gabian oil spring.

---

[1] From time to time, since the days of Selligues, attempts—more or less abortive —have been made both by French and Scotch experts to revive the *French* shale oil industry. The shale oil industry of Scotland, established some twenty years later than the Autun works, after coming through many vicissitudes, may now be said to have weathered the storm of foreign competition. The *French* industry is hardly in so satisfactory a condition.—*Tr.*

The boring, being pushed as deep down as 413 metres, yielded negative results, it is true, but, to judge by the beds traversed, it has only confirmed the existence of petroleum, in large quantities, at Gabian. Where? That is the question. As is the case with all new deposits, oil is not struck without groping or feeling one's way. Sloboda in Galicia, and Noworossirsk in the Caucasus, are cases in point. At Gabian, more than anywhere else, nature would appear to take delight in leading oil prospectors astray, so as to rob them of its treasures. Rarely indeed have we been in the presence of such disturbed and distorted strata. These strata, which belong to the miocene, are downthrown, folded back on themselves like a sheet of paper, crumpled up, upheaved by volcanic phenomena dependent on the grand neighbouring upheaval towards the east of the central plateau. To the south, and within a short distance of Gabian, stretches a narrow band of nummulitic strata which, passing Caumette, extends towards the outlying flanks of the Pyrenees, and would appear to constitute part of the geological formation of these mountains.

The nummulitic strata, in this band, is superimposed by a calcareous bed which extends to the base of the black mountain. It is the lignite limestone, which furnishes a fuel wrought in Hérault.

In these strata, petroliferous signs are met with, not only at Gabian, but notably also at VENDRES and at GABOUX.

HISTORY OF THE HEALING FOUNTAIN OF GABIAN.

The Gabian fountain has enjoyed for centuries a celebrity which attracted all the sick people of Languedoc. A long way back in the very distant past, there was constructed for their accommodation a kind of subterranean pond. The ancient buildings are still in existence. They consist of a principal tunnel, which cuts another tunnel which ends at the basin where the naphtha was collected.

The principal passage, of a pointed form and entirely of granite, runs from north-west to south-east; it is at the end of this tunnel that the spring flows. This lobby is cut by another tunnel of the same form, but of smaller dimensions, and which extends in a north to south direction, parallel with the river, and which terminates in a room, containing the basin,

9

the waters of which easily run away through a reversed syphon, whilst the liquid petroleum (it weighs 0·894) remains in the reservoir. It was collected there as far back as 1608 for pharmaceutical purposes.

DR. RIVIÈRE'S ACCOUNT.

The first work which gives us some details in regard to the Gabian Well is that of Dr. Rivière, entitled *Memoir on some Peculiarities of the Lands of Gabian, and chiefly on the Fountain of Petroleum which flows There*. This work was published at Montpelier in 1717. We read in it as follows: "The spring of petroleum oil is thus called because it issues from a rock. It is 1,000 paces from the village of Gabian, and in a valley formed by two small mountains, on the border of a rivulet. It flows by two subterranean conduits, with the water on which it floats, into a basin contained in a building, where it always keeps above the water without mixing with it. . . . The petroleum is collected every eight days, when it is put into a barrel, where it is allowed to settle for a few moments, so that the water may separate, after which this water is emptied into the basin of the fountain through a hole in the bottom of a barrel, and when the petroleum commences to flow out it is run into earthenware vessels, where the separation is completed. This spring, which is believed to have been discovered in 1608, has not always yielded the same quantity of oil; that, which has been collected during the last twelve years, is not nearly so considerable as formerly. The petroleum, which it yielded for more than eighty years, reached each year the quantity of 96 quintaux.[1] At present, the tenant of this well has only collected 4 quintaux, annually, during the last twelve years or thereabouts. It dried up, and remained dry during two whole months in the summer of 1715, a circumstance which had never occurred before. But, after the rains, it flowed just as before the drought. It is asserted that this spring yields more petroleum at the equinox than at any other time; more also in summer, during mild, humid weather, than in the cold of winter. This oil is opaque, and its colour is a deep brownish red; in the basin it would appear to have a slight greenish, strongly brown cast. It has a strong

[1] Quintal = 220 lb.

agreeable odour, similar to that of bituminous matter, and is inflammable."

RECENT DEVELOPMENTS AT GABIAN.

Later, it is the Bishop of Béziers[1] himself who praises Gabian petroleum, from a medicinal point of view, in a memoir dated 1735; afterwards, in 1752, Bovillet, a nobleman, equally vaunted the virtues of Gabian oil in his writings. In 1834, A. Hugo expressed himself as follows: "The petroleum spring of Gabian is a well on which petroleum oil floats. It would appear that its flow is getting exhausted. At one time, they collected 36 quintaux annually; in the last century, the yield did not exceed more than 10 quintaux, and now no more than 80 to 100 kilogrammes are collected." In 1844, a company was floated to prospect for petroleum, and soon commenced trial experiments. These experiments were undertaken in a manner that deserves to be recorded. Instead of making several small borings in different localities — borings which would have afforded a general idea of the arrangement of the strata, their direction, their points of outcrop, and the nature generally of the ground beneath Gabian—the work was conducted in an altogether novel fashion. This is how they set about it: They commenced their search by a tunnel round about the old works, and by this tunnel, 50 metres long, they convinced themselves by the sweatings that the sheet of petroleum was spread out under the whole of the subsoil. Such experiments could not, therefore, throw any light on the position and direction of the beds at greater depths. In fact, it was not an investigation of the Gabian beds, but, rather, that of its fountain which was made. As a matter of fact, the real experiments were those undertaken blindfolded, and neither the one nor the other of which yielded any result. But let us continue. The Gabian Petroleum Company, as soon as the first trials had been made, and the proof established that the oil fountain of Gabian was fed by a petroliferous stratum—a fact which had been already demonstrated by common-sense — desired to pass on to the working thereof. Two systems of boring were proposed to

[1] Béziers, from which this bishop takes his title, is a town 72 kilometres from Montpelier.

them ; the Canadian system by M. Zipperlen, and boring by
free fall by Dru's system.   The velocity of the Canadian
system and the delicacy of its parts frightened the company.
It chose Dru's system and commenced, in November, 1884, a
boring, 400 metres below the fountain.   Underneath is a
statement of the beds traversed :—

SECTION OF STRATA MET WITH IN BORING FOR PETROLEUM AT GABIAN OIL FOUNTAIN, COMMENCED NOVEMBER, 1884.

	Metres of 3·28 feet.
Alluvium	1·38
Blocks of yellow sandstone mingled with sand	5·86
Rounded pebbles, alluvium, yellow marls	4·67
Yellow quartzose sandstone, yellow marls and sands	12·28
Grey marl and bands of limestone	4·67
Grey and crystallised gypsum	26·16
Grey limestone	7·91
Red limestone	4·01
Red conglomerates, pudding stone	33·58
Greasy red marls	8·48
Black marls	4·80
Greenish grey marls	5·00
Black quartzose limestone	1·69
Reddish grey limestone	1·69
Greenish grey limestone	6·45
Black quartzose limestone	0·73
Grey hard quartzose limestone	5·19
Bluish sandy limestone	4·38
Flags of sandstone, limestone and grey marls	1·14
Grey limestones slightly violet	4·67
Violet shaly marls	2·90
Grey marly shales in flags	3·79
Greyish blue, sandy limestones	0·55
Green marls	0·89
Violet marls	0·37
Reddish conglomerate, pudding-stone (very hard)	0·56
Grey marly limestone mixed with sandstone	7·15
Brown, grey and blackish grey marls	2·84
Grey, very quartzose sandstone, mixed with grey sand	1·30
Grey sandy limestones (hard)	38·57

Arrived at a depth of 202 metres (662 feet) the boring was
stopped.

Summary.—The boring of this well with Dru's appliances
had an initial diameter of 0·40 metre (15¾ in.), and terminated
with a diameter of 0·25 metre (10 in.).   The well was cased to a
depth of 170 metres (557 feet).   The boring of these 202 metres
lasted ten months.

At the depth of 69 metres a bed of red conglomerates
having been reached, the sides of the well caved in, and a

mineral water spring began to flow. This water, which continued to flow intermittently, had the following composition :—

			Grammes per Litre or parts in 1000.
Carbonate of lime .	.	.	1·080
Sulphate of lime .	.	.	0·232
Sulphate of soda .	.	.	0·902
Chloride of sodium	.	.	0·105
Carbonic acid	.	.	1·910

It was during the flow of this well that M. Zipperlen wrote the following remarks :—

" Strong petroleum emanations are increasing in intensity. I estimate, as far as that is possible in a ground so upheaved by volcanic commotions as that of Gabian, the depth of the pudding-stone conglomerate at 120 or 130 metres. It appears to me that it is this impermeable bed which keeps back the petroleum and the gases, only allowing a few drops to sweat through the fissures.

" We adhere to the theory that the concurrence of three essential circumstances are necessary before any strata can become the source of liquid petroleum.

" 1. That there has been a development in the country of the organic or eruptive force which generates those hydrocarbides.

" 2. That the phenomena of oscillations of the ground have produced fissures or communications with the starting point of the gaseous strata.

" 3. That these cavities terminate in an impermeable bed capable of retaining in its interior the hydrocarbide emanations by closing all exit to the exterior.

" There is no doubt but that Gabian fulfils all these conditions.

" And here we have more than a probability, since the petroleum flows naturally, and that daily we collect a few litres from the spring.

" Moreover we have, as in the Canadian petroliferous strata, a considerable extent of shale, pyroschists with nodules of pyrites and limestones.

" We have analysed these nodules of white pyrites and mar-cassite containing as much as 54 per cent. of sulphur and 44 per cent. of iron. There would appear, therefore, to be no doubt that, with perseverance, success will crown our efforts, and that we shall be able to endow our country with a new industry, which will be a source of considerable profit to us, and save us from being tributaries to the foreigner."

This first boring abandoned—in spite of the hopes which the petroliferous signs encountered gave rise to—without having reached the petroliferous stratum, M. Zipperlen started, in 1886, a new boring at a distance of 900 metres to the north of the first boring. There he used the Canadian system, and the experiment was altogether in its favour. In fact, 413 metres (1354 feet) were bored in four months, the well, commencing with a depth of 0·30 metres (11·8 inches), was cased to the depth of 315 metres (1033 feet), with a final diameter of 0·13 metre or 5 inches.

The boring was made on rising ground corresponding to an important volcanic upheaval. After the first few metres permian beds were struck. The results in this well, as in the first, consisted of signs of petroleum.

Here are the beds met with in the boring :—

Sections of Strata Encountered in Second Boring for Petroleum at Gabian, 900 Metres North of First Boring.

		Metres of 3·28 feet.
Permian strata :	Red hard silicious conglomerates	21·25
	„ „ „ (quartzose) . . .	30·00
	Hard, marly, very quartzose limestones . . .	16·00 .
	Very hard, grey, sandy limestones with pyrites . .	28·65
	Grey sandy limestone	33·80
	Sandy limestones, with crystallisations, very hard . .	15·90
	Dolomites, with crystals of carbonate of lime, very hard .	154·40
Devonian strata :	Dolomites for 15 metres, traces of bitumen which afterwards disappeared completely, very hard strata .	113·35

A mineral spring also flowed during the boring of this well.

Cause of Failure of Gabian Borings.

Now, No. 2 well was bored in a designated spot. As far as we are concerned, we are certain that this spot where the volcanic upheaval took place was chosen deliberately for the erection of the second derrick.

Whereas in Europe, all the deposits of any importance are met with in tertiary beds. At Gabian itself, even, the well flows from decidedly miocene beds.

What are we to conclude from these facts ?

1. That the well No. 1 would have struck oil at a greater depth.

2. That, as regards well No. 2 they systematically deceived themselves by boring too far to the north, and in strata lower than the tertiary.

3. That a new boring, to have any chance of success, ought to be made between these two wells, at a point as near as possible to the tertiary strata, which can easily be determined by preliminary soundings to the depth of 15 metres on an average.

As a matter of fact, the disposition of the strata, where the two wells were bored, may be represented by the accompanying ideal section.

The position of affairs may be told in two words—one of the wells has not struck the petroliferous stratum, the other is sunk beyond it, outside the line.

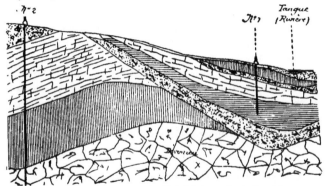

Fig. 24.—Showing how Nos. 1 and 2 Wells failed to strike the Petroliferous Tertiary Strata at Gabian.

The traces of bitumen in this well and in the Devonian, too, at a depth of about 300 metres would, in our opinion, indicate that these beds are, as shown in Fig. 24, immediately subjacent to the petroliferous stratum, and that the bitumen, far from being the product of the Devonian strata in question, is the simple result of infiltration from a neighbouring tertiary bed.

Moreover, we must note, along with M. Zipperlen, that these traces are only found in the first 15 metres of those strata. The infiltration goes no farther in this compact bed. This peculiarity shows decidedly that our assertion is the only one possible.

This argument is important. It shows clearly the existence of hydrocarbides at a great depth in the ground beneath Gabian.

The fountain of oil on the surface soil, the traces of bitumen at great depths (300 metres) ; these facts are in absolute concord ; they fit into each other admirably.

VAUCLUSE.

THE ROQUE-SALIÈRE BITUMINOUS SHALES.

The vicinity of Apt (807 kilometres from Paris and 65 from Avignon) also shows evident signs of petroleum, amongst which may be cited those met with at Roque-Salière. The place takes its name from the phenomena, which may be observed there, of a schistose rock sweating an almost liquid substance, and containing between its foliae a rather large proportion of common salt.

So far back as 1834, A. Hugo mentioned these peculiarities in these terms : "There are included amongst the natural curiosities of the department, the singular schists of Roque-Salière, near to Apt. They look like a mass of old books, which time has blackened and rolled their leaves one over the other. Maritime petrifications are found there, and especially a species of fish, two to three inches in length, some of which are perfectly complete, with dorsal and lateral fins.   These schists, they say, burn easily, and then give off a disagreeable, slightly bituminous odour."

LIMAGNE.

Limagne, Puy-de-Dôme, is a vast valley, through which the river Allier flows (the Allier rises at the foot of the Cevennes, in the department of Lozère, and after flowing 370 kilometres joins the Loire at Nevers, 270 kilometres from Paris). It forms a magnificent plain, bounded on the one hand by the mountains of Forez (Lyons), and by those of the Puy-de-Dôme on the other.

GEOLOGY OF THE LIMAGNE DISTRICT.

Surrounded on all sides by volcanic upheavals, the signs of petroleum are more generalised here than in Hérault. However, the geological structure of Limagne differs but little from the strata of the latter department. Here, as in Hérault, we encounter a tertiary basin, sharply defined and limited on all sides, except on the north, by primordial strata.

The lower portion of these tertiary beds consists, as at Gabian, of important masses of red conglomerates and marls. In the central zones we find sandstones and green foliated marls. Finally, in the upper portion, the most prevalent beds consist of a travertine, quite similar to that in the neighbourhood of Rome, and marls more or less mixed with gypsum. The Limagne travertines very much resemble certain Roumanian petroliferous strata. Like them they are characterised by the abundance of certain fossils.

These beds should be unhesitatingly classified as partly belonging to the base of the miocene, partly to the eocene. We notice here, just as at Gabian, upheaval phenomena, even volcanic fissures, which have disturbed the relative positions of the different strata quite as much as at Gabian.

LIMAGNE BITUMEN.

In regard to surface signs of petroleum, Limagne is one of those countries where naphtha has mostly manifested itself by a large number of salses, small natural springs, from which an almost liquid bitumen has oozed for centuries. "We know moreover, says Leymerie in his *Geology*, that this latter bitumen, softened by a mixture of petroleum (pissasphaltum), oozes constantly in the Pays de la Poix (Pitch country) et de Cronelles, near to Clermont, in Auvergne. It is without doubt natural to think that certain sources of bitumen result from a kind of underground distillation, caused by the central heat acting on masses of coal for example, but many other sources, and particularly the bituminous artesian wells of America, would appear to *indicate subterranean sheets of petroleum, the origin of which we are ignorant of up to now.*"

It is always the case, in a general manner, that all the travertine which constitutes the underground of Limagne is impregnated with a bitumen so far liquid that it may already be called petroleum, and which impregnates with naphtha the thin layer of arable soil which forms the cultivated surface of fertile Limagne, as readily as it would blotting paper (bibulous paper). Pardon the familiar comparison, but it is accurate.

SMALL BITUMEN "MINES".

This faculty which the soil has of pumping petroleum has been known and taken advantage of for a long time in Auvergne, and has given rise to quite a host of small bitumen mines (*minières de bitume*).

Some years ago, M. de Guasco visited one of these works. Here is the account which he gave of his visit to a meeting of the Mineral Industry Society of St. Etienne :—

"I found myself in presence of an abnormal fact, or one at least regarded as such. A field, situated in the precincts of Riom (407 kilometres from Paris and 13 from Clermont Ferrand) surrounded by ditches full of iridescent water, spotted in places with patches of grease, was *periodically* wrought for the extraction of bitumen. The surface, highly impregnated with bitumen to a depth of 30 to 40 centimetres, whilst the subsoil did not contain enough to be treated, was scummed, and the ground, after being treated, was put back in its place. After a greater or less lapse of time, the operation could be repeated, and a yield obtained sensibly equal to the previous one.

"How can this phenomenon be explained otherwise than by a physico-chemical action ; the condensation and oxidation on the surface of the soil of emanations proceeding from a subterranean reservoir of mineral oil?

"Starting with this line of ideas, I have searched for precedents which might impart to them a material existence, and during a residence of two years, 1871-2, not only have I been able to collect some, but also to attack the study of the geology of Limagne which confirmed me in the belief of the existence of mineral riches at great depths : coal or petroleum.

"Now the petroliferous stratum has been met on two occasions, in two points far removed from one another, during borings made with another end in view, and that at an epoch before Selligues had commenced his researches on the utilisation of mineral oils.

RECENT BORINGS IN LIMAGNE.

"Since, and quite recently, *La Société des Bitumes et Asphaltes Français* have made two rudimentary borings, one 25 and the other 20 metres deep ; it also has struck the petroliferous strata."

In the vicinity of Clermont Ferrand, the Tramway Company, in sinking a well to obtain water, struck a sheet of saline water, mixed with a notable proportion of petroleum.

In the Pays de là Poix (a characteristic name), a well 50 metres deep yields a very fluid bitumen containing, according to M. P. Juncker, 52 per cent. of crude petroleum.

At the present day, the search for petroleum in Limagne is being developed. Several borings on Fauvelle's hand system are in progress in the vicinity of Riom. Although this is but a small beginning, we wish good luck to the French petroleum pioneers.

LE CREDO (AIN).

The flanks of the Jura would appear to contain a certain quantity of petroleum. At Boge, on the flank of the Credo, near to Collonges, manifest traces of the presence of oil have recently been found. Up to now, the ground has not been examined but in quite a superficial manner by serpentine tunnels a few metres below the surface of the ground.

An examination of the ground would appear to prove that in one place it presents numerous analogies with that of Pechelbronn, the geological formation belongs to the cretaceous. At a depth of some metres below the surface the sand is completely soaked with bituminous oil, thick, like all surface oils, and of density 0·975.

We believe this deposit to have a grand future before it.

LE GUA (ISERE).

THE "BURNING" FOUNTAIN.[1]

Ten kilometres to the south of Grenoble (the chief town of Isère, 633 kilometres from Paris), a little to the east of Vif, there are important cement quarries. These quarries are situated in a mountainous country at the bottom of a valley, which is terminated by a circle of limestone rocks. This place, which is situated in the commune of Gua, has been known from time immemorial under the name of the burning fountain. This significant name is due to a phenomenon of nature essentially petroliferous.

" The burning fountain, situated in the commune of Gua, four and a half leagues to the south of Grenoble, is a spring issuing

[1] " La Fontaine ardente."

from a shallow excavation, which exists on a cultivated plateau at
the foot of a chain of calcareous mountains.

"The water, which flows from it abundantly, rises in columns
of flames when the receiving vessel is removed and a light brought
near to it.

"After the summer rains, the well produces, even spontaneously, flames three feet in height, and these can cook eggs,
without the water, from which they issue, having a higher temperature than the air.

"The burning fountain increases or diminishes like the other
springs in the neighbourhood. Its water is tasteless, and has no
action on blue vegetable colours. The bubbling is produced by the
gas with which it is impregnated, and which exhales an odour of
sulphuretted hydrogen or phosphuretted hydrogen. The gas
is believed to come from an iron mine in decomposition. It
is very inflammable, very easily collected, but it loses its inflammability in a few days, without losing either its weight or
its volume, when it has been enclosed in a bladder. There are
no indications in the neighbourhood of an extinct volcano. It
has been noticed for some years that spontaneous inflammation
has become more rare, but there escapes still from the stream of
water, which issues from the surrounding earth on boring to
a slight depth, a gas, by means of which light substances such as
dried leaves, straw, paper, etc., may be set fire to " (A. Hugo, 1834).

Certain cements in the neighbourhood of the burning fountain
of a brown colour and unctuous to the touch would appear to be
soaked with petroleum. The soil, in that respect resembling
that of Baku, could easily be lighted by digging down a little.
All these phenomena could not pass unperceived in this age of
industrial progress. In 1885, an engineer took in hand a boring on
the site of the burning fountain itself.

GEOLOGY OF THE "BURNING" FOUNTAIN NEIGHBOURHOOD—
GAS PRESSURE.

Any one could easily convince himself that the underground
strata presented all the most characteristic appearances of
petroliferous strata. After the first metres, the eocene was encountered; the soft burning ground, like that of Baku, tended

always to fall in under the pressure of the gas, which became formidable in proportion as the works advanced.

That was the stumbling-block ; for a boring accident, caused by this pressure, stopped operations. This unfortunate result proved the death-blow of the researches which courageous industrials had attempted at Gua. They did not persevere further.

A new company undertook operations a little below the old works.

An American engineer, Mr. Piret, undertook the systematic exploitation of the ground of the burning fountain. He had the utmost confidence in his tentative. He expressed himself as follows : " If we examine attentively, in the country, the black schists and the stone quarried for the manufacture of cement, we will find them to be slightly bituminous. The clays are black, impregnated with greasy matters, the clear limpid waters of the stream, precipitated into the well of the burning fountain, soon become black, and one sees surging, circulating, serpentining on the surface, in the bubbling provoked by the pressure of the gas, in spite of an overcharge of 4 metres of water, long traces of a greasy inflammable black substance ; and when this water is thrown on the flames, these become lively, enlarge and spread themselves over all the surface of the well with frightful rapidity, giving rise to slight explosions, finding fresh fuel in this water.

" The same observations made in America are met with here, as much amongst the schists as amongst the anthracites :

" The black schists, from a very shallow depth, show crystals of calcite, associated with slight agglomerations of carbonate of lime, and leave a certain saline taste on the tongue. The anthracites of the neighbourhood have also, in the portions broken from the veins, a certain peculiar resinous substance which forms the binding material."

" This substance, which is of a reddish-brown colour, has a vitreous lustre. It is as tenacious as calcipar, its fracture is conchoidal and has neither taste nor smell."

ANALYSIS OF THE " BURNING " FOUNTAIN GAS AND KRAFFT'S DEDUCTIONS THEREFROM.

The abundant gases issuing from the burning fountain, were analysed by M. Raoult, Professor of Chemistry to the Faculty of Grenoble. The following is the result of his analysis :—

						Per cent. by Volume.
Carbonic acid	.	.	.	.	.	0·58
Nitrogen	.	.	.	.	.	0·48
Oxygen	.	.	.	.	.	0·10
Marsh gas	.	.	.	.	.	98·81
Loss	.	.	.	.	.	0·03
						100·00

On the result of M. Raoult's analysis, M. Krafft, Professor of Industrial Chemistry, made the following observations :—

"1. Petroleum always and everywhere accompanies burning wells. Since we have here a burning fountain which has burned for centuries, I find it reasonable to suppose that petroleum may exist not far distant therefrom.

"2. Petroleum, which must not be confounded with naphtha, the product of another geological epoch, is never far distant from anthracite deposits. Now there are numerous anthracite deposits around Grenoble.

"3. Up to now, petroleum has always been met with in subterranean cavities which it fills along with saline waters. From what we find here, I do not see why it should be otherwise. There are deposits of salt and saline springs in the two neighbouring departments at, geologically speaking, insignificant distances."

"Then again we are right in the middle of the lower jurassic and it is in the jurassic that the salt mines are found."

PIRET'S ATTEMPTS TO CAPTURE AND UTILISE THE GAS OF THE "BURNING" FOUNTAIN.

Struggling against the routine of formalities Mr. Piret obtained from the Council General of Isère the suffrage "that the mining law be modified in this sense that the concessions of natural gas may be granted with the same title as that of petroleum or other analogous deposits".

The works were commenced, but Mr. Piret, instead of searching for petroleum, desired rather to capture the gas of the burning fountain, so as to use it for heating and lighting purposes. He pursued the sources of this gas on the flanks of the hill by a horizontal tunnel; *no vertical research, no boring* was undertaken. The tunnel was 1·50 metres in height, and 1 metre in width (59 × 39 in.). The gas soon manifested its presence. An explosion occurred, which led to the loss of a gasometer. However, the tunnel reached the length of 150 metres, but accidents multiplied more and more, which forced Mr. Piret to abandon an undertaking which gave rise at the outset to such brilliant expectations. Again, the burning fountain has been forsaken by industry.

Only a stone exists which attests its ancient celebrity. The following words, engraved without doubt during Roman rule, are written thereon :—

L MATERNUS OPTATUS

VULCANO AUG.

SACRUM

P.

CHÂTILLON (UPPER SAVOY).

GEOLOGICAL FEATURES OF THE DISTRICT.

The tertiary strata, which form the base of the Alpine flanks in Upper Savoy, would appear to be particularly rich in petroleum.

Signs of naphtha are abundant there; sometimes they are long bands of denuded strata, laying bare brown shales with the characteristic odour of petroleum; at other times they are inflammable gases which escape from the soil in ground which does not present any volcanic characteristic, and which everywhere melt under the feet of the observer. Often it is petroleum itself, which seems to say, " *Here I am; pick me up* "; and we, we French, remain callous, preferring to bury our capital in Italian borings.

NATURAL GAS UTILISED AS FUEL AND ILLUMINATING AGENT.

At Châtillon, and, in its vicinity, the phenomena of the burning fountain is regenerated. Carburetted hydrogen escapes, in such abundance from the soil, that several of the inhabitants lead it into tubes and utilise it for lighting and domestic purposes. All that requires to be done, in certain places, is to drive a stick into the ground, and a disengagement of gas is obtained. A simple hole, 1 metre deep, made in this manner produced such an abundant flow of gas, that when it was lighted it sent up a

shower of flames 2 or 3 metres high, which continued to burn
for five or six hours. We have estimated at 40 cubic metres
the volume of gas which may be obtained from such a hole
during this lapse of time.

"This Châtillon gas," says M. Delesse, "injures the crops
on the spots, where it escapes spontaneously from the soil.
On the other hand, it may be utilised, not only for lighting,
but also as a motive power for gas engines—*les machines
Lenoir*." Different experiments have been tried to see whether
the layer of gas-charged strata was altogether superficial. These
experiments. whilst demonstrating the depth of the strata in
question, produced an unlooked-for result. Several borings
were made ; the first, at Prêle, was only pushed to the depth
of a few metres. As they were clearing the bore-hole, they
heard a dull, heavy noise, a sort of rumbling, which was very
soon followed by a violent eruption of gas. This gas burnt
for fifteen days.

To the east of Châtillon, the result was altogether different.
Without going beyond the gas-charged strata, having reached
a depth of 30 metres (98·4 feet), *the presence of petroleum at the
bottom of the well was materially demonstrated.*

THE "PETROLEUM STONE".

We, ourselves, have collected a certain quantity of brown shaly
stones which we have heaped together in the vicinity of the Prêle
boring. These stones have an unctuous, the so-called greasy,
surface, and a well-developed smell of petroleum.

All these characters give them an evident resemblance to the
petroliferous *Ropianka Schichten*, which we have often met with in
Galicia, and we have no doubt that we are in presence of what we
"oil prospectors" call the *petroleum stone*. To establish our point
some decisive experiments were necessary ; we have performed
them ; they succeeded in all points.

1. The fragments of rocks to be observed were of a brown
colour, clean fracture and smooth surface. When we had exposed them for some time to the air they became paler, turned
to grey, and had tendencies to separate into lamellæ. When these
lamellæ are separated there is found between them a humidity,

which was drawn from the surface, to be stored in the centre of the stone. This proved to us clearly that the humidity of these rocky fragments was not due to exterior causes, depending on the state of the atmosphere, but rather to the fact that *the stone itself distilled a fluid substance* which we now have to determine.

2. With this end in view, we took some pieces of Prêle schist, crushed them, and plunged them into a vessel containing a certain quantity of boiling water. At the end of some time, when the liquid was cool and clarified, we found that the surface was iridised by a liquid, which floated on the top. The liquid distilled by the stone was therefore *a greasy liquid, or a grease lighter than water.*

3. We were already satisfied ; nevertheless, we pushed the experiment further, and submitted the stone to a further proof. After that of the water that of the fire. Having filled a sort of metallic basket with fragments of the stone, we fixed it over the flame of a gas furnace. After a little time a smoke, with a decided odour of petroleum, was disengaged from the basket. Under the action of the flame, the grease seemed to sweat from the stones ; soon a reddish, smoky flame crowned the heap.

We concluded that this flame proceeded from the schists contained in the basket, and we extinguished the gas. We had not deceived ourselves, for the flame continued to burn, demonstrating to us that it was due to a combustible furnished by the stones themselves. It was evident that *we here had to deal with a combustible stone like a coal, or with one distilling a liquid hydrocarbide.*

In order to satisfy ourselves which of these two conditions prevailed, we measured afresh the volume of the stones used in the experiment, by plunging the whole of these materials into a vessel full of water. The amount of liquid displaced in the vessel yielded us a volume appreciably equal to that which we had measured before the experiment. The stone itself, therefore, had not burned, and the combustion *could only have proceeded from the liquid which it distilled,* and whose presence we had demonstrated. This combustible liquid was a hydrocarbide. The following then are the result of our observations :—

The shales, collected on the borders of the Prêle boring, near to Châtillon, Upper Savoy, distil a combustible oil, lighter than water,

10

which can only be regarded as petroleum, of which it possesses all the properties.

> *Colour :* Bright brown, with blue fluorescence.
> *Density :* 0·782.
> *Great fluidity, which prevents it being classified with the asphaltums.*

· These facts, the result as much of our own observations as that of the boring experiments undertaken in Upper Savoy, may pass as comments.

We do not despair of seeing at the foot of our French Alps, if not fountain wells as at Baku, at least flourishing undertakings which will share part of a tract of land which, in our opinion, has all the appearances of a rich oil field.

CHAPTER XII

PETROLEUM IN ASIA.

TRANS-CASPIAN AND TURKESTAN TERRITORY.

Isle of Tchéléken—" Naphtonia, Naphtnia Gora."

THE Caucasian oil field, which terminates in the Apshéron peninsula, would appear to project its veins of naphtha-impregnated sandstone beneath the waters of the Caspian. In fact, on the eastern shore of this sea, exactly opposite to Baku itself, we come across the rich oil fields of the Trans-Caspian region. The island of Tchéléken—which the Russians have baptised with the appropriate name of *Naphtonia* or *Naphtnia Gora* (naphtha mountain)—forms the advance guard of the Trans-Caspian oil field.

Geological Structure of the Island.—This designation, moreover, is very applicable to Tchéléken, that island being of even a still more extraordinary petroliferous structure than that of the Apshéron peninsula. The soil consists of dried naphtha (*kir*), over which in all directions gush bubbling waves of liquid naphtha. Judging from this peculiar geological structure, the island itself can only be regarded as the result of a gigantic eruption of ozokerit, mixed with *kir* and sand.

Development Retarded by Situation, etc.—Owing to the distance and isolation of Tchéléken on the one hand, and the petroliferous riches of Russia on the other, the development and working of the petroleum industry has not as yet been extended to the *Naphtnia Gora.*

Tentative Borings.—Nevertheless, during the construction of the Trans-Caspian Railway, several borings were made by the Russian Government—the sole proprietor of the island of Tchéléken. One of these shallow borings yielded 1,000 barrels daily. Since that time, Nobel Brothers have acquired a concession in the neighbourhood of the Russian works.

Average Yield.—The average yield of a well at Tchéléken is 100 barrels a day.

Properties of the Oil.—The oil extracted is black and viscous. A sample, submitted to the authors for examination, had a density of 0·912. Another sample, in the almost pasty condition of asphaltum, had a density of 0·983.

Mikhailowsk.—A few kilometres from Tchéléken, on *terra firma*, is situated the town of Mikhaïlowsk. To the south of this town the prospector may find numerous signs of petroleum-bearing strata, indicating a country particularly rich in liquid fuel. This oil field extends eastwards, parallel to the Trans-Caspian Railway, in the sandy valley where the waters of the Oxus formerly flowed. The northern flanks of the little Balkan are landmarked by numerous petroleum springs.

Backward Condition of the Industry.—In spite of its riches in petroleum, the industry has made but little progress in this country, as much on account of its arid nature and its isolated distance as from fear that an over-production would lower the market price of Russian mineral oils.

A Store for Future Use.—Under these circumstances, the Trans-Caspian region may be regarded as a reserve store—and, without doubt, a formidable one—the working of which promises prodigious yields, when the uses and applications of petroleum, already so varied, shall have increased still further, and the increased consumption shall have banished all fear of over-production.

GENERAL PROPERTIES OF DIFFERENT TRANS-CASPIAN PETROLEUMS.

Place of Extraction.	Density.	Colour and Appearance.
Tchéléken	0·912	Black
,, (2)	0·983	Black pasty
Mikhaïlowsk	0·946	Black viscous

TURKESTAN.

Turkestan would also appear to furnish its contingent to the Russian production of petroleum. M. G. Mouchkétoff has published the following data :—

Fergane Oil Field.—No petroleum-bearing strata are known to occur in any place throughout the whole of Turkestan except in the single territory of Fergane. In the chain of mountains which surround Fergane, petroleum, mixed with water, appears at the the surface in several places. At the present day the following sources are known :—

1. In three places, 55 kilometres (34·3 miles) NNE. of the town of NAMANGANE.

2. To the east of these wells, in the northern part of the district of Andidjane, 25 kilometres (15·5 miles) from the village of ISBAKENT, on the banks of the Maila.

3. At a distance of 30 kilometres (18·6 miles) to the east of the town of Andidjane, on the right bank of the Kara-Daria, in front of the village of AÏM-KICHLAK.

4. At a distance of 12 kilometres (7½ miles) to the south of the village of RICHITANE.

5. Near to the road between Khokand and ISPARA.

6. At a distance of 15 kilometres (9·3 miles) to the south of the village of MAKHRAM.

Altitude, etc.—All these springs are located at a height of 700 to 1,100 metres (2,296 to 3,608 feet) above the level of the sea. The regions where the petroleum springs are found are totally deprived of forests, and often even of water. Solitude reigns supreme, although the road is practicable.

Small Yield.—The daily yield of petroleum which comes to the surface in all the known springs is very small (at the most 20 vedros, for example the Maïla spring) ; this does not, however, prove that the springs are barren. It is a well-known fact that springs, which yielded 3 or 4 vedros before being converted into wells, yielded afterwards, in the same period of time, a great number of tons.

Geological Formation. — The petroleum comes entirely from cretaceous strata, which consists of highly fossiliferous sandy limestone, red sandstones, green clays, gypsum and marl. All this assemblage of strata rests on jurassic rocks, and is covered by tertiary strata. The petroleum is found most often in the limestone with shells, or lower, at the contact of the limestone with the gypsum. As is the case with the clay, rock salt is sometimes met with. The water on which the petroleum floats is always salt. The more the petroliferous strata are disrupted, the more abundant are the springs, and *vice versâ*. More or less extensive beds of hardened petroleum *kir* are to be observed.

Probable Existence of Undiscovered Springs.—If the vast extent of land—dozens of kilometres—over which petroleum is found under the same geological conditions to the north, east and south of Fergane, as well as the scarcity of population in these districts, be taken into consideration, it may be safely said, without fear of error, that there are many more springs in existence than those actually known.

PERSIA.

Extent.—The Persian oil field extends in a line running from north-west to south-east, which, commencing at Shahku, on the Turko-Persian frontier, terminates on the east coast of the Persian Gulf.

Northern Section.—The northern portion of this oil field has its centre at KASHARASHIRIN, near to Shahku. Round this village numerous wells are sunk, the depth of which does not exceed 10 metres (32·8 feet). This deposit is situated on an eocene bed of sand and marl. The Kurds work it in a very rudimentary manner. They simply collect every four or five days the oil which has accumulated in the wells, the average yield during that time being 10 barrels per collection. The petroleum is very limpid, and has a green cast. It is refined, to a greater or less extent, on the spot.

Central Section.—In the centre of the Persian oil field, on a line parallel with the Bakhtiari Mountains, the petroliferous district of LAURISTAN extends.

Geological Formation.—This district, like that of Kasharashirin, is contained within a region belonging to the eocene, and characterised by the same blue clays which are met with in Galicia. The petroleum springs are met with here in the neighbourhood of important deposits of salt and sulphur. It is in the environs of CHOUSTER that the existence of petroleum is revealed in the most evident manner: the inhabitants collect it on the surface of the ground.

Chouster petroleum is of a peculiar nature. Its colour is a

very bright, almost transparent to a certain extent, opalised yellow. Its density is 0·773.

To the south of this station, a few kilometres from Ram Ormuz, are the natural spouting wells of CHARDIN. One of these wells has a regular output of 1 hectolitre (22 gallons) a day.

Natural petroleum springs also flow quite close to the Persian convent of Imamzade, at HAF-CHEID. These springs, which have an output of 1 barrel per day, yield a greenish oil of density 0·927.

Southern Section.—The shores of the Persian Gulf run along the southern extremities of the Persian oil field. DALIKI is the most important of the stations of this portion of the oil field. There is an establishment here which has commenced work in a rational manner. Daliki petroleum is heavy and bituminous. According to Boverton Redwood, its density is 1·016.

. A neighbouring station of Daliki, KISM, yields a greenish petroleum of a normal density of 0·837.

From this review it may be said that the petroleum oil fields of Persia have hardly been wrought. The cause of this fault is the absence of means of communication, the want of resources of all kinds and also the uncertainty of the issue.

GENERAL PROPERTIES OF THE DIFFERENT PERSIAN PETROLEUMS.

Place of Extraction.	Density.	Colour and Appearance.
Kasharasharin.	0·864	Greenish very fluid
Chouster	0·773	Bright yellow volatile
Haf-Cheid	0·927	Greenish
Daliki	1·016	Black, bituminous
Kism	0·837	Greenish

BRITISH INDIA.
BELUCHISTAN.

Beluchistan posseses an oil field which has recently commenced to be wrought. The principal district is KAHATAN, where the workings extend over a length of 40 miles, and the annual production is about 7,000 barrels. In the same region, the springs of MAGALKOT attract the attention of the prospector.

They yield in fairly large quantities, a deep yellow-coloured petroleum, charged with paraffin, and of density 0·819.

PUNJAUB.

In the Punjaub the existence of petroleum had been demonstrated many years ago.

The English Government put itself at the head of the search movement. The results were not so brilliant as had been expected, and the borings hitherto sunk have not yet led to a remunerative yield in a mountainous country where means of communication are far from being assured. The principal deposits are PUNOBA, ALLUGE and GUNDA. The latter place yields a deep green oil, with a density of 0·907 in quantity sufficient for the local illumination. The works belong to the Punjaub Oil Prospecting Syndicate. A boring 25 metres (82 feet) in depth produces a barrel a day.

ASSAM.

The Indian province of Assam, in the basin of the Brahmapootra, possesses an oil field with a fine future.

Early Attempts at Developing.—So far back as 1866, attempts were made to develop it by an English company, but without any result. At the present day the oil field is in the hands of two companies : the Assam Railways and Trading Company and the Assam Oil Syndicate.

The principal works of these companies are at DIGBOI and MAKUM.

Geological Structure.—At DIGBOI the ground, very favourable for boring, consists of :—

> Coarse-grained sandstone.
> Argillaceous sand.
> Brown shales.

The latter cover the petroliferous strata. A score of wells have been established at Digboi.

Properties of Oil.—They yield a deep brown-coloured petroleum of a density varying between 0·835 and 0·845. Traces of ozokerit have been found in this locality.

At MAKUM, the ground consists of hard sandy conglomerates and shales, equally mixed with sandy veins. Half a score of borings have been established here, the output of which varies from 2 to 8 barrels per day.

Properties of the Oil.—The oil consists of a reddish-brown petroleum of density 0·944, which, from its great viscosity, may be regarded as a real, natural lubricating oil. Its flash point is 82° C. (179·6° F.).

Number of Borings and Yield.—In 1897, 10 borings were in active operation in the Assam oil field. The production was 11,000 barrels of oil.

BRITISH BURMAH.

The Burmah oil field is the oldest known and the richest in the British possessions in Asia.

Divisions.—It may be divided into two sections which, according to their geographical position, may be distinguished as the *Coast* or *Lower Burmah* district including the Arakan Isles, and the *Interior* or *Upper Burmah* district.

Coast District—Arakan Isles.—The *coast* district includes a series of islands which have all, in a great or less degree, shown decided signs of petroleum. These are the Arakan Islands.

Geological Structure and Phenomena—Mud Volcanoes.—The geological structure of these islands resembles very much that of the Caspian strata ; like them, they are the scene of violent petroliferous phenomena. Numerous mud-volcanoes, salses, etc., are met with, which disengage, with a peculiar hissing noise, inflammable gases. Petroleum intermittently exudes from these small craters, and frequently more violent eruptions of gas upheave the ground in all directions.

Native Methods.—The natives did not observe these phenomena without attempting to utilise them. To a small extent, all throughout the Arakan Isles, and that from a very remote period, they dug shallow pits, or excavations, on the surface of the ground similar to the *Fettlöcher* of Northern Germany. They allow the petroleum to accumulate in these, during a greater or less length of time; then, when they think they have got a sufficient quantity, they collect, in buckets, the infiltrated oil, which they use for illuminating purposes, without any other previous preparation than allowing it to deposit in earthenware jars for a considerable time, so as to free it from any earthy or aqueous impurities which might be detrimental to it.

The Different Arakan Oil Fields.—The petroleum-bearing islands of Arakan are four in number. They consist of the islands of Ramri (the most southern), Cheduba, Flat and Barongah.

RAMRI.

The development and working of petroleum has been extended the most greatly in the island of Ramri. It includes the important deposit of MIMBAIN.[1] In this locality 14 wells have been bored, 8 are unproductive, but the 6 others yield a quantity of oil which may be estimated at 450 barrels per month.

At LETAUNG, a little to the south of Mimbain, a boring has been sunk which, at the depth of 124 metres, say 406 feet, has yielded 25 barrels per day of a bright brown oil of extreme fluidity, containing almost no paraffin. It may be classified with the oils rich in lamp oil. Its density is 0·826.

LIKMAU we believe has not yet been the object of systematic researches; the oil collected by the natives is similar to that of Letaung. Its density is 0·831.

KIAUK PHYN has the lightest oil in the Arakan district; its density is from 0·818.

CHEDUBA.

THE ISLAND OF CHEDUBA has also some wells excavated by hand by the natives, with quite a minimum yield, consisting of a petroleum similar to that of Ramri, of a density varying between 0·824 and 0·843.

FLAT.

THE ISLAND OF FLAT is but little wrought.

BARONGAH.

The Island of BARONGAH comprises two working centres, the West and the East. Here is the manner in which Fernand Hue (*Au Pays du Pétrole*, Paris, 1885) relates the story of the beginning of the Barongah petroleum industry :—

Fernand Hue's Description.—"In 1877, two English companies undertook the search for petroleum in Barongah. Using the gaseous eruptions as guides, they made some trial borings.

[1] Minbyin.

One of these, sunk to the depth of 260 feet, gave almost no result; the other was commenced at the bottom of a hole 30 feet broad and 30 feet deep. Oil was struck after boring 68 feet, when a violent expulsion of gas occurred, accompanied by a wave of oil. During the first week, the daily production was about 500 gallons, but it rapidly decreased, and a few days after the opening of the well the yield was almost *nil*. This was, we believe, the first attempt at boring on the American plan in this region. In 1880, other borings were made, which would appear to have afforded several flowing wells, rather an extraordinary fact, considering the nature and structure of the ground beneath.

" The wells, bored according to the American system, soon became exhausted, while the pits wrought by the natives during centuries have always yielded the same quantity of oil."

Present Production.—At the present day numerous borings have been made all over the island. Several wells have produced 8 to 10 barrels per day.

Properties of the Oil.—Barongah petroleum differs essentially from Ramri petroleum; it is of medium density, and, moreover, increases in density in crossing from east to west. Its colour is black, and it has very great viscosity. Its density varies between 0·835 in the east and 0·888 in the west.

COAST OIL FIELD (MAINLAND).

On the Continent of Asia, opposite the Arakan Isles, extends a tract or band of land, bounded on the east by the Yomadang Mountains. The indications of petroleum along this coast are altogether similar to those observed in the adjacent islands. The petroleum exudes from the soil and is hardly wrought, except in the most rudimentary manner, by the natives.

Production.—The coast production may be very approximately estimated at 6,000 barrels annually.

Properties of the Oil.—Samples of oil from this country have been examined by the authors; their density varies between 0·821 and 0·877; their colour is brown and rather inclining to bright. In contradistinction to Ramri petroleum, the viscosity

of Burmah coast petroleum is very great, and their solidifying
point is $+14°$ C.

THE IRRAWADDY OR INTERIOR OIL FIELDS ("RANGOON OIL").

It is in the interior of the country, on the banks of the
Irrawaddy, about 200 miles to the south of Mandalay, where
the *great oil field of Burmah* is found.   Methodically wrought
by the natives (by excavations) long before the English conquest, the Burmah oil field is one of the oldest in existence.
About 1850, Price & Co., at Belmont and Sherwood, in England,
used Burmah petroleum (the so-called "Rangoon oil") for
the manufacture of candles by Warren de la Rue's process.

History.—Many years ago, Dr. Robertson, the English traveller,
was surprised, as he ascended the Irrawaddy towards Mandalay, to see a number of junks which were engaged along
the river in carrying petroleum in earthen vessels.   This petroleum came from the royal mines of Tongune (Twingoung).

TWINGOUNG [1] OIL FIELD.

The property of the King of Burmah, the oil fields of
Twingoung, extended on the left of the Irrawaddy on a ridge of
denuded strata.   Hundreds of wells had been excavated by
hand along a triple range, and over a length of about ten
kilometres (6·2 miles).

Native Methods of Excavating Wells.—The following is Fernand
Hue's description of the working of these wells :—

"These wells, the depth of which vary from 50 to 400 feet,
are about 1 metre (3·28 feet) in diameter, that is to say the width
absolutely necessary to permit of a man working.   To prevent
them from collapsing, they are lined inside with boards.   'It
is really astonishing,' writes an American engineer, who visited
these districts, 'that, with all the modern processes at our
disposal, we have never been able to excavate a well to this
depth and of this width, on account of gaseous emanations.' [2]
And nevertheless the natives perform this task daily.   After
they have got to a certain depth they work in complete darkness ;
they cannot remain more than from *two to five* minutes at the

[1] Hill of Wells.—*Tr.*
[2] The assertion is controverted at least by the ancient wells of Galicia.

the bottom of the well, and moreover they are often brought to the surface insensible and half-asphyxiated.

"As soon as the oil-bearing stratum—the *kabaa*, as the natives term it—is broached, the oil appears accompanied by gas, but without any eruption ; it comes out of the sandstone in the same way as water issues from a squeezed sponge. None of the wells in this region yield a drop of water ; they are perfectly watertight, whilst those excavated outside the zone of production now being wrought, only yielded a water highly charged with sulphuretted hydrogen."

Thickness of Oil-bearing Stratum.—" The oil-bearing stratum would appear to be no more than from 4 to 6 feet thick, and to form a continuous bed fed from one source, because it has been remarked that, if oil was drawn simultaneously from the whole of the wells in the same group, the yield is much less than when only the half of the wells is wrought."

Yield.—" The yield varies from 1 to 1,000 gallons daily per well. Several yield, for a fair length of time, 500 to 700 gallons daily to the great detriment of those in the neighbourhood, the yield of which descends to a small number of gallons. One is quoted which, amongst others, had given a daily yield of 700 gallons for a long number of years ; its output then suddenly fell to 130 gallons, and for twenty or twenty-five years it has never diminished for one instant."

Present Yield of Twingoung Oil Field.—At the present moment, Twingoung comprises 441 wells, excavated by hand, the annual production of which may be estimated at 150,000 barrels.

Properties of the Oil.—This petroleum is very rich in paraffin ; it has a buttery appearance at the ordinary temperature ; it solidifies at + 22° C. (71·6° F.), is brown or almost black in colour, and has an average density of 0·814.

YENANGYAUNG OIL FIELD.

Situation.—To the north of Twingoung is a new oil field, YEN-ANGYAUNG. Yenangyaung, the characteristic name of which means in the Burmese language, *the river of stinking waters*, is situated on the right bank of the river of that name, on a barren height, on the left of, and only a few kilometres from the Irrawaddy.

GENERAL PROPERTIES OF YENANGYAUNG PETROLEUM.

(According to BOVERTON REDWOOD.)

Number of the Well.	Depth in metres.	Production in kilog.	Density of the Petroleum.	Remarks.
1	61	20	0·900	Very dark colour
2	51	15	0·869	
6	56	22	0·890	Rich in paraffin
8	77	45	0·882	,,
9	78	120	0·877	,,
10	74	22	0·860	
11	58	6	0·867	
12	70	8	0·862	
13	75	90	0·875	Poor in paraffin
14	75	105	0·890	Rich in paraffin
15	75	30	0·885	,,
17	72	75	0·872	Poor in paraffin
18	77	75	0·877	Very rich in paraffin
19	78	30	0·880	,,
20	77	67	0·880	,,
23	75	22	0·890	,,
25	25	75	0·890	,,
26	70	15	0·915	Poor in paraffin
27	60	12	0·915	,,
28	55	3	0·920	,,
31	50	12	0·925	,,
32	60	165	0·882	Rich in paraffin
33	65	3	0·925	Little or no paraffin
34	55	15	0·892	Rich in paraffin
35	50	37	0·902	Paraffin may be neglected
36	40	30	0·890	Very rich in paraffin
37	32	15	0·890	,,
38	15	240	0·883	Rich in paraffin
39	20	225	0·881	,,
40	5	150	0·900	,,
41	33	240	0·880	Very rich in paraffin
42	60	188	0·890	,,
43	7	135	0·882	,,
44	20	180	0·875	Paraffin may be neglected
45	25	210	0·880	Very rich in paraffin
46	22	225	0·890	,,
48	30	225	0·872	,,
49	2	165	0·905	,,
50	5	165	0·926	Little or no paraffin
51	10	192	0·868	Rich in paraffin
53	70	85	0·900	
54	12	232	0·875	Very rich in paraffin
55	12	210	0·872	,,
56	12	210	0·874	,,
57	12	180	0·871	,,
58	12	180	0·872	,,
59	45	210	0·874	,,
60	5	195	0·870	,,
61	22	165	0·873	,,
62	2	45	0·956	Little or no paraffin oil, very viscous
63	2	45	0·919	Little or no paraffin
64	2	45	0·950	,,
65	2	124	0·919	,,
66	105	180	0·886	Very rich in paraffin
72	40	210	0·881	,,
73	40	210	0·881	,,
74	20	210	0·890	
75	3	120	0·940	No paraffin oil, very viscous
76	2	75	0·956	,,
77	20	210	0·894	Very rich in paraffin
79	35	194	0·904	,,
80	30	220	0·890	,,
81	30	210	0·880	,,
84	15	170	0·884	Rich in paraffin
86	20	120	0·890	Very rich in paraffin
68	20	150	0·876	Rich in paraffin
89	3	180	0·880	,,

Number and Depth of Wells.—The wells wrought in this oil field are 89 in number. Their depth varies between 2 and 105 metres (6·56 and 344 feet).

Yield.—The average yield is 95 kilogrammes, say 1 cwt. 3 qrs. 14 lb. per day, per well. The annual yield may be estimated at 25,200 barrels.

Chaotic Structure of the Ground.—In no part of the globe is nature more capricious than at Yenangyaung, not only in the geological structure and conformation of the ground, but also in the composition of the oils extracted, the chemical and physical characteristics of which may be altogether different in two neighbouring wells.

Geological Horizon.—The ground in this locality cannot be assigned exclusively to any precise geological formation. It is in fact a chaos, where all kinds of rocks, all sorts of sediments, jostle and overlap each other, the glory of the geologist, but the terror of the miner. Throughout the greater part of the oil field, coarse-grained friable sandstones would appear to predominate ; they are intermingled with rather thin beds of clay mixed with sands.

Great Variation in Properties of Oil.—For the reasons given above, it is not possible to give general data which would designate the properties of the different oils of Yenangyuang ; there are as many different qualities of oils as there are wells in this locality. The authors are, therefore, pleased to be able to submit to their readers this table (per well) of the general characters of the Yenangyuang oils.

KODAUNG is one of the successful places of the Irrawaddy oil field. Recently developed and wrought, it numbers some 70 derricks, with an annual production of 17,000 barrels.

BÊMA only comprises shallow wells excavated by hand.

YENANGYAT includes an establishment of the Burmah Oil Company. The results have not in any notable way as yet fulfilled expectations. Yenangyat oil has a density of 0·823.

MIMBU, on the right bank of the Irrawaddy, shows signs of petroleum, which lead to the belief that this place will be developed before long.

Annual Production.—The Irrawaddy oil field would not appear,

in the author's opinion, to produce more than 650,000 barrels
annually.

GENERAL PROPERTIES OF THE DIFFERENT PETROLEUMS OF THE BRITISH POSSESSIONS IN ASIA.

Place of Extraction.	Density.	Colour and Appearance.
BELOUCHISTAN		
Mogalkot	0·819	Deep yellow, rich in paraffin
PUNJAUB		
Gunda	0·907	Deep green
ASSAM		
Digboi	0·835 to 0·845	Deep brown
Makum	0·944	Reddish brown, very viscous
BURMAH		
Létaung	0·826	Bright brown, very fluid
Likmau	0·831	„ „
Kiauk Phyu	0·818	„ „
Cheduba	0·824 to 0·843	„ „
Barongah (east) . . .	0·835	Black, viscous
Barongah (west) . . .	0·888	„
Arakan (coast)	0·821 to 0·877	Brown, viscous; solidifies at + 14° C.
Twingoung	0·814	„ „ „ at + 22° C.
Yenangyaung	0·860 to 0·956	„ „
Yenangyat	0·823	„ „

CHINA.

Chinese Petroleum an Unknown Factor. — China is a mining
country, which, in addition to rich deposits of coal, possesses
important iron, copper and tin mines. In Kouang-Si, quite close
to the frontier of the French-Indo-Chinese colonies, petroleum
deposits are also known to exist. The exact locality of these
deposits, their importance and value are so many unknown
factors, as far as we are concerned. What we do know is the
fact that the crude petroleum, brought to the light of day by rope-
boring—the system used in China for thousands of years—is
utilised by the Chinese, not only to caulk their junks, but also to
coat the roofs of their houses and for illumination, being burnt,
for the latter purpose, in pottery lamps.

Huart's Memoir.—M. C. Imbault Huart, French consul at
Canton, has recently sent to the Paris Society of Commercial
Geography a very interesting memoir on the mining industry of
China. It is from this privileged gentleman, who has had the
opportunity of visiting and observing, that a good part of the
facts which follow are borrowed.

Native Hostility and Superstition.—" The natives have a horror of divulging their underground secrets. The inhabitants of Canton are, moreover, peculiarly excitable and very hostile towards strangers. The system of transport between the mining centres and the waterways, which form the only means of communication in the province, is vile. Finally, the local superstitions have to be fought against, especially the wind of happiness (*Fong-Chouei*), wind and water, a speculation which presides over the fate of towns, houses, localities, families, individuals, etc. The mining industry is not in favour with the country people, who fear that the earth, robbed of its treasures, would have its revenge by producing no more harvests."

As already stated, in spite of this apathy, petroleum has been wrought, very imperfectly it is true, in Kouang-Si.[1]

Taï-li-Chen Oil Field.—Near to the southern frontier of the province of Kouang-Si, at TAÏ-LI-CHEN, some wells were bored by native hand labour for the purpose of obtaining brine. These wells yielded petroleum; they were visited by M. T. A. Mathon-Jozet, who was good enough to send to the authors a sample of the petroleum obtained, which consisted of a viscous, greenish oil, very rich in paraffin; of density, 0·881.

Yield.—At Taï-li-Chen, Mathon-Jozet tells us, the oil produced by 8 wells collectively does not exceed 6 barrels per day. The depth of the deepest well is about 150 metres (492 feet). According to the inhabitants, other borings situated in the east, and especially in the neighbourhood of Haï-tha, produce a greater yield.

Boring Tools.—The Chinese boring implements, which have not varied for centuries, consist of a boring system of bamboos terminated by a conical bit. The petroleum or gas pipes are also made of hollow bamboos.

Production does not Meet Home Consumption.—As will be seen, on account of the absence of data, the petroleum production of China cannot be estimated. This production, in spite of the richness of the deposits, which would appear to correspond with those of Burmah and India, does not seem to exceed the exigencies

---

[1] Sainte Clare Deville has given us the principal properties of a sample of petroleum from this district. The oil in question came from Foo-Choo-Koo. It was fluorescent and but slightly coloured; its density at 0°C., the point at which it solidified, was 0·860.

of local needs.   In no locality do we believe petroleum to be transported far from its point of origin, and China imports notable quantities from the United States.

The English scent Rich Oil Fields.—We have already said that "wherever petroleum is met with Englishmen are to be found". In virtue of this axiom, for a long time past, the British subjects of Hong-Kong, scenting rich oil fields, have asked the authority of the Chinese Government to make trial borings.   This authority has been invariably refused them.

French Concessions and Treaties.—At the present day, China, however, would appear to be disposed to make some concessions in favour of France.   Very soon a French railway will penetrate into the Celestial Empire, in fact into this province of Kouang-Si, the petroleum riches of which we have described.   Is there no bold pioneer to be found, who will take to heart that (in virtue of recent conventions, a clause of one of which runs as follows) "China, for the development of its mines in the provinces of Yun-nan, Kouang-Si and Kouang-tonng, may apply in the first instance to French industrials and engineers?" (Act of 20th June, 1895.)

GENERAL PROPERTIES OF DIFFERENT CHINESE PETROLEUMS.

Place of Extraction.	Density.	Colour and Appearance.
Foo-Choo-Koo	0·860	Slightly coloured fluorescent solid at 0° C.
Taï-li-Chen	0·881	Greenish, viscous, rich in paraffin

CHINESE THIBET.

In the western parts of China, in the but little explored provinces of Chinese Thibet, there would appear to be no reason opposed to the supposition that the Turkestan oil field (the Fergane deposits) extends towards the east.   No geological investigation has been made into this subject.

Natural Gas.—The fact remains that these provinces, as well as those of Se-Tchouen, possess numerous *tse-liou-tsin* (fire wells). These wells, which furnish in notable quantity the inflammable gas required to crystallise the salt in the brine of the salt springs of the district, can only be classified with the natural gas wells of Pennsylvania and Galicia.

SE-TCHOUEN OIL FIELD.

Geological Formation.—In Se-Tchouen the ground is composed of beds of sandstones of a different enough nature, but belonging almost entirely to the tertiary formation.

In the northern part of Se-Tchouen, Tchouen-pé, are the gas wells of CHEE-KONG, PONK-KI, SAN-TAY-LE-TCHE, YEN-TIN, TCHONG-KIANK, SU-LIN and GAN-YO.

In Southern Se-Tchouen and Tchouen-lan the principal natural gas workings are at TSE-LIOU-TSIN, KONG-TSIN and HO-TSIN.

R. P. Coldre, an Apostolic missionary, who visited this region, has described (*Ann. des Mines*, 1891) all the melancholy grandeur of this country, in which the stems without leaves (the derricks) project their black silhouettes against the horizon. He retains an indelible impression of the song of the borers as they leap on the bascules, to work the Chinese cross beam ; and of that song, mingled with a dull, solemn, continual murmur—the voice of the earth, the song of the gas of a well of fire.

JAPAN, FORMOSA AND SAGHALIEN.

JAPAN.

Ancient Knowledge of Petroleum.—Petroleum has been known from antiquity in Japan. It was, say the chroniclers, discovered under the reign of Teniyteno, in the seventh century A.D.

Producing Centres.—The producing centres are the districts of Echigo, Ugo, Schinano, Toutomi, Akita-ken Tosan, Nügata, Némuro, Ishikari, Iburi, Hidaka, Oshima, Aomori, Yamataga and Yokoyama.

Production most Minimum.—Up to the present, the production has not gone beyond a most minimum figure. Notwithstanding a great number of wells, a bounteous producing vein has not as yet been found.

Primitive Methods of Excavating Wells.—It is true, however, that the Japanese methods of working are still in their infancy, and that the boring of the wells has only quite recently made its appearance in the Japanese oil fields.

Description.—The Japanese method is at the present day still the most universal, and, as has been said, it is one of the most primitive. The reader can form his own opinion of it from the following description by Fernand Hue :—

" The wells excavated by the natives do not, in any way, resemble those which we have seen in America. These wells, reaching a depth of 160 to 200 feet, are square-shaped, the sides of the square being one metre. They are dug with the pick and shovel. Their width does not admit of more than one man working in the hole at a time. The well is lined inside with boards, four large beams fill the corners, and are joined by cross pieces, which serve as a ladder for the workmen. At the bottom of each of those wells, four horizontal tunnels are con- structed, one leading from each face, by means of which, at a certain distance from the main hole or shaft, borings are made to ascertain the extent of the deposit. These tunnels are short, 12 metres (39·36 feet) at the most, on account of the darkness reigning at this depth, and the difficulty of ventilating, so as to fight against the emanations of gas.

" This, we may add, is the way the excavation of a well is carried out :—

" Above the place where the well is to be made, a hut of earth and turf, 10 feet in height, is constructed, the top of the roof of which is exposed to the sky. Above this opening the reflector to enable the workmen to see is fixed. This reflector consists simply of a bamboo window sash frame, 5 feet long and 3½ feet wide, on which is fixed a piece of yellow paper soaked in oil. The window frame forms an angle of 45° with the horizon. In spite of its simplicity, this artifice enables the excavation to be carried on for six hours daily. The work is done by two men, the one works from 9 A.M. to noon ; the other replaces him from noon till 3 P.M. The earth and stones are drawn up in a net of twine of very narrow meshes, attached to a rope passing over a pulley fixed to the top of the hut. When the net is full, it is raised up by three men hauling the rope ; two of them work on opposite sides of the well, and the third in a hole dug near to the hut, 3 feet deep by 2 feet wide."

Air Machine.—" Whilst one of the men is at work in the bore hole the other works the air machine, a kind of bellows, which continually blows fresh air into the bottom of the well to counter-act the gaseous emanations.

" The bellows is of the most primitive description ; it consists of a large box, 6 feet long and 3 feet wide, laid horizontally. It is divided longitudinally by a plank placed vertically. Above the box, a plank of the same length is fixed see-saw ways (*en bascule*) on the vertical division ; the plank is dressed in such a manner as to fit exactly into the box. The man who works the pump walks alternately from one end of the plank to the other, so as to make it see-saw under his weight. When he reaches each end he strikes a hard blow on the end of the lid so as to close by the concussion a small valve, which opens when the plank is in the air. At each see-saw movement of the bascule the air is driven out first from the one end of the box, then from the other, into two bamboo tubes, fixed to the ends of the box, and which, uniting in a third, descend to the bottom of the shaft. The blast is continuous, and sustains the worker unin-terruptedly.

Oil raised in Buckets.—" When the well is sunk to a sufficient depth, and oil having been struck, the oil is extracted by means of buckets fixed to the end of a rope, which are lowered to the bottom of the well and raised when they are full. In certain districts this operation is only performed as required and to such an extent as only to meet the wants of the inhabitants. But in districts where the works have been developed, the extraction of oil goes on continuously, and in many of the wells they have commenced to adopt the use of the pump, the pipes of which are made of bamboo."

Progress impeded by Want of Capital and cost of Drilling Machinery. —There are two causes which would appear to militate against the progress of the Japanese petroleum industry. The first is that, on the one hand, the wells are wrought by numerous com-paratively insignificant companies, whose capital, in many in-stances, is less than 10,000 yens (the yen = 5·15 francs). On the other hand, American boring plant runs very dear, from the fact that its price is doubled by freight and custom house dues. However, according to recent reports, a syndicate for the working and development of Japanese petroleum by modern methods is on the point of being formed at Tokio. This first initiative will no doubt be the means of the Japanese business people—so adroit and enterprising—adopting a rational plant and system of boring.

Oldest District.—The oldest wrought district is KOZODZU (Province of Echigo). Kozodzu signifies, in Japanese, *the water that burns*. Here, and at MIYÔHÔJI, according to Lyman, in addition to abandoned, there are 178 productive wells. The average production of the Kozodzu and Miyôhôji wells collectively is 4⅔ barrels daily. By dividing this figure by the number of wells, an average production of 4 litres per day, per well, is obtained.

Best Well.—The best well of the region is at MACHIKATA. Its daily yield is 1½ barrels. In this locality the wells are generally productive; in fact it is related that, in 1871, in this same district, another well yielded for several years 19 barrels daily; then the yield fell to 8 barrels, and this yield was maintained for a comparatively long time.

Summary.—Summing up, we are far from the fountains of Baku, and the production of Echigo does not tend as yet to compete to any great extent, even in Japan, with Russian or American oil. The province includes 522 productive wells, the deepest of which is 732 feet, and the best of which yields 1½ barrels per day.

Yield.—The daily production is about 22 barrels, which gives an annual production of 8,000 barrels, the selling price of which forms a total of 31,500 yens.

Property of Oil.—The oil extracted has a density of 0·831.

In the district of SCHINANO are 32 productive wells, the deepest of which is 342 English feet; the best has a yield of 2½ barrels per day. The average production per day, per well, is, in this district, 40 litres.

Schinano puts on the market 1,900 barrels annually, the selling price of which gives a total of 6,250 yens.

Properties of Oil.—The oil produced is greenish black. Its density is 0·839.

At MABANA, a small quantity of a translucid white petroleum is extracted.

In the district of Ugo, the yield is almost insignificant.

At ORGUNI, very deep dark greenish brown petroleum is extracted with a density of 0·840.

Present Progress.—At the present time, however, the petroleum industry, owing to the exertions of foreign companies, is progressing in Japan. Improved plant has been introduced.

All efforts are concentrated towards the district of Tosan. There are 56 boring machines working at the present time. At the end of the current year, 1899, there will be 84. The first trials were made at Tosan, in 1894 ; in 1895, the petroleum production of Japan leaped from 1,561,760 gallons to 5,481,656 gallons.

Production.—The following table of Japanese production will show the progress effected in this country up to now, better than any discourse :—

Year.		Gallons.
1882	. . . .	708,000
1883	. . . .	814,000
1884	. . . .	859,000
1885	. . . .	246,000
1886	. . . .	290,000
1887	. . . .	535,000
1888	. . . .	350,000
1889	(Boring adopted for the first time). .	1,429,000
1890	. . . .	1,960,000
1891	. . . .	2,017,000
1892	. . . .	2,241,320
1893	. . . .	2,915,720
1894	. . . .	1,567,160
1895	(District of Tosan developed). .	5,481,656
1896	. . . .	4,874,920
1897	. . . .	6,131,400

The production of Japan in 1897 was therefore 145,985 barrels.

FORMOSA.

Traces of petroleum are reported to have been met with in Formosa, in the island of HAÏ-NAN. Some years ago flourishing works even were spoken of as existing in Formosa. These reports, spread by Americans, have not been confirmed. A few borings only, without giving remunerative results, have confirmed the presence of oil.

SAGHALIEN.

In the island of SAGHALIEN a few surface borings, undertaken in rather a tentative way, have given at depths not exceeding

10 metres, 80 to 120 litres of petroleum per day. These small results at the surface of the ground, may be the sign of a fruitful oil field.

PROPERTIES OF JAPANESE PETROLEUM.

Place of Extraction.	Density.	Colour and Appearance.
District of Echigo	0·831	Brown
„ Schinano	0·839	Greenish black
Orguni	0·840	Very deep greenish black
District of Tosan	0·882	Deep brown, viscous

CHAPTER XIII.

PETROLEUM IN OCEANIA.

SUMATRA, JAVA AND BORNEO.

PETROLEUM is found in notable quantity in the Dutch East Indies.

The deposits of Sumatra, Java and Borneo, although but recently developed, are at the present time amongst the most productive. The systematic realisation of this rich oil field goes no further back than half a score of years. Nevertheless, the natives knew of the existence of petroleum, and used it in the preparation of the leather which they sold to Europeans.

Geological Structure.—The group of islands, which form the Asiatic archipelago, consists of a volcanic mass, the recent elevation of which has displaced and upheaved, almost to the surface, the rich oil-bearing strata of the tertiary formation. In spite of the presence of active volcanoes, and of phenomena of volcanic nature everywhere throughout the Asiatic archipelago, petroleum has been found in the different strata of the tertiary formations, in which it would appear to have originated. In order to confirm, moreover, the theory which we have propounded, of a petroliferous geological line of strike or trend without other breaks of continuity than those which owe their existence to plunges, faults or immersion, the tertiary strata of Sumatra can only be regarded as those which we have already seen in the isles of Arakan on the shore of Burmah. " *The Sumatra chain certainly continues, but with more regularity, the Indo-Chinese range of mountains of Arakan, which forms Cape Negrais, to the east of the Irrawaddy, and which describes afterwards the prolonged curve of the Andaman and Nicobar Islands.*" (Elysée Reclus.)

SUMATRA.

Towards the north-west point of the island of Sumatra, in the point designated by Elysée Reclus as being opposite to Nicobar, the tertiary strata crop out at the surface and bring with them

FIG. 25.—The Oil Fields of the Asiatic Archipelago.

Bassins exploités = Oil fields being wrought.   *Indications pétrolifères* = Signs of petroleum.   *Mer de Chine* = China Sea.
Mer de Java = Java Sea.

traces of petroleum similar to those of the Arakan region, and which have given rise to fruitful researches.

Langkat Oil Field—Geological Structure and Horizon.—In this region is the rich oil field of Langkat. The geological structure of this island we just mentioned as being decidedly tertiary. A zone of this formation—longer than it is wide—ends the modern madreporic strata of the coast, and extends, towards the interior of the country, the important volcanic upheaval of Sumatra. The petroliferous zone identifies itself in a very sharp, decided manner with the tertiary formation, without deviating in the slightest extent to the one side or to the other. It occupies a north-west to south-east direction, making an angle of 22° with the meridian (north to south). This inclination corresponds exactly with the general direction of the great petroliferous line of strike of Europe and Asia. The oil field of Langkat is situated between the rivers Sekoendar and Lépan at a minimum distance from the coast. The level, at which the oil is found, crops out in several points, and consists of bluish, coarse-grained sandstones. In several localities these sandstones contain bands of limestone or marl.

In those localities where the petroliferous formation does not crop out at the surface, it is generally covered by a rather important bed of clay shale, comparable to the beds of the same nature met with in Galicia.

Development of Oil Field.—It was in 1884 that the district of Langkat commenced to be wrought and developed in a rational manner by Warren, who had obtained a concession for the purpose. At the present time several companies, as many English as Dutch or German—no French establishment is in existence—are at work in the chief localities of this oil field. From among these places the following may be mentioned :—

TELEGA TOENGAL.

TELEGA TIGA, whose wells produce over 500 barrels daily of an oil of density 0·765. Telega Tiga is connected with the refinery of Balbalan by a pipe 7 kilometres long, say 4·34 miles.

More to the south numerous borings with a remunerative yield have had a successful termination.

Properties of Langkat Petroleum.—The oil collected is of excellent quality ; its density decreases, according to what would appear to be a general rule, in accordance with the increase in depth of the spot from which it was collected.

The following are, according to Boverton Redwood, the general characteristics of the oils of the southern portion of the Langkat oil field :—

At SUNGIE REBAH, in the argillaceous schists, according to the depth of the wells, the density of the petroleum is 0·945, 0·943, 0·940.

At SUNGIE SICHINO, in the coarse-grained sandstone, oils are met with of a density of 0·897 to 0·843. This latter oil comes from a well, situated towards the south of the concession and in beds of clay shale.

At SUNGIE PENANTI, in an important bed of clay shale, the oil has a density of 0·800 and 0·798. At the extreme south of this concession, the oils found have proved to be of great fluidity and remarkably light—0·777 and 0·769.

To the south-east of those places which have just been mentioned, towards BELOE TELANG, extend some oil fields which, as yet, have been but little wrought or developed. This locality, however, possesses a well 48 metres (157·44 feet) deep; the petroleum of which spouted to a height of over 3 metres.

Properties of Oil.—The oil yielded in large quantity by this well had a density of 0·771 to 0·789. Its colour was bright brown. The surface petroleums in this district are reddish brown, and of density varying between 0·855 and 0·868.

A deposit of bituminous coal prolongs towards the south and in the same direction the Langkat oil fields.

At the southern extremity of the island of Sumatra, near to PALEMBANG, are beds similar to those of the Langkat district. Of later years numerous works have been grouped together on this oil field. Several English and Dutch companies, and amongst them the important *Dort'sche Petroleum Industrie Maatschappij* have realised magnificent profits.

JAVA.

The Island of Java would appear to be as rich in petroleum as the neighbouring island of Sumatra.

Geological Structure.—As is the case in the latter island, the oil is met with in the tertiary strata. These beds are everywhere more upheaved by volcanic action than those of Sumatra.

Division into Oil Fields.—They are divided into two portions, those on the *south coast* of the island and those on the *north*.

South Coast Oil Fields. — The oil fields of the southern coast extend the whole length of that coast parallel to the volcanic upheavals, which form the backbone of the island. They occupy the western part of the country, and include the well-developed oil fields of TJELATJAP, NGAWI and POERWADI. This latter place has been conceded recently to the Dutch East Indian Petroleum Company, who have just started there, several borings which are so far advanced at the present time as to warrant the best hopes being entertained.

Northern Coast Oil Field. — The courses of the River Semarang and the Rio Proge, which seem to separate the volcanic crater of Java form the boundary of the south coast petroleum oil field. The petroliferous line of strike would appear, starting from this section, to have been thrown back into the northern coast, which it occupies in the whole of the eastern part of the island.

The first works in this oil field are met with in the residency of REMBANG. In the environs of this town rise the derricks of the *Dort'sche Petroleum Industrie Maatschappij* at ·PANOLAN, where there are 15 flowing wells. The last well opened in this district in January, 1898, has a daily production of 2,400 barrels.

Mud Volcanoes. — To the west of this deposit, in the direction of Japara, is the mud volcano of Grobagan, of which Dr. Horsfield said: "When you have approached near enough for the view not to be obscured by steam, you see a spherical mass, of about 16 feet wide, consisting of black earth, mixed with water and viscous substances; it rises to about 30 feet in the surrounding air, in a uniform manner, as if it were propelled by a subterranean force. At this height it bursts, projecting in all directions its debris of stinking mud. After a few seconds of repose the eruption recommences, and a fresh ball ascends to burst once more. The air is impregnated with a strong odour of petroleum."

The development and working of the TINAWEN oil field has just commenced. Several borings are in progress.

In the residency of Rembang, at Plœtœran, and also at

Semarang, the *Dort'sche Petroleum Industrie Maatschappij* has built oil refineries and paraffin manufactories. Pipe line conduits connect these with the oil wells.

The oil wells of the TOEBAN district are likewise very prosperous and have a considerable output.

The deposits of the residency of SOERABAJA are the richest on the island. It is there that the *Dort'sche Maatschappij* commenced its operations in 1888. The first undertakings of this company, which were destined to assume such extensive proportions, were located in the village of DJABA KOTA. By 1893, the *Dort'sche Maatschappij* had as many as 30 highly productive borings in this locality at a depth of 75 to 225 metres. An oil refinery was built at Wronoko, and connected with the wells by 6 kilometres of pipe line. At that time the production of the company might be valued at 100,000 barrels annually.

The company now possesses in the same district numerous petroleum locations, amongst which may be mentioned :—

LIDAH, with 27 flowing wells, amongst which a boring finished in the month of May, 1898, yields 1,400 barrels daily.

KOTÉÏ, which likewise possesses flowing wells, has over 15 productive wells.

MATATŒ, discovered only a few months ago; which nevertheless already includes two flowing wells, one of which produces 400 and the other 1,000 barrels per day.

GOGOR, with a well of natural gas.

All these works (with the exception of Matatœ, still very recent) are connected with the Wronoko refinery.

In the residency of Soerabaja, once more extend the deposits of GOENOENG SARIE, belonging to a Chinese proprietor, who had the good luck, some few years ago (in 1894), of finding a flowing well at the depth of less than 100 metres (328 feet), the grandeur of the output of which was more than 2,500 barrels per day.

To the north of Soerabaja, the island of MADOERA likewise yields petroleum. This district is nevertheless, as yet, but little wrought or developed.

PROPERTIES OF JAVA PETROLEUM.

The density of Java petroleum varies between 0·876 and 0·898. That of Djaba Kota is brown red fluorescent, with a not unpleasant smell, and density of 0·878. It contains a rather large proportion of paraffin.

BORNEO.

Until lately, the working of petroleum in the Island of Borneo would appear to have been less prosperous than in Sumatra, and more especially Java. Recently, a lucky hit in boring has just demonstrated that the large island has no reason to be envious of its neighbours.

Geological Structure and Horizon.—It is also in tertiary strata, and relatively near to the coast, that petroleum is most generally met with in Borneo.

On the northern coast, in the English territory of Sabah, the petroleum-bearing ground of SEKNATI extends. In that locality, a well, excavated by hand, after having traversed a bed of clay, encountered ferruginous sands, intersected by petroliferous shales. The almost bituminous oil extracted, in a minimum enough quantity, is of great viscosity.

The independent territory of BRUNEÏ possesses petroliferous deposits, conceded to the English. The working of these . is rather modest, and hindered by an inclement climate. In the island of LABUAN, matters are in the same condition. MALOODA likewise shows traces of oil.

On the south coast, in the territory belonging to the Dutch, some works are found disseminated at KOTEÏ, TANDJONY and MARTAPOERA. In the north of this latter place, back in the interior, is Amoentaï, which is the centre of a rather extended petroliferous district.

Balangan Spouting Well.—Until lately, working was confined to wells of a minimum yield. These springs it is true were very numerous ; the country seemed impregnated with petroleum but nowhere had it been discovered in large quantity. In the month of May, 1898, news was received from BALANGAN, that a well, bored in this locality by a Dutch engineer, Van der Made, had suddenly spouted. The spouting was as violent as it was unforeseen.

During three days, the rivers Negara and Martapoera were entirely covered with oil, and the water rendered undrinkable over a course of several hundreds of kilometres. The oil fountain of Balangan, surpassing that of Java, had, we are told, a yield of 6,400 barrels per day.

Other borings are being made in the neighbourhood.

ISLE OF TIMOR.

The southern portion of the island of Timor, situated in the prolongation of Java, possesses petroliferous deposits. These deposits are partly situated in limestone beds of the tertiary period, partly in carboniferous schists. At Timor, a well dug to a depth of 10 metres struck a pocket of oil, which became exhausted after having yielded 120 barrels of petroleum of a deep brown fluorescent colour, of density (according to Boverton Redwood) of 0·825.

SUMMARY.

Rapid Growth and Development of the Dutch-Australasian Oil Fields.
—Few petroliferous countries have had such a rapid growth and development as the Dutch-Australasian possessions. This is owing to several causes : in the first place to the excellence of the boring plant imported from America ; and secondly, to the rational manner in which the enterprise was undertaken by big companies with ample capitals, who were able to tide over, without getting discouraged, the numerous trial borings and the prolonged disillusions of the periods of researches.

In order to judge as to the correctness of the facts which we have stated, the reader is referred to the following tables.

Production.—The Dutch-Australasian possessions produced :—

Year.	Barrels of Petroleum.
1890	18,800
1891	21,000
1892	40,500
1893	68,300
1894	88,400
1895	198,000
1896	320,500
1897	468,000

The production is thus twenty-six times greater in 1897 than it was in 1890.

EXPORTS.

The exportations have followed this progressive movement.

Year.	Barrels of Petroleum.
1890	281
1891	232
1892	18,500
1893	55,500
1894	81,000
1895	181,700

Anticipated Future Progress.—The authors firmly believe that, thanks to the fortunate results of the borings already undertaken, and also to the equipment of the recently discovered deposits, these figures will be notably increased during the course of 1898.

Moreover, in the preceding tables of production, that of Borneo has not been taken into account, nor of Timor, whose production up to the end of 1897 might be regarded as insignificant. Commencing with 1898, as will have been seen, Borneo will also furnish its contingent to the production to be tabulated.

GENERAL CHARACTER OF THE PETROLEUM OF THE DUTCH-AUSTRALASIAN POSSESSIONS.

Place of Production.	Density.	Colour and Appearance.		
Sumatra	0·771 to 0·785	Bright brown (according to Boverton Redwood)		
Langkat	0·765	,,	,,	,,
Sungie Rebah . .	0·945	On the surface	,,	,,
,, ,, . .	0·943	,,	,,	,,
,, ,, . .	0·940	,,	,,	,,
Sungie Sichino . .	0·897	,,	,,	,,
,, ,, . .	0·843	,,	,,	,,
,, Penanti . .	0·800	,,	,,	,,
,, ,, . .	0·798	,,	,,	,,
,, ,, (south)	0·777	Very fluid		
,, ,, . .	0·769	,, ,,		
Beloe Telang . . .	0·771 to 0·789	Bright brown		
,, ,, . .	0·885 to 0·868	Surface petroleum, red brown		
Java	0·876 to 0·898			
Djaba Kota . . .	0·878	Brown red, fluorescent, rich in paraffin		
Timor	0·825	Deep brown (according to Boverton Redwood)		

PHILIPPINE ISLES.

The Philippine Islands exhibit numerous signs of petroleum. Unfortunately, this country has for some years been in the throes of political convulsions, and has therefore been unable to participate in the industrial progress of the neighbouring

12

countries, consequently the petroleum industry there is still in its embryonic stage.

1. ISLE OF CÉBU.

Surface Wells.—The more decidedly petroliferous strata extend over the island of CÉBU, situated to the south of the large island of Luzon.   Some years ago, two shallow surface wells were excavated there, and a few barrels of oil have been extracted from them.   This oil, of a brown colour and strongly fluorescent, had a density of 0·809 ; it was rather rich in paraffin.

2. PANAY ARCHIPELAGO.

Traces of petroleum lead to the belief that rather rich oil fields exist in the islands of LEITE DE BOHOL, and throughout the whole of the Archipelago of small islands which surround the Isle of Panay.   There is no doubt whatever but that these tertiary beds are a continuation of the Borneo deposits.   At the present day, American prospectors are inspecting and prospecting it.   On the other hand, an English company of Manilla is in course of formation for their development and working.

Inertia of French Industry.—As is everywhere the case, the inertia of French industry is remarkable, and, in this connection, let us recall the remarks, so just and so to the point, of our colleague, M. René Menant.

European Population.—" The number of Europeans now in the Philippine Isles may be estimated at 25,000 without including, let it be well understood, the troops in occupation.   Of this number, about 12,000 reside in Manilla, the capital of the Archipelago, the centre of the colonial Government ; of these, thirty-four Frenchmen alone represent French commerce in this rich country : moreover they are not engaged on the large scale, either in the export or in the import trade."

Monopolies.—" All the products of these islands are monopolised by powerful British, German or Spanish firms, who advance to the producers considerable sums on their crops so as to secure them to themselves.   This way of doing business evidently requires immense capital, and is liable to some risks, covered in every case, however, by enormous profits.   Why then, we French, do we not adopt it ?

"Is the want of money the stumbling block, or is it the want of business knowledge ? No, it is the want of boldness or nonchalance, perhaps it may even be indifference which keeps us at home, and causes a great number of merchants or small holders in the funds to take their money or their savings to the first bank, or to sink them in some gold mine, rather than invest them in some great and profitable foreign enterprise " (*Bulletin de la Société de Géographie commerciale*, tom. xix., No. 4, p. 242).

New Zealand.

M. Fernand Hue (*Au Pays du Pétrole*, p. 146) expresses himself thus in regard to New Zealand :—

"In the northern isle of New Zealand, between the volcano of Tongarino and the sea, extends the lake district, the geysers, the solfataras, the fumeroles and the mud volcanoes. The latter, numbering more than a thousand, are, according to Dr. Hochstetter, the most active in the whole world."

Mud Volcanoes.—" The mud volcanoes which we have seen in the island of Trinidad, which we meet with throughout the whole of the Asiatic Archipelago, in Burmah, in the Caucasus and in Italy, are continuous or intermittent eruptions of asphaltum and petroleum mixed with fused substances, boiling water and mud. This phenomenon is always a sure sign of petroleum in the subjacent strata of the neighbourhood.

" In fact, not far from these volcanoes, petroleum has made its presence manifest, especially in the district of Taranaki. There have been, in certain points, attempts made to start operations. A squad of workmen and a full equipment of plant were brought from America, but the contractors were apparently deceived in their attempt, and so matters remain. Everything, however, points to the supposition that a well-directed undertaking would yield good results."

At the present date, a big company, the New Zealand Petroleum and Iron Syndicate, has resumed the old works. They have dug a well, 915 feet deep, the yield of which, though not enormous, is already encouraging, having risen to 4 barrels per day.

Geological Structure.—Contrary to what the reflections of our esteemed *confrère*, M. Fernand Hue, would lead us to suppose, the petroliferous strata of the Taranaki district are not in any way of a volcanic nature; they rise up directly from the lower tertiaries, partly miocene, partly upper eocene. The beds traversed, rich in marine fossils and in traces of fucoids, are grey clays and sandy shales. The oil-bearing bed is generally a coarse-grained, friable, brownish sandstone.

The principal deposits wrought in the district at the present time are SUGAR LOAVES (site of the New Zealand Petroleum and Iron Syndicate), which produces 1,000 barrels annually of a bright brown (nut brown) petroleum of density 0·840 very rich in paraffin. The surface petroleum, produced by exudation, is very different; of a deep brown, its density 0·971 is much higher than that of the deeper beds.

Boring at Sugar Loaves is rendered much more difficult by the presence, in contact with the petroliferous strata, of several water-bearing beds which, not being without absorbent properties, carry away a portion of the petroleum. There, more than elsewhere, hermetically sealed casing is to be recommended.

To the south of this place are numerous tracts where surface indications are numerous. The petroleum collected on the surface of the ground is greenish and has a density of 0·966.

At POVERTY BAY, numerous companies have followed each other without finding petroleum in sufficient quantity for profitable working. The oil produced by exudation is very dark, and has a density of 0·864 to 0·873.

At MANUTAHI, the oil produced in minimum quantity by the natives is brown, and of density 0·829.

In the opinion of the authors, the presence of petroleum in New Zealand is the result of a geological accident. In fact, this island is situated well to the south of the belt which runs from the Atlantic to the Pacific, across Europe, from Scotland through Asia and Oceania to New Caledonia, its extreme point, and which includes almost the whole of the important deposits of these countries.

At this point we terminate the study of petroliferous Oceania, leaving purposely on one side the deposits of New Caledonia, which will be examined in the chapter dealing specially with the French colonies.

GENERAL PROPERTIES OF NEW ZEALAND PETROLEUM.

Place of Extraction.	Density.	Colour and Appearance.
Sugar Loaves (surface) . .	0·971	Dark brown
,, ,, ...	0·840	Bright brown, rich in paraffin
,, ,, South . . .	0·966	Greenish
Poverty Bay	0·864 to 0·873	Dark green
Manutahi	0·829	Brown

CHAPTER XIV.

HISTORY.

The Mound Builders and Petroleum.

THE United States of America and Canada possess an aggregate of petroliferous deposits only exceeded in importance by those of the Caucasus and Baku.

If American petroleum cannot lay claim (and for a good reason) to the descriptions which the authors of antiquity have given of the petroleum of our continent, nevertheless the knowledge of this substance over there is very ancient, and the latter centuries have left undeniable testimony in regard to this point. Moreover —and this is altogether to the credit of American genius—the petroleum industry, as it exists at the present time, first saw the light of day in the United States.

Long before the appearance of the races which, to-day, people the United States, perhaps before the settlement in these regions of the present race of Indians, an ancient civilised people, who, according to the buildings which they have left, would appear to have been pretty far advanced, inhabited the vast tracts of country to the south of the great lakes. That race, the most ancient known of North American races, has entirely disappeared. The Americans of the present day distinguish them under the name of the *mound builders*, on account, no doubt, of the gigantic piles of stones which they have left which, it is supposed, were used by them as tombs. These cyclopean tombs form striking landmarks on the vast plains of the United States, and recall to the modern Yankee the grandeur of the defunct race.

They are brought back to mind in an equally forcible manner, and to us "oil men" in a more particularly interesting way, by excavations of equally gigantic proportions—the deep wells, 7 to

8 metres deep and 2 to 3 metres wide—scattered through the oil region between Titusville and Oil City. These wells, carefully built of masonry, or lined inside by strong boxing with boards, have braved the ravages of time. Saturated with petroleum, and preserved by it from the inroads of moisture, they descend side by side with the modern wells, like so many ancestors, surrounded by a numerous progeny.

The Red Indians and Petroleum.

To satisfy the unbelievers, the sceptics who would have no faith in the remote antiquity of these subterranean works, we shall invoke the testimony of the Indians who themselves preceded the Anglo-Saxons in the country.

From father to son, the Indians have left a legacy of legends in regard to the oil pits, which they attributed either to the spirit of the earth, or to a highly-civilised race who occupied the ground long before them, and who centuries ago had returned to this same mother earth.

As if the testimony of the Indians were not in itself sufficient, nature has taken upon itself the duty of bringing before us arguments which cannot be refuted. In fact, from the bottom of these wells, there rise up trees, hundreds of years old, the age of which may give the exact epoch when the oil pits were abandoned. It would seem to coincide with the epoch when the mound builders became defunct.[1]

Later on the Indians, who replaced the mound builders, were not ignorant of the sources of mineral oil. The Indians are great naturalists, or rather great observers of the phenomena of nature. Completely deprived of what we understand by the benefits of education, they nevertheless possess a science, based on their own faculties of observation, and on an instinct of deduction, ever on the alert.

In the plains of Canada there then existed a small lake, which presented two peculiarities : the first was that the surface, already darkened by the foliage of the surrounding forest, appeared absolutely black—hence its name of the Black Lake.

[1] See footnote at end of chapter, p. 201, for an explanation of these evidences of ancient civilisation.

The second consisted in the attraction which this black lake appeared to exercise over the animals of the region who came there in flocks to quench their thirst.

The Indians were not long in observing these peculiarities; they, moreover, found out that the black colour of the lake was due to a brown, greasy liquid of a penetrating odour. They concluded that if the beasts of the forests drank this oily substance, they did so because it had no injurious effect upon them; because, perhaps, thanks to their instinct, they foresaw a draught of benefit to their constitutional system. The Indians tried this funny cure, which was neither more nor less than crude petroleum. They found it do them good, and sought for it on the surface of the ground, even digging shallow holes to facilitate its advent.

By-and-by its use became universal amongst the Iroquoise tribe; they knew no other medicine. Petroleum possessed extraordinary curative virtues according to the Indian magicians. Taken inwardly, it was a drastic purgative—a specific for inflammation and flatulency, and destroyed the worms or serpents which, according to the aforesaid magicians, might choose to make a domicile in the human body. Outwardly it cured pain and skin diseases; it warded off from the warrior the bites of reptiles, and, —perhaps,—rendered him invulnerable in combat.

It will be readily understood that all these qualities, the one more precious than the other, increased the renown of the oil of the Black Lake and its neighbourhood. The Seneca Indians, who dwelt in the region, were ardent in their search after the oil. They learned to excavate—like the Germans in Hanover—deep trenches, in the bottom of which they spread woollen stuffs which became soaked with the petroleum brought in by the infiltrated waters.

At this time, the inflammable properties of petroleum would appear to have been known to the Senecas. In fact, in the first part of the eighteenth century the tribes, when at war the one against the other, made use of the oil as an incendiary agent. Forest tracts were burned down by inflamed petroleum, which brought calamity and desolation in its train.

The Early French Settlers and Petroleum.

It was towards the middle of the eighteenth century that Europe commenced to concern itself with petroleum.

It contributes to our renown for us Frenchmen to have been the first to give to science, exact, though yet vague, ideas on Seneca oil.

So far back as 1627, Delaroche, the missionary, speaks in his memoirs of a "fountain of bitumen" which he saw, issue from Lake Ontario.

Father Joseph, a French monk, attached to the Canadian missions, mentions American petroleum in the description of his travels in the Iroquoise country. He describes the virtues attributed to them by the Indians, and would not appear to have doubted their efficacy.

Later on, in 1750, the commandant of Fort Duquesne despatched to France a report describing a *fête* given in honour of him and his troops by the Senecas, friends of France. During this *fête*, at nightfall, the Indian chief gave an order, and some of his followers, torchbearers, applied their torches to the surface of the small river, near to which the gathering had taken place. A gigantic flame suddenly ran along the waves, spreading rapidly over the whole river. This festal day concluded with an alliance treaty, sealed in the light of the gods, the expense of which was borne by petroleum.

This affair was mentioned to the commander-in-chief of the French forces in New France. The Marquis de Montcalm,[1] general of the Canadian army, desired to study this phenomenon *de visî*. In 1757, Montcalm sent a report dealing with this subject to France. With a rare clairvoyance, the hero of the Canadian war had, in passing, seen the use which could be made of the new illuminating agent. He described the appearance of this substance in regard to which the Indians were so enthusiastic. He detailed its properties, its use and the manner in which it was collected.

---

[1] The French general despatched, 1756, to defend the French North American colonies. He was mortally wounded under the walls of Quebec, 1759, in the battle in which France lost Canada.—*Tr.*

But the King of France[1] was "as regardless of this as he was of Canada," and the *rôle* of the pioneer Frenchman in the New World was played out. The French chronology of American petroleum closes at this point.

The Trappers and Petroleum.

The war of independence, then the evacuation of Canada by the French,[2] left but two Anglo-Saxon peoples in the country— the British in Canada and the Yankees in the United States. These two peoples are adventurers, practical men and, above all, traders. Commerce accordingly took possession of the dis- covery of the Senecas ; it was sold in the drug stores under the name of *Seneca oil*. In 1791, an article in the *Massachusetts Magazine* describes, with forcible details, the oil springs which surround the river, already baptised, even at this period, with the name of *Oil Creek*. It boasts of their therapeutical virtue, and relates, in support of its statements, that the American soldiers, as they passed by, stopped at this spot and rubbed their weary limbs and joints with the oil which they collected ; the fatigue dis- appeared ; moreover, pains and rheumatics were permanently cured.

In spite of this publicity, petroleum was only collected on a small scale. The odour of the new remedy was obnoxious to sick people, its sale was but very small, and the produc- tion of the natural springs was quite large enough to meet the wants of a minimum consumption.

The Anglo-Saxons, less advanced than Montcalm, less familiar with the substance than the Indians, had not yet attempted to utilise it as an illuminating agent.

As in previous centuries, once more, animals were the cause of the development of petroleum, just as they had been of its discovery. History repeats itself.

In fact, in the beginning of this century, the fur trade was the principal North American industry. It was the age of the trapper, the age of Cooper's heroes. In their wanderings

[1] Louis XV., great grandson of Louis XIV. (le grand).—*Tr.*

[2] Canada was evacuated by the French *troops*, not by the French *colonists*, who, at the present day, are amongst the most loyal of British subjects, *e.g.*, Sir Wilfred Laurier, the present Prime Minister.—*Tr.*

through the forests, in the same way as the Indians of a previous epoch, the trappers remarked that the deer congregated together round springs of a peculiar nature. The water which flowed from those springs was highly charged with common salt. Salt, being a prime article of consumption, its sale was assured. The trappers (the more so as game became rare), from being hunters, changed their occupation to salt-makers.

The Salt-makers and Petroleum.

For the first time, after the *mound builders* became defunct, wells were excavated on the American continent.

Unfortunately—at least the salt-makers thought so—a great number of these wells yielded more substances than brine; the brine, in fact, was accompanied by a black, oily liquid, with a particularly disagreeable odour, and which they would have recognised as " Seneca oil " if they had read the advertisements of certain apothecaries who, in an honest way, strove to impregnate with it the organisms of their fellow countrymen.

The salt-makers were hale, hearty men ; they had no traffic with doctors, and moreover dwelt far away from towns. Petroleum and its virtues were unknown factors to them ; they only knew the grease that spoilt their brine.

So long as the brine wells were shallow, petroleum only came in minimum quantities, and was of little or no inconvenience to the salt-makers. However, when the greed of gain led to the sinking of deeper wells, and to the sinking of real artesian wells, very often the striking of oil caused them to despair, because, in large quantity it rendered the salt absolutely unfit for culinary purposes.

Nevertheless, this salt industry rapidly developed in Pennsylvania and Ohio ; Tarentum, near Pittsburg, became the working centre.

In 1814, this industry abated ; the *bad wells*—the oil wells— multiplied. Many, in face of this " misfortune," abandoned the business.

DR HILDRETH'S FIRST USE OF PETROLEUM AS AN ILLUMINANT.

In 1826, however, a forerunner, Dr Hildreth, saw, in his day, the proper use to make of this petroleum which ruined the proprietors of brine wells.

In the American *Journal of Science* Dr. Hildreth wrote as follows : " A man excavated a well on the bank of the Musk-ingum, near to Marietta, Ohio, in order to procure brine; on reaching a depth of about 400 feet, when he believed he had struck the brine bed, he met with naphtha, commonly called *Seneca Oil ;* and the last turn of the drill caused a prodigious quantity of brine, oil and gas to spout from the well, all of which ascended with incredible force."

" This product," remarked Dr. Hildreth, " offers great resources as an illuminating agent, and has commenced to be asked for in earnest in shops and factories ; it gives a brilliant, bright light and will certainly become of great utility in lighting the future villages of the Ohio ".

Dr. Hildreth preached in the wilderness—his voice was not heard. In fact, four years after these lines were written, in 1830, petroleum spouted from a Kentucky brine well. Great was the consternation of the owner, the more so as his brine was entirely spoiled. Let us think of it — 2,000 barrels of black, stinking oil issuing from the well daily ! There was no other course open but to abandon it ; at least such was the opinion of the salt-makers who evidently had not read Dr. Hildreth's remarks.

The well was abandoned, the salt-makers fled before the *plague,* and the petroleum continued to ·spout; it covered the country, flowing towards a river, which it rendered nauseous. Moreover, children at play set fire to the crude oil . . . and very soon a torrent of flames rolled over the country, bringing frightful calamities in its train.

EARLY THEORIES AS TO ORIGIN OF PETROLEUM.
SILLIMAN'S OPINION.

This fortuitous event drew the attention of men of science to the origin of naphtha. In 1833, Benjamin Silliman, an American chemist and geologist, desired to satisfy himself as to the facts related, touching the existence of petroleum. He visited the oil fields of Pennsylvania. The American *Journal of Science* published his observations in the same way as it had published those of Dr. Hildreth. But Silliman's opinion differed altogether

from that of his predecessor. After describing the appearance
of the district which he had visited, the celebrated American
savant expressed a very strange opinion which deserves to be
recorded. In his opinion, petroleum was of no great importance
by itself, its use as a medicine not being within his province ;
the only thing he saw in the presence of this liquid was a
sure sign of the existence of vast beds of anthracite in the
subjacent strata. It was therefore desirable to bore in the
spots where the oil exuded to bring these deposits to the
light of day. The miners of this period were not yet men
of science ; possibly the American *Journal of Science* was unknown to them. Without bothering with the anthracite of
which Silliman spoke, they continued to prospect for salt.

CRUDE PETROLEUM PUSHED AS A MEDICINE.

In 1845, Lewis Peterson, Esq., of Alleghany county, Pennsylvania, bored 2 wells to obtain salt. In 1847, instead of
salt, the 2 wells yielded petroleum to the extent of 40 gallons
per day. Lewis Peterson faced fortune with a stout heart ;
he sold his oil to the apothecaries who, faithful to ancient
traditions, re-sold it—whether good or bad—to their . . . patients
(we were going to say their victims).

This trade, as has been mentioned, made but little progress.
A man was, however, found who tried to push it by unlimited
publicity. This man was a druggist of Pittsburg, a man
named Kier, who opened in his native town a sumptuous
establishment, and by great dint of advertising, tried to render
general the medicinal use of petroleum. Elegant small half-
litre bottles were exposed for sale ; they bore the following
label : —

KIER'S

Petroleum or Rock Oil, celebrated for its wonderful
curing power.

A NATURAL MEDICINE

Pumped from a well in the Alleghany Country, Pennsylvania
400 feet below the surface of the ground.
Samuel M. Kier, 363 Liberty Street, Pittsburg.
Price : 58 cents.

The *natural medicine* had decidedly too bad a taste ; the sick people of free America could not take to it. *The stock of bottles got larger.*

Kier did not despair ; so as to draw the attention of the public, he caused prospectuses to be printed, imitating the appearance of the bank notes of the Union. His shop and his warehouses were plastered with these notes. The public did not respond to his ingenuity, and *the stock of bottles continued to grow.*

PETROLEUM FAILS AS A MEDICINE, BUT SUCCEEDS AS AN ILLUMINANT.

As the Americans strenuously refused to swallow the petroleum which was served to them so liberally at the price of 25 francs the litre, Kier turned his ingenuity in another direction. Perhaps he called to mind Dr. Hildreth's memoir, perhaps without having read it he had a lucky intuition, or perhaps he was hampered by his stock of bottles. At any rate, he said his panacea could very well be used as an illuminating agent, and sold his purgative for lamp oil.

This thick oil, contaminated with impurities, burnt badly. In 1852, Kier bethought himself of distilling it, and extracting from it the most volatile portions. *The stock of bottles began to go.*

The Shale Oil Industry paves the Way.

The American people took the more readily to this innovation from the fact that, for some time previously, the mineral oils extracted from lignites and anthracites, according to the process of Selligues, the Swiss chemist, were in current use.[1] The celebrated engineer, Gessner who, moreover, later on was to become one of the historiographers of petroleum, had introduced shale oil into the United States, the first attempts having been made at his premises on Prince Edward's Island in 1846. Later on, this method of illumination spread to Halifax, Nova Scotia. A company was not long in being formed to provide the United States with shale oil. This was the North American Kerosene Gas Light Company, whose business was already in a very prosperous condition, when Kier first thought of selling distilled petroleum

[1] See pages 4 and 129.—*Tr.*

In this country, where petroleum already covered the soil of Pennsylvania and Ohio, it is curious to notice how the trade in shale oil had necessitated the erection of works. There were already 5 more than those already mentioned. These were :—

Brooks, at Zanesville (Ohio).

Carbon Company, New York.

Empire State Company, New York.

Enon Valley Company, at Enon Valley (Pennsylvania).

Ritchie Company, at Ritchie (Virginia) and at Boston.

The town of Pittsburg, the birthplace of Kier, was the first to use distilled petroleum for illuminating purposes. The consumption soon became so great as to use up all the products of the old salt wells of Tarentum.

THE PENNSYLVANIAN ROCK OIL COMPANY.

In 1853, as we have just seen, the consumption equalled the production, and new deposits became the object of search. A company, the Pennsylvanian Rock Oil Company, was founded for this purpose by Messrs. J. G. Eveleth and George Bissel, of New York. From 1853 to 1859, the Pennsylvanian Rock Oil Company passed through many vicissitudes. It is a romance, but an American romance, where the leading motive may be summed up in the one word *business*. We hand the pen to our friend, the eminent popular writer, W. de Fonvielle (*Le Pétrole*, Paris, 1888).

Its Vicissitudes.

" Messrs. Brewer, Watson & Co., millers, of Titusville, entered into partnership with Mr. Angier, a proprietor of this locality, who used a slight modification of the Indian method for abstracting the greasy matter floating on the surface of Oil Creek.

" Alongside the oil river, ditches were excavated. These led into a reservoir, from which the water was raised by means of a pump driven by a mill. The water fell into troughs, and the oil which it brought in its train was collected by the aid of flannels, over which it passed, and which were afterwards pressed. In spite of all these improvements so small a quantity of oil was obtained that the company was dissolved.

" But in the summer of 1854, Dr. Brewer, son of the mill

manager, having occasion to visit relations, who dwelt in a town
of New Hampshire, where he had been educated, took with him
a sample bottle of petroleum to his old professor, Dr. Crosby of
Dartmouth, for the purpose of asking him whether nothing could
be done with a substance possessed of such extraordinary properties, and so similar to the shale oil of which everybody spoke
so enthusiastically, and which was easily sold at the rate of
1 fr. 20 c. the litre. (say 4s. 3d. the British gallon).

Bissell's Introduction to Petroleum.

" Dr. Crosby, some time afterwards, received a visit from a
young New York lawyer, named Bissell, to whom he showed the
bottle of oil which he still had in his laboratory; he even sacrificed
a part of the liquid to afford him the pleasure of seeing it burn in
a lamp brought from Europe. Quite enthusiastic over the bril-
liancy of the flame, Bissell betook himself to Titusville with Mr.
Crosby's son, inspected the source, and founded a General Petroleum Company, with a capital of 1,250,000 francs, divided into
10,000 shares.   Two thousand shares and the sum of 5,000 dollars
were paid to Messrs. Brewer, Watson & Co. as the price of the
source which they held."

Early Difficulties of the " Oil Kings ".

" But the reign of the ' oil kings ' did not commence without
the founders of these dynasties—more numerous than the Pharaohs
—having to encounter astounding difficulties.

" Notwithstanding the favour with which a substance endowed
with such remarkable illuminating power was received by the
public, the quantity of oil collected, by skimming the oil either
off the river water or off the brine, brought up from the bowels
of the earth, was so slight that it was not available for any
economical or industrial use.

" The process which we have described, of multiplying the
surfaces of contact by connecting—with cross ditches—the impregnation pits along the oil river, did not yield any material
result.   The only industrial method which really had a future
in it was that of boring wells, by means of which the oil could
be brought to the surface free from brine.   In a word, the
problem was to bore in a locality where one of those subterranean

crevices is met with, in which the oil accumulates naturally. Benjamin Silliman, when consulted by Bissel the lawyer, had given this advice; but it was necessary to find shareholders bold enough to put it in execution.   Messrs. Pierpont and Havens, persuaded, and rightly so, that these precious natural deposits ought to be found in the very spots where the trenches had been opened, proposed to the company to lease the petroleum sources for 15 years, and to bore there, at their own risk and peril, all the artesian wells which would be necessary for the development of the industry, and in exchange to pay a royalty of 15 centimes per litre of oil obtained during the currency of the lease.   Unfortunately, these far-seeing men were not able to get the necessary sum for commencing operations together within a year, and in virtue of a saving or reverting clause the bargain became null and void.   Bissel, who had become director of the company, tried to find another *concessionaire*.

Colonel Drake.

" He cast his eyes on a certain Mr. Drake, a but poorly educated New Yorker, but a very active man, who was bordering on fifty, and from his youth had led a varied, wandering life in different parts of the Union.   For one period a commercial traveller, for another employed in a hotel, and for some time a railway conductor, he had saved a few crowns.   Drake arrived at Titusville, decorated with the title of "colonel," with which he had been dressed so that he might gather together the necessary funds more easily.   But the first act of this *colonel of industry*, in taking command of his regiment, was to insist on the royalty being reduced to 5 centimes, the third of that in the first contract.

" But the speculators, who higgled with such greed, over the aliquot part of the royalty, who tore their hair, with such desperation, knew full well the necessity of keeping the secret, and to avoid attracting the attention of competitors, they made the necessary legal publications in the newspaper of one of the smallest villages of the country.

" It soon became apparent that the company were short of money, for funds did not arrive at Titusville, with all the regularity desirable.   The borers employed on the job, had no

13

confidence in an enterprise which appeared to them on the high road to bankruptcy, and they deserted every time that a salt manufacturer proposed to open a small brine well."

First Use of Iron Tubing to " case " the Bore Hole.

" From a technical point of view, matters were not going on well. The ground in which the oil well was being bored was sandy, and fell or caved in. They had tried in vain to box it up; the woodwork was driven in. The undertaking was then in a desperate condition, when Drake was seized with an ingenious idea, *viz.*, to sink in the shifting sand an iron tube, by means of which he could reach solid rock, and through which he could pass the rods by means of which he turned his auger, thus enabling him to continue work."

Uncle Billy Smith.

" The first sign of a change of fortune was meeting with a faithful workman, whose name history has preserved. He was a certain man named Smith, who, having already wrought for Kier, was not altogether a stranger to the business, since that quack sought a brine mingled with oil for the purpose of 'skimming it'. Towards the end of the month of August, 1859, one Saturday evening, old Smith was able to pass Drake's tube through 56 feet of shifting sand, which would have presented an impassable barrier with the old methods. Then he commenced to turn the drill by means of the handle, which he held in his hand, an operation which no longer presented any difficulty, and the boring progressed regularly at the rate of 1 metre per day.

" All at once, seeing that the rod sunk of its own accord, old Smith thought that an accident had occurred, and that the drill was broken. He gathered his tools together hurriedly and went to bed, leaving, until Monday morning, to see what had happened. However, as the night brings wisdom, he said to himself that perhaps he had struck one of those pockets of oil which the colonel was looking for, and which he had never heard mentioned without shrugging his shoulders. Consequently, instead of going to church, he went back to the well, and with a hand shaking with emotion he lowered a tin pail and raised it again. It can only be left to the imagination to appreciate with what joy he saw that it was filled with precious liquid, which had accumulated quite alone in a subterranean cavern, the roof of which had been pierced by the drill!   The prophecy of the man of science, Silliman, was realised in the most brilliant manner."

The First Oil Well.—The Rush to the Oil Fields.—Bissell and the Oil Kings.

" Although Smith was not what is termed garrulous, the news was known in the village before the close of the day.   On the Monday morning, there was a crowd of people round the well, where old Smith and his two children were drawing oil in bucketfuls.   They were not long in filling several barrels, before the eyes of the stupefied rustics, who commenced to understand instinctively that chance had made them witnesses of some great event.

" As the electric telegraph had by this time assumed great developments, it may, without exaggeration, be said that the news of Drake's discovery spread over the whole of America like a flash of lightning.   People ran there from all parts of the Union— bravadoes, *déclassés*, adventurers, etc.—all decided even to work if they could not make their fortune otherwise.   They all ran a race to the same goal, analogous to that which took place to California on the discovery of gold, and which was reproduced when the diamond fields of South Africa were discovered.

" Bissell was one of the most active, as well as one of the most fortunate, of these speculators.   He rapidly acquired an immense fortune, and placed himself at the head of the " oil kings ".   He represented them at New York and Philadelphia when the politicians proposed to create local taxes which would have paralysed the prodigious development of an industry to which these two great cities owed the greater part of their prosperity.   Besides, he organised the first railway company, laid down with the view of connecting with the American network a country where riches, equal to the treasures described in the *Thousand and one Nights*, were acquired with a rapidity which to-day would appear to us fabulous."

Colonel Drake and Speculation.

" . . . As soon as they knew of the success of Colonel Drake, a host of speculators fell on the country, as we have described.   Their unique occupation was to rent, on the best possible terms, for a very long time, all the farms which they made haste to corner, so as to impose tribute on the real oil prospectors.   But Colonel Drake would not hear of this kind of operation.   He confined himself to working his well—that is to say, to pumping his oil.

"Very soon he began to see, to his great grief, that the source diminished, and at the end of the year 1859 his production was reduced to a score of barrels per day.

"During this time, his neighbours had adopted the process which Colonel Drake had omitted to patent. 'Derricks' arose on all sides. The oil flowed in waves, and enormous fortunes were made in front of his face.

"The idea of seeing himself ousted by neighbours, to whom he had set the example, disgusted him with mining.   In 1860, he was nominated justice of peace of the town of Titusville—an office worth about 15,000 francs a year—on account of the activity which the oil fever had imparted to all transactions in movable and immovable property.

"As American magistrates are not forbidden to engage in trade, the colonel became the representative of an important New York firm engaged in the oil trade, and thus brought his income up to 25,000 francs per annum. This sum was so moderate, when compared with what the "oil kings" gained daily, that the colonel decided to withdraw from the game with his gainings (*faire Charlemagne*), and to go back to New York. He settled up his affairs completely, by realising all his property, and returned to his native town with a capital of 150,000 francs, the whole of which he devoted to speculation on the Petroleum Exchange."

Colonel Drake Retires Ruined.

"Unfortunately, there happened to him that which is almost always the lot of those who change their profession at the end of their career. The small fortune was swallowed up in one opera-tion, and he very soon found himself wandering, without resources, in the great metropolis of New York.

"One of his friends offered him a refuge in a cottage, to which he retired with his wife and four children. The money, which this unfortunate woman gained by her needle, was the sole resource of the unlucky family to whom American society owed so many millions.

"The colonel was in such miserable circumstances that it was often impossible for him to go to New York, for want of the money to pay his fare, which came to 4 francs.

"One day he decided to make this sacrifice, so as to obtain a situation for his eldest son. Having been unsuccessful, he was on the point of despair, and was walking along Broadway with a feeble step, when he met one of his friends, who was struck by his lugubrious air and grim countenance.

"Pressed with questions, the colonel confessed, with tears, that he had not been able to come to New York with enough money to get anything to eat, and that he had gone without his dinner. His friend forced him to partake of a good dinner on the spot, forcibly placed a few dollars in his pocket, and assured him that he would straight away endeavour to ameliorate his lot."

Colonel Drake Pensioned by a Grateful Country.

"This gentleman wrote forthwith to Titusville in regard to what had happened. Holding a meeting, the "oil kings" of the country assembled and opened a subscription in the colonel's favour. In a few days it produced a sum of 100,000 francs—a poor fraction of what he would have gained by the granting of licenses if he had taken out a patent for the sinking of his tubes.

"The Pennsylvanian Legislature held it a matter of honour to show that ingratitude is not, as people believe, one of the virtues of Republics. In 1873, it passed, enthusiastically and unanimously, an Act conferring on the colonel as a national recompense, a life pension of 1,500 dollars, with reversion to his wife."

The Fate of American Petroleum Industry Decided by Colonel Drake's Success.

Coming back to the Pennsylvania wells, it was therefore Drake's success and also that of his workman, *Uncle Billy Smith*, who, in 1859, decided the fate of the petroleum industry. In fact, the oil was found by Smith on the 28th of August, 1859, and just a year afterwards in September, 1860, 55 wells yielding 25,000 gallons daily were opened in United States territory,

and 14 refineries had been established to treat the extracted
oil. On the banks of Oil Creek, the price of land had risen
in the proportion of 1 to 1,000. The acre (rather less than
half a hectare) had risen from an average of 10 dollars to the
fabulous price of from 9,000 dollars to 10,000 dollars. The
owners of the ground on the oil fields, made their fortune at
a single stroke. They did so, the more suddenly, as some of
them—the most intelligent—had the lucky thought of making
some sacrifice on their selling price, so as to retain an interest
in the working. And this working succeeded beyond their
expectations—the spouting wells brought them millions.

THE FIRST SPOUTING WELL—THE " FOUNTAIN ".

The first well which, to the great astonishment of the " oil
men " projected its petroleum into the air in the form of a
heavy shower of liquid, was the " Fountain ". Bored on the
Funk property, the " Fountain " was in the midst of accumulated
difficulties when the boring had reached 460 feet. The well
spouted on the 1st June, 1861 ; Funk, finding that his well
yielded 300 barrels daily, was nearly mad with joy. The
boring had cost him very dear, and if he had struck a " dry
hole," it meant ruin ; 300 barrels per day was a fortune ; at
the then price of petroleum (nearly 1 franc the litre), it was
an assured income of 50,000 francs (say £2,000) a day.

The event became noised abroad in the district. People
declined to believe in such a marvellous yield. The " Fountain "
was nicknamed the " *Oil Creek humbug* ". In spite of the remarks of the incredulous, when the well stopped flowing, its
output had produced more than 12,000,000 francs (say £480,000)
to its lucky owner.

OTHER SPOUTING WELLS.

Moreover, before this time other flowing wells had been opened.
The first was the " Empire " which, at the depth of 450 feet,
exceeded in its unexpected proportions, the flowing of the
" Fountain ". The " Empire " gave 2,500 barrels per day. The
spouting which took place on the 3rd September, 1861, was
unexpected. No vessels could be found in which this formid-

Pl. IX.—Oil Creek in 1867. Photograph communicated by MM. Desmarais (Vol. i., to *face p.* 198).

able quantity of petroleum could be collected which, during the first day or two, flowed into the river.

The " Phillips " Well.—Two months afterwards in November, 1861, a well, situated on the Torre Farm, exceeded this yield. This was the " Phillips " which, with a depth of 491 feet, spouted with a production of 3,000 barrels daily. The " Phillips " did not stop spouting until 1873, after a flow of 12 years, producing 1,000,000 barrels yearly.

The price of petroleum began to lower. Nevertheless, work in the neighbourhood of Oil Creek did not slacken. Then there spouted in succession the " David Wherlock " which produced 1,500 barrels daily, the three " Densmore " wells, the yield of which was 600, 400 and 500 barrels, the " Maple Shade " of Hyde and Egbert which, in August, 1863, supplied its contingent of 800 barrels daily, the " Noble and Delamater " with 3,000 barrels per day and a total production of 3,000,000 dollars.

In the neighbourhood of Pithole, starting in 1865, there followed each other in quick succession the " Twin " 800 barrels, " No. 54 " 800 barrels, the " Grant " 450 barrels, the " Eureka " 800 barrels.

MUSHROOM TOWNS.

It was a real deluge of oil, the increasing production of which brought quite a cosmopolitan population of workmen and, in their train, dealers and speculators of all kinds. The country began to live a feverish life which gave rise, in a few months, to roads, railways, towns; all this prosperity, all this civilisation being based on the flow of oil which seemed to animate the region, and at the mercy of any diminution in that flow.

Pithole : Its Rise and Fall.

When the " Twin " commenced to spout in 1865, Pithole, a village consisting of but a few huts, was assailed by strangers.

" On the first of May, Pithole was only a small clearance in the heart of the forest—90 days afterwards it was a town with about 16,000 inhabitants. Its mailbag was the third in Pennsylvania, and ranked after Philadelphia and Pittsburg. Hotels, theatres, clubs, music halls, were counted by the score ;

it had a fire brigade, a municipality, mayor and aldermen. A railway was projected for improving its communication, when all at once ruin came with frightful rapidity ; the wells no longer yielded oil, they were abandoned, and the 1st of January, 1866, dawned on an uninhabited town, on silent streets and deserted dwellings. Pithole, which, according to the American expression, ' went up like a rocket and came down like a stick,' was no more. ' Moreover, if any Pitholian was born there,' says M. Simonin, ' this good citizen will some day find it rather difficult to re-discover his native town.' "

Cherry Grove.

" We may also mention the two towns in the district of Cherry Grove after the boring of well 646. As soon as the figure of its production was known, " oil men " flew there. Six weeks afterwards, two cities were founded—Garfield, named after the assassinated president, and Farnsworth, the name of the owner of the ground where this marvellous well was bored. Three months afterwards, the two towns, which numbered 20,000 inhabitants, were two more deserted cities to add to the long list of dead towns.

" Others, the existence of which was not dependent on the issue of such and such an oil field, have endured to now. Sustained by the numerous industries which are connected with petroleum, they have grown and are become the most important centres of engineering manufacture—the branch lines of pipe lines or refineries."

Oil City.

" The oldest, and also the most important, Oil City, is situated at the confluence of Oil Creek and the Alleghany ; it is the city of oil *par excellence*, with its dirty, muddy streets. One breathes an atmosphere poisoned by the smoke of the refineries and petroleum emanations. The town is well built ; it is the seat of the United Pipe Lines Company ; it possesses the largest petroleum exchange in the whole world " (Fernand Hue).

All Pennsylvania, and also all Western Virginia, were smitten with this oil fever. Borings multiplied, wells spouted, towns were born.

CLARION AND BRADFORD DISTRICTS.

In 1863, the rich deposits of the Clarion district were discovered. The district of Bradford commenced to be wrought in 1863.

In Virginia, operations begun in an old salt spring at Burning Spring yielded 50 barrels daily. A short time afterwards, a neighbouring well, the "Llewellyn," yielded, at a depth of 100 feet, a daily product of 1,000 barrels.

So far back as 1866, the year where we conclude this history, the petroleum industry had been immensely developed. The oil production of Pennsylvania, which in 1859 was 2,000 barrels annually, had risen from this figure to the respectable amount of 3,732,000 barrels; and Pittsburg, the oil metropolis, built around the old Fort Duquesne, from whence the first reports regarding petroleum were despatched to France, counted hundreds of thousands of inhabitants and thousands of manufactories.

According to advices from Washington, under date 15th November, 1900, a startling story of ancient archives unearthed at Pekin is current there. It is said that documents tending to prove that America was discovered by the Chinese, centuries before Columbus, were found after the occupation of the Chinese capital by the allied forces. Information of the discovery is said to have been given to the State Department by Mah Twah Win, a Chinese student. The documents are said to record that a party of Mongolian missionaries crossed the Pacific and landed in Mexico about 499 A.D. They spread the doctrines of Confucius, erected a number of temples, taught the natives many arts and crafts, thus affording an explanation of the evidences of Asiatic civilisation shown in the works of the Aztecs which have long puzzled antiquaries..—Tr.

CHAPTER XV.

PHYSICAL GEOLOGY AND GEOGRAPHY OF THE UNITED STATES OIL FIELDS.

THE United States include a number of rich oil fields, which we shall classify according to the States in which they exist.

EASTERN BASIN (*course of the Ohio and Mississippi*).—States of New York, Pennsylvania, Western Virginia, Ohio, Indiana, Kentucky, Tennessee.

NORTHERN AND CENTRAL BASIN (*course of the Missouri and the Upper Mississippi*).—States of Wisconsin, Minnesota, Dakota, Illinois, Missouri, Kansas, Texas.

SOUTHERN BASIN (*banks of the rivers of the Mexican Gulf*).—States of Alabama and Louisiana.

WESTERN BASIN (*Rocky Mountains*).—States of Wyoming, Utah, Colorado, California.

GENERAL TREND AND DIP OF THE PETROLEUM-BEARING STRATA.

Geologically speaking, the United States petroleum-bearing beds would appear to follow a line of strike conformable to the mountainous system of the country, and parallel with its great folds.

In *the north, east and southern basins* this line strikes from north-east to south-west, appearing to be in absolute conformity with the direction of the Alleghany mountain chain. This line at first sight would appear to be perfectly perpendicular to that which we have studied on the old continent. Contrary to what we have seen in Europe, the petroliferous strata of these regions form a very regular series of stratified rocks, consisting of several oil levels which tend to dip generally towards the south-west.

GEOLOGICAL FORMATIONS.

All these beds belong to older formations than those to which the petroliferous strata of our continent belong, and in our opinion

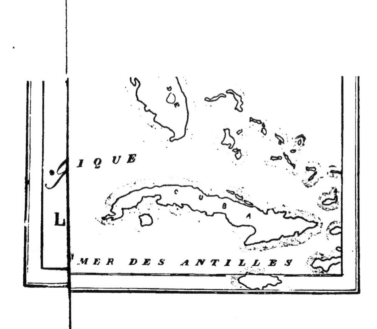

IQUE

L

MER DES ANTILLES

no relation whatever exists between either of the two. In fact, whilst in Europe, Asia and Oceania it is always the *tertiary* strata (*pliocene, miocene* and *eocene*) which are considered by the geologist as the typical petroliferous formations ; on the other hand, in the great northern, western and southern basins of the United States, as well as those of Canada, which are but an offshoot, the oil-containing beds are found in the geological level of the *transition* [1] *beds* or in the *carboniferous strata.*

SILURIAN PETROLIFEROUS STRATA.

Owing to the general dip which has been mentioned, it is towards the north, in the neighbourhood of the great lakes, that the older rocks come to the surface, such as the *Silurian*, which is found along the St. Lawrence and over a large belt on the shore of Lake Erie. This formation, which may be looked upon as that of the great lakes, is that which constitutes in the greatest meas-ure the oil fields of the State of New York (between the lakes and Albany) and those of Northern Ohio (districts of Frémont, Find-lay, Lima and Corey).

DEVONIAN PETROLIFEROUS STRATA.

A more extended oil-bearing zone, at least in the United States, is that in which *the Silurian is covered by the Devonian strata* (districts of MacKean, Alleghany, Warren, Venango, Clarion, Lawrence, Beaver, Armstrong and Butler in the States of New York and Pennsylvania, and the districts of Washington in Western Virginia and of Nobel in Ohio). Tennessee and Ken-tucky belong to a large extent to this formation.

As is the case in the lower Silurian system, the Devonian beds are especially rich in marine fossils, in the the same way as in China, Galicia and the Caucasus salt springs abound ; these are points of resemblance, almost of contact, with the beds of the ancient continent, arguments which may be added to those

[1] Transition beds is a loose phrase of geological terminology applied by the older geologists to the Cambrian and Silurian as being the transition stage between the stratified and metamorphic rocks. Here it is evidently meant to include the Devonian, whereas it is now used in a much more restricted sense. The rocks forming the crust of the earth have been divided into (a) Palæozoic or Primary, (b) Mezozoic or Secondary, (c) Kainozoic or Tertiary. The oil fields of the Old World are found in the Kainozoic or Tertiary, whilst those of the Eastern oil fields of the U.S.A. and Canada belong to the Palæozoic or Primary. See note p. 32.—*Tr.*

which we have already formulated in regard to the organic and
marine origin of petroleum.

CARBONIFEROUS PETROLIFEROUS STRATA.

The carboniferous strata in their turn lie upon older beds
which, in virtue of the rule which we have already enunciated,
dip towards the south. It is thus, in the south of Pennsylvania,
that the carboniferous zone is found ; it occupies a rather irregu-
larly outlined space in the environs of Pittsburg. It is, we believe,
the least rich in oil of any of the districts just enumerated in
detail. The carboniferous strata again crop up in Ohio (Wellsburg and Maksburg districts) and in Kansas.

FIG. 26.—Geological Section of the Canadian and Pennsylvanian Oil Fields
(after a drawing in Fuchs and De Launay's *Traité des Gîtes Minéraux*).

WESTERN OIL FIELD.

The western basin of the United States would appear to present special geological conditions. Its direction is rigorously
influenced by the Rocky Mountain system, and would appear
to be perfectly parallel thereto.

CRETACEOUS AND TERTIARY PETROLIFEROUS STRATA.

In the western basin, the strata would appear to be in general
of much more recent nature than in the other American basins,
viz., those of the Alleghany, Ohio and Mississippi. The cretaceous beds found in Utah and the rich deposits of California
belong to tertiary strata. This basin, from the nature of its beds
and the direction of the strata, would appear to be in more
intimate geological relation with the South American basins
than with those of the remainder of the United States and
Eastern Canada.

Details of the different deposits of the four great basins into
which we have divided the United States are given below.

NEW YORK AND PENNSYLVANIAN OIL FIELDS.

Geographically, since 1890, the New York and Pennsylvanian deposits have been divided into five principal districts. These are :—

1. Richburg district (New York State).
2. Bradford ,, on the north-east.
3. Middle Field district.
4. Lower Field ,,
5. Washington ,, on the south-west.

Geologically, an ideal section of these beds collectively would yield the following classification :—

IDEAL SECTION OF NEW YORK AND PENNSYLVANIAN PETROLIFEROUS STRATA.

Paleozoic (Primary) Rocks.	Permian Strata.	Waynesburg series. Washington ,, Monagallia ,,
	Carboniferous Strata.	Pittsburg coal. Barren beds. MAHONING SANDSTONE (*petroliferous*), wrought at Dunkar Creek (Green county, Washington district). Lower coal measures (Alleghany series). POTTSVILLE CONGLOMERATES (*petroliferous*), (Washington district.—Lawrence and Beaver county, in the Lower Field district).
	Anthracitic Beds.	Mauch Chunk red schists, with some beds of limestone. POCONO AND BUTLER SANDSTONES (*petroliferous*), natural gas level of Butler county (Lower Field district) ; a restricted quantity of petroleum in same region.
	Devonian Strata.	Catskill red sandstone, including the petroliferous level of VENANGO SANDSTONES, very rich beds, comprising the greater part of the deposits of Venango county (Lower Field) and the southwestern part of Warren county (Middle Field). Chemung series, comprising the *Warren sandstones* (Warren Middle Field), and lower down, the MACKEAN sandstones (MacKean, Bradford county).
	Silurian.	CANADIAN SANDSTONES AND LIMESTONES (*petroliferous*), Enniskillen formation. In the north of Bradford district, towards Richburg.

STRATA ENCOUNTERED IN ACTUAL BORING NEAR PITTSBURG.

(According to Fuchs and De Launay.)

In a boring made in the neighbourhood of Pittsburg, Washington county, the following beds were encountered :—

(a) Carboniferous Strata.

1. *Productive Coal Measures.*—The productive coal measures of Pittsburg ; then

2. *Barren Coal Measures.*—Fifty metres (say 164 feet) of barren coal measures ; then

3. *Lower Alleghany Coal Measures.*—One hundred and thirty metres (say 426·4 feet) of the Lower Alleghany coal measures, with a dozen beds of coal, terminated at the base by a character-

istic bed of ferruginous limestone, from which starts the anthracitic beds.

4. *Anthracite Beds : Crawford Beds.*—Below the ferruginous limestone come the anthracite beds, generally sandy, with a variable thickness according to the locality, consisting of the Mauch Chunk, and Pocono sandstones of the preceding section, and called the Crawford beds, from the name of a petroliferous county to the south of Lake Erie.

The anthracitic beds comprise in descending order :—

(1) The *Mauch Chunk shales*, which may attain a depth of 900 metres (2,952 feet) ; shales with *imprints of waves and raindrops, traces of reptiles, labyrinthodontes,* etc.

(2) *The Anthracitic Sandstones.*—Six hundred metres (1,968 feet) of anthracitic sandstones. Towards Pittsburg this formation thins out to 120 metres (393·6 feet) of sandstone, with petroliferous impregnations, and 140 metres (459·2 feet) of the lower shales, which may be observed in Crawford county.

(b) Devonian Strata (Continental Samennien).

At last the Devonian strata, which corresponds to the European Samennien, comprises the three principal zones of the Pittsburg region. It comprises in its upper part :—

1. *Sandstones, Shales and Conglomerates.* — One hundred and twenty metres (say 393·6 feet) of compact sandstones, shales and conglomerates, beneath which there is found the first petroleum-bearing stratum wrought in the counties of Warren, Clarion and Budlew, producing in 1884, 17,000 barrels per day.

2. *Schistose Impermeable Strata.*—Then there is traversed another 100 metres (say 328 feet) of schistose, impermeable, absolutely barren strata; and there is met with in

3. The *Warren sands*, a second petroleum-bearing stratum, reaching a depth of 100 metres (328 feet), and which, in 1884, gave a maximum production of 16,000 barrels per day.

4. *Argillaceous Barren Shales.*—Still another 100 metres (328 feet) of argillaceous barren shales, containing only some *rare beds of slightly impregnated sandstone,* that are passed in boring.

5. *Petroliferous Sands.*—Below the barren argillaceous shales a third petroleum level is reached, which is by far the most

important, although only 25 metres (say 82·2 feet) thick. It consists almost exclusively of petroliferous sand, and produced in 1881 as much as 83,000 barrels per day. At the present time it still yields 34,000 barrels per day.

(c) Silurian Strata.

Lower down the Devonian argillaceous shales no longer contain petroleum, and the Silurian is reached (corniferous), in which a boring has revealed two petroliferous horizons, the lowest of which corresponds to the petroliferous horizon of Canada (Enniskillen).

1. THE OIL FIELDS OF NEW YORK STATE—RICHBURG DISTRICT.
GEOLOGY.

The deposits which belong to the State of New York, between Syracuse and Jamestown, on Lake Erie, are the most northern of the eastern basin ; consequently, and in accordance with the geological principles which we have enunciated, petroleum is found there in the lower horizons similar to those which crop out on the other side of Lake Erie, at Enniskillen. This type of strata (the Silurian), characterised by quartzose particles, is especially met with in the west of Albany, towards SYRACUSE. Salt springs are numerous in this region, but petroleum deposits are rather scarce, and not very fertile in yield.

The yield increases a little in the districts of RICHBURG and JAMESTOWN. There the geological formation answers to the Devonian, and with the exception of slight outlying tracks of the petroliferous limestone of Canada of Silurian formation (the Trenton limestone), which may be observed to the north of Richburg, the petroleum-bearing bed consists of fine-grained black sand, or of a chocolate-coloured sandstone, which contains mineral oil in rather restricted quantity. This formation corresponds with the MACKEAN SANDS, and forms the northern extremity of this horizon, which would appear to become richer in petroleum as it dips towards the south. The principal deposits of the region, besides Pittsburg, are :—

NILES, rather a poor locality.

WIRTH, where the working of natural gas appears to specially tempt the owners.

HARDING AND O'CONNOR.

ANNUAL PRODUCTION OF NEW YORK STATE OIL FIELD.

The deposits of New York State may be counted amongst the oldest wrought in the United States. Their yield has been subject to great fluctuations. A few figures will give an idea. In 1889, the production was 541,634 barrels. In 1891, two years afterwards, it reached 1,121,547 barrels—that is to say, more than double.

The wells would now appear to be getting exhausted, and the present yield may be estimated at 450,000 barrels.

PROPERTIES OF THE OIL.

The oil yielded by the petroliferous region of the New York basin is generally dark in colour, verging towards black, its average density being between 0·807 and 0·833.

2. BRADFORD DISTRICT. GEOLOGY.

The Bradford district only extends for a few miles to the south of the Richburg district. It includes a few deposits in the territory of New York State, but the principal oil fields are situated in Pennsylvania. It further comprises a part of Cattaraugus and Potter counties, and the whole of MacKean county. The petroliferous bed of the Bradford oil field is wholly situated in the MacKean sands (Silurian). The petroliferous sands there are of the same nature as in the part of New York. However, as has been previously observed, they would appear to be much richer in petroleum.

Divisions.—The district is divided into different mining regions, the details of which are as follows :—

1. THE GREAT BRADFORD FIELD, which extends from Eldred to KINZUA on the one hand, and from Bradford to Colegrove and to SMETHPORT.

2. KINZUA, which is situated to the west of Bradford, almost on the line which separates MacKean county from Warren county. Kinzua possesses a number of productive wells of especially long continued yield.

3. WINDFALL RUN, a relatively new locality (1892), of which great things were expected, which have only been partially realised.

Properties of the Oil.—The density of the petroleum of this district varies between 0·769 and 0·817.

History of Bradford District and the "Barnsdall" Spouting Well.

The development and working of the Bradford oil field has received more than one check. The first well was bored there in 1862. This well was, however, abandoned at a depth of 200 feet as unproductive. The owners covered it with planks and, leaving the country, went to seek their fortune elsewhere. Whether they did not find it, or whether they were stricken with remorse for having shut up their "Barnsdall" is a moot question. The fact remains, however, that, four years after having left the country, the owners returned and re-opened the "Barnsdall," and boring was continued unceasingly to a depth of 850 feet. There the owner stopped, thinking that the well was decidedly a "*dry hole*," and that the best plan was to carry the money lost and the time spent to profit and loss. Happily God watched over the future of Bradford district. He watched in the person of Samuel Ficht, the foreman, who said to his master that, as there was coal still remaining on hand for the engine, and that as the workmen would have to receive their month's wages, in full, if they dismissed them forthwith, it would be more to the purpose not to lose the work due, and the fuel already paid for. Samuel Ficht was listened to, and two days afterwards, at a depth of 875 feet, the "Barnsdall" spouted.

No sooner did this occur than "oil men" arrived from all points of the horizon. Wells were bored, and in 1867 the "Gilbert" well of the Foster Oil Company, then the "Butts" well, then others, spouted, yielding considerable returns. Briefly, in 1868, 400 wells were grouped round the old "Barnsdall"; in 1875, 2,536 were counted, whilst to-day the figures are 6,800 productive wells.

Annual Production of Bradford Oil Field.

The district of Bradford is one of the richest in Pennsylvania. At one time it furnished almost one-third of the production of this State. However, during the last few years, the general yield would appear to be diminishing.

It is also from Bradford that the New York section of the pipe lines starts, and here likewise are located the immense reservoirs, in which the conduits leading from all parts of Pennsylvania and bringing the oil to follow this section, terminate. We shall return to this subject in a special chapter on the organisation of pipe lines.

3. The Middle Field District. Geology.

The district of Middle Field touches that of Bradford, of which it forms the south-west continuation. It comprises Warren and Forest counties. Two different geological horizons, but both belonging to the *Upper Devonian*, form the Middle Field district. To the north, the *Warren group*, and to the south the *Venango group*. The *Warren level* consists of 5 or 6 beds of fine-grained, compact, rather bright-coloured sandstone.

The thickness of the productive zone varies between 175 and 215 metres (574 and 705·2 feet). The *Venango group sands* are of very different texture. Its sandstones are white, greyish white or yellowish, friable and coarse-grained, and often mixed with quartzose particles. This group is one of the richest in Pennsylvania.

Localities in Middle Field.

Petroliferous localities are numerous in this district—the following may be mentioned :—

(1) In Warren County.

Sugar Run, a place with a poor enough yield.

Dew Drop.

Warren, irregular beds, numerous pockets, production of very limited duration.

Wardwell is a place with the same physical configuration as Warren. The same remarks apply.

Clarendon, whose productive and durable wells yield a petroleum of very light gravity.

Garfield has a large number of wells, the yield and duration of which are above the average.

Cherry Grove, whose rich deposits are celebrated. Cherry Grove was the first place wrought and developed in Warren county ; and it is from the day on which its celebrated well, " 646," then unique in that region, spouted, that the fortune of Warren county dates. This oil field—the expansion of which has been very rapid—is nevertheless one of the last arrivals amongst those which bring their contingent to swell the Pennsylvanian production. In fact, "646" spouted on the 17th May, 1882.

THE HISTORY OF CHERRY GROVE.

"The circumstances which marked the discovery of this deposit, and the consequences which it had on the petroleum industry, deserve that some details should be recorded of the first wells bored.

"Cherry Grove, in Warren county, was still, in the month of April, 1882, an almost desert locality. In a vast clearance, in the midst of a forest of birch trees, rose some miserable farms; a dozen inhabitants, mostly woodcutters, comprised all the population of this hamlet, which had been decorated with the rather pretentious title of hamlet. Ten miles from there is the little town of Clarendon—a railway station on the Philadelphia-Erie Railway, and the centre of a small oil field. For a long time the "oil men" were certain that if borings were made to the south-west of Clarendon, petroleum would be found in large quantity, but, nevertheless, none of them had made the attempt. In the month of April, 1882, four men, more convinced or more adventurous than their comrades, arrived in the Cherry Grove clearance, got a concession for a few acres of ground, and commenced boring. It would appear that as soon as the enterprise was known, people had a presentiment of the consequences which would follow. The brokers and the speculators sent out spies to keep them informed of the progress of the work, but the owners exercised the greatest surveillance around the derrick; they kept good watch during the night, and, to drive off the curious, they fired several gunshots in all directions. Nevertheless, a young man, smarter than the others, succeeded in penetrating under the planks of the scaffolding; he remained there, they say, seventeen hours, got away unperceived, and announced to those who had sent him that the well "646" was a spouting well, the like of which had never been seen. The name of "646," by which this boring was known, was given to it from the fact that the plot of ground on which it was bored, bore this number on the survey register.

"The fortunate proprietors could not hide the result of their operations for long, and it was soon known that the well yielded 4,000 *barrels per day.*

". . . It was on the 17th May, 1882, that "646" spouted for the first time. Before the end of June, 321 wells were bored in the neighbourhood, and all had a considerable output. This prosperity did not last long; rapidly exhausted, the petroliferous bed was not long in being drained dry. In the month of October of the same year, the production was almost *nil*—the greater number of the wells were abandoned and the clearance again became a desert.

"The fortunes of the thousands of individuals who had founded the greatest of expectations on Cherry Grove, had taken wings with the riches of the district" (Fernand Hue).

What may surprise the reader is that the thousands of individuals in question had not been so badly advised in founding the highest expectations on Cherry Grove.

The State of Affairs since Improved.—Since 1885, the time at which the lines just quoted were written, the situation at Cherry Grove has altogether changed. Cherry Grove is situated on the *Warren Oil Fields.* Now this oil field has several productive superimposed beds; by boring, lower deposits, at least as rich as those previously wrought, were discovered. At the present day, Cherry Grove has resumed its ancient splendour; 8,000 barrels are produced there daily.

The deposits of COOPER and SHEFFIELD are not nearly so rich. Pockets, the yield of which is very limited, constitute the foundation of this district. STONEHAM and TIONA are to-day places of secondary importance, producing an oil of remarkably low gravity, almost that of a lamp oil. There are further met with in the county, the deposits of KANE and Grand Valley and the old works of TIDIOUTE on the Alleghany. Tidioute is one of the most ancient localities in which oil was found at the most shallow depths. At the present time, this deposit is almost exhausted. The petroliferous formation of Tidioute belongs to the *Venango group.*

County Warren oil is of an amber colour, and its density varies between 0·788 and 0·801.

(2) FOREST COUNTY LOCALITIES.

The principal oil fields in the district of the Middle Field in Forest county are BALTOWN and WEST HICKORY, the beds of which are superimposed on the Venango sands.

A little heavier than the Warren oils, the Forest oils have an average density of 0·800.

4. THE LOWER FIELD DISTRICT.

The Lower Field is the oldest wrought district in Pennsylvania. It comprises a part of Crawford county, and those of Venango, Clarion, Butler, Armstrong, Lawrence and Beaver. Situated to the south-west of the Middle Field district it extends in the valleys of the lower Alleghany, the Oil Creek, the French Creek, the Clarion, the Red Bank Creek, the Mahoning, and is bounded on the south by the left bank of the Ohio. The petroliferous strata of the Lower Field includes three different geological horizons.

1. *The Venango Group*, already described, in the county of Venango (in the northern part of the district); this formation is rich in gas in the southern portion of the county.

2. *The Pottsville Conglomerates* (carboniferous strata) form the south of the Pittsburg basin, and are met with in the south-west of the district, in Beaver and Lawrence counties.

3. *The Butler Sandstone* (anthracite beds) to the south of the district in Butler county. The level is not very rich in oil, but produces much gas.

(1) CRAWFORD COUNTY.

Starting from the north, following Oil Creek valley in Crawford county, we come across the deposits of CHURCH RUN and those in the neighbourhood of TITUSVILLE. The last are the *doyens* of the Pennsylvanian deposits. We have seen that Drake's well dates back from 1859; at the present time they are almost exhausted. It is to be remarked, however, that Drake's well, twice deepened, continued to yield until 1888—a very good performance indeed. The average density of the petroleum in this district is 0·803.

(2) Venango County.

Still following the course of the Oil Creek into Venango-county, the oil fields continue from the south of Titusville up to the confluence of the Oil Creek with the French Creek at FRANKLIN. The principal deposits are those of OIL CREEK, RENO OIL BELT and OIL CITY, which yield a relatively heavy oil, of an average density of 0·818.

In the Oil Creek district the wells are very close to one an-other; very often two neighbouring wells, working the same vein of petroleum, have been the occasion of much annoyance to their owners. Two anecdotes, related by W. de Fonvielle, may be quoted in support of this statement.

"The owner of a flowing well, which spouted in great abund-ance, had a neighbour who only extracted salt water from the ground, as one of his wells had a salt vein, which hampered him. This nabob, thinking he would do his neighbour a service in regard to what he was in search of, tried to divert the inconvenient jet of brine by pumping in his neighbour's well. The operation succeeded very well indeed, but far beyond the intentions of the owner of the opulent oil artery. He, in fact, produced a regular subterranean revolution, in consequence of which the sheet of petroleum changed its course. It deserted the rich man, and, by an apparently inexplicable caprice, it passed over bag and baggage to his poor neighbour, who immediately became an "oil king," whilst his old patron fell back once more into the ranks of the plebeians.

"The dethroned monarch invoked the aid of the law, but the district judge, after hearing the case, decided in favour of the *nouveau riche*. Since that decision, the 'oil kings' resigned them-selves to their fate and put up with being embarrassed with brine, whilst at the same time they take good care not to pump it up through the wells of their less fortunate neighbours.

"Two wells, sunk one after another at some distance from each other, give another example of these secret—perhaps adulterous —liaisons. When the junior well was being wrought, the senior sank to such an extent as to stop spouting, and it became necessary to use the pump to extract the oil. Instead of quarrelling, the

Pl. XI.—General View of Oil City.   Photograph communicated by MM. Feuaille and Despeaux (Vol. i., *to face p.* 214.)

two proprietors came to an agreement. It was settled that the one who was the first to occupy the ground should work his well during the six working days; whilst the other, who did not, nevertheless, get the smaller share, reserved the day when God rested after having created the world for the working of the second well. As is generally the case in Protestant countries, the Sabbath is very much respected, this well very soon became celebrated far and near under the name of the 'Sunday' well."

PITHOLE, the grandeur of which is greatly diminished, is in this district. Beginning in 1865, spouting wells followed each other with incredible rapidity; the "Twin," then the "Grant" and the "Eureka" made this oil field a place of the first rank. In a few days, where only a few huts existed previously, a town was formed which, in a few months, had attained considerable commercial importance, and a population of 20,000 inhabitants. Alas! a few months afterwards, the yield of the wells had decreased, the population migrated to more fortunate localities, and of the deserted town a Pennsylvanian proverb says: "Pithole went up like a rocket, but came down like a stick".

The other localities in the district are SHAMBURG, PLEASANTVILLE, CASHUP, WALNUT BEND and HENRY'S BEND. Near to the confluence of French Creek are SUGAR CREEK, COOPERSTOWN, MACCALMONT FARM, SUGAR AND FRENCH CREEK, UTICA and FRANKLIN.

In the latter place an oil is extracted from comparatively shallow wells of an abnormal density for Pennsylvania, viz., 0·885 to 0·898.

Two petroliferous lines of strike run from the north-west to the south-east, extending into the counties of Clarion, Butler, Beaver and Armstrong:—

(8) CLARION COUNTY, ETC.

The first, the most important, commences to the north in Clarion county, about a score of miles to the west of the town of that name. This line, very rich in oil, is mapped out or delineated by the deposits of EMLENTON, FOXBURGH, PETROLIA and PARKER.

The rather deep wells of this district, 250 to 450 metres (820 to 1,476 feet), increase in depth as we descend towards the south-west. The average density of the oil collected is 0·804.

To be mentioned amongst the principal localities of this line of strike are CRANBERRY, GAS CITY, SCRUBGAS, BULLION, etc.

(4) BUTLER COUNTY.

The second line of strike is less extensive ; it is wholly included in Butler county, and is determined by the deposits of BALDBRIDGE, the very brilliant production of which some years ago would appear to be decreasing. WILDWOOD, HARMONY, EVANS CITY, ZÉLIENOPLE, and a little to the south, GLADE RUN. Other new localities are BUTLER CROSS BELT, RUMBAUGH FARM, WHISKY RUN.

Properties of Butler Petroleum. — According to Halphen and Riche, the density of the oils of Butler region would appear to be as follows :—

Zélienople	0·772
Glade Run	0·783
Harmony	0·771
Evans City	0·790

(5) BEAVER COUNTY.

In Beaver county, keeping to Ohio State territory is the deposit of SMITH'S FERRY, which produces an amber coloured oil which may be used for illumination without previous treatment, the density of which is 0·788.

5. WASHINGTON DISTRICT.

The Washington district is the most recently developed in Pennsylvania. From 1860 to 1865, borings were undertaken there, but without success. These old works were completely abandoned. It was only in 1884 that the Citizens' Fuel Company undertook fresh researches in this country. These researches were very soon crowned with success ; in fact, in August, 1884, the "Gordon" well spouted, followed in March, 1885, by the "Pew and Emerson Manifold" and the "Thayer," which in April of the same year yielded the remarkable quantity of 2,000 barrels per day. In August, 1886, the yield of the district rose to 14,000 barrels per day. At the present time, Washington

Pl. XII.—Group of Oil Wells in Butler County. Photograph communicated by MM. Fenaille and Despeaux (Vol. i., *to face p.* 216).

district furnishes a quarter of the total production of Pennsylvania.

Except in the northern districts, in the environs of Pittsburg, where petroleum is found in the Venango sands, of the Devonian formation, the productive beds of the country are wholly situated in the coal measures, *Pottsville conglomerates* to the south-west, and *Mahoning sandstone* to the east in Green county.

The regions may be divided into three oil fields :—

1. THE ALLEGHANY POOL.
2. WASHINGTON COUNTY.
3. GREEN COUNTY.

(1) ALLEGHANY POOL.

The Alleghany Pool possesses a few works, with a poor enough yield, in the neighbourhood of Pittsburg. The number of unproductive wells is greater than in the remainder of Pennsylvania. The principal deposits are BRUSH CREEK and SHANNOPIN, both in the Venango horizon. To the north-east of Pittsburg extends the county of Westmoreland, which yields but little petroleum, but where gas wells are frequent and coal mines abound. It is this abundance of different combustibles required in industry that has made Pittsburg so prosperous, and which gave birth to the important city on the very spot where the ancient Fort Duquesne stood.

ELYSÉE RECLUS ON PITTSBURG.

"Subterranean petroleum lakes and deep accumulations of combustible gases have successively developed the outbursts of local industry. In 1812, the first manufactories were opened, and now the whole town, with its Alleghany annexe, on the other side of the river of the same name, and other suburbs, forms one immense manufactory, with numerous chimneys. It is the *Fire City*. Some assert that the nickname *Smoky City*, which was also given to it, has ceased to have any justification, since the conduits of natural gas bring this smokeless combustible to a great number of factories. Pittsburg, however, still remains (1891) a town with a black sky, with an almost unbreathable atmosphere, a 'hell' on the banks of the Ohio. In front of the confluence only two smoky avenues can be dis-

tinguished, and the bridges appear as if in a dream. When, after
having crossed either of the rivers, by one of the twenty bridges
which connect the town with the suburbs, we scale the opposite
abyss by one of the numerous cables which bring foot passengers,
carriages and loaded waggons up the abrupt slope, we no longer
see the town at our feet ; the smoke rolls in black billows above
the cupolas and the chimney stalks " (Elysée Réclus).

2. WASHINGTON AND GREEN COUNTIES.

These possess three rich oil fields. At the head of these
comes MACDONALD POOL, discovered in 1891, the production of
which at that time reached 27,000 barrels per day ; MAC CURDY,
WASHINGTON, TAILORSTOWN and SISTEVILLE POOL, 137 miles
to the south of Pittsburg.

So that the reader may realise the rapidity with which this
country has developed, it may be mentioned that in August,
1892, 36 wells had been sunk ; and in April, 1894, 1,266
derricks had sprung up as if by enchantment, constituting the
richest oil field of North America.

ELK COUNTY.

To the north-west of the Pennsylvanian oil fields extend
the, as yet, but little developed oil fields of Elk county.

GENERAL REMARKS ON PENNSYLVANIAN PETROLEUM.

Density.—We have mentioned that, compared with European
oils, those of Pennsylvania are of a much lower density (0·808
on an average).

Pipe Lines.—These oils are admirably adapted by their quite
low viscosity for transport by pipe lines. Besides, this method
of transport has multiplied in such inconceivable proportions
that a special chapter must be devoted to its description, to
which the reader is referred (see page 470 *et seq.*).

Depth at which Oil is Struck.—Petroleum is met with in the
different geological levels already indicated, at a depth varying
between 200 and 370 metres (784-1,215 feet), the greater depths
being towards the south of the oil field.

Ratio of Barren Wells to Productive Wells.—The number of
barren wells, in proportion to the productive wells, would appear

to increase. The greater number of the Pennsylvanian oil wells, if not exhausted, are at least becoming worn out. From time to time, however, new localities bring their contingent to swell the production, thus preventing a decline in the yield of this country. In virtue of this principle, the producing centres would appear to move towards the south, alongside the new deposits of Washington and the borders of the States of Ohio and Western Virginia.

The following table gives the relative proportion between productive wells and dry holes. This table confirms our remarks :—

TABLE SHOWING INCREASE OF BARREN COMPARED WITH
PRODUCTIVE WELLS.

Year.	Number of Borings Sunk.	Of which were Barren.
1880	4,203	148
1881	3,848	167
1882	3,263	178
1883	2,949	263
1884	2,193	256

PENNSYLVANIA.

TABLE SHOWING THE ANNUAL PRODUCTION OF PETROLEUM IN
PENNSYLVANIA FROM 1859 TO 1895; BOTH INCLUSIVE.

Year.	Production. Barrels of 42 gallons.	Observations.	Year.	Production. Barrels of 42 gallons.	Observations.
1859	2,000		1878	15,866,000	
1860	200,000		1879	19,827,000	
1861	2,110,000		1880	26,048,000	
1862	3,055,000		1881	27,238,000	
1863	2,610,000		1882	30,460,000	The spouting wells
1864	2,130,000		1883	23,128,000	of Cherry Grove
1865	2,721,000		1884	23,772,000	opened up in
1866	3,782,000		1885	20,776,000	Warren county
1867	3,588,000		1886	25,798,000	Opening up of the
1868	3,716,000		1887	22,356,000	Washington oil
1869	4,351,000		1888	16,484,000	field
1870	5,871,000		1889	21,487,000	
1871	5,531,000		1890	29,000,000	
1872	6,357,000		1891	34,000,000	Spouting of the
1873	9,932,000		1892	27,149,084	wells of Macdon-
1874	10,883,000		1893	19,283,122	ald oil field
1875	8,801,000		1894	18,077,559	
1876	9,915,000		1895	18,231,442	
1877	13,043,000	Development of MacKean district			

DECLINE OF PENNSYLVANIAN PRODUCTION.

As far back as 1893, the American mining annuals registered the decline of the yield of the old oil fields. According to these, the total yield of Pennsylvania in 1891, 34,000,000 barrels, was only an abnormal fact, resulting from the extraordinary yield of the new oil field of Macdonald Pool. As will be seen further on, other American States produce as much as, if not more than Pennsylvania, and that the oil reservoirs of the New World are not yet nearly exhausted.

WESTERN VIRGINIA OIL FIELDS.

Recent Rapid Development.—Within the last few years the surface extension and the yield of the oil fields of Western Virginia has been altogether extraordinary.

For a long time, however, the petroleum industry in this locality remained in its infancy, the region producing a yield altogether quite insignificant to that produced each year by Pennsylvania. It was only on the diminution of the yield of this latter oil field that suddenly, like a premeditated act, the yield of the Western Virginian oil field rose in a single year (1893) from 2,000,000 to 8,000,000 barrels.

It was in 1860, that capitalists, who had heard of Drake's fortune, re-opened a salt spring at BURNING SPRING, in Virginia. This initiative attempt was crowned with great success. The well spouted in a short time, yielding 50 barrels per day. A short time afterwards the boring of the " Llewellyn " was taken in hand, which, at a depth not exceeding 100 feet, yielded more than 1,000 barrels per day, and that during several months. Very soon, the " oil men " began to flock thither, and derricks were erected at BURNING SPRING, COW CREEK, STILLWELL, HORSE NECK, WHITE OAK.

The petroleum industry, however, languished. There was a saying that the Virginian oil wells were only very modest *pumping wells*, and that it was folly to spend time oil-prospecting in that country. So well was this propagated that gradually only a few industrials in real earnest remained. Later on, the Standard Oil Company established itself in Virginia, but then was the time when Pennsylvania produced each year a yield of nearly 30,000,000 barrels. The Standard, fearing without doubt the effects of an over-production on the American market, husbanded their resources, and only pushed their operations forward in a very prudent manner. As already said, it was not until the Pennsylvanian production had fallen to a less formidable total that suddenly Virginia burst forth in such a graudiose style.

GEOLOGY OF WESTERN VIRGINIA OIL FIELD.

The geological formation to which the oil fields of Western Virginia belong is superimposed on the Devonian. It is included in the basins of the rivers Kanawha and Hughes. This country, known under the name of the *Great Oil Belt*, consists of highly disturbed strata. Divided by ravines and narrow passes, at the bottom of which torrents flow, its geological structure is of the most irregular description. Western Virginia is divided into four oil fields.

1. TURKEY FOOT.

2. Volcano, situated in carboniferous strata ; the most productive of the Virginian oil fields.

3. MOUNT MORRIS.

4. BURNING SPRING.

The density of Western Virginian petroleum is comprised between 0·806 and 0·895. The average figure being 0·863. This density is much higher than Pennsylvanian petroleum.

The progressive increase in production has been as follows :—

TABLE SHOWING THE ANNUAL PRODUCTION OF PETROLEUM
IN WESTERN VIRGINIA FROM 1878 TO 1894 ;
BOTH INCLUSIVE.

Year.	Production.	Observations.	Year.	Production.	Observations.
	Barrels of 42 gallons.			Barrels of 42 gallons.	
1878	180,000		1887	145,000	Stagnation period
1879	180,000		1888	119,000	
1880	179,000		1889	544,000	
1881	151,000	Stagnation period	1890	492,578	
1882	128,000		1891	2,404,000	
1883	126,000		1892	3,810,086	Progressive period
1884	90,000		1893	8,445,412	
1885	91,000		1894	8,533,046	
1886	102,000				

OHIO OIL FIELDS.

Their Important Production—Resemblance of their Petroleum to that of Canada.—The Ohio deposits have become so important that at the present time they more than counterbalance those of Pennsylvania. The petroleum of Ohio—of a quality which is perhaps less esteemed than that of the last-named State—approaches, in chemical composition and physical properties, that of Canada. For a long time, this peculiarity was a drawback to the petroleum industry of Ohio. However, during the last few years, with improved methods of treating oils, the prejudice against Ohio oils has completely disappeared.[1]

GEOLOGY OF OHIO OIL FIELDS.

The oil fields of Ohio extend from the south-west of the Pennsylvanian oil fields, which they touch on the border of Beaver county. They are in direct connection with the *Great Oil Belt* of Western Virginia, from which they are only separated by the Ohio river; thus their geological structure does not differ from that of the neighbouring States. The *Silurian* formation is present in the north, extending along Lake Erie, and presenting numerous petroliferous horizons of the Canadian oil fields (Enniskillen formation) in the oil fields of Sandusky, Frémont, Findlay, Bowling Green and Corey.

Devonian strata are present, to the south-west of Pennsylvania, in the prolongation of the Middle Field; it may be observed in the central portion of Ohio, to the south-west of the Silurian formations of Neff, Wellsburg and Macksburg.

The territory of Ohio is divided into two oil fields :—

1. Macksburg (to the east).
2. Lima (to the north-west).

1. MACKSBURG OIL FIELD.

The first of these oil fields is situated along the Ohio; one half on the Pennsylvanian frontier, and the other on that of Western Virginia. The geological formation is Devonian, and the oil extracted from it differs but very little from the *Great Oil Belt* petroleum or that of Beaver. It is a little heavier than the average of Pennsylvanian oils.

[1] A description of several of these methods is given in the *Oil and Colourman's Journal* for October, 1900.

LOCALITIES OF MACKSBURG OIL FIELD.

The deposits which are met with in the districts of Macksburg are :—

DUCK CREEK (old works).

MECCA, which yields a viscous oil, which may be used for lubricating purposes in the crude state, with an average density of 0·888.

BELDEN, WASHINGTON, NOBEL, HARRISON.

More to the south, on the borders of Ohio, in the environs of Marietta, there are a few old works with a very restricted output.

2. LIMA DISTRICT.

The district of Lima (belonging geologically to the Enniskillen beds of the Silurian formation) extends from the north-west of Ohio State. It is bounded on the north by Lake Erie, and on the west by the State of Indiana. The petroleum met with there is of variable density, sometimes light (about 0·790), sometimes higher (between 0·835 and 0·850). It is dark green in colour, and gives off a disagreeable smell similar to that of the Canadian oils, which proceeds from dissolved organic sulphur-containing compounds.

The Lima deposits have only been wrought since 1875, since which time they have sent a continually increasing contingent to swell the production of the United States.

SUBDIVISIONS OF LIMA DISTRICT.

The principal subdivisions of this oil field are the rich basins of ALLEN, AUGLAIZE, HANCOCK and FINDLAY, where important sources of natural gas are found, NEW BALTIMORE, ST. MARY, GIBSONBURY, UPPER SANDUSKY and SPENCERVILLE. The average yield of productive wells in this district is 40 to 70 barrels daily. The maximum was 1,750 barrels. In the richest portion of the oil field it has been calculated that a hectare of ground would produce from 10,000 to 15,000 barrels.

Localities of less importance are : MERCER, WYANDOT, SENECA and VAN WERT.

ANNUAL PRODUCTION.

Progress.—The production of the State of Ohio is progressively increasing. The following table shows the progress :—

TABLE SHOWING THE ANNUAL PRODUCTION OF PETROLEUM
IN OHIO, FROM 1875 TO 1895.

Year.	Production.	Observations.	Year.	Production.	Observations.
	Barrels of 42 gallons.			Barrels of 42 gallons.	
1875	200,000		1886	1,782,000	First spouting wells at Lima
1876	31,786		1887	5,018,000	
1877	29,888		1888	10,010,868	
1878	38,179		1889	12,741,000	
1879	29,112		1890	15,000,000	
1880	36,940		1891	14,500,000	
1881	33,867		1892	15,169,507	
1882	39,761		1893	18,643,804	
1883	47,682		1894	16,792,154	
1884	90,081		1895	19,545,233	
1885	650,000				

Barrels of 42 Gallons.

In 1895 The yield of Ohio was . . . 19,545,233.
Superior to that of Pennsylvania by 1,313,791.
That of Pennsylvania being only . 18,231,442.

INDIANA.

The oil field of Indiana is also of quite recent development. Some years ago it did not rank on the market ; at the present day it holds, however, a very honourable position thereon. The two oil fields of Indiana are :—

1. MONTPELLIER, especially rich in gas.

2. TERRE HAUTE, which, in 1889, had a daily production of 1,000 barrels, then fell to 75 barrels, to mount up again very soon afterwards.

Here are some figures which will show the rapid growth of the petroleum industry in Indiana. The figures are quite eloquent enough to speak for themselves :—

Barrels of 42 Gallons.

In 1890 32,700
,, 1891 124,900
,, 1892 698,068
,, 1893 2,385,298
,, 1894 3,688,666
,, 1895 4,386,132

Nowhere has progression been more rapid ; in six years it has increased more than one hundred-fold.

KENTUCKY.

The State of Kentucky is one of the oldest wrought oil fields of the United States; the inhabitants even contend with pride that it is the most ancient oil field in the Union; and, in support of their contention, they quote the "American Oil Well," mention of which was made in *Niles's Register* so far back as 1829. The American oil well is situated in Cumberland county. It would appear to have flowed as far back as the beginning of the century, and covered the surface of the Cumberland river, which was also set on fire. However, to be just, it must be added that in spite of the report in *Niles's Register,* and in spite of the affirmations of the inhabitants of Kentucky, the "American" was never bored in seeking for petroleum. Its owners only dug it in the hope of finding brine. The "American" was therefore a common salt spring.[1]

It was in 1862 that the search for petroleum began in Kentucky.

LOCALITIES.

The localities where borings have been sunk are numerous. The following may be quoted: PULASKI, WAYNE RUSSEL, CLINTON, CUMBERLAND, RIG SANDY, BURKESVILLE, which has the fame of possessing the "American Oil Well". Up to now the production of Kentucky has been but a very minimum one; it oscillates between 10,000 and 15,000 barrels annually.

TENNESSEE.

Tennessee possesses a few very deep wells in Dickson district. The yield, whether in oil or gas, would not appear to be very remunerative.

WISCONSIN.

"But little petroleum and much gas." Such might be the summary of the production of OAK CREEK and TOWNSHIPPE. The annual production of Wisconsin does not exceed 35,000 barrels.

MINNESOTA.

Some not very encouraging researches have been attempted in the environs of FREEBORN. The yield is insignificant.

[1] No matter what these people were in search of, they appear to have been the first to strike oil in quantity. That they were in search of brine and found oil does not alter the fact.—*Tr.*

DAKOTA.

As in Minnesota, the results obtained in Dakota have been
rather negative, and that in spite of the presence of signs of
petroleum and several gas wells.

ILLINOIS.

A petroleum of an abnormal density in the United States,
0·924, is likewise found in very small quantity (1,500 to 2,000
barrels) in Illinois.

ALABAMA AND LOUISIANA.

Alabama possesses surface traces of petroleum, which we
believe have not as yet been the object of any operations. In
Louisiana, a well 175 feet deep, yielded a very viscous oil, but
in very minute quantity.

WYOMING.

Traces of petroleum are abundant, and manifest themselves,
in a violent manner, in the valleys of the territory of Wyoming.
The production may as yet be neglected, but the state of the
ground leaves the prospector with the hopes of a rich oil field.

UTAH.

The territory of Utah produces very little petroleum, but in
the neighbourhood of Salt Lake City, a few years ago, an important
deposit of mineral wax was discovered. This discovery was not
made without causing a discussion in the scientific world. Whilst
Prof. Newburry regarded this American wax as true ozokerit,
similar to that of Boryslaw, Prof. H. Wurtz declared that it was
Zietriskisite, which is distinguished from ozokerit by being insoluble
in ether.

The question rests there.

COLORADO.

It was only in 1889 that the petroleum deposits of Colorado
began to be wrought.   Since 1866, however, trial borings had been
attempted, but with little success, in the valley of the Arkansas,
to the south-west of Denver (Frémont county).   These old works,
at the present time in full production, are conducted by numerous companies, the principal of which are the United Florence
Rocky Mountain and the Triumph.   The deposit is situated in

a synclinal basin, and from above downwards the following beds may be observed:—

1. CRETACEOUS (a) Colorado group (*Dakota, Benton* and *Niobara horizons*); (b) Montana group (*Pierre* and *Fox Hill horizons*); (c) Lanmie group.

2. JURASSIC.

3. TRIAS.

4. CARBONIFEROUS.

5. SILURIAN.

The Pierre horizon is the most productive of the petroliferous beds of the cretaceous system. The oil-bearing bed consists of greyish clays, having sometimes a tendency to become sandy, with lenticular limestones. A little oil, no doubt of infiltration origin, is found at the bottom of the jurassic. The petroleum extracted has a great resemblance to that of Galicia. The average density is 0·874.

The production of Colorado was:—

	Barrels of 42 Gallons.
In 1889	316,000
,, 1891	326,000
,, 1893	665,000
,, 1895	831,482

MISSOURI, KANSAS AND TEXAS.

MISSOURI.—The production of Missouri is almost *nil.* Only some signs of petroleum necessitate the mention of this country. Annual production about 20 barrels.

TEXAS.—The same remark applies to Texas.[1] This latter State, however, since 1898, has been the object of researches which require rather many borings.

KANSAS.—In Kansas during the last two years rather productive deposits have been discovered. The production of this State may be estimated at 100,000 barrels per annum.

CALIFORNIA.

The first Europeans who established themselves in California were not slow to remark the frequent indices of the presence of petroleum in a great number of localities situated,

---

[1] Several prodigious "gushers" have recently (Feb.-Mar., 1901) been discovered about four hundred miles from Dallas at a place called Beaumont.—*Tr.*

some in the interior, others along the sea coast. They recognised even that the sources of asphaltum must be continued underneath the bed of the ocean, because this substance, rising up out of the deep, floated on the surface of the waters. Moreover, asphaltum constituted an important factor in the domestic economy of the first inhabitants of the country. In many localities on the sea coast, formerly inhabited, asphaltum has been found. The natives employed it as a substratum for a kind of ornamental mosaic in beads; they coated baskets with it so as to render them as impermeable as pottery, and made use of it for a thousand other purposes.

The natives of the neighbouring islands got their store of asphaltum from the sea. At the present day, the rocks are still coated in many places with asphaltum, which has condensed there. This is particularly visible after an easterly wind, which indicates that there is, at the bottom of the Santa Catalina channel, a vast superficies from which asphaltum is disengaged.

LOCALITIES.

At REDONDO (Los Angeles county) asphaltum exudes through the sand, and is deposited there. Between Saint Monica and Los Angeles there are undoubted deposits, and several other considerable ones to the north of SANTA BARBARA.

SANTA BARBARA.

The Santa Barbara deposits extend almost under the sea, and the scaffolding of the works which have been erected there, looking like so many windmills without sails, have been multiplied in a very short time outside the narrow gorge where the liquid had been primitively found, and have invaded the river banks in the direction of Santa Barbara. At first they rose along the slopes of the hills, which in this locality border on the ocean, but gradually they have been advanced towards the sea until one of the prospectors, bolder than the others, has installed his scaffolding in the midst of the waves.

SUBMARINE OIL WELLS.

The work of establishing this latter well has been undertaken at the extreme limit laid bare at low water, so that at high water, the scaffolding appears twenty or thirty feet from the shore, as if it were floating on the sea.

Certain wells, even at low water, have the base of their outside works in the water, and at high water are completely submerged. The workmen have platforms within their reach, placed at different heights, and on which they mount when they require to work when the tide flows. The scaffoldings, erected in the sea, do not appear as yet to have been put to the test of a south-west wind, and certain people are afraid of their being carried away by a high tide.

CONVEYANCE OF OIL TO THE SHORE.

An engine works the pumps which brings the oil to the shore; the fuel used is the oil itself. A single engine is sufficient for several wells.

At the present time the wells at the greatest distance from the shore lie about six feet under the sea at high water, and the rumour is current that other borings farther out will be made without delay.

It is perhaps the only locality in the world where petroleum is wrought under the waters of the sea.

SUBMARINE SPRINGS.

There is no doubt whatever but that the whole of the neighbouring shore rests on the petroleum-bearing bed. Half a mile from the locality known as More's Wharf, the oil rises to the surface in several places. In the same locality, a jet of fresh water rises with such velocity that when pumped up it has only a very slight taste of salt. A similar marine spring is known on the coast of Florida, where ships can get a supply of fresh water in the middle of the ocean.

LOS ANGELES.

Los Angeles commenced to be developed in 1889, and its extension has been rapid. In 1894, 100 wells had been bored there; at the present time they number about 400. The oil is of very variable density, but rather high, varying between 0·832 and 0·924. It has a disagreeable odour, with a dark brown opaque colour, and contains in suspension a large proportion of bituminous sediment.

One of the most extraordinary spectacles arising out of the working of petroleum is exemplified in Los Angeles. The petroleum was discovered in the western part of this town, which was the favourite residence of the highest society, and which was transformed, as if by magic, into a forest of scaffolding similar to that of the oil fields of Pennsylvania.

The wells have since advanced in a north-easterly direction, and would appear to be arrested, at the present time, by the great Catholic cemetery, which rests on the petroliferous strata. Not far off, is the river Los Angeles, which will in all probability be dammed up and diverted, so that the riches covered by its bed may be realised.

The other deposits of California are SANTA PAULA, OJAI, SESPÉ, TORREY CANNON, PUENTE, SANTA CLARA and VENTURA.

The production of California was :—

	Barrels of 42 Gallons.
In 1889	303,290
„ 1891	350,000
„ 1893	452,000
„ 1894	705,929
„ 1895	1,208,482
„ 1896	3,010,124

TOTAL PETROLEUM PRODUCTION OF THE UNITED STATES.

In order to sum up this review of the United States deposits, we place before the eyes of the reader the following table of their production, by means of which he will be able to realise with a single glance of the eye the condition of the petroleum industry in the different States of the Union.

TABLE SHOWING THE TOTAL ANNUAL PRODUCTION OF PETROLEUM IN THE UNITED STATES AND THE DIFFERENT STATES OF THE UNION FROM 1859 TO 1895 ; BOTH INCLUSIVE.

Year.	Pennsylvania and New York.	Western Virginia.	Ohio.	Indiana.	Colorado.	Kentucky, Tennessee and other States.	California.	Total American Production.
	Barrels.	Barrels.	Barrels.	Barrels.	Barrels.	Barrels.	Barrels.	Barrels.
1859	2,000	—	—	—	—	—	—	—
1860	200,000	—	—	—	—	—	—	500,000
1861	2,110,010	—	—	—	—	—	—	2,113,000
1862	3,055,000	—	—	—	—	—	—	3,056,000
1863	2,619,000	—	—	—	—	—	—	2,611,000
1864	2,180,000	—	—	—	—	—	—	2,160,000
1865	2,721,000	—	—	—	—	—	—	2,797,000
1866	3,732,000	—	—	—	—	—	—	3,791,000
1867	3,788,000	—	—	—	—	—	—	3,847,000
1868	3,716,000	—	—	—	—	—	—	3,833,000
1869	4,351,000	—	—	—	—	—	—	4,420,000
1870	5,371,000	—	—	—	—	—	—	5,673,000
1871	5,531,000	—	—	—	—	—	—	5,715,000
1872	6,357,000	—	—	—	—	—	—	6,531,000
1873	9,932,000	—	—	—	—	—	—	9,978,000
1874	10,883,000	—	—	—	—	—	—	10,950,780
1875	8,787,514	3,000,000	200,000	—	—	—	175,000	12,162,514
1876	8,968,906	120,000	31,736	—	—	—	12,000	9,132,669
1877	13,135,475	172,000	29,888	—	—	—	13,000	13,350,363
1878	15,163,462	180,000	38,179	—	—	—	15,227	15,396,868
1879	19,685,176	180,000	29,112	—	—	—	19,858	19,914,146
1880	26,027,631	179,000	38,940	—	—	—	40,552	26,286,123
1881	27,376,509	151,000	33,867	—	—	—	99,862	27,661,238
1882	30,053,500	128,000	39,761	—	—	—	128,636	30,349,897
1883	23,128,389	126,000	47,632	—	—	—	142,857	23,444,878
1884	23,772,209	90,000	90,081	—	—	—	262,000	24,214,290
1885	20,776,041	91,000	650,000	—	—	—	325,000	21,842,041
1886	25,798,006	102,000	1,782,970	—	—	225,000	377,145	28,285,115
1887	22,350,198	145,000	5,018,015	—	—	51,817	678,572	28,249,597
1888	16,484,668	119,448	10,010,868	—	—	310,612	690,333	27,615,929
1889	21,487,000	544,000	12,741,000	—	316,000	116,476	303,290	35,279,989
1890	29,000,000	1,717,300	15,000,000	32,700	300,000	100,000	350,000	46,500,000
1891	34,000,000	2,404,000	14,500,000	124,900	325,000	86,080	452,000	51,891,980
1892	27,149,034	3,810,086	15,169,507	678,068	500,000	1,182,489	2,000,000	50,509,136
1893	19,283,122	8,445,412	16,643,804	2,335,293	665,000	2,140,085	2,000,000	48,412,666
1894	18,077,559	9,000,000	16,792,154	3,688,666	550,000	530,037	705,000	49,344,516
1895	18,231,442	8,000,955	19,545,233	4,386,182	831,482	780,000	1,208,482	52,983,526

CHAPTER XVI.

CANADIAN AND OTHER NORTH AMERICAN OIL FIELDS—HISTORY, GEOGRAPHY AND GEOLOGY.

EXTENT.

THE Canadian deposits form the summit of those of Pennsylvania and Ohio. Belonging to the lower beds of the same formation, they extend from the north-west to the south-east from the Gaspé peninsula in the province of Quebec to the shores of Lake Superior in the province of Ontario.

Owing to their immediate proximity, the history of the Canadian oil fields is intimately linked with those of Pennsylvania.

HISTORY.

The first European colonist who, in the middle of this century, interested himself in Canadian petroleum, was a farmer of the Black Creek region—John Rows—who was struck with the analogy which the substance used by the Indian magicians to cure pains presented to the petroleum, which had begun to be used in the United States. Rows communicated his observations to business men of the country, and shortly afterwards the first wells were bored at Enniskillen to the south-west of Toronto.

The first spouting well was discovered in the month of August, 1861. The well had a depth of 67 metres (say 230 feet); its yield varied in the space of fifteen months from 2,904 hectolitres (63,888 English gallons) to 2,178 hectolitres (47,856 English gallons). It belonged to a Mr. Shaw.

SPOUTING WELLS.

The Famous " Shaw " Well.—" The comparatively recent, but already legendary, story of the famous ' Shaw ' well in the Enniskillen district, is still told by the *raconteurs*," says M. Albert

Duhaigne in his little book on petroleum (Paris, Victor Palme, éditeur, 1872).

" John Shaw, blessed with small capital and a strong will, had purchased a small lot in the better quarter of the concession. He wrought from morning to night, boring and pumping without ceasing; months were thus spent and his capital also. His neighbours' wells overflowed, but his constantly deepened without reaching the desired vein. By-and-by he had spent his last farthing. He then lived upon his credit, dieting himself stintly, but always boring.

" One day in the month of January, 1862, he bored, he pumped and paddled about in the stinking, slippery mud, thanks, amongst other things, to boots resewn and restuck by cobbler's wax every day; he went about his work with the more grim determination, because he had had a difficulty in the morning of finding bread on credit. All of a sudden, after a special exertion, he felt a sharp feeling of cold; the two soles had come quite off both his boots— the misfortune could not be remedied.

" There was only one hope left to him. He had a neighbour— a shoemaker—who perhaps knew and esteemed his activity in working, his sobriety, his economy. He enters the shoemaker's shop humbly, and proffers his request. . . . But in America, like everywhere else, a ruined man, with empty pockets, covered with clods of mud, gets no credit in rich shops. John Shaw only met with a disdainful refusal.

" Desperate, he returned barefoot to his well, protesting that, if he did not strike oil before night, he would leave that inhospitable country that very evening. He seized his tools and struck his blows with redoubled force.

" All at once the rock gives way; a gurgling sound is heard. The oil comes; it fills the pipe, then the well, which, in its turn, begins to overflow. John Shaw, in readily understood emotion, promptly adds a beck of boards, but the oil still overflows. Quick! casks, cases, vats, anything that can be got! But everything is full; the overflow continues; it is a river, what do I say? a torrent, which runs across the valley and precipitating itself into the river speeds towards the lakes.

" The fame of the extraordinary occurrence has, however,

become known. The neighbours fly to the spot. The fortunate possessor of the marvellous well is overwhelmed with congratulations. He is no longer 'old' Shaw, as he was in the morning. He is 'Mr.' Shaw as large as life (*gros comme le bras*). But look! there is some one running; the crowd makes way for him. But why? It is the shoemaker, who comes to humbly salute our 'dear Mr. Shaw,' and to offer him the best pair of boots in his shop. John Shaw could well have paid him back in his own coin with the disdainful refusal he got in the morning, but . . . . he was barefoot, and the oil continued to flow; and, like a true Yankee, he put on the boots without a word, and set to his task again.

"The well produced 2 barrels of 180 litres in a minute and a half—say at the lowest calculation of 2 centimes the litre, 5 francs per minute, 300 francs per hour, 7,000 francs per day, 2,000,000 francs per annum.

"There is no example, neither in legendary lore nor in fairy tales, of a more extraordinary nor of a more sudden change of fortune than that of this man; at the lowest depths of poverty and misery in the morning, and a millionaire thirty times over in the evening.

"But he was not destined to enjoy this changed life brought about so suddenly for any length of time. One day in April, 1863, fifteen months afterwards, the rich owner of the great spouting well, always active and economical, directed operations himself, when the end of a pipe fell into the widened opening of the well which served as an oil reservoir. Shaw, who happened to be nearest the side, seized the chain ending in a stirrup which was used to lower the buckets in the oil, and made the signal to be lowered. Coming suddenly, at a depth of 5 metres (16·4 feet), into the midst of a deadly atmosphere, he reached the pipe, and gave the signal to be raised up again. But very soon he was seen to gesticulate with great efforts; respiration failed, his eyes closed, he dropped the chain, fell and disappeared in the oil.

"His corpse was found by emptying the reservoir. . . ."

The Seed-Bag : Its Discovery.—The second spouting well (14th April, 1863, the day it is said on which Shaw died) was the great well of Messrs. J. Piero & A. Grovier. The well had a depth of 90 metres. The spouting was sudden and impetuous.

The owners, in search of an expedient for capping the well, invented the seed-bag (a bag of linseed). The following is the account of M. Gauldrée-Boileau, ex-consul of France in Canada, who was present at the spouting and the discovery of this method of capping the well:—

"When the oil spouted for the first time from this well, a liquid column rose to a height of a score of feet above the orifice, and it flowed at the rate of at least 5,000 barrels per day. To stop the overflow of the well, the hole was plugged by a bag filled with linseed, through which there passed a pipe of smaller diameter than the mouth of the hole. As linseed swells very much in contact with oil, it forms a hermetic seal. By closing the second pipe with a new sack, in which a third pipe is introduced smaller than the second, and by repeating this operation several times, they were at length enabled to convey the petroleum into a pipe of but an inch in diameter, and the flow is thus controlled by means of a simple tap. It is a method which can easily be put into practice, which is perfectly successful, and which is now employed in all the oil fields. Before it was thought of, enormous quantities of oil were lost for want of the power of controlling the fluid masses which were shot out by the wells."

Since that time numerous spouting wells have contributed their quota to the Canadian production of this region. Here is the list of the principal, with the figures of their daily yield:—

YIELD OF CHIEF CANADIAN SPOUTING WELLS.

Well.	Yield.	Well.	Yield.
	Barrels of 42 gallons.		Barrels of 42 gallons.
Solis	600	Sambon and Shannon	2,000
Purdy	1,000	Campbell and Forsyth	1,000
Evoy Brothers	600	Wilkes	2,000
Jenery and Evoy	800	Bradley.	3,000
Fairbanks	500	Webster and Shepley	6,000
Campbell	200	Liannworth	500
Bennetts Brothers	500	Curler	200
Chandler	100	Allen	2,000
Jenery and Evoy	2,000	Barnes	800
Sifton, Gordon & Bennett	150	Pettit	3,000
J. W. Sifton	800	George Graff	150
Wanless	200	Hohnes	500
MacLane	3,000	MacColl	1,200
Ball	250	Sivan	6,000
Rumsey	250	Black and Mathewson	7,500
Whipple	400	Fiero	6,000

GASPÉ OIL FIELD.

In the neighbourhood of Gaspé the entries in the petroleum register are not so brilliant. The intermittent working of this district has from 1859 given but rather poor results.

This latter deposit being nevertheless geographically the head of the Canadian basin, we shall commence with it the study of the geology of that basin.

Gaspé forms, on the southern side of the embouchure of the river St. Lawrence, a tongue of mountainous land, which, packed between the river, the State of Maine and New Brunswick terminates in a peninsula towards the east.

It is on the extremity of this peninsula, in the neighbourhood of the Bay of Gaspe, that the most notable petroliferous manifestations occur. We there find the formation termed the Gaspian, consisting of upper Silurian limestones and Devonian sandstones. These decidedly petroliferous beds dip towards the south, to be covered by lower carboniferous strata. The same general arrangement and direction of the strata, as in Pennsylvania, also hold good here.

In the Gaspian formation, petroleum has been demonstrated to exist, especially in the vicinity of the anticlines, where the Silurian limestones, similar to those of the Enniskillen formation, crop out.[1]

Petroleum is met with, in the Gaspé district, in the following localities :—

SANDY BEACH, where two rather shallow wells were excavated about twenty-five years ago. One of the two yielded, for a certain time, about 20 barrels daily. In 1886, a United States company, The International Oil Company, was established at Sandy Beach. Several borings were made, without any great result.

Another English company, The Petroleum Oil Trust, in 1889, likewise undertook borings. It managed to strike oil, but up to now in very restricted quantity. It was only in the month of February, 1898, that the news reached us that this company had at last found a spring with a great output.

HALDIMAND, numerous exudations and signs of petroleum.

[1] According to several oil experts, the Gaspé sand and gravel are not suited to oil—the dip of the rock being altogether wrong, instead of being 50 feet to the mile it is about 1,000 in most places where wells have been sunk.—*U.S. Consul Report.—Tr.*

DOUGLASTOWN, where there is an old surface boring, 30 metres (98·4 feet) deep.

BIG FORK, on the river St. John.

M. J. Obalski, mining engineer, cites moreover traces of petroleum at CAP AU GOUDRON (Tar Cape) in a dyke of eruptive trap, which would appear to traverse this region in an east to west direction, and would seem to have given birth, with other similar eruptions, to the anticlinal folds of this part of Gaspésia.

As Gaspé is approached, petroliferous signs are numerous in the neighbourhood of the York river, amongst which may be mentioned the SILVER RIVER, the LITTLE FORK, the BIG FORK, the BARREL. The International Oil Company have established works on the Silver stream.

"The district covered by the Gaspé sandstone," says M. J. Obalski, "extends around the Gaspé oil field, especially in the valleys of the York, St. John and Dartmouth rivers. It extends also into the interior of the peninsula, covering large tracts of unexplored land, and stretching even to the Inter-Colonial Railway line towards the river Casupscul.' This sandstone is supposed to be of considerable thickness, which may, in certain cases, entail very deep boring before striking the limestone. The experts, who have visited these districts, have found them present certain analogies with the petroliferous regions of the United States ; but they are new, but little explored, and but few borings have yet been made. It is therefore to be hoped that the rich and workable tracts which certainly exist in this district will be got at by further operations." [1]

Properties of Gaspé Oil.—The petroleum got in the Gaspé district is, as far as surface oils are concerned, of a dark colour and of a but comparatively slightly fluorescent nature. At a certain depth, it is a pale green, becoming brown by transmitted light. Its percentage of sulphur compounds is not considerable ; it may be taken as between 0·09 and 0·20 per cent.

---

[1] From 1889 to 1900 the Petroleum Oil Trust dug 33 wells from 2,500 to 3,800 feet deep; oil was found in all but one, but it seems to be soon exhausted. The Canadian Petroleum Company, who have taken over the Petroleum Oil Trust, has 4 wells under way and intends drilling 10 or 15 more forthwith. It has laid a 12-mile pipe line and is building refineries.—*Tr.*

Boverton Redwood has given the following densities :—

Big Fork	0·871
Tar Cape, near to the trap dyke	0·939
Near to Patawagia brook	0·949
Silver stream	0·894
Gall's brook	0·921
Well No. 1 at 2,057 feet	0·853
,, 2 ,, 906 ,, ...	0·877
,, 5 ,, 1,946 ,, ...	0·861
,, 5 ,, 2,361 ,, ...	0·847
,, 5 ,, 2,661 ,, ...	0·847

The southern part of the valley of the St. Lawrence possesses numerous gas wells. Petroleum, however, has not been found in workable quantity. Traces, however, and even small pockets of oil have been met with at POINTE AUX TREMBLES, at RIVIÈRE A LA ROSE ; at L'ILE DE LA TRAVERSE, situated on Lake St. John, the inflamed oil has burnt even for several weeks. The geological formation of this country is Silurian ; the petroleum and gas would appear to be derived from immense beds of limestone which crop out near Quebec.

Speaking of this formation, Sir W. Logan [1] said in 1853 : " We must not lose sight of the possibility of meeting with petroleum in profitable quantity in some points of the Trenton formation ". M. J. Obalski, who made a special study of the mineral deposits of this region, expresses his opinion as follows : " We can assert that the formation included between Quebec and Montreal, cropping out over several miles of the north coast and extending over 50 to 60 miles on the south, is apt to contain gas and petroleum " (*Mines et Minéraux de la province de Québec*, 1889-90).

ONTARIO. GEOLOGY.

The rich oil fields of the province of Ontario, in themselves alone, furnish almost the whole of the Canadian production. The deposits are found grouped together in the county of Lambton, upon that kind of a peninsula which is situated between the Lakes Ontario, Erie and Huron ; their centre would appear to be the town of Enniskillen. The strata belong to the Silurian formation, and consist of a bed of limestone of the greatest importance, which

---

[1] Director-General of Geological Survey of Canada, discovered the fossil *Eozoon Canadense*, the nature of which, whether plant or animal, still remains disputed. —*Tr*.

crops out on this shore of the great lakes, dips towards the south to lose itself in Pennsylvania below the *Chemung* Devonian beds. Wrought to the north of the great lakes (Canadian oil fields), and to the south (Bradford district), and the northern part of Ohio, this bed is most easily got at in Lambton county, where, in the greater number of the spouting wells mentioned on page 235, oil has been struck at a depth of less than 100 metres (328 feet). The most extensively wrought level is, on account of this easy access, situated between Enniskillen and Dereham ; it extends generally, at no considerable depth, in a long anticline of corniferous limestone surmounted by beds of clay shale.

One of the peculiarities of the petroliferous strata of Canada is the profusion in these beds, and even in the oil, of an organic debris of a marine nature (*crustaceae, molluscs, fucoïds, algae* and traces of a powerful *marine vegetation*). These facts speak for themselves, and are, perhaps, of such a nature as to throw some light on the vexed question of the origin of petroleum, which has been the subject of so much controversy. According to Ste. Claire Deville, the density of the petroleum of this region varies between 0·828 and 0·857.

The principal deposits on the shores of the great lakes are :—

ENNISKILLEN.

OIL SPRING, whose fortune commenced in 1862 with the "John Shaw" oil well.

PETROLIA, whose prosperity dates from the spouting of the famous "King" well in 1867.

COLLINGWOOD.

BOSANQUET, which is endowed with a rather important petroliferous deposit in a bed of Devonian strata.

SHERKSTOWN.

Of less importance are TILSONBURG, DEREHAM, and further on towards the west, SAINTE MARIE and CAPE SMITH, in the island of Manitouline on Lake Huron.

WESTERN CANADA.

In the west of Canada, petroliferous districts have recently been discovered. Two basins, differentiated very distinctly by altogether different geological formations, extend from each side of the rocky mountains. The first, on the eastern watershed, is situated in the territory of Alberta; the second, on the western watershed, is situated entirely in British Columbia.

The Alberta basin lies on cretaceous strata; it consists of outcrops of sandstone saturated with oil. Gaseous manifestations are frequent in this locality. This petroliferous district, but little wrought hitherto, extends close to the rivers Peace and Athabasca.

BRITISH COLUMBIA.

The British Columbian oil field is situated near to Kootenay Pass. The most notable traces of petroleum have been found near to a brook, a tributary of the Flathead river. The petroleum, which exudes from the soil, is of a bright yellow colour; it comes from the outcrop of a silicious sandstone, rich in magnesian limestone, and which would appear to belong to the upper Cambrian formation.

TOTAL CANADIAN PRODUCTION.

The petroleum production of Canada, without approaching the enormous figures of that of the United States, is, however, very respectable. The annual production has been as follows :—

TABLE SHOWING THE TOTAL PRODUCTION OF PETROLEUM IN CANADA FROM 1875 TO 1894; BOTH INCLUSIVE.

Year.	Production.	Year.	Production.
	Barrels of 42 gallons.		Barrels of 42 gallons.
1875	220,000	1885	250,000
1876	312,000	1886	250,000
1877	312,000	1887	424,000
1878	312,000	1888	597,000
1879	575,000	1889	591,000
1880	350,000	1890	633,000
1881	275,000	1891	610,000
1882	275,000	1892	648,000
1883	250,000	1893	664,100
1884	250,000	1894	686,900

NEWFOUNDLAND.

Traces of petroleum have been found on the coast of Newfoundland.

Nova Scotia.

A few shallow wells have been dug, without any success, in Nova Scotia, and that notwithstanding surface indications which should warrant the most favourable anticipations.

Alaska Territory.

The following note is extracted from the *Moniteur des Pétroles* of the 28th February, 1898 :—

" There has recently been discovered in Alaska (North America) a lake of almost pure petroleum, five or six miles long by three or four miles wide, and of unknown depth. This lake, barely two miles from the sea, is surrounded by hills whose flanks conceal coal and asphaltum in abundance. Samples of this petroleum have been brought to Seattle, where a company was quickly formed to work this lake, which would appear to be inexhaustible."

The authors have not been able to confirm the absolute correctness of this statement, which was likewise published by the American journals.

CHAPTER XVII.

ECONOMIC DATA OF WORK IN NORTH AMERICA.

UNITED STATES.

Wages.

IN the United States, and especially in the rich oil fields of Pennsylvania and Ohio, the cost of manual labour has a higher tendency than in the European oil fields.

A foreman borer receives per month 50 to 150 dollars.
A mechanic   ,,   ,,   50 to 90   ,,
A labourer   ,,   ,,   30 to 70   ,,

The *Rig Builder*, who is the contractor to whom the erection of the derrick and the machines is generally entrusted, receives 50 to 225 dollars per boring installed ready to start work. The carpenters are paid by the day; they are most generally casual hands, who receive, during the time they are employed, from 1·25 to 3 dollars per day.

If these wages be compared with those given for manual labour at Baku, we have the following data for the staff of an ordinary boring per month:—

BORING IN RUSSIA AND AMERICA—COMPARATIVE MONTHLY
WAGES BILL.

	North America.	Russia.
1 Mechanic	70 dollars	50 roubles
1 Stoker	50 ,,	25 ,,
1 Foreman Borer	100 ,,	100 ,,
2 Assistant Borers	100 ,,	50 ,,
4 Labourers	200 ,,	60 ,,
1 Smith	70 ,,	25 ,,
	590 dollars	310 roubles
Say in francs	806 fr. (£32 5s.) in Russia and 2,950 fr. (£118) in the United States	

Comparative Cost of Drilling in America, Baku and Galicia.

There is a great difference between these prices, but it may be stated that the costs of working come pretty close to each other by the fact that the higher wages paid for manual labour is compensated in America by the higher price got for American crude oil, and also, in most cases, by the ground being more favourable for boring, and by the progress, whilst not involving such complicated plant—as in the Baku oil field —being almost always much more rapid.

The cost of plant in the United States is much the same as in Europe.

In addition to the windlasses and boring system, 500 dollars is generally allowed for a boiler, 200 dollars for the engine, 500 dollars for the derrick and buildings, and 100 dollars for sundry repairs during the course of boring.

Summary.—The initial charges amount to 3,220 dollars (say 16,000 francs or £640) against 20,050·2 francs or £802 in Europe.

Almost the whole of the boring processes are done through the intermediary of a contractor. The latter, to whom the plant is supplied (with the exception of the windlasses and the boring system), as well as the manual labour, is paid at the rate of 1·75 to 2 dollars per foot.

The manual labour amounts, on an average, to 15 dollars per 24 hours.

Generally a well of 300 metres (about 984 feet) is finished in 2 or 2½ months. It costs about 2,400 dollars for the contractor, and about 1,100 dollars for the manual labour, say a total of 3,500 dollars (17,500 francs, say £700 or thereabout).

We have seen that at Baku the boring of a well to the same depth by a contractor costs 28,350 francs, and that in Galicia it may be estimated at 29,125·20 francs. It will be observed, from this example, that in spite of the high price of labour the contract work in the United States is carried on in a much more practicable manner and in a way more profitable to the owner of the undertaking.

Depth of Wells.

We have here taken the depth of 300 metres as a point of comparison, but we must hasten to state that if this depth be not often reached in America, yet it is often also exceeded there.[1]

In the Northern oil fields (Canada-Franklin, Mecca Belden) the wells are very frequently of a restricted depth and above this level. On the other hand, in the south-west of Pennsylvania and Western Virginia depths are often reached of from 400 to 500 metres (1,312 to 1,640 feet).

Yield and Duration of Flow.

The yield and duration of the wells is, naturally, not uniform. The average yield in Pennsylvania may be taken as 15 hectolitres (say 330 gallons) per well per day, or 1,200 kilogrammes per 24 hours. The average duration of the yield would appear to be 2½ years.

Minimum Remunerative Yield.

For a well to be remunerative it should produce more than 7 hectolitres (154 English gallons) per 24 hours. This yield is considered as covering normal expenses.

Ratio of Barren to Productive Wells.

Moreover, the proportion. of barren wells to productive wells is most variable. In Pennsylvania it would appear to have increased, and, in our opinion, that is not one of the least causes of the lowering of the production figures of that region.

> In 1880, of 4,203 wells bored, 143 were barren.
> In 1881, „ 3,848   „   167   „

Ten years later on, in a slow but steady manner, this proportion had notably increased. In fact,

> In 1890, of 7,306 wells bored, 1,069 were barren.
> In 1891, „ 4,562   „   664   „

Land Purchase—United States.

With regard to the purchase of lands and their development, the States of the American Union are very liberal.

---

[1] The two last clauses are evidently contradictory in terms. What the authors mean is apparently that although the general run of wells in America are below 300 metres, yet there are not a few wells of greater depth in that continent.—*Tr.*

These operations do not occasion any administrative formality. In the principal oil fields of the United States, the rights to make trial borings are conceded, by the proprietor of the ground, for a period of 30 years, to the " oil man " who proposes to work and develop it. In the general run of contracts, the proprietor, in addition to a fixed sum, which may vary from 250 to 15,000 dollars, according to the district, retains for himself a *land interest* (say from 10 to 15 per cent.) on the total production. The land interest is generally 8 per cent. In almost all agreements the settler is bound to commence work within 2 months ; sometimes even he binds himself to finish within a specified time.

In the State of *Pennsylvania*, the laws provide a penalty for the owners who leave an unwrought well open. These wells, for fear that the surface water should reach the petroliferous stratum and influence its yield, should be carefully filled up with sand.

CANADA.

Initial Cost.

In Canada manual labour, as well as plant and land, are on a much more moderate scale than in the United States.

Cost of Rig.

The wells there are likewise of less depth, and the plant less complicated. The total cost of a Canadian rig, for a well of 300 metres (984 feet), amounts to 3,000 dollars in the Enniskillen district, say 15,000 francs, or £600 sterling.

The cost of a boring of 300 metres (984 feet) rarely exceeds 2,500 dollars, say 12,500 francs, or £500 sterling.

Mining Legislation.

The mining legislation of Canada is rather complicated. The rights of making trial borings and working petroleum may be conceded to individuals on paying a tax of 5 dollars per acre. The ground landlord has the preferential right over every would-be buyer of purchasing the mine discovered on his land. The mine may constitute a property quite distinct from that of the surface ground.

With the owners of the ground the contracts are almost analogous with those current in the United States.

AMERICAN EXPORT TRADE.

Crude Oil.

Whilst in Russia but little crude oil is exported, in America a good portion of the production is exported, without having been subjected to any preliminary treatment. This method of despatch opens many channels for American products, especially in countries which, like France, have established a tax on the refined products and allows the crude to enter free of duty. In this way France has become the best customer of the United States.

In 1894, the American exportation was as follows :—

	Gallons.
France	84,434,953
Germany	4,877,593
Spain	15,176,034
Other European countries	2,009,747
Mexico	8,026,169
Cuba	6,865,549
Other American countries	584,304
Other countries	2,000
Say a total of	121,926,349

New York Crude Oil Reservoirs.—These are shown in Plate XIII. (opposite).

Pl. XIII.—Crude Oil Reservoirs near New York. Photograph communicated by MM. Fenaille and Despeaux (Vol. i., *to face p. 246*).

✦

CHAPTER XVIII.

PETROLEUM IN THE WEST INDIES AND SOUTH AMERICA.

I. WEST INDIES.

Asphaltum Deposits.—The West Indian archipelago possesses numerous bituminous deposits, amongst which some, from the nature of their products, of a liquid consistency and of a lower density than that of water, may be regarded as producing petroleum.

(1) CUBA.

In the island of CUBA, numerous deposits of a peculiar kind of asphaltum are met with on the surface of the ground. Although known to the Spanish settlements at Havanna, these natural riches have been but little wrought or developed, and this fact will surprise no one. One only requires to reflect on the greatly disturbed history of this unfortunate island to understand the abandoned condition of its mineral resources. There can be no doubt that the political changes which have but lately supervened, whilst ensuring peace to Cuba, will enable these deposits to be realised.

Cuban Asphaltum.—We have said that the asphaltum of Cuba was of a peculiar nature, and this statement requires explanation. The asphaltum which comes from Cuban deposits, and to which the inhabitants of the country give the name of *chapafote* or *chapapote*, is differentiated from other asphaltums by the fact that it is completely soluble in naphtha and in spirits of turpentine. It has a pasty consistency ; sometimes it is even liquid. Its colour is black. The consistency of *chapapote* is not strong enough for it to lend itself to the ordinary purposes for which bitumen is used, but, on distillation, it furnishes a notable quantity of products analogous to those of petroleum. It is therefore more often used at the close of the distillation, and, from this fact, it can only be classified with the mineral oils.

Products from the Destructive Distillation of Chapapote.—About 120 gallons (about 550 litres) of crude oil, of density 0·892, are obtained from a cubic metre of *chapapote* (55 per cent.). This oil, with a rather profitable yield of lamp oil and paraffin, has a disagreeable smell, which as yet it has been found impossible to remove entirely.

The crude oil of *chapapote* yields on distillation :—

	Per Cent.
Gaseous products	1
Light „	5
Burning oil	44
Oil rich in paraffin	48
Heavy residuum	2
Total	100

The paraffin-charged oil yields 6·39 per cent. of solid paraffin. The principal deposit of the island is situated at BANES.

(2) BARBADOES.

Barbadoes Asphaltum.—BARBADOES likewise possesses deposits which, in their nature, are closely allied to those of Cuba. Experience has proved that, in this country, the pasty masses of asphaltum which are found on the surface of the soil—in virtue of the principle that the density of the oils decreases with the depth—may change into true petroleum. In fact, a well, dug by hand to a depth of 30 metres (98·4 feet), at ST. ANDREW, yields on an average 2 barrels of petroleum daily, with a density of 0·952. The most fluid exudations on the surface of the same place have a specific gravity approaching that of Cuban *chapapote, viz.,* 0·978.

The authors firmly believe that deep borings in the West Indies would lead to the discovery of mineral oils, with a normal density of 0·850 to 0·875.

(3) TRINIDAD.

Configuration of the Island.—The island of TRINIDAD is of a rectangular form. The principal petroliferous manifestation of the island is in the south-west corner, about 40 kilometres from Port of Spain. It is a lake, a real shallow lake of about 3 miles in circumference. This lake is upheaved by masses of islets of an eruptive nature, and which are formed by the continual flow of a kind of fluid asphaltum, which gradually solidifies. Several of these islets attain a length of 20 metres, others assume a conical form. "Between these islets," says M. Abraham Gessner, who was the first to utilise the product of PITCH LAKE, "flow shallow currents of clear and transparent water, in which live several species of beautiful fish. The sea along the shore is covered by a considerable quantity of petroleum coming from submarine sources, thus causing the surface to become iridescent."

According to Fernand Hue, "gases escape continually from all the points of the lake, impregnating the atmosphere with a strong smell of bitumen. They burn with a pale yellow flame. Their temperature is 36° C. The average heat of the water and of the liquid asphaltum does not exceed 35° C.

"In spite of the pronounced taste and peculiar smell spread by the gas, animal life was quite well developed on the shores of the lake; before its being wrought, numbers of birds made their nests in the crevices of the dried pitch; fishes frequented the small channels. A traveller asserts that he even saw an alligator there.

"The soil round about the lake consists, in great measure, of limestone. The shore is covered with cakes of more or less pure asphaltum, which the waves throw back unceasingly, and the shore itself consists of a black, ferruginous sand.

"It is probable," continues the same author, "that these deposits owe their existence to vegetable matter so abundantly prevalent in the neighbourhood, substances whose presence may be demonstrated in the asphaltum."

Products of the Destructive Distillation of Trinidad Asphaltum.— The following are the results of experiments made about 1860 by Abraham Gessner :—

Density, 0·882.

	Gallons.
Crude oil by first distillation	71
Oil by second distillation	62 per cent.
Paraffin	1·75 per cent.

The proportion of burning oil was 42 gallons per barrel of 160 litres, and 11 gallons of lubricating oil.[1] The colour of Trinidad

[1] The figures given in this sentence are incomplete. Presumably 42 per cent. of illuminating and 11 per cent. of lubricating oil were finally obtained from the crude asphaltum.—*Tr.*

asphaltum is greyish; its odour is disagreeable. According to Abraham Gessner, "this bitumen is grey, rather friable. It, however, melts more or less in summer by the heat of the sun.

Shipmasters refuse to accept Asphaltum as Cargo.—"A cargo of solid lumps of this bitumen agglomerated in the hold of a ship to such an extent that it had to be dug out in order to be discharged. They now refuse to accept a freight of it because they believe it strains their ships."

Pitch Lake is the property of the British Government. Petroleum is likewise found at BRÉA, a sulphuretted oil of density of 0·971, and at ARIPERO (a lighter oil of density 0·938).

There are no petroleum works in Trinidad worthy of the name.

II. SOUTH AMERICA.

(1) COLUMBIA.

Near to the Rio Arboletes, numerous natural springs are encountered—small kinds of craters from which petroleum exudes, sometimes rather abundantly.

On the borders of the River Iquana, one of these springs yields a large quantity of oil, which comes to the surface, with a temperature of about 40° C. (104° F.).

In the plain of MEDINA, pits have been dug by the inhabitants, who collect in them, for their own use, a certain quantity of a brown coloured oil of density 0·926.

(2) VENEZUELA.

Extent and Boundaries of Oil Field.—The Venezuelan oil field would appear to extend from north-east to south-west from the end of Lake Maracaïbo in the neighbourhood of the Zulia lagoon, to the borders of Columbia. The productive zone, as yet but little studied, would appear to follow the direction of the railway line from San Carlos to Villamina.

On the shores of Lake Maracaïbo, it would appear to crop out in the neighbourhood of a bed of sand, from which respectable quantities of petroleum already exude continuously. The output of these natural springs may be valued at 12 barrels daily. The petroleum, which is collected by the natives, is bituminous, of a brown almost black colour, with a density amounting to 0·947.

(3) ECUADOR.

About 200 kilometres to the south of the port of Guayaquil, in the neighbourhood of SANTA ELÉNA, there extends a wide petroliferous region, which would appear to be the head of the rich Peruvian oil fields. In spite of a position on the Pacific coast very favourable for its development, this oil field has not yet been realised. Excavations from 2 to 6 metres (6½ to 19½ feet) in depth, alone represent the work of man in this locality. The principal producing centres of this district are SANTA PAULA, SAN RAYMONDO and ACHAGIAN. The petroleum collected on the surface varies in density between 0·971 and 0·984. At San Raymondo, at a depth of 8 metres, they have been able to collect a few barrels per day of a greenish coloured oil of lower gravity, viz., 0·933.

(4) PERU.

Extent and Boundaries of the Oil Field.—The Peruvian oil field would appear to be one of the richest of South America. It extends along the Pacific Coast from the borders of Ecuador to Séchura Bay, that is to say, over a length of 300 kilometres (186 miles). The extent to which this oil field penetrates into the interior of the country has not yet been determined.

Reclus' Delineation of the Peruvian Oil Fields.—"The chief petroliferous deposits, spread over about 1,000,000 hectares (say 2,500,000 acres), are distributed in the mountains, and along the whole line of coast from Tumbez to Séchura, and notably exceed in extent the famous oil region in the upper basin of the Alleghany, in Western Pennsylvania. Asphaltum is found at an average depth of 30 to 120 metres above various strata, sands, sandstone of marine origin, calcareous deposits which formerly consisted of a mass of shells, and shales more or less impregnated with oil. In many places, the oily mass rises by filtration through upper strata, and even gas and oily substances rise up to consolidate at the surface. To the south of Sechura, some hills, analogous to the Amotape mountains, are perhaps richer in subterranean petroleum lakes; and in the plains near to the sea the swollen ground rises in blisters as large as 10 metres (32·8 feet) in height and 200 metres (say 656 feet) in circumference, like so many miniature volcanoes, from which asphaltum (*la pierre de goudron*) escapes in the liquid state, mixed with salt water, and very soon consolidates on the surface of the ground.   The plains of La Garita and Reventazon, in the neighbourhood of the sea, are dotted by hundreds of these miniature mountains of hardened tar, submarine currents of petroleum are scattered far and wide, and the iridescent oil is often seen to shine brilliantly on the surface of the water" (Elysée Reclus, *Nouvelle Géographie Universelle*, vol. xviii.).

History.—According to W. de Fonvielle, the history of petroleum in Peru goes back to a very remote period.

"If we believe," says that author, "the traditions, which include nothing that is not very plausible, if we consider the love which the Incas had for nature, the Indians were acquainted with petroleum and even used it as a fuel.   They were not even unacquainted with the art of hardening it by evaporating its most volatile constituents.   They used this kind of tar to line the inside of the earthen vessels which they manufactured.   This product was even so widely distributed that the Castilian Government would appear to have made it the subject of a monopoly.   But the possibility of establishing a fiscal measure was the sole reflection which the sight of this extraordinary substance suggested to the conquerors.   After the downfall of the Spanish domination, Peruvian petroleum ceased to receive attention.   It was only in 1863 that an American engineer, employed by the National Government, called the attention of the natives to the existence of riches which, less intelligent than the docile subjects of the enlightened despotism of the Incas, the ignorant and haughty citizens of an anarchical Republic had completely forgotten the use."

The American engineer to whom W. de Fonvielle refers was A. Ruden who, in 1863 bored, at the bottom of a square pit, long ago dug by the natives, a well which he stopped at a depth of about 40 metres (131 feet), and from which he collected a certain quantity of oil.

In 1867, a well bored at Zorritos by another Pennsylvanian engineer, Mr. Prentice, yielded a daily production of 60 barrels per day.

From this time forward the petroleum industry of Peru dates.

Mr. Prentice became the *concessionaire* of the oil fields then known, that is, the northern part of the oil field of the present time. The works multiplied, but did not really extend until the discovery, in the south of the Zorritos district, of new and rich deposits.

At the present day, the petroliferous region of Peru is included within the province of Piura, the farthest north of the coast provinces of this country. The Peruvian oil field—the direction of which is parallel with the coast—may be divided into three districts :—

1. TUMBEZ, on the north, the oldest wrought.

2. PAÏTA, the central oil field.

3. PIURA, the southern oil field.

Petroliferous Strata correlated with those of California.—In these three districts the petroliferous strata are of a similar geological formation to that of California, and rest on beds of the tertiary formation.

1. TUMBEZ OIL FIELD.

History.—The district of TUMBEZ is, as we have said, the most northerly and the oldest wrought of the Peruvian oil fields. It was at ZORRITOS, to the south-west of Tumbez, that Mr. A. Ruden made his first trial borings in 1863. Mr. Prentice was able in 1867, after having bored half a score of surface wells in the neighbourhood of Zorritos and QUEBRADA, to secure for himself the monopoly of the extraction throughout the whole of that region.

The Faustino Piaggio Company.—This monopoly he shortly afterwards conceded to an Italian company, the Faustino Piaggio Company. This company made a lucky start. In the very first well it bored, it struck such a productive bed of oil that the petroleum spouted to a height of over 20 metres. This spouting was only stopped by an accidental cause—a collapse of the casing (tubing) of the well. Some time afterwards, another well, a neighbour of this latter, yielded, for more than two years afterwards, a daily production of 600 barrels. The wells of the Faustino Piaggio Company have a depth which varies, on the average, between 200 and 300 metres. At the present time the company possesses, in the neighbourhood of Zorritos, 216 hectares of ground, on which some 100 wells have been sunk. These wells have all been more or less productive. No case of absolute barrenness has been pointed out. To-day, 20 borings are at work, and the annual production of the establishment is 75,000 barrels.

Properties of the Oil.—The petroleum collected is deep brown in colour, and varies in density between 0·810 and 0·840, according to Weinstein. This oil is converted into two kinds of kerozene, and the remainder into lubricating oil.

A *refinery* has been constructed by the company at Zorritos; it can turn out 9,000 cases of refined petroleum daily.

In the neighbourhood of Zorritos more modest establishments are encountered. Amongst these, mention may be made of TUCILLAL, two and a half miles from Zorritos, where 2 wells were bored from 1891 to 1892. One of these wells produces 20 gallons (say about 1 hectolitre) per day of a dark brown petroleum with a density of 0·940 (according to Boverton Redwood).

At QUEBRADA, some wells with a modest production yield a dark coloured oil with a density of 0·859 (according to Boverton Redwood).

At LA CRUZ, a well bored to a depth of more than 500 metres (1,640 feet) has given no result.

In the interior of the country at MANCORA, some rather important works have recently been taken in hand.

2. PAÏTA OIL FIELD.

The oil field of PAÏTA is situated to the south of the Tumbez district. Like the former, the PAÏTA district extends along the coast of SICHÈS, where surface exudations of a dark·coloured fluorescent oil of density 0·920 (according to Boverton Redwood), are encountered.

The principal locality wrought in Païta district is NÉGRITOS, six and a half miles to the south-east of Talara.

The London and Pacific Petroleum Company.

From the time of the Spanish Conquest, the territory situated between Talara and Négritos had been regarded as a mining property. This concession was designated under the name of " *La Mina Bréa y Pariñas.*" In 1888, the Peruvian Government granted it to Mr. H. W. Tweddle, who conveyed it in 1889, for a period of ninety-nine years, to an English company, the London and Pacific Petroleum Company. The ground, which this company soon started to work, consists of an immense tract of land which covers a superficial area of more than 15,360,000 hectares, or say a length from north-east to south-west of 30 miles, with a width of 20 miles, say 3,840,000 acres. This is one of the largest existing concessions.

Sixty wells have been bored by the London and Pacific, and the annual and yearly-increasing production amounts to 50,000 barrels. This production has been, up to now, voluntarily reduced for fear of over-production ; it is therefore evident that the figure given is only a minimum, and that millions of wells could be bored on the ground of the London and Pacific, and in a single day the yield of this magnificent concession could be increased tenfold.

As a matter of fact, the ground wrought at Négritos is particularly productive. Amongst the 60 wells already in existence not one is barren, and, if no spouting has been seen, a well there has been observed to yield, in ten hours, 750 barrels, owing to having burst its "kalpack". The wells reach a depth of 150 to 200 metres. There is no doubt that their greatest depth does not strike the very deep beds whose existence may be determined as the result of geological experience. The extension of trade in the refined article in South America, and a protectionist régime in these countries, will be the signal for a gigantic development of the works of the "London and Pacific".

The petroleum produced by this concession is brown and very fluorescent, and varies in density between 0·834 and 0·848. Its trade constituents are :—

Distillation Products of London and Pacific Crude Petroleum.

	Per Cent.
Benzine	10
Lamp oil	30
Mazout (used as fuel)	50
Loss and gas	10
	100

Pipe Line.—The works are connected with Talara, which is the company's seat, by a pipe line 10 kilometres long. The petroleum flows to that point by its own weight, the difference of level between the point of departure and the point of arrival being sufficient to do away with the necessity of pressure pumps.

Headquarters. — The administration buildings, the refinery—regulated to treat 1,000 barrels daily,—and a port which includes a depot with storage space for 130,000 barrels, are likewise situated at Talara.

One of the singular things connected with the working of this concession is the want of water and timber. This double inconvenience has been obviated by the use of sea water for the boilers and mazout for fuel. For domestic purposes, the sea water is distilled.

3. PUERTO GRAU.

Compagnie Française des Pétroles de l'Amérique du Sud.—In the same district, near to PUERTO GRAU, are the works of the *Compagnie Française des Pétroles de l'Amérique du Sud.* Although of younger date than its forerunners, as its establishment only dates from 1895, this company, whose works are directed by a capable engineer—our friend, De Clercy—has not been slow in producing very encouraging results.

Extent of Concession.—The concession of the *Compagnie Française* extends over a surface of 220 hectares (5,434 acres). The boring plant was imported from France, and De Clercy has tried a bold innovation which is justified by the want of water and fuel in the country, *viz.,* the use of compressed air as a motive power. The authors have not had the opportunity of examining the working of this new agent, and its value for boring purposes must be left for the future to decide.

Tank-Steamer.—Since its establishment at Puerta Grau, the *Compagnie Française* have built depots and a refinery, and they have bought a tank-steamer, the *Madeleine.* In 1896, an English consular report, estimated the production of the concession at 15 English tons per day, say about 7,800 barrels per annum.

State of the Wells.—According to a note from M. de Clercy, under date of the 21st October, 1897, 12 wells were opened on the concession. Of these 12 wells, only 6 could be regarded as productive. The other 6 varied in depth, between 6·1 metres and 23·4 metres (20 to 76·7 feet).

The following is the state of these wells on the 21st October, 1897 :—

					Depth in Metres.	
Well No.	1	.	.	.	.	324·84 (struck oil at 200 metres)
,,	2	.	.	.	.	120·00 (yielding)
,,	3	.	.	.	.	250·00 (struck oil at 75 metres)
,,	4	.	.	.	.	55·00 (yielding)
,,	5	.	.	.	.	66·00 ,,
,,	6	.	.	.	.	125·62 ,,
,,	7	.	.	.	.	22·84
,,	8	.	.	.	.	6·10
,,	9	.	.	.	.	10·20
,,	10	.	.	.	.	10·55
,,	11	.	.	.	.	6·20
,,	12	.	.	.	.	6·20

Production.—According to this same report, the daily production was 1,500 hectolitres, say 93 barrels, which gives an annual yield of about 34,000 barrels.[1]

Richness of the Puerto Grau Oil Field.—It is to be remarked that this production is proportionately much greater than that of the old companies who wrought this oil field. In fact, if we consider the 60 wells of the London and Pacific with their 100,000 barrels, the 100 wells of the Zerritos Company and their 75,000 barrels, and, finally, the 6 wells of the *Société Française* with their 34,000 barrels ; and, if we moreover bear in mind that the depth of the productive wells of the *Société Française* is but a minimum one, it will be acknowledged that the ground of Puerto Grau is of an exceptional character, or that the production, shown in the report of the company, of 1,500 hectolitres per day, can only be the result of a temporary production which cannot be taken into account in any reliable statistics.

4. PIURA OIL FIELD.

The district of PIURA forms the south of the Peruvian oil field. By the arrangement of its deposits, which would appear to deepen or dip towards the interior, it would appear to be the least capable of being developed. It is also the least wrought.

Total Production of Peru.—The total production of Peru may be estimated, by including the small works, at 180,000 barrels per

[1] There is something wrong with these figures. 1,500 hectolitres at 2 hectolitres to the barrel would give 750 barrels daily, or an annual production of 273,780 barrels.—*Tr.*

17

annum.   The European capital engaged amounts to about
10,000,000 to 12,000,000, of which France contributes 3,000,000.[1]

Price of Peruvian Crude Oil.—The average price of crude petroleum in Peru fluctuates between 35 and 55 francs (say 28s. to
44s.) the ton.

BRAZIL.

The province of Bahia possesses several petroliferous springs,
which yield a small quantity of good quality oil, of density 0·888.

BOLIVIA.

Petroliferous traces have been reported in the south of Bolivia
at CUARAZUTE, PLATA and PIGUERENDA.   M. Fernand Hue speaks
even of *rivers of oil*.   No works worthy of the name are developing
the deposits reported to exist in Bolivia.

CHILI.

Neither has any serious attempt been made up to now to
realise the petroleum deposits which exhibit themselves by evident
traces in Chili.

ARGENTINE REPUBLIC.

Petroleum has been reported in several localities in the territory of the Argentine Republic, especially in the north of the
province of MENDOZA, where a sample has been taken and analysed
by Boverton Redwood, who found it to have a density of 0·935.
In the province of JUJUY, several German works have been
started to realise the oil fields, but without any great success.
The same thing has happened in the province of SALTA, in spite of
a partial success, consisting in the discovery of a vein which, for a
certain time, yielded 300 barrels daily.

[1] ? Francs.—*Tr.*

CHAPTER XIX.

PETROLEUM IN THE FRENCH COLONIES.

ALGERIA.

THE BLACK SPRING.

Mythological History.—In the mountains, in the midst of a fertile tract of land, lived, loaded with years and riches, Mohamed ben Harousch ben Lakdar. He was as rich as he was avaricious, and hard to the unfortunate. In spite of the teachings of the Koran, he kept his heart hardened against the poor, and his door closed on the famished traveller.

One day, when Ben Lakdar looked with pleasure on the spring of fresh water which constituted the richness of his fields and fertilised his patrimony, a traveller—a poor shepherd—followed by four goats, passed by. The heat was great, the shepherd was thirsty; he approached the fountain to drink. But Ben Lakdar coveted a portion of the goods of the poor shepherd. He explained to him that he would not allow him to drink at his fountain, unless a goat—and he pointed out the finest one—was given to him. The shepherd refused to part with the goat, and Ben Lakdar called his sons, who drove away the man who had desired to drink without their father's consent. They pelted him with stones and took from him a part of his property, the goat coveted by their father, who was greatly pleased therewith.

Old Ben Lakdar stopped, with the goat, near the spring. But gradually its water darkened, and suddenly blackened. He called his sons to show them that the water of the well had become black and that it exhaled an abominable odour. Then they all lamented, because they saw in this the result of their evil deed.

They searched everywhere for the shepherd, so as to beg

him to restore its clearness and limpidity to the water of the
well.   They never saw him again . . . neither him nor his goats.

And the days ran on, whilst the fields of Ben Lakdar
perished, and the herbs and pasturages faded and died, killed
by the deadly smell, given off by the well. Ben Lakdar died
with vexation, and his sons were forced to go further afield,
fleeing from the cursed place which the vapours of the poisoned
spring had changed into a desert.

And this place was called Aïn Zeft (*the black spring*).

Such is the origin that the Arabs ascribe to petroleum.
Aïn Zeft is, if not the principal deposit of Algeria, at least
the oldest wrought, and, as far as we are concerned, we are
almost grateful to Ben Lakdar for the evil deed which endowed his country with a fortune. Let us add that we are
equally grateful to the Arab *raconteurs* who enliven and adorn
with an unpublished story the dryness and commonplace nature
of our subject.

Algerian Pirates come to it for Tar.—The petroliferous spring of
Aïn Zeft is found on the shore of Chèliff, not far from the Mediterranean coast. ˙ This situation explains the fact that, a long time
before the conquest, the Algerian pirates came to the *black
spring* to get a supply of tar with which to caulk their ships.

The Spaniards and the " Black Spring."—Later, the Spaniards,
numerous in this part of Algeria, saw a source of profit in working
the well.   In order to increase the output of mineral tar, they dug
pits and even shallow galleries, into the flanks of the hill. Such
was the nature of the first works undertaken at Aïn Zeft.

Geological Horizon of the Algerian Oil Fields.—At the present
time, when the petroleum industry would appear to be rapidly
developing in Algeria, it seems to us to be desirable, before examining each deposit individually, to take a collective glance at
the geological structure of the regions in which they are contained.

The petroliferous region of Algeria would appear to extend
parallel to, and at a short distance from, the Mediterranean coast.
The Tellien Atlas possesses numerous mineral oil springs in its
tertiary strata, and, as is the case, in Europe, this formation
would appear to be that in which the petroleum of the district
originates.   It would appear to be contemporaneous with the

upheaval of the Alps in the tertiary epoch; it follows the same general trend, and exhibits convincing analogies in the folds of the strata.

Depth at which Oil is Struck.—The depth at which petroleum in large quantity would appear to be encountered in Algeria would seem to be rather considerable. In the department of Oran, over 400 metres (1,312 feet) have to be bored (see the results obtained at Aïn Zeft). In the department of Alger, no surface indications have given rise to any trial borings; in that of Constantine the explorations have all been quite superficial.

| ◎ *Gisements de Pétrole·* | *Échelle en Kilomètres* |
| Oil Fields. | (Scale in kilometres of 0·62 miles.) |

FIG. 27.—Oran Oil Field (Algeria).

The Oran Oil Field.—The most decided signs are found in the department of Oran ; it is in that department also that the search for petroleum has been the most eager. In that district, the oil fields occupy a vast tract which, in its explored part, extends over a length of 135 kilometres (say 94 miles) from south-west to north-east. Parallel with the folds of Mount Atlas, this basin would appear to be bounded on the north by the shores of the Mediterranean, and on the west by the course of the Oued Houenet, on the south by a line passing through Mascara, Palikao and Ammi-Moussa, the general boundary of the department of Alger. The oil fields ought to stretch outwards from the border of this department, but no explorations have been made. In the Oranaise district of this region, four physical divisions, quite distinct from one another, are to be differentiated :—

1. The eastern part of the plain of Sig.
2. The eastern part of the mass of the Beni Chougran.
3. The western part of the mass of the Ouarsenis.
4. The mass of the Dahra.

Lines of Strike.—Two petroliferous lines of strike, each corresponding with an important anticlinal fold, have been observed in the Oran oil field.

1. The first, the most northerly, extends in an almost rectilinear direction, from Port-aux-Poules, in the heart of the Gulf of Arzeu, to Mazouna on the right bank of the Cheliff, close to the boundary line of the department. This line is landmarked by the deposits of PORT-AUX-POULES, CHABAT-HARMELA (Sidi Brahim), AIN-ET-KAALA (Beni Zenthis), AÏN ZEFT and TARRIA.

2. The second follows in a direction almost parallel to the first, but situated entirely on the south of the Cheliff, in a mountainous country. It has been less studied up to now. Its characteristic deposits are TILIOUANET and RELIZANE.

Each of the deposits mentioned now come under examination.

1. *Deposits on First Line of Strike.*

PORT-AUX-POULES.

PORT-AUX-POULES is situated about 3 kilometres from the sea, near to the embouchure of the Sig. For some few years back, attempts have been made to work and develop this locality, which shows numerous, rather considerable, surface exudations of mineral oil. The works, executed at Port-aux-Poules, cannot be regarded as a real attempt at working. Several excavations have been dug out, and a few shallow wells sunk, which have gone no further than to confirm the petroliferous indications of the surface. The gases evolved from these wells are very abundant, and give rise to the highest expectations.

SIDI BRAHIM.

SIDI BRAHIM is the name of a mineral concession, which for the greater part extends along the right bank of the Cheliff, and covers a superficies of 9 square kilometres.[1] The centre of this concession is situated near to CHABAT-HARMELA, 34 kilometres (say 21 miles) to the east of Mostaganem, and on the right bank of the Cheliff at less than a kilometre from the course of the river. Moreover it was in this locality that the first signs of petroleum were observed, twenty-five years ago. On the surface a feeble exudation was visible in beds of gypsum, mixed with rare crystals of sulphur.   Quite recently a tunnel, 5 metres (16·4 feet long) was dug into the beds of gypsum and marl, at the spot where the exudations occurred.   Although this tunnel can only be regarded as a preliminary experiment, the fact cannot be got over that the very feeble exudation, observed on the surface was, at a *depth of 5 metres* (16·4 feet), changed into a permanent flow of mineral oil, the regular yield of which was estimated at 29 litres (nearly 6·2 gallons) per day.

As has been already mentioned, the beds which were traversed by this tunnel and by the trench which conducted thereto, consist of shales intermingled with marls and gypsum. The traces of sulphur are only to be seen on the surface. These strata would appear to belong to the Swiss miocene.

Four borings have up to now been undertaken by the Sidi-Brahim concession.

Well No. 1 did not strike the oil-producing bed, owing to its having been started with plant insufficient for the great depth. In this boring, traces of oil were encountered at a depth of 8 metres, and in a constant manner below this depth. The chief petroliferous manifestations were observed at the following levels : 17 metres, 128 metres, 168 metres (say 55·76, 419·84, 531·04 feet). Boring was first stopped at 212·95 metres (about 700 feet) in consequence of an accident, then, after boring having been resumed, the depth was not carried beyond 290 metres (751·2 feet) owing to insufficient plant, 1895-7.

Well No. 2 was bored in the neighbourhood of the preliminary prospecting tunnel.   Oil was successively encountered in small quantity at 50·96, 56·0, 91·50, 130·14 and 228·40 metres (say at 167, 184, 299, 426 and 749 feet). At this depth work was interrupted, then resumed some time afterwards to be pushed down as far as 332 metres (1,089 feet) on the 29th January, 1897.

[1] Say, 5½ square miles.

Well No. 3 was commenced on the 30th March, 1890, and was completed on the 19th February, 1897, at a depth of 361·37 metres (say 1,184 feet). Petroleum had been encountered, but not in workable quantity, at a depth of 150 metres (492 feet).

Well No. 4 was taken in hand towards the beginning of April, 1897. The boring system adopted was that of Fauvelle, with circulation of water. At the moment this is being written, this well has reached a depth of 127 metres (416 feet). The beds passed through consist of silicious sandstones (pliocene). This last boring has been made with much more important plant than that of the first three wells, which could not go beyond 400 metres (1,312 feet).

Properties of Sidi-Brahim Petroleum.

According to Mr. Alfred Evrard, the oil flowing from the tunnel at Sidi-Brahim is slightly sulphuretted, very viscous, of a deep brown colour, with a density of 0·980.

BENI ZENTHIS.

BENI ZENTHIS is an important concession, situated like that of Sidi-Brahim, on the right bank of the Cheliff, and at a distance of 10 kilometres (say 6·21 miles) from Mostaganem. The surface signs of AÏN-EL-KAALAH, the exploration of which has led to the search for oil in the concession, consisted of exudations from the clays which are found in contact with an outlier of gypsum (Sahelien bed).

The first boring was put in hand on the 20th February, 1897. It passed through the Sahelien (selenitic marl). At 80·80 metres (265 feet), a disengagement of a very strong smelling gas was encountered. At 90·03 metres (295 feet) traces of oil were met with ; at 168 metres (551·04 feet), gas was again abundantly evolved ; at 190 metres (623 feet) and again at 193 metres (633 feet) traces of oil. At this depth, an accident in boring has stopped the works for some time. At the present time, boring is continued regularly in the petroleum impregnated ground. The highest expectations are entertained.

AÏN ZEFT.

AÏN ZEFT, which includes AÏN-DELLAH, VIEUX JARDIN and AÏN LOUISE, is up to the present the chief place of the Algerian oil field. As previously mentioned, the petroliferous fountain of Aïn Zeft has been known for a very long period of time, but it is only within recent times that it has been studied in a scientific manner.

As far back as 1877, the surface oil was the object of some operations, which consisted of some few rudimentary works—as many pits open to the skies as tunnels. A certain quantity of oil has been from that time forward collected in this locality.

In 1881, an official report of M. Baille gives the most precise information in regard to Aïn Zeft and its neighbourhood.

It was not, however, until 1887, that M. Jules Delecourt Wincquz, a Belgian engineer, drew the attention of the industrial world to Aïn Zeft and demonstrated the profits which would accrue to a well-equipped company working in a rational manner.

Evrard's Report on the Geology of the Aïn Zeft Oil Field.—" The mountainous mass, on the back of which lies the petroliferous works of Aïn Zeft," says Alfred Evrard (*Génie Civil* tom. xxix., an. 1896, p. 236), " constitutes part of an anticlinal mass highly dislocated and folded by volcanic action. There can be no doubt that, in its centre, this mass is cut by the plane of fracture which, over a length of 120 kilometres (say 75 miles), has given rise to the sulphurous, gaseous or bituminous emanations of Port-aux-Poules, Sidi-Brahim, Aïn-Dellah, Aïn-Louise, Taria and Mazouna, marshalled in such remarkable alignment.

" These manifestations on the exterior of subterranean phenomena—the energy of which may not be extinct—are only visible in isolated spots, where the disintegrated gypsum has enabled them to penetrate easily, or in the vicinity of central radial crevices, rejoining the plane of fracture, and which have not been completely plugged by the decay and flow of the superficial marls which enter so abundantly into the composition of the tertiary beds of this region. As a matter of fact, the system of fractures exists throughout with remarkable continuity, and it warrants the assertion that there exists a great flow of petroleum underground.

" But, in order to constitute a workable oil field, another indispensable condition must be fulfilled : the presence in the tertiary strata, at a reasonable depth, of a permeable zone, which may serve as a sufficiently vast reservoir to condense the hydro-carbons coming from the interior.

" We see by the general section that this second condition is fulfilled in the 40 to 50 metres (say 131 to 164 feet) of sandstones, sands and conglomerates which exist at the base of the Sahelien. It is because they ignored this local geological structure, that the explorers, who have succeeded each other at Aïn Zeft, have been for many years stopped half-way.

" This double confirmation is of such a nature as to encourage the *Société d'Aïn Zeft* in the pursuit of its efforts. But, at the same time, it imparts an incontestable value to the rights of search beyond the rather restricted zone reserved to this company, which have been granted to other explorers on the prolongations of the line of fracture, and in districts where there exists at the same time an identical geological structure and exterior emanations, similar to those encountered by the first explorers of Aïn Zeft."

We shall now pass in review the different works undertaken in the Aïn Zeft concession.

Borings at Aïn Zeft.—The search commenced at Vieux Jardin in the boring, by the Spaniards, of long trenches from which they collected the oil. The quantity of it must have been rather considerable, because it follows from the report of M. Baille that, during the fifteen years which preceded his visit (in 1881), the flow of petroleum might have been estimated at 56 cubic metres, (say 12,320 gallons).

Later on, in 1892, the works assumed a more practical direction. A tunnel, 100 metres long, was pierced into the rock ; this tunnel, directed from south-east to north-west, encountered several exudations of thick, bituminous oil in the gypsum and sub-Apennine marls which dip towards the north at an angle of about 20°. In front of the cutting a permanent flow of bituminous matters, mixed with water, was encountered.

An æration shaft, 14 metres (say 46 feet) deep was also excavated in the impregnated beds.

It was likewise in 1892 that the first boring was made at Aïn Zeft. This boring was executed by the Canadian system, under the management of Mr. Armitage. He was unable to push the depth further than 300 metres owing to the insufficiency of the original diameter of the well. The numerous collapses of the soft beds in the strata underneath Aïn Zeft necessitate special precautions as demonstrated by this first experiment. Oil was encountered, but in restricted quantity, at a depth of 130 metres (426½ feet) in this well. At 210 metres (689 feet) a rather important disengagement of gas was produced.

Well No. 2, also bored in 1894 by the Canadian system, was stopped at 276 metres (say 905 feet), without having produced appreciable results.

Well No. 3, bored in 1894, did not go beyond 215 metres (say 704 feet).

At the same time as these three wells were being bored, tunnelling work was being carried on at Aïn-Dellah and at Aïn Zeft.

At Aïn-Dellah, in the west of the concession, a tunnel cut at the bottom of a ravine met, in the selenitic marl, with a flow of oil, which demonstrated the persistency of petroliferous indications in the direction of Sidi-Brahim. This persistency of indications was an acknowledged fact, on the east of the concession ; in fact, in the ravine of Aïn-Louise, near to the northeastern boundary of the concession, on the coast of Tarria, numerous signs of petroleum and rather important exudations can be observed.

At Aïn Zeft, in the centre of the concession, on the line which joins Aïn-Dellah and Aïn-Louise, a tunnel of 180 metres (590½ feet) was made. In one of its branches, leading from east to west, an important flow of petroleum was brought to light. This spring produced at once the remunerative yield of 5 hectolitres (110 gallons) per day during its earliest days. This yield gradually sank to 1 hectolitre (22 gallons), a figure which it seems to be going to maintain.

Well No. 4 was begun towards the end of 1894. The plant employed had been planned to reach great depths, because it was judged that if, in the first three wells bored, oil had only been met with, in but small quantities, this drought was only due to the impossibility of striking the oil-bearing bed, owing to the want of the means for so doing. The success obtained in the No. 4 well has shown that the works' engineers were in the right, and has closed the mouths of those who asserted that petroleum existed but in small quantity in Algeria. Oil was struck at a depth

of 250 metres (820 feet), but a yield of a few hectolitres per day
was not going to satisfy those who had for such a long period
concentrated their efforts in the Aïn Zeft beds. They continued
boring. At 296·70 metres (say 972 feet) a shale level was again
reached ; several pools of petroleum were passed through and
neglected. Violent disengagements of gas frequently occurred,
which on several occasions gave rise to great falling in or collapses
of the ground. However, in spite of these obstacles, on the 26th
of June, 1893, when the bit struck a sandstone bed (Sahelien
strata), a column of petroleum burst into the well. . . . The
depth was 416 metres (1,404 feet).

Before the pumps could be set to work, the oil reached the
orifice of the bore-hole and began to spout. The yield amounted
to 202 hectolitres (125 barrels) per day.

This spouting did not surprise the works' engineers—they expected it ; all their efforts had been concentrated towards this
goal. They were troubled by another matter : had they struck
a pocket of oil or a really productive bed, comparable with those
of America and Baku, which yield petroleum for several months ?
Time would tell. In fact, at the end of 1895, more than six
months after the first stroke of the pump, the well produced 70
hectolitres (1,540 gallons) of petroleum daily, and with such
regularity as augured well for the duration of the flow. At that
time the total output of No. 4 well might be already estimated
at 12,500 barrels in six months.

It may be remarked in passing that in Pennsylvania the average yield has been estimated at 15 hectolitres (330 gallons) per
twenty-four hours, and the average duration of a well at two
years, which gives a total output for the two years of 6,900
barrels.

"Therefore," says M. Evrard, "the first boring at Aïn Zeft,
which went beyond a depth of 400 metres (1,312 feet) met with
a petroleum-bearing bed exceeding in daily output the average
of 60 hectolitres (1,320 gallons) yielded by the wells of the famous
oil field of Baku.

" This result was obtained as soon as they had penetrated 1·30
metres (say 4·26 feet) into a sandy bed, which was able to act,
by its natural permeability, as a reservoir for the petroliferous

emanations.   Up to then the work of searching had been constantly pursued in gypsums or in marls, that is to say, in but
slightly permeable or quite impermeable media, and the accidental
meeting with some traces of oil was generally due to the presence
of small beds of shale or limestone, which offered rather less resistance to the penetration of the oil.

" Now, if we glance at the general section of the strata passing
through Sidi-Slimann and the Aïn Zeft works, it will be seen that
the permeable bed of soft silicious sandstone and conglomerates
into which boring No. 4 penetrated to the extent of 1·3 metres
(4·26 feet), is not a simple local accident.   It will be found in

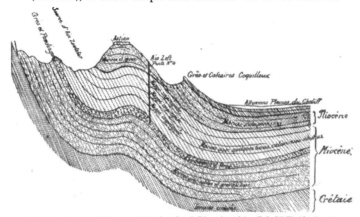

Fig. 28.—Geological Section in North to South Direction of Aïn Zeft Oil Field, Algeria.

EXPLANATION.—*Grès et poudingue* = sandstone and conglomerate. *Source* =
spring. *Puits* = well. *Nappe imprégnée* = impregnated level. *Marne et gypse* =
marl and gypsum. *Grès et calcaires coquilleux* = sandstone and shelly limestones.
Alluvions Plaines de Cheliff = alluvial Plains of Cheliff. *Marnes sub-appenines* =
sub-Appenine marles. *Marnes avec quelques bancs calcaires schisteux* = marls with
some bands of shaly limestone. *Marnes avec sable et grès à la base* = marls with
sand and sandstone at the base. *Terrains crétacés* = chalk beds.

reality to have slightly broached a *permeable aggregate of* 40 *to* 50
metres (say 131 to 164 feet) *in thickness*, the outcrops of which are
found to the north at rather a great height above the level reached
by boring No. 4.

" It is very possible—it is even probable—that by penetrating
deeper into this permeable zone, boring No. 4 would still further
increase its daily output."

New Wells.—At the present time the boring of 2 new wells has been taken in hand at Aïn Zeft—No. 5 and No. 6. The latter opened on the coast, 390 metres (say 1,279 feet) to the north-west of wells Nos. 4 and 5, is stopped for the moment in consequence of an accident at a depth of 350 metres (1,148 feet).

Properties of the Oil.—The oil yielded by the Aïn-Zeft tunnel is dark in colour, with a sulphuretted odour, and a density of 0·924.

Distillation Products.—According to Mr. Armitage, this oil consists of the following ingredients :—

	Per cent.
Lamp oil	15
Intermediate oil	28
Lubricating oil	34
Paraffin, vaseline, etc.	23
	100

The percentage of marketable products indicates a viscous oil highly charged with paraffin.

Tarria.

Evrard's Report.—" The perimeter of Tarria," says M. Evrard, " is situated to the east of that of Aïn Zeft, of which it is the continuation, and which it touches in its south-west point (S.-A.-E.-K. Zegnoun). This perimeter has a shape approaching that of a quadrilateral, which would be crossed diagonally from south-west to north-east by the course of the Oued Tarria, and from the north-west to the south-east by the prolongation of the great rectilinear fracture passing through all parts of the Dhara region, where bituminous, sulphuretted or gaseous emanations have manifested themselves on the exterior.

" In l'Oued itself, in ascending the course of the Tarria, after the Mahommedan priest's olive tree (*l'olivier du marabout*) had been passed, bituminous emanations, with deposits of sulphur, were discovered. Near to this legendary olive tree, in contact with the Astien and the sub-Apennine marls, a spring flows, to which the Arabs come to draw potable water for their daily use.

" Only insignificant excavations have been made on the perimeter of Tarria, which have laid bare gypsums and bands of marl, blackened with bitumen, or dotted with numerous spots of native sulphur.

" The nature of these exudations are, as a whole, the same as those of Aïn Zeft, with, perhaps, a more sulphuretted aspect. It may be presumed that if a boring were made near to l'Oued, which yields a sufficiency of water for the purpose, petroleum would be found in the same way as in No. 4 well at Aïn Zeft if it were pushed down to the sandy zones which, in this district, it would appear, should be met with at a less depth than at Aïn Zeft."

2. *Deposits on Second Line of Strike.*

It has been stated that the meridional line of strike of the

FIG. 29.—Boring near Relizane, in the Oran Oil Fields, (Algeria).

department of Oran was manifested by the oil fields of TILI-OUANET and RELIZANE.

At TILIOUANET, it was only in the month of May, 1897, that important efflorescences of petroleum were discovered. These efflorescences are met with 18 kilometres (say 11 miles) to the south of Hillil. The presence of petroleum is manifested in a very decided manner, and abundant exudations hold out much hope for the future.

To the south of RELIZANE efflorescences similar to those of

Tiliouanet are encountered. Borings were undertaken a few months back by a French firm in this region.

Amongst the signs of petroleum in the department of Oran, the following facts may be registered: near to OUED OURIZANE a well was bored in 1894, by M. Lainé. This boring was made for the sole object of finding water. When the boring had reached a depth of 74 metres (243 feet), a hot spring, with a temperature of 40° C. (104° F.), spouted to a depth of 15 metres (49·2 feet). This spring was intermittent, and its flow was accompanied by the disengagement of a very large quantity of combustible gas, which, when set fire to burned with an enormous flame. Since then, several wells have been bored in this region, each of them producing combustible gas, like the wells of Pennsylvania. This gas is present in such quantity that it could easily be used for industrial purposes."

DEPARTMENTS OF ALGER AND CONSTANTINE.

It has been stated that there are no signs of petroleum in the department of Alger. In that of Constantine they are numerous, and extend in the same direction as those which characterise the Oranais basin, from the mountains of Little Kabylie in the west, to the Medjerda mountains in the east.

The places where these traces are revealed are: BENI-SIAR, 15 kilometres (9¼ miles) to the south-east of Djidjelli. Exudations of a bituminous petroleum extend over comparatively a rather restricted area corresponding with the outcrop of a bed of shale belonging to the marine series of the pliocene.

FERDJIOUAH, about 40 kilometres (say 25 miles) to the south-west of Milah, exhibits a fissure filled with bitumen, produced by the drying of mineral oil which exudes from the shales by which it is surrounded. The geological formation of this deposit would appear to lie on suessonien beds.

CHEBKA DES SELLAOUA, situated 43 kilometres (say 26½ miles) SSW. from Guelma, on the front western flanks of the Medjerda mountains, exhibits to the observer abundant signs of petroleum. These indications consist of flows of bitumen which exude from fissures in beds of cenomanian shales and limestones.

None of the oil fields of the department of Constantine are wrought.

Future Hopes.—We conclude this short study of the Algerian oil fields by expressing the hope that the day will come when the oil fields of French Africa will give to its general industry an impetus comparable to that which has made Baku one of the principal towns of Russia.   To private initiative belongs the duty of profiting thereby ; let us hope that it will not fail to do so.

TUNIS.

The authors have not been able to procure precise documents on the mineral oil industry, which, in Tunis, is still more in its infancy than in Algeria.

Researches have been undertaken, but too recently to even discount their results.

It was, we understand, on the 1st May, 1896, that the first authority to search for mineral oils in Tunis was granted. This authority was given to M. Assereto over a concession situated between DJEBEL-SERRA and L'OUED EL KSOB, as well as in the valley of L'OUED CRATTOUNA.   On the 13th July, 1896, a similar authority was granted to MM. Henri and Gaston, of Vêsine-Larue, in respect of a concession situated at DJEBEL BOU KOURNINE.

MADAGASCAR.

Explorers' Accounts.—Several explorers, in their accounts of their travels, speak of petroliferous ground in Madagascar.   These oil-bearing beds would appear to be more frequent to the southwest of Tananarivo, at a medium distance from the coast, in the valley of Manambolo.

Gautier's Account. — Gautier, the explorer, says in regard thereto :—

" It would, unfortunately, appear that the coal bed of Ambaratobé to the south of Nossi-Bé is unworkable.   Perhaps compensation may be found on the coast of Ankavandra ; I there saw sources of bitumen, although it would be premature to speak of petroleum, as borings have not yet been made " (*Bulletin de la Société de Géographie commerciale*, tom. xvii., *an*. 1895).

M. Grosclaude, in the narrative of his expedition across Madagascar, relates the following facts which go to support the statements of M. Gautier.

18

Grosclaude's Account.—"As far as Rocheron and myself were concerned, we proposed to start in search of the bituminous, perhaps petroliferous, shales, whose presence in the region was reported by Gautier, the explorer.

"Their situation is localised in rather a vague manner in the neighbourhood of Ambohitsalika; but we wanted exact information as to the village, which some said was two days' march on the other bank of the Manambolomaty, and the very existence of which the others vehemently contested. Cruel enigma! The difficulties with our guides start afresh more embarrassing than ever. . . .

"In the morning we were furnished with information in regard to the petroleum springs, in search of which we had sent one of our guides the previous evening, escorted by some men from the village. They returned with an oily odoriferous earth in a preserve box.

"After some hours' walking, we see some huts: it is the village of Yankeli, whose population, informed beforehand by our emissaries, receive us with sympathy, and conduct us to the springs, which are about a kilometre (say 0·62 miles) to the south-east of the village. It is a flow of bituminous exudations, where we made some summary excavations to collect a few samples of the earth, the rock and the mineral oil which burns marvellously."

Mining Legislation.—The mining industry of Madagascar is subject to a particular administration, of which the following are the principal regulations:—

Each individual, every firm, other than the proprietor of the ground, who is desirous of searching for minerals, must ask permission to do so, either from the mining department of Tananarivo, or from the resident of the province.

The cost of the permit to search is fixed at 25 francs (£1).

The permit to search is valid for one year; it may be renewed if the administration deem it advisable.

The explorer may dispose of the products of his search, provided he pre-informs the mining officials, and that the mines are not deteriorated in the process. The maximum superficies of land to be prospected must not exceed 2,500 hectares.[1] This is also the

[1] 6,175 acres.

maximum extent of the concession which is granted to whoever asks for it, if he fulfils the requisite conditions, and if there be no opposition.   In the latter case the government determines, arbitrarily, to whom, and to what extent, the concessions should be granted.

Every concession, every mining company of the same category must be authorised by the administration.

Taxes.—A tax of 4 per cent. is levied on every deed of concession or company formation.

Annual Rent Charge.—Each mine is subject to a fixed annual rent charge of :—

```
          s.  d.
1 franc  = 0   9·6  per hectare up to 200 hectares (494 acres).
2 francs = 1   7·2  per hectare up to 500 hectares (1,235 acres).
3 francs = 2   4·8  per hectare, over and above up to 1,000 hectares (2,470 acres).
4 francs = 3   2·4  per hectare, over and above up to 1,500 hectares (3,705 acres).
5 francs = 4   0·0  per hectare, over and above up to 2,500 hectares (6,175 acres).
```

This rent is not levied after the expiry of the second year of the concession.

Moreover, the extracted products pay a proportional royalty of $2\frac{1}{2}$ per cent. *ad valorem.*

NEW CALEDONIA.

New Caledonia, by its geographical situation, would appear to form the prolongation of the great petroliferous line of strike which runs across the old continent and the Straits Archipelago.

On the coast of the island of Wagap there extends a thin band of tertiary strata, and it is in this geological formation that traces of petroleum were discovered in 1896.

Pools of oil are formed on the surface of the ground, notwithstanding no works have been established.

Properties of Oil.—A sample of petroleum from the Neo-Caledonian basin presents the following characteristics : Density, 0·930; colour, very deep brown, almost black; viscosity, very great; percentage of paraffin, high.   This oil would appear to yield a large percentage of lubricating oil and a lamp oil perfectly suitable for lamps constructed to burn heavy petroleum.

GEOLOGICAL CLASSIFICATION OF THE DIFFERENT
OIL FIELDS.

Formation.	Characteristic Fossils.	Place of Origin of Petroleum.		Strata in which the Deposits are encountered.
		Oil Fields.	Deposits.	
I. RECENT STRATA	Human remains in recent alluvial deposits. *Débris* of species now living. Madreporic (coral) deposits.	EGYPT	Djebel Zeit. N.B.—*The presence of petroleum in recent formations would—except in the above instance — appear to be due to infiltrations of oil, the source of which is to be found in lower beds of a different nature.*	Alluvium. Coral deposits
II. QUATERNARY STRATA	*Débris* of several extinct races. Mammoth. Cervus giganteus.	NORTHERN GERMANY	Wietze, Steinforde, Weenzen, Verden, Linden, Hänigsen, Heide, Mehldorf. (*According to Hans Höfer*)	Diluvium Sand
		CANADA	Enniskillen. N.B.—*Often also in these beds the presence of petroleum is only the result of simple infiltratum.*	
III. TERTIARY STRATA	Pliocene	Ursus minutus. Megatherium cuvieri. Pithecus maritimus. Balena. Turtles. Sharks and other dog fish. Voluta Lamberti Lymnœa palustris.	NORTHERN ITALY ISLE OF JAVA JAPAN NEW ZEALAND	Province of Emilie de San Colomboat Faenza. [igo. Province of Ech-Waipava.	Sandstone and marls
			CALIFORNIA ALGERIA ROUMANIA	De Santa-Clara à San Diégo. Aïn Zeft. Colibatzi, Ploesti, Buseo, Tintea, Bacau, Telega Campina, Dragonese, Pacuretzi.	Pliocene sandstones Sandstone beds mixed with clay
			TRANSCASPIAN TERRITORY GALICIA	Isle of Tcheleken. Boryslaw, Dwiniaz, Starunia. (*Ozokerit and petroleum charged with paraffin.*)	
	Miocene	Mastodontes. Coniferae. Cypris. Fishes. Paludina. The latter are very numerous in the petroliferous strata. Nummulites.	ROUMANIA	Moïnesti, Campeni, Taslau, Comonesti, Maïnesti, Tirgu Okna, Solanti.	

GEOLOGICAL CLASSIFICATION OF THE DIFFERENT OIL FIELDS—(Contd.).

Formation.	Characteristic Fossils.	Place of Origin of Petroleum.		Strata in which the Deposits are encountered.
		Oil Fields.	Deposits.	
Miocene	Mastodonts. Coniferæ. Cypris. Fishes. Paludina (the latter are very numerous in the petroliferous strata). Nummulites.	RUSSIA	Baku and the Apshéron Peninsula. Tamansk Peninsula.	Sandstones and sands
		ALSACE	Hirbach, Ste-Croix, St. Bilt Echery, Hirsingen, Roderen, Mutzig, Mohlsheim, Lobsann, Biblisheim.	
			Pechelbronn, Schwabwiller.	Sandstones and sands
		NORTHERN ITALY	North Watershed of the Apennines between Padua and Bologna.	Clays and sandy shales
		FRANCE	Gabian.	
		TRANSCASPIAN TERRITORY	NaphthniaGora, Buja Dagh.	Beds of sand
		VENEZUELA	District of Maracaibo and of Punto d'Acaja.	Clay Shale, and sandy limestone
		TRINIDAD		Disintegrated shales
		ALGERIA	Deposits of the department of Oran and of the department of Constantine.	Shales
Upper Eocene	Dicotyledonous plants. Nummulites and Paludina. Lymnisæa caudata. Cypris. Fucoïdes.	GALICIA	Bobrka, Pohar, Koziowa, Schodnica.	Clay shales
		ROUMANIA	Cobalescu.	
		GERMANY	Petroliferous indications of Tegernsee.	Nummulitic beds
		ITALY	District of Plaisance, — Vergato, Pietra Mala. District of Abruzzi—Tocco, Roccamorice, Albatagio.	Limestones
		GALICIA	Bobrka, Schodnica, Sloboda Rungurska.	Sandstone
		FRANCE	Limagne.	
Lower Eocene	The Nummulites are characteristic of these strata. Nummulites levigata. Nummulites exponens. Nummulites scabra.	TURKEY IN ASIA	Petroliferous traces at Hit (government of Bagdad).	
		PERSIA	The chief deposits.	
		BRITISH INDIA	Kahatan (Beluchistan), Puncoha, Gunda, Lundigar, Doula, Chorhot, Ruta Otur, Digboi, Assam district.	Nummulitic beds
		BURMAH	Arakan Islands, Yenangyaung, Beme, Yenangyat, Twingoung.	Sandstone and clays / Sandstone

III. TERTIARY STRATA (Continued).

Eocene

TECHNOLOGY OF PETROLEUM.

GEOLOGICAL CLASSIFICATION OF THE DIFFERENT OIL FIELDS—(Contd.).

Formation.		Characteristic Fossils.	Place of Origin of Petroleum.		Strata in which the Deposits are encountered.
			Oil Fields.	Deposits.	
IV. CRETACEOUS STRATA	Upper Cretaceous	Reign of the ammonites. Belemnites in the upper beds. Absence of mammalia. Saurians and mollusca.	GALICIA	Ropianka, Mrasznica, Kimpolung, Putna, Krasna.	Chalk
			RUSSIA	South Caucasian Oil Field.	
			NORTH GERMANY	Heide, Mehldorf, Henning, Darfeldt (according to Hans Höfer).	
			SOUTHERN ITALY	Manopello, Tocco, Chieti, Gardagreli. Rionero di Molise, Triolo, Squillaci, Gerace, Sacarite.	
			FRANCE	Credo.	
			SPAIN	Huidobro.	
			TURKESTAN	Namangane.	
			SYRIA	Traces of petroleum on the Watershed of the Djebel el Dahr.	
			CANADA	Alberta district, Rocky Mountains.	Sandstone
			PALESTINE	Dead Sea and source of the Jordan.	Limestones
	Neocomian	Lymnoea. Gigantic saurians.	GALICIA	Ropianka.	Ropianka shale
			NORTH GERMANY	Klein Oedesse, Oelheim. Traces of petroleum at Horsdorf (according to Hans Höfer).	Limestone and sandstone Black clays
			PORTUGAL	Traces of petroleum in Estramadura.[1]	
			ARGENTINE REPUBLIC	District of Selta. District of Jui-Jui.	
			UNITED STATES	Colorado and Nevada deposits.	
V. JURASSIC STRATA	Upper	Ostrea deltoides.	NORTHERN GERMANY	Hânigsen, Linden.	Marl, limestone and marl
	Middle	Many ammonites. Ammonites biplex. Ammonites cordatus. Ammonites anceps. Disaster ellipticus.	,,	Reitling. Wietze, Steinforde, Verden.	Sandy clays Deep green clays
	Lower	Commencement of Belemnites. Reign of the ammonites. Ammonites spinatus. Ammonites Bifrons. Ammonites serpentinus.	,,	Schoningen-Sehnde, Wietze. Steinforde, Weenzen.	Clays
			ARGENTINE REPUBLIC	Mendoza (district of Selta)	Deep green clays
			UNITED STATES	Signs in Colorado.	

[1] This Portuguese province must not be confounded with the province of the same name rich in minerals in central Spain.—Tr.

GEOLOGICAL CLASSIFICATION OF THE DIFFERENT OIL FIELDS—(Contd.).

Formation.	Characteristic Fossils.	Place of Origin of Petroleum.		Strata in which the Deposits are encountered.
		Oil Fields.	Deposits.	
VI. TRIASSIC STRATA			No petroleum deposits, much bitumen.	
VII. PERMIAN STRATA [1]	Reign of trilobites and regular brachiopods.[2]	RUSSIA	Traces of petroleum at Sukkowo (government of Kazan). Traces of petroleum at Michailowska, Kanischki, Schugorowa, Nowosemeckino (government of Samara).	
VIII. CARBONIFEROUS STRATA		GERMANY	Traces of petroleum at Wetlin (Saxony).	
		PENNSYLVANIA	Pittsburg district.	Limestones and black or brown coarse grained sands
		OHIO	Neff, Wellsburg, Magsburg.	
		KANSAS	Gas Wells of Sola, Fork Seot, Kansas City, Rosdale.	Clays and shales
IX. TRANSITION STRATA [3] — Devonian	First reptiles. Trilobites.	PENNSYLVANIA AND NEW YORK	Districts: Alleghany, MacKean, Warren, Venango, Clarion, Laurence, Beaver, Armstrong, Buttler.	Fine grained compact sands
		OHIO AND WESTERN VIRGINIA, TENNESSEE AND KENTUCKY	Districts: Nobel, Washington.	Shales and clays
		CANADA	Gaspé.	
		FRANCE	Gabian (according to Zipperlen).	
IX. TRANSITION STRATA [3] — Silurian	First imprints of fishes. Fucoïds and algae.	OHIO	Districts of Fremont, Findlay, Bowling Green, Lima, Corey.	Coarse grained sandstones with quartzose particles
		NEW YORK	Near Albany.	
		CANADA	Manitouline Island, Packenham, Rivière à la Rose. Districts of Enniskillen, oil spring and petrolia.	Limestones

[1] From the province of Perm in Russia.—Tr.

[2] This remark would appear not to refer so much to the Permian itself, but rather to the strata below the Permian. As a matter of fact the *trilobites* first appear in the Cambrian, become very numerous in the Silurian, less so in the Devonian and practically disappear in the upper Carboniferous.—Tr.

[3] See note p. 203.—Tr.

PART II.

EXCAVATIONS.

CHAPTER XX.

WORK OF EXCAVATION: HAND-DIGGING.

First European Method of Oil Well Sinking.—Manual labour, with the help of the pick and the shovel, was the first means employed in Europe in seeking for petroleum.

Excavation the Slowest and most Costly Method.—This process is very simple, and its mechanism has but few complications. Just as it is the slowest, it also requires much less capital than the others, but it is the most costly in proportion to the amount of work done. That is why we do not recommend it but in exceptional cases, either for digging out a surface well, or in soft ground.

Plant. — Quite rudimentary plant suffices for excavating a well by this method. The following is, moreover, the method of working :—

Method.—After having covered the spot to be excavated with a shed to shelter the workmen, operations are commenced to excavate a well about 1·5 metres (say 5 feet) square. In soft surface ground (gravel or clay) the shovel will be the tool which will play the chief *rôle*. The rubbish is drawn up the well by a sheet-iron bucket, which is suspended from a steel-wire cable of at least 0·007 metres—about 0·275 inch—in diameter, and which rolls round a windlass modelled after the simplest and most primitive plan. This bucket also serves to raise and lower the man working in the well.

When the soft ground permeable to water has been passed, and the sides of the well have been carefully boarded up, digging is continued, but within a more restricted space. The width of the well should then be 0·8 metre to 1 metre (say from 2·6 to 3·28 feet) square. The object of this operation is first of all to form, by means of the boarding of this second portion, together with that of the first, a kind of gutter (Fig. 30 G), which will retain the

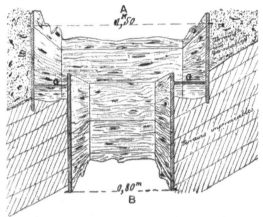

FIG. 30.—Section of a Well excavated to the Point where Surface Infiltration ceases.
A. Surface Well.   B. Well being continued, but only 2·6 feet square.

infiltrated water, the superabundance of which might hinder the work. This gutter is emptied either by means of a small pump, or by an overflow tunnel (Fig. 31), if the natural position of the

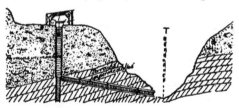

FIG. 31.—Well and Outflow Tunnel.

well lend itself thereto, that is to say, if the well has a ravine below it whose channel (T) is below the level of the gutter, so that the tunnel has a sufficient slope to allow the water to flow away easily.

Necessity for Diminishing the Dimensions.—Again, another thing

which especially necessitates the restriction of the dimensions
of the well, is the diminution, by this same restriction, of the
rubbish to be raised, and consequently of work to be done. Moreover, when descending and ascending the well, the workmen ought
always to have the sides of the well within reach of their arms.

Aeration of Excavated Oil Wells.

Beyond a depth of 20 to 30 metres (65·6 to 98·4 feet),
respirable air must be forcibly injected, by means of a ventilator,
down to the bottom of the well, to counteract the action of the
deleterious gases.

The excavation of the well is continued, under these conditions, until petroleum is reached.   But we again repeat that
this method is long, and what is more, it is a dangerous one
to employ.   We shall now, in addition, explain in a summary
manner the work of the mining labourer in the way still practised
in Galicia—ozokerit wells of Boryslaw.

When the well has been carefully aerated by means of some
species of air propeller, the safety lamp, locked by the foreman,
is attached to the handle of the bucket which is lowered slowly
into the well.   If the flame be extinguished, stifled by the gases,
it is dangerous to let a man descend.   If, on the other hand,
the lamp can be lowered to the bottom and brought up again
alight, the workman can descend, provided that the air propeller
does not stop working for one minute.   In this case, the miner
clothes himself with braces, the straps of which pass over the
shoulders and between the legs, the whole joining into a sound
belt.   To this safety belt, which the workman should never remove
before having been raised up again, is attached a strong hemp
rope, 24 millimetres (1 inch) thick at the least, which is rolled
round a windlass independent of that for the bucket.   The miner
puts his right foot in the bucket ; the left remains hanging, and
enables him to obviate rocking or overbalancing.   The lamp is
held in the right hand, the arm of which will press the steel
cable of the bucket against the body ; the left hand is free, and
can at will seize the rope of the bell used as a signal.   When he
has got to the bottom, after having hooked his lamp to some
projection, he attacks the rock with his pick, and when he has

broken up enough he fills the bucket with his shovel, which is raised and lowered quickly so that he may refill it.

Work Hard and Dangerous.—There is no need to remark that this is very hard work, and that the workmen have to be accustomed to live in an atmosphere in which we could not breathe; but, in spite of being used to it, even the best seasoned amongst

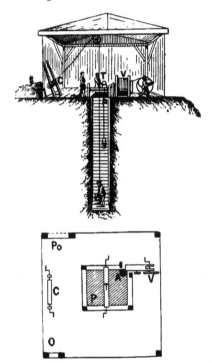

Fig. 32.—Section and Plan of a well being excavated by Hand. A, Ventilation Pipes; V, Ventilator; P, Well; T, Windlass; C, Safety Windlass; Po and O, Doors.

them can hardly remain more than two hours in a properly aerated well.

The dangers to which the workman is exposed are many and varied: explosion, fire, inundation, asphyxia, landslips and many others.

Tools, etc.—The equipment of a well to be excavated by hand, in addition to picks, spades, shovels, buckets, etc., consists of an air propeller.

Air Propeller.—This ventilator is wrought by hand, and supplies pure air through white-iron tubes, fixed firmly into the corner of the well. This air is propelled by the rapid rotation of a wheel, with paddles, in a drum, at the bottom of which is the orifice of the aeration tubes. The wheel with paddles is driven by a grooved wheel, having a diameter of about 1½ metres (say 5 feet), which a workman turns by hand. It is this grooved wheel which drives, by means of a belt, another wheel of 0·20 metre (say 8 inches) in diameter, on the axis of which are the paddles or blades.

The pipes of the air propeller ought to be firmly fixed to the sides of the well, or rather suspended by an iron cable, so that they may be drawn up and replaced as occasion may require. In Eastern Galicia white-iron tubes are employed; but the authors have sometimes noticed, in the Western oil fields, aeration tubes consisting of a rectangular chimney of carefully adjusted boards. This latter kind of pipe has the enormous advantage of not being liable to suffer from shocks, of being unsoldered, nor of being dented in or buckled; but they are difficult to handle when it is desired to draw them up. The pipes of the air propeller should always descend to within 2 metres (6·56 feet) of the bottom.

Lowering a Man into an Insufficiently Aerated Well for Urgent Repairs.—When it is necessary to lower a miner for urgent repairs, when the well is not sufficiently aerated, the greatest precautions are absolutely necessary. He does not take the lamp, and it is desirable that he should cover the mouth and the nose with a sort of gag (or aspirator), soaked in strong vinegar. It often happens that, in spite of the greatest precautions, a semi-asphyxiated workman faints in the well.

Remedy for Asphyxiation.—When this takes place, he is first raised by means of the safety rope, carried into the open air, made to respire ether, or strong vinegar, and in those cases, where the fainting is prolonged, artificial respiration should even be resorted to. Milk is the best antidote for the toxic effects of carburetted gas. But the surest plan is to well aerate the well. Accidents, often painful, will thus be avoided.

SIGNALLING AND ALARUM BELL.

An instrument which has its importance as a preventative against these accidents is the bell fixed to the mouth of the well, the rope of which hangs down to the bottom. Three pulls of this bell signify that the bucket or the workman is to be raised; two pulls to lower; one pull to stop at once raising or lowering; finally, four pulls and more indicate alarm—in this case the miner ought to be raised as quickly as possible.

DAVY LAMP EXPLOSIONS.

The lamp employed is an improved Davy lamp, which should be locked with a key. When the workmen sees the flame of his lamp has become blue, or if the gas burns in the inside, he should be careful *not* to blow to extinguish the lamp, for the lamp may have a flaw, and the blowing, by driving the flame outside, might set fire to the gas in the well. When the lamp becomes inflamed, there is only one way of putting out the flame, and that is to deprive it of air, by placing it under the clothing. One would do well to examine the lamps often and carefully, because upon a hidden fault may depend the death of several men.

Well on Fire.—In case of the well taking fire, if no explosion has occurred (when it has been sufficiently aerated), the only resource is to deprive the fire of air. The well should therefore be covered with planks, and earth thrown over this covering. When an explosion does not occur forthwith, it is no longer to be feared, and the work indicated above can be gone about tranquilly.

SMITHY TOOLS.

A small smithy equipment, consisting of an anvil, a bellows, a small vice, a set of hammers and spanners, ought to be found on the works. It is used for sharpening, by reforging, the blunt picks (*the work of a single well blunts on an average a score per day*), as well as for urgent repairs to windlasses, buckets, etc.

CARPENTRY AND JOINERY WORK.

It is also advisable to have a few axes and saws on hand for boarding up purposes. The boards used for this purpose have generally a width of 0·30 metre (11·8 inches) and a thickness of 0·04 metre (1¼ inches). They ought to be dressed in such a manner as to fit into one another to form a continuation of horizontal frames. In the wells, the woodwork should be kept in place by battens nailed to it in a perpendicular position to that of the boards forming the sides—that is to say, vertical.

In certain localities where the ground easily falls in or land slips, the thickness of the woodwork varies considerably. At Boryslaw, boards of 0·1 metre (say 4 inches) in thickness are used. The boards used for the woodwork are pine deals. The authors recommend them to be steeped in crude petroleum for three or four days before use, so as to render them so far—even if only to a slight extent—impermeable.

Fig. 33.—Arrangement of Battens intended to consolidate the Boarding of the Wooden Casing.

USE OF DYNAMITE.

Finally, we may further mention, although it is only used in a small number of wells and its use also requires special permission from the authorities, that redoubtable but formidable agent—dynamite. At the present day, this explosive is not so much in vogue as at the time when Colonel Roberts extolled its discovery in the oil fields of the New World.

A dynamite cartridge should only be used in those beds where the rock presents great resistance. In such a case, a mine hole is bored about 0·30 metres (11·8 inches) in thickness (*the authors advise that this hole should be bored in the softest part of the stone, this method of working being favourable to the dislocating effect*). The cartridge should be introduced into the mining hole; and a fuse, dipped in sulphur, inflamed and furnished with a capsule of fulminate of mercury at the extremity (next to the cartridge) will cause the explosion of the latter after a lapse of time proportional to the length of the fuse. To light the latter, ardent or glowing coals without flame are used, it being understood that any flame would inflame the gas and might blow up the well. It is useless to add that the workman—generally the foreman—who lights the fuse, ought to get himself raised up as quickly as possible.

In certain works, the dynamite is exploded by electricity. This method is safer, and on that account is to be preferred in every respect. We may add that the use of dynamite can only be attended with beneficial results when the ground is very hard and all other means have been tried and failed. The dislocating effects of this substance are almost *nil* in soft ground, and, as after it has been used it is indispensably necessary to spend two hours at least in aerating the well, the operation in this case balances itself by an appreciable loss of time.

ALLOCATION OF WORK.

Each man working in a well has his job assigned to him beforehand. The staff consists of three men, who descend to the bottom in turn, the two who remain on the surface being employed in the raising of the rubbish. A fourth workman works the air propeller; when the working miner ascends or descends, he likewise works the hand winch of the safety rope. As will be seen, the staff of a well is not numerous, and the expenses connected with it are not heavy. But as the proverb says, the *cheapest market is often the dearest.* One will do well to bear it in mind.

SPOUTING WELLS.

It is rare, we will say almost impossible, that spouting can occur in a well excavated by hand labour, for the reason that the dimensions of these wells are much too great. For the oil to overflow an enormous quantity of liquid would be required. Besides, the oil does not often come suddenly; one is always warned beforehand of its proximity by an average yield, which gradually becomes greater as the depth increases. The disengagement of gas is likewise redoubled.

Under such conditions it becomes, in the majority of instances, impossible to continue to work by hand, and if it be desired to go to greater depths, resort must be had to one of the methods of perforation, which are about to be passed in review in the chapters devoted to boring.

In any case, if a spouting, or rather, looking to the diameter of the well, an overflow occurs against all expectation, the old Canadian method should be preferably adopted. This method consists in plugging the orifice with a sack filled with linseed. These seeds, as every one knows, expand considerably in contact with moisture, and in the act close the well. There is then introduced into this improvised cork, a pipe provided with a tap, which is opened at will according as required. Let it be well understood that this method is only provisional, but it allows time for the well to be capped in a more substantial manner. (*See* p. 234-5.)

Well Cap.—A hermetic covering of steel plate, furnished with a tube closed by a tap, is the most simple covering, the most practical and the most substantial. (*See* p. 442.)

ESTIMATE.

ESTIMATE FOR SINKING, ETC., A WELL, 150 METRES (492 FEET) DEEP, TO BE EXCAVATED BY MANUAL LABOUR.[1]

1. *Construction of the Shed covering the Well.*

	Francs.	£	s.	d.
Beams	8	0	6	4·8
150 thin boards at 0·20 francs each	30	1	4	0
1 lock	1	0	0	9·6
1 pulley	14	0	11	2·4
4 men, 2 days at 1 franc each per day	8	0	6	4·8
	61	2	8	9·6

[1] Cost in Galicia.

2. *Cost of Sinking the Well, etc.*

	Francs.	£	s.	d.
Wood for making windlasses and lining well . .	6	0	4	9·6
Steel cable 0·007 (say 0·275 inch) diameter and				
150 metres (492 feet) long	22	0	17	7·2
2 buckets for drawing water at 4 francs each . .	8	0	6	4·8
2 „ strengthened for transporting solid sub-				
stances and raising and lowering the miners,				
at 6 francs each	12	0	9	7·2
1 hemp cable, 3 centimetres (say 1·18 inch) diameter				
and 150 metres (say 492 feet) long . . .	104	4	3	2·4
2 safety braces at 14 francs each . . .	28	1	2	4·8
1 bell	2	0	1	7·2
150 metres (say 492 feet) of rope for bell . . .	8	0	6	4·8
An air propeller complete with belt	80	3	4	0
200 metres (say 656 feet) of aeration piping at				
1 franc per metre	200	8	0	0
20 picks at 1·2 francs each	24	0	19	2·4
4 shovels at 1·5 francs each	6	0	4	9·6
4 spades at 1·5 francs each	6	0	4	9·6
1 lever	8	0	2	4·8
2 dies for drilling at 2·4 francs each . . .	4·8	0	3	10·08
Pearwood for handles	4	0	3	2·4
4 safety lamps at 10 francs each	40·0	1	12	0
Lamp oil and wick	40·0	1	12	0
Lamp glasses and renewal chimneys . . .	4	0	3	2·4
200 metres (say 656 feet) of boarding, 500 boards at				
1 franc each	500	20	0	0
	1101·8	44	1	5·28

3. *Cost of Building and Furnishing the Smithy.*

	Francs.	£	s.	d.
200 sacks of charcoal at 1·8 francs the sack . .	360	14	8	0
1 bellows	50	2	0	0
A small anvil	40	1	12	0
„ „ vice	30	1	4	0
A set of hammers	20	0	16	0
3 spanners	10	0	8	0
Trunk of tree (bed for anvil)	4	0	3	2·4
Bar iron and iron plate	90	3	12	0
Building of shed to cover smithy same as for well .	61	2	8	9·6
100 bricks for furnace	8	0	6	4·8
Building the furnace	14	0	11	2·4
1 window	3	0	2	4·8
	[1] 690	27	12	0

General total for the well and the smithy, construction and equipment,[2] 1,852 fr. 8 (or £74 2s. 2·88d.).

When a second well is opened the same expense will be in-

[1] The authors make this add to 680 francs.

[2] The authors give this total as 1,892·80 francs or £1 12s. more (say £75 13s. 7·2d.). The figures given by the translator correspond with the totals of the different items.

curred, with the exception of the smithy, which will not figure therein.

Freight.—The cost of freight, being very variable, cannot be included in this account, nor in any other of the same nature.

Wages.—We have said that it required 5 men to staff a well. These men are paid at the following rates:—

	Francs.	s.	d.
The 3 miners per day 1·4 francs each . .	4·20	3	4·32
The labourer who works the air propeller. .	1·00	0	9·6
The foreman	2	1	7·2
	7·20	5	9·12

Progress and Cost.—Again, from 1 *metre up to* 100 *metres* in depth, 6 metres may be excavated monthly.

The expenses for these 6 metres (19·68 feet) are therefore :—

	Francs.	£	s.	d.
For the staff	222	8	17	7·2
„ charcoal	18	0	14	4·8
„ the smith	40	1	12	0
„ aeration tubing	6	0	4	9·6
„ boarding	42	1	15	7·2
„ lamp oil	4	0	3	2·4
„ repairs and sundries	40	1	12	0
	372	14	19	7·2[1]

From 101 *to* 200 *metres*, no more than 5 metres per month can be done, and that at the rate of 80 francs the metre (about £1 per foot).

These general indications are naturally only approximative. The miners divide the strata into rocks of greater or less hardness. The following is a table which may be made in regard to the conditions of sinking a well of a square metre of bottom :—

	Monthly Progress.	
	In Metres.	In Feet.
1. Hard granites and quartzose rocks (use of dynamite)	2	6·56
2. „ „ with veins of sandstone	3 to 5	9·84 to 16·4
3. Ordinary granites	6 „ 10	19·68 „ 32·8
4. Clay shales or soft micaceous shales . . .	8 „ 12	26·24 „ 39·36
5. Sandstone and sands petroliferous	5 „ 15	16·4 „ 49·2

From 201 to 300 metres, barely 4 metres will be excavated in a month, at a minimum cost of 100 francs per metre (say 24s. 6d. per foot).

[1] Say 15s. the foot.

It may, therefore, be reckoned that it will take 5 years to excavate a well of 300 metres (984 feet) at a cost of 32,000 francs (say £1,280 or 26s. per foot).[1]

As to the cost of the equipment, it is represented by the plant on the spot.

Daywork versus *Piecework*.—The authors do not recommend working with men paid by the day, because, unless an almost impossible surveillance be exercised, the work will progress much more slowly than the rate they have indicated, and will cost more than the average amounts just quoted. It is better to get the excavation done at so much the running metre, keeping as near as possible to what has been indicated by the authors.

[1] This evidently includes interest on capital.—*Tr.*

PART III.

METHODS OF BORING.

CHAPTER XXI.

GENERAL IDEAS AND HISTORY OF BORING.

Differentiation from Coal and Ozokerit Mining.—In order to reach petroleum it is not advisable to excavate wells tunnel fashion, and consequently, to have to send men below ground who, as we have seen in the previous chapter, are exposed to a thousand dangers. On the other hand, the oil is not extracted, like ozokerit or coal, by the aid of men lowered into the pit, who load it into waggons, but by means of a few arrangements working above ground, and by the use of buckets or pumps, unless the well happen to be a flowing well; in which case all that has to be done is to collect the petroleum in reservoirs.

Classification of Different Systems of Boring.—The lowering of workmen can, therefore, be dispensed with, and the excavation of the well is reduced to simple boring. Several systems of boring have been utilised. These are:—

Shock systems of boring	1. Boring with the rope (by hand).
	2. Free-fall boring (by hand and by steam power).
	3. Canadian system (steam power).
	4. Combined system ,,
	5. American system ,,
	6. Hydraulic system, with the bit (hand and steam power).
Rotation systems of boring	7. Auger system.
	8. Diamond system.

Boring involves greater Initial Outlay but is the most Rapid, and finally the Cheapest and most Efficient System of Well Sinking.—Boring, no matter what the system may be, is more rapid than excavating with the pick, but it always necessitates a more complete equipment of tools, and a much more considerable initial outlay.

The False Economy of Excavation.—People did not realise in the beginning that a work executed rapidly is always cheaper than another which lasts ten times longer. This was one of the causes which weighed against the growth of petroliferous works in Galicia and in Roumania. The first "oil men" who searched for petroleum in these countries obstinately refused to use any other form of excavation than the pick and the shovel, which appeared to them the most economical. These rapacious proprietors, whose researches, moreover, were but rarely crowned with success (it has been previously explained above how difficult it is to reach great depths, and consequently to strike oil, in wells excavated by hand). Far from understanding their own interests, they ruined themselves, retarded the industrial progress of the Carpathian oil fields, and only just missed stopping it altogether. At the outset there were, in fact, more wells excavated by the pick and shovel in that country than there ever had been in America or Russia; and it is only a few years since that boring has dethroned the older processes.

History—Boring an Ancient Process.—As far as age is concerned, we must remark that boring would appear to be, however, much more ancient than the other methods of excavation. The ancient processes are, in reality, boring processes!

Was it not by means of the drill that Drake, the inventor of the "casing" pipes, was the first to reach a bed of petroleum in the United States in 1855? Was it not by means of a rudimentary boring that the Chinese sought for salt springs, and that at a time when Julius Cæsar went to conquer the Gauls. The Chinese boring tool, which is described in the narratives of several explorers who travelled at the beginning of the last century, and which, according to the translation of the ancient chronicles, consisted of a horizontal balancing pole (walking-beam) moving round an axis, and acting as a lever which was used to lift an angle iron, the bit, suspended by a rope from its extremity. At the other end, a workman, by his own weight, basculed this lever, then allowed it to resume the horizontal position, thus producing an alternate rise and fall movement. It was the blows of the angle iron which gradually disintegrated the rock. Here was the principle of the present system of drilling, and almost of boring, by the rope, which Drake made use of twenty centuries afterwards. As will be seen, there is nothing new under the sun.

The Jesuits and the Chinese System of Boring.

In order to obtain proof of what we state, one has only to read the numerous narratives which the Jesuits published in the seventeenth century on this method of boring. One of these is still preserved at Amsterdam. The author there describes, with undisguised admiration, the ingenuity of the Chinese, who bore deep holes in their salt springs " by an iron hand hung from the end of a cord ". In our own day we have taken a step to the rear. By boring the ground with a pick, we have gone back almost to the time of the Troglodytes (cave dwellers).

Introduction of Canadian Methods of Drilling.

In Europe, up to a few years ago, only a few borings were to be seen ; the greater part were wrought according to the free-fall system (Fabian's system) and wrought by hand. It was in 1884 that an intelligent man imported from Canada steam engines and perforators on the Canadian system. This was Mr. Mac-Garwey, who has since made a colossal fortune—the deserved recompense of his initiative.

Enhanced Results Obtained by Canadian System.

It was reserved for his machines to strike the most important oil-bearing beds hitherto discovered in the Carpathians. Two wells belonging to Mr. MacGarwey, together yield more than 16,000 hectolitres (352,000 English gallons) of petroleum per day, representing a value of 95,000 francs (£3,800).

Comparative Percentage of Productive Wells Obtained in Galicia (a) by Drilling, (b) by Excavation.—These figures speak for themselves. We shall only add that, according to the most recent statistics, the owners of the wells *bored* in Galicia have found petroleum in large quantities in 71 per cent. of the wells, and that this proportion becomes reduced to 19 per cent. in the case of *excavated* wells—it having been found impossible to continue the greater part of the latter on account of numerous difficulties.

This remark sums up and ends our chapter.

CHAPTER XXII.

BORING WITH THE ROPE.

History.—The first method of boring adopted in America was boring with the rope. We only mention it as a reminder, and simply as a matter of history.

Impracticable in European Strata.—This method, derived from the Chinese system, was but rarely employed in the European oil fields, where, moreover, its use is rendered altogether impossible, on account of distorted strata caused by the upheaval of the beds, which form the ground beneath; but, in the earlier days of the discovery of petroleum, boring with the rope was very often used on the other side of the Atlantic, and thanks to the few difficulties met with in boring in America, it was, in numerous instances, successful. However, we repeat that the chances of success, with perforators without rigid rods, are more than uncertain, both in the Caucasus and in the Carpathians.

Advantages and Disadvantages.—Boring with the rope presents this advantage. It does not necessitate but comparatively cheap plant, and, afterwards, the time is saved which is lost in raising or lowering the rigid rods, which must each be screwed together at least each 10 or 15 metres (32·8 or 49·2 feet); but, on the other hand, there is the risk of boring slantwise and losing the well, of locking it up (*de l'enclouer*), in the language of the trade. That is a point which the manager of the works ought always to bear in mind.

The Drilling Tackle.—The first drilling tackle which was used in the search for petroleum was almost similar to the apparatus which the inhabitants of the celestial empire formerly used in their search for salt.

Bit.—It consisted (Fig. 34) of a bit, *r* (a tool which at that period was made of a conical form), and which was suspended by a

hemp rope, c, to a horizontal beam, P, capable of oscillating round
an axis, A, and which we shall designate by the name which is
generally given to it, that of *bascule* or "walking beam". This
"walking beam" is suspended at the axis, A; about its middle
length, from one of its extremities, hangs at the end of a cord, a
stirrup, e, in which a workman places his foot, and, in this position
imparts a continuous movement of alternate oscillation to the
"walking beam."

Drilling.—As we have said, the bit is suspended by a cord
between the point of attachment of the stirrup and the axis of the
"walking beam". So as to limit the oscillations of the "walking
beam," care is taken to fix it, by means of a rope, through the end
opposite to that of the stirrup, to a post, P, fixed firmly in the
ground.

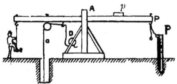

FIG 34.—First method of Rope Boring.

Moreover, in order that all the strength of the workman may
be utilised, there may be fixed to the "walking beam," if need be,
on the side opposite to his one, a weight, p, which will, as far as
possible, balance that of the drill, and thus regulate the oscillatory
movement which the workman imparts to the "walking beam".
It is this movement, communicated to the drill, which causes it to
strike the rock and perforate it.

Sand Pumping.—When the disintegrated materials are bulky
enough to deaden the blow of the drill on the rock, it is necessary
to draw them up. The rope, which supports the drill, is wound up
on a hand winch, D (Fig. 34), of the simplest construction,
and the drill is raised out of the bore hole and then replaced at
the end of the rope by a "sand pump" (Fig. 35), the object of
which is to remove all the detritus from the bottom of the

well which would impede the work, and bring it to the surface.
The sand pump, which is most generally used for hand labour,
consists of a galvanised sheet-iron cyclinder, rivetted or soldered,
in the case of wells of very small diameter, about 2 metres (say 6½

feet) in length, and a diameter equal to that of the
hole being drilled. This
cylinder is provided at the
bottom with a valve open-
ing from without inwards.
When, instead of the drill,
the sand pump has been
lowered down to the bot-
tom of the well, the workman draws up 1 or 2
metres (say 3¼ to 6½ feet)
of the rope from which
it hangs, and allows it to
fall freely several times.
This movement causes the
cylinder to strike forcibly
against the sides of the
bottom where the rock
debris has accumulated.
The debris opens the valve
of the cylinder, and gradually enters it.

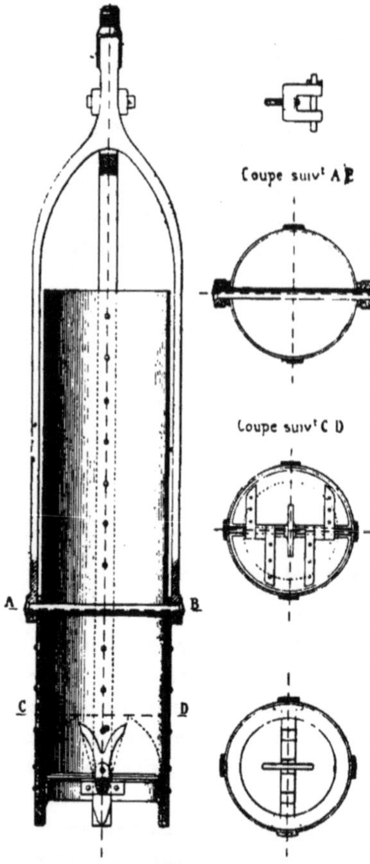

When the workman
thinks the sand pump is
full, he raises it to the
surface, and recommences
the operation as often as
he thinks it necessary.

The number of sand
pumpings necessary in a day vary according to the more or less
friable nature of the rocks traversed.

Examination of the Detritus.—It is very desirable to carefully
examine the rock detritus brought to the surface by the sand

Coupe suiv' A B

Coupe suiv' C D

A   B

C   D

FIG 35.—Sand Pump.

pump. The nature of the beds traversed will thus be ascertained and valuable deductions drawn of the proximity to, or distance from, petroleum, as well as the greater or less number of petroliferous veins contained in the debris, and, moreover the number and the force of the balls of gas which escape from its surface.

Advantage of filling Bore Hole with Water.—We must remark here, and the remark holds good in all systems of drilling, that the bore hole ought always to be filled with water, to the extent at least of a column, 15 metres high (49·20 feet). This is of the highest utility because :—

1. The petroleum, floating on the surface of this layer of water, will not be taken up by the sand pump along with the detritus, as if it were lying quite on the bottom of the well.

2. The water assists the action of the sand pump and prevents to a certain extent the wear of the drill.

3. Finally, certain clays and certain shales, without this expedient, would collapse and fall in, and, if this precaution be not sufficient to stop collapses, it often prevents them.

LANDSLIPS.

Lining with Timber.—In the infancy of drilling, an insurmountable difficulty, or one which was regarded as such, was encountered — landslips. It had been attempted, and almost even accomplished, but at what a cost! to line the bore holes with wood. But, in addition to the fact that this method very quickly reduced the diameter of the wells to a minimum, the woodwork did not last; it could resist neither the pressure of the earth nor the gas pressure; it shattered or broke. It meant the loss of the well! What was to be done in front of this state of affairs? No one found and nobody could reach a sufficient depth to get anything like a sufficient yield.

Colonel Drake's Discovery.—It was at this time that Colonel Drake thought of the plan of lowering retaining pipes of sheet iron, rivetted in the cold, into his wells of a thickness of 0·001 to 0·003 metres (say from 0·0393 to 0·118 inch) according to the greater or less stability of the beds traversed.

Casing the Well.—These pipes are driven into the loose part of the ground liable to slip and shift. As each tube is driven in, the width of the bit should be diminished. It is, therefore, only for fear of the ground falling in that recourse should be had to lining the well with the sheet-iron casing referred to.

Casing with Cast-Iron Pipes.—However, if the sides of the well, even although tubed with sheet-iron casing, be bulged by the weight of the clays, shales or sands composing the ground kept back by the pipes, there should be no hesitation in driving a new sheet-iron pipe in its place, and even, if need be, to use tubes of cast iron screwed together.

In general, the nature of the petroliferous beds are such that the bore hole has to be carefully and entirely cased.

Lowering the Casing.—The pipes are lowered into the well attached to a fastening strong enough to bear the weight of the pipe, but not sufficiently strong to resist a sharp shock or pull.

Snapping the Fastening.—It will be understood that the strength of this fastening can only be determined by experience. It is when the tube is in its place and on raising the rope suddenly and rapidly that the concussion is given which should rupture the fastening.

Plant, etc.—The implements required for drilling with the rope are easily procured. To inaugurate a boring according to this system, neither engineers, expensive implements, nor derricks are required.

System Impracticable and Uneconomical. — Unfortunately, we repeat it, this system is impracticable in the majority of cases. Besides, it is not economical, except in appearance ; in fact, if it only requires 4 workmen for the first 30 metres (98·4 feet), this number is increased by a fresh workman every 30 metres (98·4 feet), so that following this progression, when a depth of 300 metres (984 feet) is reached, each equipment requires 13 men.

There is no system of boring, either by hand or by steam, which necessitates such a numerous staff.

In spite of this, however, this system (improved, it is true) is still used in America. It will be described later on under the title of the American system.

IMPROVED ROPE-BORING SYSTEMS.

1. *Jobard's System.*—Drilling with the rope has been the object of some improvements invented by M. Jobard, of Brussels. Although these improvements have not yet been used in seeking for petroleum, we nevertheless think it right to mention them here. M. Jobard took out a patent for fifteen years for his apparatus, as simple as it is ingenious. The patent is now expired, and is therefore public property.

There is no walking beam in Jobard's apparatus; the drill consists simply of a gin raised above the bore hole provided with a transmission pulley, P (Fig. 36), and a windlass on which the rope is rolled. A bell-ringing movement is imparted to the latter by means of several smaller ropes which are hauled on by the workmen.

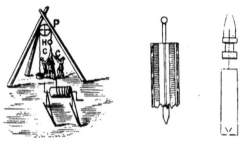

FIG. 36.—Jobard's System of Rope Drilling, Cylindrical Ram and Ram Sand Pump.

These smaller ropes are fixed to the larger main rope by a ring which can be made to slip up and down the rope at will in accordance with the depth of the well.

The drill is raised a height of 0·30 to 0·60 metres (say 1 to 2 feet) according to the resistance of the ground traversed.

The rope which sustains the drill may be made indifferently of aloes, or consist of a cable of iron wire with a hemp core.

Description of Drills.—The drills used are of two kinds :—

1. In hard grounds a cylindrical ram or channelled pile was used, *cannelé en coquille* (Fig. 36).

It was terminated in its lower part by an iron rod with a steel point, destined to pierce and always to fall back into the same hole and in the same direction.

The mud, consisting of the crushed material, ascended the channels and collected in the upper part of the ram, hollowed out for this purpose, in the form of a cone.

2. In soft grounds, a sand pump, with a valve similar to that which has been described above, but surmounted by a ram which, following the movements of the rope, by gliding along the length of a rod placed on the top of the sand pump, drove the latter into the ground by repeated blows.

Goullet-Collet System.—Following the same train of thought, two Frenchmen—Goullet and Collet—have been able to undertake rope borings with their special apparatus with extraordinary economy.

With a drill, consisting of a cylinder of iron plate, 2 metres (say 6½ feet) long, and provided at its base with a steel ring or band and two knives in the form of a cross, each forming a diameter of this circle (Fig. 37), Goullet - Collet have bored more than 100 wells in Champagne, with plant costing no more than 500 francs (£20).

Staff and Progress.—Each boring requires 2 men, and advances 8 metres (say 26·24 feet) daily.

Cost per Foot.—Goullet-Collet fix their price at 9 francs the running metre (say 2s. 3d. the running foot).

But, let us hasten to say, the sands of Champagne and the chalks of the Marne basin are far from presenting the difficulties to the drill which the petroliferous strata, with which we have to deal, do.

FIG. 37.—Goullet-Collet Drill.

This is the reason why, in doing all justice to the apparatus of both Jobard and Goullet-Collet, we confess that we have never seen them used in any borings for petroleum, and that we would not be tempted to use them for that purpose.

We finish this notice of drilling with the rope by these few appreciations of one of our most competent French borers—M. Lippmann.

Lippmann on Rope Boring: " Is there anything more simple than such an equipment ? Why, then, since numerous successes may be quoted, does not the borer take the trouble to impart to this system all the improvements necessary to make its application general? It is because these successes have been and only are obtained in those countries where the homogeneity and solidity of the rocks, the slight dip of the beds, the absence of sections liable to fall in, etc., do not solicit or entice the deviation of the boring implement, do not jam it, and do not necessitate the casing of the bore hole; because that, alongside of the successes obtained, no mention is made of failures, which often necessitate the abandonment of the boring, in which the plant engaged, and the sums expended would not represent a value comparable to the expense which would have to be incurred in clearing the bore hole; because there is no borer, even coming to Jobard and Lechatelier, the most indefatigable partisans of the system, who has had the idea of reverting to the attempts of his predecessors, who has not had to acknowledge the necessity of having, alongside the rope-boring plant, a rigid drill with all its accessories, and the arrangements which its working renders necessary; whether it be to disengage a jammed tool, or to draw up a broken cable, or to traverse clay beds with the auger, or to extract samples, etc."

CHRONOLOGICAL NOMENCLATURE OF THE DIFFERENT SYSTEMS OF ROPE BORING.

Chinese.	17th Century.
Jobard	1826
Combes	1832
Sello	1833
Selligue	1835
Frommann	1835
Alberti	1838
Thomson	1852
Kolb (1st flat drill)	1863
Hattan (Steam)	1825 (? 1865)
Gaiski (free fall)	1868
Goullet-Collet	1868
Sonntag (free fall)	1869
Köbrick	1870
Klerity	1871
Stracka	1872
Sparre (free fall)	1873
Fauck (free fall)	1873
Noth	1873
Mather	1874
Rugins	1875
Siskerlé (free fall)	1876
Benda (free fall)	1877
American	1877
„ improved by Fauck	1883
Neuburger	1895

CHAPTER XXIII.

DRILLING WITH RIGID RODS AND A FREE FALL — FABIAN SYSTEM.

Advantages and Disadvantages.—The system of drilling with rigid rods and a free fall, whether by hand or by steam power, is very much practised at the present time. Moreover, it is an excellent system, which can, in many points, claim superiority over the Canadian and American system. These, however, have an advantage over it, in being more expeditious and lighter. But, on the other hand, the free-fall systems, and more especially the Fabian system, are superior to all others for wells of wide diameter, and constitute the safest and the most economical method of hand boring.

First Introduction.—The first drilling machine used in Galicia (at Bobrka) was on the free-fall system of Fabian. The owners of the well to be drilled—Messrs. Klobassa and Lukasieweiz—realised that boring with the rope did not suit the Carpathian underground strata. They did not have to repent of having eliminated this process of boring, because they found at the outset comparatively difficult ground, and they had to take great precautions not to deviate, by following the direction of the soft beds, even with rigid rods.

Difference between Free-Fall and Rope System.—The principal difference between the free-fall and the rope system consists essentially (without referring to the replacement of the rope by a system of iron rods screwed solidly the one to the other) in the manner in which the drill disintegrates the rock; it is raised by the engine or the workmen who draw the extremity of the bascule I (Fig. 38) towards the ground (*this movement will be easily understood when it is known that the rods are suspended at the opposite extremity* A). When it is at the end of its course or play, instead of following the descent more or less slowly, as in rope boring the drill weighted with a rod heavier than the others is set free in space by a work-man releasing a hook, and falls freely by its own weight to the bottom of the well.   During this time, the rods descend again, and, by an inverse movement of the foreman borer, the drill is again seized, brought up again to a height varying between 0·4 metre to 1 metre (16 inches to 3·28 feet) and again released, and so on.

Each of these differences present a certain advantage :—

1. The force acquired by the drill in its fall.

2. The comfort of the workmen, who have only to lift the drill.   The rods descend by their own weight when the bascule or " walking beam " is well proportioned and adjusted.

3. The safety incidental to the use of iron rods, and the almost certainty there is of keeping to the vertical.

Greater Initial Outlay and Staff.—This system involves a greater outlay of capital than rope-drilling.   The staff is more numerous at the beginning than at the end of the boring (it diminishes in

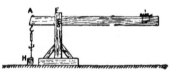

Fig. 38.—" Walking Beam " independent of the Derrick.

direct proportion with the width and weight of the drill).   This staff consists of seven workmen and a foreman borer (*Bohr-meister*).   We only refer just now, let it be well understood, to working by hand power.

The Work of the Foreman Borer.—In regard to the foreman borer, it must be observed that, in order to obtain the maximum effect, he ought to unclutch at the precise moment when the labourers have brought the drill to the end of its journey.   In order to fix this moment precisely, the workers strike the extre-mity of the bascule or " walking beam," I, on a block fixed for the purpose.   The unclutching ought to coincide with the shock, the more so as this shock, causing the drill to spring, eases the work of the foreman driller, who ought to take great care that the drill performs a continuous movement of rotation in the bottom of the well, so that this bottom may be perfectly hori-zontal.   This rotation is made by the wrist which works the

20

clutch. The foreman borer has thus two movements to perform by hand :—

1. A turning movement, from right to left, which effects the unclutching, and also the turning of the drill.

2. An inverse turning movement which reclutches the drill. (This second movement should not be so forcible as the first for fear of inversing the rotation of the drill and bringing it backwards.)

The rotation of the drill should not always be effected at a uniform rate. The foreman should dwell particularly on the hard sections of the bottom of the well, which would otherwise very soon project; and it ought not to be forgotten that deviation from the vertical almost always proceeds from the bottom of the well not having been horizontal.

Sand Pumping. — Sand pumping is done by the sand pump mentioned in the previous chapter. This sand pump is suspended from the end of a cable, so that it may be raised and lowered more quickly than if it were suspended from rods. The latter are raised by means of a strong steel cable, 25 millimetres in diameter (say 1 inch), rolled on to a very powerful winch. As they can only be drawn up by portions of 5 metres (16·4 feet) at the least, their use involves the construction of a drilling tower, otherwise called a " derrick," which supports a pulley at least 6 metres (19·68 feet) from the ground, which is used for lowering the rods. Let us now pass to the description of the various organs of the free-fall system of drilling.

DERRICKS.

Derivation of the Word.—The term derrick, given to these constructions, is almost synonymous with power. In fact it owes its origin to *Theodoric*, the name of a celebrated English hangman, who lived in the sixteenth century.[1] We have no desire to discuss this etymology, in spite of its incongruity, and pass on at once to the technical question.

Precautions to be Observed.—The builder of the derrick ought always to have two principal points in view, *viz.*, the perfect equilibrium (and consequently the stability) of the derrick, and also the lightness of its construction. The resistance to the wind

[1] W. de Fonvielle in his volume, *Le Pétrole*, quotes Ogilvie's Dictionary as his authority for this etymology.—*Tr.*

must also be taken into account, as well as the firmness of the ground covered, and if the derrick be raised over a well which has already been excavated by hand, care must be taken that the pulley for raising the rods is in a perfectly vertical line with the bottom of the well.

Minimum Height of Derrick.—We have said that the iron rods of the free-fall system are nearly always 5 metres (16·4 feet) in length, and that the drill and the auger stem, plus the mechanism of the clutch, are together always at least 6½ metres (21·32 feet). The derrick ought therefore to have its pulley placed at a minimum height of 8 metres (say 26·24 feet) above the opening of the well; unless by special arrangements, which for the most part can only be defective, its total height cannot be made less than 9 metres (say 29·52 feet).

Long Section Rods save Time in Unscrewing.—It will be readily understood that if the rods could be raised by only unscrewing them every 10 or 15 metres (32·8 or 49·2 feet), that is to say, by sections of two or three rods, a great saving of time would be made, which would be very advantageous in working. Let us reflect, in fact, that each unscrewing of a rod requires, on an average, 75 seconds; if, for example, a depth of 100 metres (say 328 feet) has been reached, the time spent in unscrewing, if the rods are taken in sections of 5 metres (say 16·4 feet), will be $20 \times 75 = 1,500$ seconds, or say 25 minutes. If the unscrewing be done every 15 metres (49·2 feet), the time spent in unscrewing will be $7 \times 75 = 525$ seconds (say 8¾ minutes). That makes 17½ minutes gained each time of raising the tackle out of the well, and each time of lowering it down the well. Supposing that 5 of these ascents and 5 descents are made in 24 hours, we get a saving of 2 hours 55 minutes per working day. This represents a gain of more than 10 per cent. In fact, a working hour costs about 4 frs.—2 hours 55 minutes represent, therefore, a loss of 11·90 frs., or 9s. 5·24d. per day.

This little calculation clearly shows that dividing the rods into large sections is the most practical. The staff required is the same as that necessitated by small sections, with the exception of the derrick, which requires to be of greater elevation.

Types of Derricks.—The types of derrick in most general use

are those of a height of 10 metres (32·8 feet) and of 16 metres
(52·48 feet). The highest which have been seen are 17½ metres
(say 57·4 feet). We do not refer here to the tripod derrick, formed

FIG. 39.—Tripod Derricks used for Pumping Purposes at Zagorz, Galicia.

of three pieces of timber, joined at the top and supporting the
pulleys. These, at the present time, are simply a souvenir, and are
no longer used in boring. They may be used at the most for the
installation of a pump to be wrought by a bascule or "walking
beam" in petroleum - yielding wells,
from which the drilling derrick has
been removed for use on another
boring (Fig. 39).

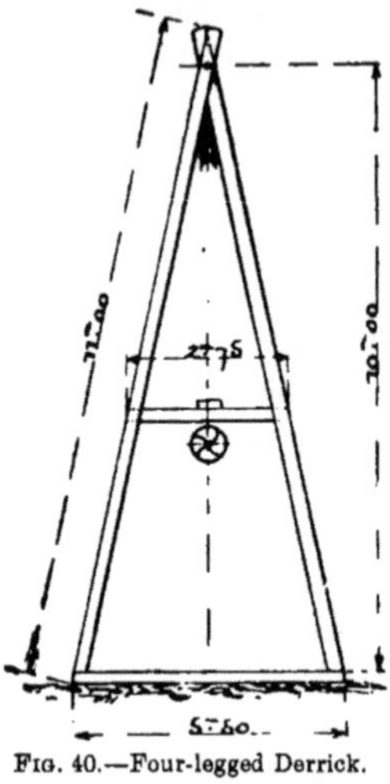

The model of a derrick (Fig. 40),
which we describe in the first instance, is also old fashioned. It is
still constructed, but it has but few
good points.

Its diminutive height 10 metres
(32·8 feet) prevents it from lifting
more than one rod at a time, and its
massive build renders its construction
rather heavy. This plan of derrick
might be selected by a capitalist, but
never by an engineer.

FIG. 40.—Four-legged Derrick.

Its carpentry work consists of four
long pieces of timber joined at the top, and fixed at the bottom
to the four angles of a square base, formed by four beams. The
four long pieces of timber are joined half-way up by four beams
forming a framework or stage, and assuring the stability of the work.

ESTIMATE.

Cost of Building a Derrick according to the above Plan.

	Francs.	British Equivalent (Approximate).		
		£	s.	d.
CARPENTER WORK.				
4 poles forming the derrick, properly so called; length 11 metres (say 36 ft.), width at thick end 0·3 × 0·3 metre (say 12 in.) square . .	24	0	19	2·4
4 beams forming the framework of the base of the derrick; length 5½ metres (say 18 ft.) and 0·25 metre (say 10 in.) square	12	0	9	7·2
4 beams forming the top frame of the derrick; length 2¾ metres (say 9 ft.) and width 0·25 × 0·25 metre (say 10 in.)	6	0	4	9·6
COVERING IN THE DERRICK.				
This consists of 520 boards, 4 metres (say 13 ft.) long, 0·15 metre (say 6 in.) wide by 0·0015 metres (say ½ inch) thick. Price per plank, 0·08	83·20[1]	3	6	4·8[1]
Cost of nailing up	14	0	11	2·4
LABOUR.				
1 carpenter, 4 days at 2 francs = 8 francs . } 10 men, 4 days at 1·4 francs each = 56 francs . }	64	2	11	2·4
Total . . .	203·20	8	2	4·8

We may remark here that, if we have only shown 520 boards for covering in the derrick, it is because we never cover in but from the ground to the smaller framework, and raise above that a roof pierced with holes for the rope used to raise the rods. Moreover 520 boards, 4 metres (say 13·12 feet) long, 15 centimetres wide, and 15 millimetres thick, are exactly what is required to cover the derrick alone. Moreover we do not refer here to the lean-to shed of the derrick used as a shelter for the workmen, which is reserved for consideration farther on.

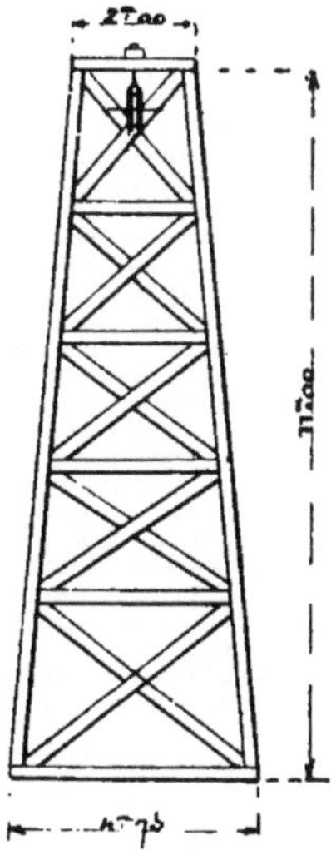

AMERICAN DERRICK.

Description.—The second model of derrick which we shall describe is an excellent one, although rather heavy and not capable of being elevated to very great heights—16

FIG. 41.—American Derrick.

[1] At rate given would appear to be exactly half this amount.—*Tr.*

metres (52·48 feet) and 17·7 metres (58 feet). It is generally known
as the American derrick, because it was employed in the United
States in the beginning of the history of petroleum.   It is like the
derrick described in the preceding—a tower of four uprights—but
these uprights, instead of joining at the top, terminate in a rather
narrow frame, which we shall call the top frame, in contra-distinction
to the bottom frame.   At an equal distance, generally every 2 metres
(6½ feet), the tower is supported by other frames, which we shall
call intermediate frames; these are connected from each of their
side ends with the opposite end of the frame immediately above
them by stays, which, on account of their crossing each other, are
termed crosspieces.   Like the preceding derrick, this one need not
be covered in all over.   It will gain in lightness and cost less, two
advantages which have their importance.

ESTIMATE.

The following is the estimate for the construction of this model
of derrick :—

	Francs.	British Equivalent (Approximate).		
		£	s.	d.
CARPENTRY.				
4 beams (derrick uprights); length 11¾ metres (say 38¾ ft.) × 0·3 metre (say 12 in.) square at thick end	24	0	19	2·4
4 beams (framework of base); length 4¾ metres (say 15½ ft.) × 0·25 metre (say 10 in.) square .	12	0	9	7·2
16 beams (intermediate frames); length of the first frame 4 metres (say 13 ft.), second frame 3½ metres (say 11 ft. 5 in.), third frame 3 metres (say 9¾ ft.), fourth frame 2½ metres (say 8 ft. 2 in.); each beam 0·20 metre (say 8 in.) square.	32	1	5	7·2
4 beams (top frame); length 2 metres (say 6½ ft.) × 0·20 metre (say 8 in.) square . . .	6	0	4	9·6
1 beam (to support pulleys); length 2·2 metres (say 7 ft. 2 in.) × 0·25 metre (say 10 in.) square .	1·5	0	1	0
8 beams (stays); length 4·70 metres (say 15 ft. 3 in.) × 0·15 metre (say 6 in.) square . .	16	0	12	9·6
8 beams (stays); length 4·3 metres (say 14 ft.) × 0·15 metre (say 6 in.) square	14	0	11	2·4
8 beams (stays); length 3·9 metres (say 12 ft. 8 in.) × 0·15 metre (say 6 in.) square . . .	14	0	11	2·4
8 beams (stays); length 3·4 metres (say 11 ft.) × 0·15 metre (say 6 in.) square	12	0	9	7·2
8 beams (stays); length 3 metres (say 9 ft. 9 in.) × 0·15 metre (say 6 in.) square . . .	10	0	8	0
COVERING IN DERRICK.				
This consists of 340 boards 4 metres (say 13 ft.) long × 6 in. wide × ½ in. thick	54·4	2	3	4·34
Cost of nailing	20	0	16	0
LABOUR.				
1 carpenter, 5 days at 2 francs = 10 francs . } 10 men, 5 days at 1·4 francs = 70 francs . . }	80	3	4	0
Total . . .	295·9	11	16	4·34

CANADIAN DERRICK.

The Canadian derrick presents a certain resemblance to the American, but it is more practical than the latter; and the authors specially recommend it to well owners.

Advantages. — If it be rather more costly than the other derricks, it largely compensates for that by the many advantages which it possesses. Notwithstanding its considerable height — Canadian derricks are to be seen 17½ metres (say 57½ feet)—it is extremely light, and the freight of the material of which it is constructed is much less costly than that of the other sorts of similar construction.

Description. — The Canadian derrick, instead of having long and heavy beams as uprights, is sustained by woodwork of planks. The 4 feet or uprights of the tower are replaced by a construction of planks; the frames are constructed in the same way. The stays, which are simple and do not intersect (they are only four for every stage), are likewise made of planks.

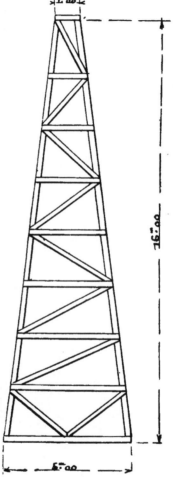

Fig. 42.—Canadian Derrick.

The derrick is entirely covered, in the first place for the sake of stability, the planks

acting as stay crosspieces and rendering all the pieces of the
derrick firm; and, secondly, to preserve it from the weather,
the planks of which it is composed being more sensible to the
action of the latter than the beams of the ordinary derrick.

ESTIMATE.

The following is an estimate of the cost of erecting a derrick of
this kind :—

	Francs.	British Equivalent (Approximate).		
		£	s.	d.
CARPENTRY.				
4 beams (framework of base), length 4·75 metres (say 15 ft. 6 in.) × 0·25 metre (say 10 in.) square	12	0	9	7·2
104 planks (forming the uprights, frames, and single stays of the derrick), length 5 metres (say 16 ft. 3 in.) × 0·25 metre (say 10 in.) wide × 0·08 metre (say 3¼ in.) thick (the dimensions of the parts of the framework and stays the same as in the American derrick), 16 metres (say 52½ ft.) high, and the top framework 1 metre (or 3 ft. 3 in.)	81	3	4	9·6
500 battens, 4 metres (say 13 ft.) long, 6 in. wide, and ½ inch thick	80	3	4	0
Cost of nailing	30	1	4	0
LABOUR.				
1 carpenter, 5 days at 2 francs = 10 francs . ⎫ 10 men, 5 days at 1·4 francs = 70 francs . ⎭	80	3	4	0
Total cost . .	283	11	6	4·8

SPECIAL KIND OF DERRICK.

We have sometimes seen used, especially for "*instrumentation*," very low derricks (Fig. 43) composed of 4 uprights joining

FIG. 43.—"Instrumentation" Derrick.

2 frames. The extra height required for unscrewing the rods
being obtained by excavating a cavity to the depth of 2 or 3 metres

in which the foreman borer stands. The work of the well is therefore by this arrangement carried on at a depth of 2 or 3 metres (say 6 to 10 feet) below the surface of the ground.

The cost of these derricks is but very little, and as they are of variable construction, we refrain from giving an estimate for them.

TRANSPORTABLE DERRICKS.

Sometimes low transportable derricks are used, principally for boring with the drill and hydraulic boring. These derricks are generally made of metal, and may be bought finished in the

FIG. 44.—Fauck's Plant without Derrick.

factory. We shall return to this subject in the chapters devoted to the systems of boring in which they are used.

INSTALLATIONS WITHOUT DERRICKS.

M. Fauck, of Vienna, reports the use of a very simple arrangement which dispenses with a derrick and which is easily transported. Although the cost of this apparatus is rather high, and the mounting of it rather delicate, we have thought it right to mention the innovation of M. Fauck. (Fig. 44).

BORING MACHINES.

THE "WALKING BEAM" OR BASCULE.

Description and Function.—In the majority of the systems of drilling, the "walking beam" is the transmitter of the movement imparted by the workmen to the rods, and, consequently, to the drill.

As is well known, it consists of a pine beam ; its general dimensions being : length, 6 metres (say 19½ feet) by 0·30 metre (say 10 inches square).

The axis of the "walking beam" is generally situated at ⅝ths of its length.

How Wrought.—The labourers work the drill at the extremity of the "walking beam" farthest removed from its axis. From the

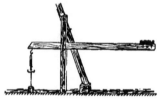 

Fig. 45.—" Walking Beam " fixed to Woodwork of Derrick.

Fig. 46.—Excavation in which Foreman Borer works to escape blows of " Walking Beam "

other end, which ought to be exactly in the axis of the bore hole, the rods are suspended.

Point of Support.—The "walking beam" is sustained by its support, which may, as a matter of indifference, either be portable or form part of the woodwork of the derrick.

Counterpoise.—On the "walking beam" there is fixed a movable counterpoise, capable of being placed at a greater or less distance from the axis, just as the collective weight of the rods and the drill is greater or less. This counterpoise considerably helps the action of the labourers in balancing the "walking beam". But this arrangement has the disadvantage of obstructing the foreman borer in his work. This defect is remedied by excavating a pit 1·5 metres (say 5 feet) deep, in which the foreman borer works without fearing the blows of the end of the " walking beam " (Fig. 46).

Double System of " Walking Beams ".—We may further quote another method, which consists in constructing a double system of " walking beams," as seen in Fig. 47. The upper beam, joined to the lower beam by bars of iron articulated to it, after the fashion of a Roberval's balance, is altogether sound and firm, and it can be placed at such a height as the stature of the foreman borer demands. This point will generally be at $2\frac{1}{2}$ metres (say 8 feet 2 inches) from the orifice of the well.

In this case, the support of the " walking beam " ought to have a soundness and stability proportionate with its height. The best plan is to construct it so as to form part of the body of the wood-work of the derrick.

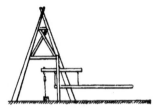

Fig. 47.—Double " Walking Beam " ; its support
forming part of Woodwork of Derrick.

RODS, TEMPER SCREW, ROTATING HANDLE.

Rods.—The rods, employed in the Fabian system, have, generally a length of 5 metres (say 16 feet 5 inches), the width and thickness being equal, and varying between 0·020 and 0·026 metre (say $\frac{13}{16}$ to 1 inch). The weight of the rod in the atmosphere is calculated per metre and square centimetre of section, at 0·8 kilogramme. The cubic centimetre of the section of the rod is therefore equal in weight to 8 grammes, which in water loses the weight of the liquid displaced, say 1 gramme. [In the original this is given as 0·008 grammes and 0·001 grammes respectively, which is altogether wrong and shows that the use of the decimal system is liable to lead to serious errors.—*Tr.*]

The upper end of the rod is formed by an enlargement which may vary from 33 to 40 millimetres (say $1\frac{5}{16}$ to $1\frac{5}{8}$ inches), in diameter in which a female screw is drilled equal in width to the thickness of the body of the rod. The upper end of the latter is furnished with a male screw fitting the thread of the female screw. This screw is cylindrical or conical in form (Fig. 48). The conical screw has the advantage of facilitating the speed in handling, and, on that account, would appear to be preferable; nevertheless, it is contended that the cylindrical screw presents greater security.

About 0·10 metre (say 4 inches) from the screw, there is an enlargement, H, of the rod; the object of which is to hold them firmly in the hand when screwing together or unscrewing, whilst they hang from a clamp or retaining key, which, from its shape, is termed a fork (Fig. 49). This enlargement is called the *safety enlargement*.

We have mentioned that the rods are generally 5 metres (say 16 feet 5 inches) in length. It is the length most generally adopted, but longer rods are also used, and we ourselves prefer the use of rods 10 metres long (say 32·8 feet) as being economical, saving time in screwing, and an economy of 10 to 16 frs. (8s. to 12s. 9·6d.) per screw suppressed.

Lengthening the Drilling Tackle.—But smaller rods must always be kept in stock, so that the length of the system of rods and drill may always be within

FIG. 49.—Fork or Rod Retention Key.

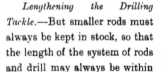

FIG. 48.—Rod of Drilling Tackle.

about 0·01 metre (say ⅖ths of an inch) from the bottom of the well. So that this may be effected, each installation ought to

FIG. 50.—Function of Retention Key. I, Rod. H, Safety Enlargement or Shoulder of Rod. J, Fixed Beams. O, Bore Hole.

have an assortment of adjusting rods, of which the following is a list:—

2 rods of ½ metre (say 19⅝ inches),
1 rod of 1 ,, (say 39¼ inches),
1 ,, of 2 metres (say 6¼ feet),

and 1 rod of 5 metres (say 16·4 feet) if boring be carried out with rods 10 metres (say 32·8 feet) in length.

In this manner the depth of the rods can always be regulated to within about 0·50 metre (19⅝ inches).

Temper Screw : Function.—As rods of 1 centimetre (say 0·4 inches) cannot be had, nor of a dimension less than ½ metre, and as a rod cannot be added every centimetre of increased depth (which would lead to an enormous loss of time), an instrument called the temper screw is used for lengthening or shortening, to a slight extent, the system of rods and drill.

Description.—The apparatus is hung directly from the walking beam by its upper part, A, a screw about 65 centimetres (say 25½ inches in length), which screws into and completely enters a sheath or lower part, B, and it is from this sheath that the rods are hung.

Working.—The working of the apparatus will be readily understood. Every time that the system of rods becomes a little short, which is known from the oscillations of the "walking beam" becoming too great, the temper screw is unscrewed a turn or two until the length of the system of rods and drill appears to be equal to the depth of the well.

When the whole of the temper screw has been unscrewed that signifies that 65 centi-metres (say 25½ inches) have been drilled. A rod, 0·50 metre (say 19⅔ inches) long, is now added, and the temper screw is screwed up thus far, and the boring continued.

The temper screw is provided with a safety nut, E, which prevents the unscrewing of the temper screw when it is not in use.

FIG. 51.—Temper Screw.

Another Method. — Another method of lengthening the system of rods consists in the use of a chain, which supports the rods, the other end of which is wound on to a pulley furnished with a crank and fixed to the "walking beam". When it is desired to lengthen the drilling system, so much chain is unwound.

Disadvantage. — But this arrangement, even although it is cheap, and only consists of a small windlass fixed on the "walking beam," has the grave disadvantage of requiring great precautions in handling, and necessitates the stopping of the work of drilling during the process of lengthening, by which a considerable time is wasted.

Working Bar.—Immediately below the temper screw and adjusted on the first rod is the working bar. It consists of a bar of iron of about 45 centimetres (say 18 inches), which is adjusted in a horizontal position on the rods.

The foreman borer uses it as a sort of lever, so as to move the drill in the desired direction so as to cause it to fall on each point of the bottom of the bore hole in turn or to " dwell " upon any particular spot.

FIG. 52.—Elongation Winch. A, " Walking Beam". S, Support of " Walking Beam ". M, Crank. F, Spring Regulating Elongation. T, Suspension of Rods.

FIG. 53.—Working Bar.

FREE-FALL INSTRUMENTS.

Various Systems.—The *free-fall instrument* is suspended to the rods. These kinds of instruments are numerous, and each system has its particular merits. We shall only quote here as a memento the instruments of Kind, those of Zobel or Sparre, as well as automatic fall instruments.

We shall revert to this subject farther on. For the moment we shall limit ourselves to the Fabian system, which we have taken for our model.

Fabian Free-fall System. - The free-fall system of Fabian is not only the oldest and the simplest, but also the soundest and the most reliable of instruments of this kind.

Merits and Function.—Besides its numerous merits, it has the great advantage of being adapted for all kinds of ground and every size of bore hole. As its name indicates, the object of the free-fall apparatus is to release the drill by a movement of the foreman borer and to catch it again auto-matically.

Description—Upper Part.—It consists in its upper part of a steel sheath of 1½ metres (say about 5 feet), screwing itself by its upper part, which is terminated in the form of a cone, to the rods of the drilling system.

Lower Part.—Within this sheath moves the second piece, generally called the lower part. It is a rod of the same length as the sheath, likewise in steel, which ought to sustain, either by means of a collar in the

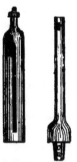

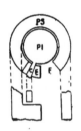

FIG. 54.—Upper and Lower Parts of Fabian's Free-fall Instrument.

FIG. 55.—Section of Fabian's Instrument at the Shoulder.

case of holes of a wide diameter or by means of a screw, the auger stem or the drilling rod, the weight of which is added to that of the bit to increase the disintegrating effects. This lower piece, PI, moves in its sheath, PS (Figs. 55 and 64). It is provided in its upper part, by a collar, C, which, in following its natural movement, glides into the vertical hollow, F, running through the sheath, and shuts firmly at the bottom of this sheath, so that, in spite of the free fall, the rod, PI, cannot separate from the upper part, PS, of the instrument and is suddenly arrested in its fall by the collar, C.

FIG. 56.—
Fabian's
Instrument
modified by
Fauck.

The upper part of the hollow, F, of the sheath, ends in a catch ledge, E (Figs. 55 and 57), which acts by holding the collar, C, and, consequently, the rod, T, the auger stem, as well as the bit during the raising of these instruments.

Clutching. — In the Fabian free-fall system the clutching is automatic. It does not include any aid like Kind's system. It operates simply by a slight bending inwards of the upper part of the hollow (Fig. 57), which, causing the collar, C, to turn, leads it into another closed hollow, where it is retained firmly by the weight of the auger stem and drill, and from which it cannot slip, unless by an abrupt movement of rotation which the foreman borer imparts by his working lever. By this movement the collar is sent back into the axis of the hollow and is then at liberty to fall freely as before.

Fig. 57.—
Plan of the
Unclutching
of Fabian's
Instrument.

Moment to Unclutch. — We here repeat that the foreman borer should only unclutch at the exact moment when the drill is at the end of its course —that is to say, when the workmen strike the extremity of the "walking beam" against the post. If this movement were done too soon, the unclutching, not being facilitated by the rebound caused by the shock from the "walking beam," would be in danger of not acting.

It is, moreover, easy to comprehend the working of the instrument. When the drill has fallen, the rod is nearly quite out of the sheath, being held to the latter only by the collar, C, arrested by the end of the hollow. In proportion, as the workmen lower the rods, the rod, T, enters the sheath, and the collar, C, arrives at the top of the groove and "clutches" itself. It is after this that the workmen raise the rods, which, thanks thereto, sustain the drill. At the top of its course the master borer unclutches,

Fig. 58.—
Auger Stem
Guide.

and thus determines the fall, and the work is recommenced as indicated.

DRILL AND AUGER STEM.

Auger Stem.—The auger stem which holds the drill by a screw or collar is screwed or fixed by a collar to the lower part of the free-fall instrument.

Description and Function.—The auger stem is 5 to 10 metres (16·4 to 32·8 feet) long, generally cylindrical in form, and of a thickness proportional to the diameter of the bore hole, between the sides of which and the auger stem there ought always to exist a space of 2 centi-

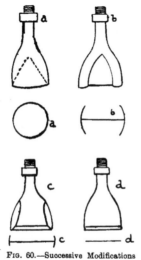

FIG. 60.—Successive Modifications in Shape of Drills.

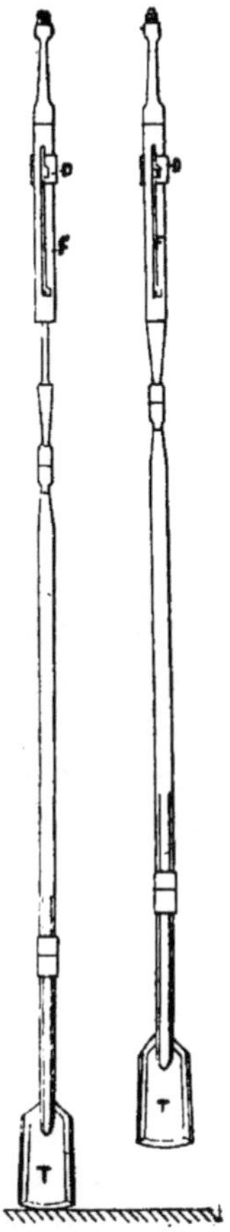

FIG. 59.—Free-fall Instrument, Auger Stem and Drill.

metres (say ¾ of an inch). The *rôle* of this auger stem is to add the force of its own weight to the force acquired by the drill in its fall.

Auger Stem Guide—Its Function.—If the auger stem is much smaller than the diameter of the bore hole, it is covered with a guide (Fig. 58) made of 4 bars of iron to increase its girth so as to obviate the want of thickness of the stem and prevent it from vacillating—in a word, to guide it and to render it impossible for the drill to deviate.

21

THE DRILL.

Transformations and Modifications.—The latter (Fig. 60) has undergone many transformations. Its form has changed so as to be unrecognisable ; from the form of a cone, which it had at the beginning, it afterwards became an inverted cone with circular hinges. In two words, it became a kind of bell (*a*). Then this bell was split on two sides, and the two-fourths of the circle (*b*), which remained of the old circular knife were joined by a straight knife.

FIG. 61.—Blades of Bit with Lugs.

FIG. 62.—Bit. TR, Body of the
Bit. CO, Neck.

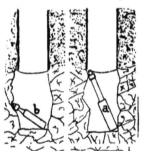

FIG. 63.—Difficulty of raising short-
necked drills when accidentally detached.

Present Shape.—At the present time, the straight knife has prevailed and the circular knives have almost disappeared from the drills (*c*), of the Fabian system, and completely from those of the Canadian system (*d*).

Free-Fall Drill.—But let us return to the bit generally used in drilling by the free-fall system. It consists of two parts, the bit, properly so-called, TR (Fig. 62), and the shoulder, CO. The bit affects the form of an isosceles triangle, whose base forms the blade, and the apex, the extremity which joins the shoulder to the bit. The bits of the Fabian system have a blade terminated, at the two extremities, by two other small curved blades, about 6 centimetres (say $2\frac{3}{8}$ inches) in length, which form, so to say, a part of the circle of which the big blade would be the diameter.

Function of Small Blades.—The object of these small blades is to impart great uniformity and regularity to the bore hole, and to avoid projections on its sides, which would cause an obstruction when " casing ".

Length of Neck.—The neck of the bit ought to have a length of about 30 centimetres (say 12 inches), so that when an accident occurs, (everything must be provided for), arising from the rupture of the point of attachment of the bit to the auger stem, the bit with a long shoulder remains in its position approaching to the vertical (Fig. 63 *a*), and can be brought to the surface of the ground, whilst a bit, *b*, with a short shoulder, falls flat on the bottom of the bore hole (Fig. 63 *b*), and there is great risk of never being able to get a hold of it again.

Screwed or Collared Connections.—Here a question arises : Should the connections of the drill and the auger stem between themselves and with the system of rods be screwed or collared ? Both ; collared in the case of bore holes of great diameter, and screwed in the case of bore holes narrower than 0·25 metres (say 10 in.). This is due to the fact that screwing and unscrewing become very difficult in the case of bits often weighing more than 200 kilogrammes (say 4 cwt.).

Tempering of the Blades. — An advice to finish : The bits having tempered blades, each time that they are re-forged—an operation which may be renewed several times in one day—they ought to be tempered again, but to do that they must be allowed to cool a little, and it is only at the moment that they have become cherry red, that the tempering process can be proceeded with, otherwise they would be too brittle, and their fragile nature might occasion the loss of the instrument, and an accident in drilling.

Fig. 64.—Free-fall System of Drill.

N.B.—In the trade of a miner there is no such thing as trifling precautions.

MACHINES FOR RAISING RODS, WINDLASSES, "LADDER WHEELS".

It is often necessary to raise to the surface of the ground the whole of the drilling system, either to sand pump the well, or to change the bit, or for any other reason.

Hoisting Winch.—The machine most frequently employed for this purpose is a winch (Fig. 65), round which an iron cable is rolled, 50 metres (164 feet) in length, and 25 centimetres (say 10

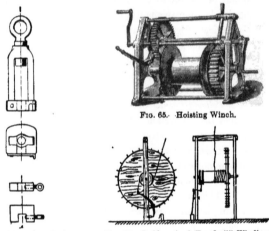

FIG. 65.· Hoisting Winch.

FIG. 66.—Safety Hook.   FIG. 67.—Ladder-wheel (Treadmill) Windlass.

inches) in diameter. At the end of this cable is fixed a safety closing hook called the hoisting key (*clef de relevée*) (Fig. 66). This hook ought naturally to be in the axis of the well, a position which is obtained by an arrangement of pulleys.

, *Ladder Wheel.*—There is also used for raising purposes an apparatus the cost of which is more moderate than that of the ordinary winch. It is the ladder wheel. The ladder wheel consists of a wooden wheel, 3 metres (say 10 feet) in diameter, the circumference of which is furnished with ladder rounds, arranged perpendicularly to its surface. The prolongation of the axis of this wheel is fixed to a drum, which is concentric with it (Fig. 67), and on which the cable for hauling up the rods winds itself. The wheel is wrought by hand, or by the foot by a workman.

Mechanical Advantage.—The size of this wheel multiplies the force of the men tenfold, and thus a cheap windlass is obtained.

Faults of Ladder Wheel.—However, it is but right to state that this windlass has three great faults :—

1. It takes up too much room.

2. Lowering can only be done very slowly through its agency.

3. Finally the fixing of it in its proper position requires a certain amount of technical skill, for the drum on which the raising rope is wound ought to be perfectly perpendicular to the imaginary line joining it with the bore hole.

SAND PUMPS.

Function.—After boring has been in progress for a greater or less length of time, the mass of detritus, formed by the rock debris deadens the blows of the drill and impedes the progress of the boring.

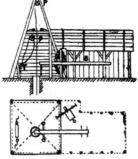

FIG. 68.—Section and Plan of Ladder-wheel Hoisting Installation.

The well must then be sand pumped.

Description.—The instrument used for this purpose is the sand pump, as described on page 298 (Fig. 35).

Sand-pump Cable.—It is suspended from an iron cable, with a hemp coir, of diameter ·007 metre (rather more than ¼ inch).

Sand-pump Windlass.—This cable is raised and lowered by a windlass of the simplest plan of construction without gearing. A drum furnished with two cranks answers very well.

The arrangement of pulleys used to place the sand pump in the axis of the bore holes is the same as that used to raise the rods.

American Sand Pump.—There is also used for this purpose, especially in America, a sand pump (Fig. 69) used in Pennsylvania. In the interior of this apparatus a piston moves, whose *rôle*, so to speak, is to pump the mass to be elevated. But the old form of sand pump is the simplest and cheapest, and moreover may be used under no matter what circumstances ; for which reasons it is preferred by the authors.

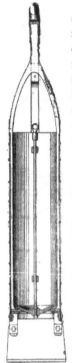

Sand-pumping by Drilling Tackle (Rods).—It is sometimes advantageous to lower the sand pump, not hanging from its own cable, but screwed to the end of the whole system of rods. This happens when masses of argillaceous debris have to be raised, the introduction of which into the sand pump necessitates pressure. In this case, there would be every advantage in using heavy sand pumps with the ordinary means of lowering ; the weight of the apparatus by itself ought to conquer all difficulties.

CASING THE WELLS.

Object : Prevention of Landslips.—When a certain depth has been bored, which may vary between 5 metres (16·4 feet) and 100 metres (328 feet), it may happen that the sides of the bore hole, shaken by the blows of the drill, break up and fall in. The energies of the staff engaged in drilling are now concentrated on the debris which has fallen in, which must be at once raised by the sand pump, an operation which may be followed by fresh landslips.

Use of Water.—The way to avoid landslips is to leave a height of 30 to 40 metres (say 98·4 to 131·2 feet) of water in the bore hole. The pressure of the column of water on the sides will maintain them as long as possible, as well as help the disintegrating effect of the drill and prevent its rapid heating. But if water retards landslips,

FIG. 69.—Piston Sand Pump.

it does not prevent them, and the only remedy is to case the well.

Method.—Casing is done almost always by means of the tubes already described.

Thickness of Tubes.—These tubes are made of sheet iron, ·001 metre to 0·005 metre thick (say $\frac{1}{25}$ to $\frac{1}{5}$ of an inch).

Rolling and Riveting.—They are rolled by a special roller, then joined and riveted together by two rows of rivets, 1 centimetre thick. Each row of rivets is separated from the other by at least 1 decimetre (say 4 inches), and the rivets in each row are at least 3 centimetres (say $1\frac{1}{5}$ inches, Fig. 70) apart, from which it follows that for a tube say 30 centimetres (say 12 inches) wide, the number of rivets required for the two rows forming the joint of the two sheets of iron will be 10 per row, or say a total of 20 rivets in all.

Holes or Eyes for Raising—Tools' Hooks.—Finally care must be taken to pierce holes or eyes in the tubes of about 3 centimetres (say $1\frac{1}{5}$ inches) in diameter. This is a precautionary measure, so as to afford a hold to the instruments in case that from one cause or another, it may be necessary to lift them.

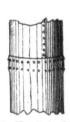

FIG. 70.—How Casing Pipes are Riveted and Joined together.

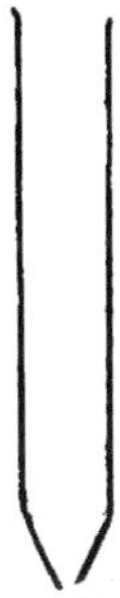

FIG. 71.—Provisional Point; Shape of Lower End of Casing.

FIG. 72.—Deviation of too short casing in A, whilst in B Long Casing retains Vertical position.

Lowering the Casing.—When this is accomplished, the lowering of the tubes into the well is to be proceeded with.

Precautions.—1. Before doing so, however, care must be taken to slightly widen the upper part of the casing to be fixed, so that afterwards the drill does not get caught on the obstruction formed by the top of the tube.

2. It will also be necessary to make the lower part point-shaped, so that the tube throughout its descent passes down easily. This point is only provisional. It will, later on, be cut away by the drill. It may therefore be conveniently made of thin sheet iron (Fig. 71).

3. The tubes ought to be at least 3 metres (10 feet) longer than the part to be tubed, because without this precaution they might be placed in a bad position, and deviate into the cavities produced

by the landslip (Fig. 72), which would lead to very serious inconvenience, as we shall see in the chapter relating to accidents.

All these precautions having been taken, nothing more remains to be done than to lower.

Lowering by Sand-pump Cable or by the Rods.—If *short*, they are lowered by means of the cable used to lower the *sand pump*, but in the case of *heavier* tubes they are lowered by means of the *rods*. The tube is, therefore, attached to the rods by its upper part, by means of two or three iron wires just strong enough to support it

during its descent (Fig. 73), but not strong enough to resist a powerful wrench. It will therefore be desirable to test their resistance inside the derrick. After this test, lowering is started. When the tube has reached the place which it is to occupy, that is to say, when its lower part touches the bottom of the bore hole, two or three strong wrenches from below upwards are given to the rods or to the cable of the sand pump, and the difference in weight will easily be felt when the wires have yielded. From this time the tube is termed *tube perdu* (lost casing).

FIG. 73.—Method of Suspending Casing when being Lowered.

Resumption of Drilling — Reducing the Size of Drill.—The rods are now raised and the drilling continued, and naturally with drills of a lower diameter than the tube just sunk. These will in general have a diameter of 4 centimetres[1] (say 1½ inches) less than those used before casing the bore hole.

Weight of Column of Casing to be Used.—In the lowering of tubes, and so as to avoid accidents, it is necessary to know the weight of the column to be lowered. This weight is ascertained by calculating from the weight of the square metre of sheet iron, which is as follows :—

Millimetres in Thickness.	Inches.	Kilogrammes per Square Metre.	Lbs. per Square Foot.
1	0·0·393	7·78	1·68
2	0·0·786	15·56	3·36
3	0·1·171	23·34	5·04
4	0·1·572	31·12	6·72
5	0·1·965	38·90	8·40
6	0·2·358	46·68	10·08
7	0·2·751	54·46	11·76

[1] The authors give this as 0·04 centimetres—obviously 100 times too small.—*Tr.*

Casing a last Resource.—Whatever advantages may result from casing the well, it should only be resorted to as a last extremity when the progress of the boring has become almost impossible through landslips, otherwise there will be great risk of not being able to sink the well so far as may be necessary to reach the petroleum-bearing bed, the smallest dimension of drill used being 6 centimetres (say 2⅜ inches) ; again, as the widest is generally about 46 centimetres (say 18 inches), it follows that there are only 13 sizes of drill at disposal,[1] and a depth of 350 metres has often to be traversed.

As one ought always to provide for the worst, this ought to be borne in mind.

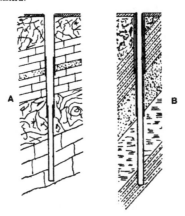

FIG. 74.—A, Partial Casing (*en colonne perdue*) ; and B, Complete Casing.

Disadvantages of "Lost" Tubes or "Tubes Perdus".—The following is another point to be considered : When ground, excessively liable to landslips, is being dealt with, and the deformation of the tubes is to be feared, owing to the pressure of the ground in question, "lost" tubes (*tubes perdus*) should not be used, otherwise there would be a liability to accidents difficult to remedy. In such a case it is therefore necessary to completely tube the well at each new portion of tubing (Fig. 74), in such a manner that the one tube enters into the other and forms a sort of telescope.

Lippmann's Opinion.—We refer to Lippmann in support of our contention :—

[1] From the data given by the authors only 11 sizes of drills are at the most available.—*Tr.*

" I take, for instance," says this engineer, " a boring in process
of execution. . . . A tube must be lowered. Shall we only give
to it a total length of 11 or 12 metres (35·48 or 39·36 feet), so as
to just ' case ' the boring to the extent of the portion liable to
slip ? Or, shall we make it 60 metres (196·8 feet) long, so that
it may come to the surface ? My question may appear *bizarre ;*
because, if the work can be carried on with a tube of 11 to 12 metres
in length, there need be no hesitation in doing so, rather than
expend with a blithesome heart a sum six times greater.

" However, this is a serious question, as to which there is
great disagreement between borers.   Some advise ' casing ' *en
colonne perdue,* that is to say, abandoned in the boring ; others
prefer ' casing ' right up to the mouth of the bore hole. I am one
of the latter, and here are my reasons.

" I have supposed the shifting bed to be 10 metres thick (say
32·8 feet). As soon as I attack it, I must place the casing, which
must be lowered afterwards in proportion to the rate of deepening
of the well, until the solid rock which comes after it is struck,
but, as is frequently the case, either by the effect of undulation
or by any other geological phenomena, the solid rock on which
I counted exists here much farther down, or is even no longer
present. I may, therefore, find myself with too short a casing,
and it will be necessary for me either to attempt with great
labour to raise the ' lost ' column, if there be yet time, so as to
lengthen it, and during this time I allowed the hole to cave
in again, or to decide to lower another tube of smaller diameter,
so that it may pass through the preceding, and what length must
I give it this time so as not to be again left in the lurch ?

" But let us admit that matters were as foreseen, and that
the *colonne perdue* was well laid on a solid bed ; the boring is
continued for 5 or 6 metres, and a fresh shifting bed is come
across. With a complete column which is held by the head, it
can be raised slightly for the purpose of being enlarged, so as to
pass through the solid bed and be caused to descend by prolonging
it so as to ' case ' all the new shifting bed, whatever may be its
thickness ; whilst with the lost column another must be lowered,
which, among other inconveniences, diminishes the diameter of
the bore hole, a serious inconvenience in trial borings, where
the reduction of the diameter may be the cause of the well

being abandoned before its object has been obtained. And, when we are traversing these thick shifting beds, through which we cannot pass the pipes, except by the exercise of great pressure on the head of the column, on account of the friction, at the same time that a boring tool disengages the bottom, is it convenient to go on striking the column, some hundred metres from the surface, whilst leaving a free passage for the drill? And when a rod breaks, when a drill falls, even when the boring system is lowered daily, is there not a chance of injuring the sheets of this telescopic tube? And all that trouble for what? To economise all the lengths of sheet iron, which you are to see doubled, tripled, quadrupled according to circumstances, which oblige you even when using complete columns to place 2, 3, 4 casings, one within the other!   But you will economise nothing! You create difficulties for yourselves; you increase the chances of an accident; you lose your diameter, and I even go farther: your tubes, for ever sunk in the ground, constitute an expense that in the great majority of instances, you would reduce very sensibly if their upper extremity rose to the surface."

In fact, in proportion as the number of tubes increase, the larger may be withdrawn to serve in the boring of other wells. In this way this method of casing becomes cheaper. But care must be taken to leave a sufficient quantity to resist the pressure of the shifting beds, and above all never to use "lost" tubes successively in the same well, if one is not absolutely sure of the resistance of the ground at the point reached.

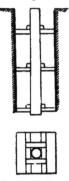

Surface Well.—Before commencing to bore the well, a surface well, 8 to 10 metres deep (26·24 to 32·8 feet), is generally excavated by hand. This well serves to contain the drilling system (consisting in the aggregate of the free-fall apparatus, the auger stem, and the drill) in such a manner that its upper part may be adapted to the temper screw, which is about 1 metre (3·28 feet) from the surface. This surface well is "tubed," and so that the casing may not deviate, it is supported by woodwork inside the well (Fig. 75). The same rule

FIG. 75.—Supports of Casing Pipes in Surface Wells.

applies to wells excavated by the pick and shovel to a greater or less depth, and which it is wished to continue by drilling.

Casing Surface Well.—Several engineers advise, for the sake of economy, that this kind of well should be cased by an octagonal tube of boards. The authors have had the sad experience of this makeshift, and that is the reason why they formally denounce it.

FABIAN SYSTEM—HAND POWER—PRACTICAL HINTS AS TO ITS EMPLOYMENT.

Staff.—The staff working the "walking beam," including the assistant borer, consists generally of 7 men, who should be under the absolute command of the foreman borer, and obey him in everything. In fact, let us consider how great is the responsibility of the latter, and that it is to him that is entrusted the duty of drilling—often difficult—without accidents.

Rests or Spells from Working.—It goes without saying that, if there be not a double shift, the periods of boring are broken by rests, so as to enable the workmen to regain their breath. These rests consist of an interval of five minutes every quarter of an hour, or at the end of 150 oscillations of the "walking beam". (The average of the oscillations in average ground, and under ordinary conditions, is 10 per minute.) In the case of a rest after a certain number of oscillations, these are counted by each of the workmen in turn.

Signs of Necessity of Lengthening Rods.—When the system of rods and drill becomes too short, which the master foreman recognises by the increased number of oscillations of the "walking beam," and, consequently, by the too deep sinking of the rods which do not always reach far enough to catch the drill, he ought to lengthen his system a little by means of his temper screw.

Signs of Necessity for Sand-Pumping.—As soon as the foreman borer feels that the drill only responds with difficulty to his shifting movement, that is a sign that the well is blocked, and that the obstacle preventing the free movement of the drill being nothing further than rock debris, it is necessary to sand-pump the well.

Duration of and Intervals between the various Operations (a) in Drilling.—It will be readily understood that the length of time

occupied in drilling is very variable.   It depends on the nature
of the ground drilled.

(b) *In Sand-Pumping.*—Sand-pumping is done on an average
after 0·40 metre (say 16 inches) has been drilled. Sand-pumping
lasts 15 minutes.

(c) *In Raising the Rods.*—The time occupied in raising the

FIG. 76.—Foreman Borer Drilling.   Free-fall System.

rods is, according to our observations, 40 minutes per 100 metres
(328 feet).

(d) *Lowering of the Rods.*—The lowering of the rods takes 25
minutes, say in all 1 hour 20 minutes per 100 metres (328 feet)
in depth of the well.

Scrutinising the Drill.—When the foreman borer has raised his

drill above ground he will examine the bit to see if it is uniformly
worn on its edge, which would, were it the case, prove that the
surface forming the bottom of the well is equally hard throughout.
If the bit be worn on the sides, that shows that on one side of the
borehole there is a bed harder than the others.

Scrutinising the Debris and Drill Simultaneously.—A consecutive
examination of the drill and the nature of the debris will enable
the nature of the beds and their arrangement to be easily ascertained.

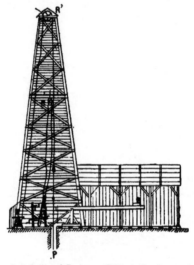

Fig. 77.—Surface Installation of Boring on Fabian's Hand-power System—P, Bore
Hole. B, "Walking Beam". R, Sand-pump Pulley. C, Sand Pump. T, Sand-
pump Windlass. T', Rod-hoisting Windlass. R', Rod-hoisting Pulley.

Daily and Monthly Progress.—With Fabian's system, wrought
by hand power, we have seen 2·8 metres bored in a day (9·2 feet).
That is the maximum which the authors have observed. However, in good ground and a good drilling squad, 45 to 60 metres
(147·6 to 196·8 feet) may very easily be drilled per month.

Average Daily and Monthly Progress.—The average in the borings
which we have observed is 30 metres (98·4 feet) per month. We
therefore give this average as a general average: upon average

ground, with good plant, with a good working squad, 300 metres (984 feet) may be drilled in 10 months at the furthest.

ESTIMATE.

FABIAN SYSTEM, HAND-POWER ESTIMATE.

Cost of Derrick.

	Francs.	£	s.	d.
Cost of construction of derrick (*see* details, p. 312).	283	11	6	4·8

Cost of Equipment of Well.

	Francs.	£	s.	d.
Temper screw	190	7	12	0
Do. suspension for	24	0	19	2·4
Windlass	680	27	4	0
Pulley for hoisting cable, 63 centimetres (say 24¾ in.) diameter	180	5	4	0
Pulley for sand-pump cable, 34 centimetres (say 13⅜ in.) diameter	38	1	10	4·8
Steel cable, 50 metres (say 164 feet, including the safety hook)	390	15	12	0
Sand-pump cable, 330 metres (say 10,824 ft.), and 1 centimetre (say ½⁹ of an inch) diameter . .	150	6	0	0
2 bits, 36 centimetres (say 14·15 in.) wide . .	500	20	0	0
2 ,, 30 ,, (,, 11·8 ,,) ,, . . .	440	17	12	0
2 ,, 25 ,, (,, ·9·8 ,,) ,, . . .	380	15	4	0
2 ,, 21 ,, (,, 8·2 ,,) ,, . . .	356	14	4	9·6
2 ,, 17 ,, (,, 6·7 ,,) ,, . . .	272	10	17	7·2
2 ,, 14 ,, (,, 5·5 ,,) ,, . . .	252	10	1	7·2
Auger stem, 5 metres (say 16 ft. 3 in.) long and 18 centimetres (say 7 in.) wide	500	20	0	0
Auger stem, 6 metres (say 16 ft. 7 in.) long × 18 centimetres (say 7 in.) wide	276	11	0	9·6
4 free-fall instruments of different maxima diameters of 16, 14, 12 and 8 centimetres (say 6¼, 5½, 4¾ and 3½ in.)	1400	56	0	0
1 adjusting piece to connect rods and free-fall apparatus	36	1	8	9·6
300 metres (say 984 ft.) of rods, 2·9 centimetres (say 1⅛ in.) wide	1400	56	0	0
Walking beam and its support, made on the spot .	30	1	4	0
4 sand pumps of different diameters, *viz.*, 30, 20, 15 and 10 centimetres (say 11·8, 7·86, 5·9 and 3·9 in.), made on the works	460	18	8	0
(Spanners, safety forks and different other tools are made on the spot).				
Sand-pump windlass made on the spot . . .	24	0	19	2·4
Total . . .	7,928	317	2	4·8

Cost of Equipment of Smithy.

	Francs.	£	s.	d.
100 sacks of coal at 1·8 francs (say 1s. 5·28d.) . .	180	7	4	0
1 cylindrical bellows	180	7	4	0
1 anvil	160	6	8	0
A set of hammers	80	3	4	0
A set of spanners	60	2	8	0
Making a small wooden crane	70	2	16	0
1 vice	160	6	8	0
1 set of dies	110	4	8	0
1 roller, wooden cylinder, constructed on the spot .	180	7	4	0
1 rail (old), 5 metres (say 16·4 ft.), for making casing tubes	40	1	12	0
Files and other small tools	60	2	8	0
Grindstone	14	0	11	2·4
Trunk of tree for anvil	10	0	8	0
Sheet iron and rivets for casing . . .	300	12	0	0
Bar iron and wrought iron	200	8	0	0
Nails and tacks	50	2	0	0
2 saws, planes, jack plane, 2 hand saws, 4 axes .	60	2	8	0
Pearwood for handles	8	0	6	4·8
Solder, acid and a soldering lamp . . .	24	0	19	2·4
Lamp store, 2 lamps for smithy, 1 lamp for the well, 2 Davy lamps, burning oil, 50 litres of petroleum, wicks, glasses, etc. . . .	56	2	4	9·6
Spades, picks, wheelbarrows, crowbars . .	50	2	0	0·0
2 buckets for smithy	8	0	6	4·8
Digging a receiver for tempering . . .	16	0	12	9·6
Construction of furnace, 200 bricks . . .	36	1	8	9·6
Erection of forge, workshops, stores, a shed 20 metres (say 78·6 ft.) × 10 metres (say 32·8 ft.) .	300	12	0	0
	2,412	96	9	7·2

General Total.

	Francs.	£	s.	d.
General total for the derrick, the well, the smithy (construction and equipment)	10,623 [1]	424	18	4·8

WAGES.

We have said that it requires seven men, in addition to the foreman borer, to man a well. The same number will be required for the night shift; they are paid as follows:—

	Francs.	£	s.	d.
12 men at 1·4 francs per day = 13·44d. . . .	16·8	0	13	5·28
2 assistant borers at 1·8 francs per day = 17·28d. . .	3·6	0	2	10·56
2 foreman borers at 3 francs per day = 28·8d. .	6·0	0	4	9·60
	26·4	1	1	1·44

The tubing comes to 2·4 francs the metre (say 7d. the foot).

[1] The authors give this total as 16,623—evidently a misprint.—*Tr.*

Cost of Smithy.—As to smithy expenses and smiths' wages, consisting of one foreman smith at 70 francs (say £2 16s.) per month, and three assistants at 50 francs (say £2) per month, it amounts to 220 francs (say £8 16s.) for the men, and to 140 francs (say £5 12s.) for coal and sundry tools, say a total of 360 francs, or £14 8s.

Sundries.—Sundry repairs and partial renewal of plant will amount to 80 francs (or £3 4s.) per month.

Fuel, Illumination.—The lighting of the works, the fuel and lodging for the men (when a house is built for them), increase the monthly budget by 40 francs (£1 12s.).

The total monthly expenses rise, therefore, in the case of a well which is being bored, under normal circumstances, 30 metres (98'4 feet) per month, to 1,291·20 francs (say £51 12s. 11·52d., or 10s. 7d. the foot).

Total Cost of Boring a Well 984 Feet.

It may, therefore, be estimated that a well of 300 metres (say 984 feet) may be drilled according to Fabian's system (hand power) in ten months, and that its cost will amount to 12,912 francs (say £516 9s. 7·2d.).

As to the price of the equipment, it is represented by the plant on the spot.

Fabian's System—Method of Commencing to Drill Without a Surface Well.

In those cases where ground of great resistance is encountered in the first few feet, such as granite or other hard rocks, the boring of a surface well is dispensed with by attaching a strong cable to the end of the " walking beam ". This cable will, by the aid of a system of pulleys (Fig. 78), sustain the drilling plant (free-fall instrument, auger stem and drill), which will be entirely within the derrick, and the workmen will work the " walking beam " as usual.

Position of Foreman Borer.—The foreman borer will stand on the platform of the derrick at the height of the working lever, and will proceed as in the ordinary course.

22

Provisional Guiding Casing.—It goes without saying that in this kind of work the greatest precautions are necessary to prevent the drill from deviating.   A provisional casing is, therefore,

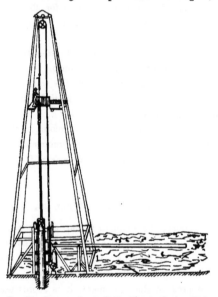

Fig. 78.—Commencing to Drill without a Surface Well.

placed in the axis of the future well, kept firmly in position by a woodwork of beams. The object of the casing, adjusted in this way, is to guide the drilling tools.

CHAPTER XXIV.

FREE-FALL DRILLING BY STEAM POWER.

THE substitution of drilling for excavation constituted a great step in advance in the petroleum industry. But drilling was not really developed until the great force of steam lent its powerful aid to those in search of oil, and enabled them to drill wells 300 metres (984 feet) deep in fifteen days, as in 1884 at Uherce, in Galicia.

As will be seen, the system of drilling always remains the same, with the exception of slight modifications. The motor only is changed, and also the form of the instruments for hoisting the rod and the sand pump; in a word, the windlass.

The methods of casing show some modifications in certain systems, but the principle of the work is the same.

Steam Engine—Unsuitability of Locomobiles.—In spite of the employment (becoming, moreover, less and less general) of traction engines (locomobiles) as motors, we shall not speak of these machines, because we consider their use as absolutely out of place and incompatible with the petroleum industry.

Its Use Ruinous.—In fact, in order to produce an effective result the traction engine must be placed at the farthest 10 metres (say 32·8 feet) from the bore hole. Now, it constitutes in itself a fiery furnace, and it is but very rare that it does not ruin the imprudent man who uses it.

Its Use Interdicted.—Fortunately, a decree of the Board of Inspection (*conseil de surveillance*) has recently forbidden these engines both at Baku and Galicia. Traction engines can no longer be used except it be for the purpose of working transmission pumps. How many accidents would have been avoided if a strict law had forbidden their use from the time steam made its first appearance as auxiliary to those in search of oil.

The engines employed for drilling will, therefore, consist of two distinct elements :—

1. *The Boiler.*— The boiler, which will be placed at a distance of 20 metres (say 65 feet) from the well, and which will be connected with steam pipes with the second part, *viz.*,

2. *The Engine.*—The engine, which can be placed at no matter what distance from the mouth of the well, its use presenting no source of danger.

Kind of Engine Required.—The engine ought to be as light as possible, in a word, transportable, if the locality for which it is intended be situated far from any line of railway; its furnace ought to be large enough and arranged for burning wood, which abounds in the greater number of petroliferous deposits, whilst coal is absent there, and because its cost is very often doubled by the freight from the railway station to the works.

Wood Used as Fuel.—The wood used as fuel for heating the boiler so as to raise steam is almost always fir mixed with one-third of its weight of beech. When the wells yield sufficient combustible gases this gas may be brought as far as the furnace by conduits, an arrangement which results in a great saving of wood.

Crude Oil as Fuel.—Finally steam can be raised with nothing else, except crude oil, or failing it, with heavy oils, the products of distillation. For this purpose an injector, consisting of two concentric pipes, is used. Oil passes through the outside pipe, and in the interior pipe condensed steam or combustible gas, the *rôle* of which is to pulverise the oil in such a manner that it is entirely burned and consumes its own smoke. The play of the apparatus is regulated by two stop cocks.

Calorific Intensity of Crude Petroleum Oil, etc.—We ought to mention here that, according to experiments made in America, 1 kilogramme of oil may convert 22 kilogrammes of water into steam (or say 1 lb. converts 22 lb. or 2¼ gallons), whilst the same quantity of coal only vapourises 10 kilogrammes of water (say 1 lb. converts 1 gallon or 10 lb.). Moreover, the same weight of oil occupies three times less space than the same weight of coal.

Selection of Specially Designed Engine.—It is preferable to buy an engine specially designed for the oil fields, and constructed for that purpose.

Various Uses which Engine has to Serve.—In fact, the uses to which such an engine may be put are very variable. It ought at one time, by a movement on the part of the foreman borer, to slowly put forth its whole energy (wrenching movement), at another increase its speed, then suddenly reverse (filling the sand pumps), finally stop suddenly. It will be seen that our engine is far from being a tranquil factory motor, the principal point of which is regularity; it should be docile and execute all the different movements of hoisting, drilling, "fishing" (repairing accidents and searching for objects which have fallen into the bore hole), without, for all that, requiring repairs after a few months. In a word, it should possess the following merits: lightness, strength, docility, simple construction, all repairs having to be done on the spot when the oil works is at a distance from the great centres.

Steam Engine Manufacturing Countries.—The countries engaged in the manufacture of engines are France, Britain, Germany, Austria and the United States.

Merits and Defects of Engines from Different Countries.—We shall now pass in review the merits and defects of the engines constructed in these different countries.

Engines of French Construction.—To our regret we must eliminate France, which does not build engines specially designed for searching for petroleum; she has no market for them at home, and it must be acknowledged that she sells too dear for exportation. French engines, however, are models, marvels of precision; and if France built special engines for boring, we would be pleased to enumerate their merits, being certain of only stating the truth.

Suitable French Boiler.—However, we recommend, as a French made machine Boulet's (formerly Hermann La Chapelle's) horizontal boiler of 20 to 25 h.p.; but only the boiler, the engine does not fulfil the necessary conditions.

Pumping Engines.—For pumping the traction engine (locomobile) of the same factory (compound system) is an economical machine without an equal, and for the same purpose we may mention the traction engines of MM. Chaligny and Guyot-Sionnets, the semi-fixed engine (compound system), constructed at the Cail engine works; this latter, resting in the front part on a cast-iron sole, and in its back portion on the base of the furnace, does not require special foundations. All that is required to fix it is a resistant soil.

Engines of American Construction.—The American engines used in searching for petroleum are models of the kind. Small, light, of simple mechanism but sound construction, they possess all the necessary merits. Unfortunately, they run too dear. Let us quote, amongst these engines, the *Buffalo* and *Electric* motors, which present in the highest degree the merits enunciated.

Engines of British Construction.—The British engine is almost always too heavy; its massive construction, however, presents all the guarantees of soundness which may be required. For

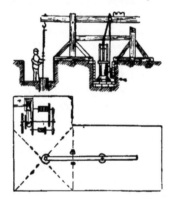

Fig. 79.—Old Steam-power Surface Installation.   Elevation and Plan.

pumping we shall mention Clayton & Shuttleworth's locomobile as a model.

Engines of German and Austrian Construction.—The German motors, as well as the Austrian, are simply copies of the American. Amongst the best we recommend the engines of G. Hammer, of Brunswick; those of M. Fauck, of Vienna; Bredt, of Ottynia, Galicia; and especially those of M. Seeger, of Czernowicz (Buko Vina).

We do not speak of boring tools and tackle (rods, drills, free-fall arrangements, etc.), which remain the same as in the Fabian *hand* system.

DRILLING, DRILLING CRANES OR WINDLASSES AND HOISTING WINDLASSES.

The " Walking Beam ".—Drilling by the aid of steam power is done by means of a " walking beam " in the same way as by hand power.

Driving Rod.—This " walking beam " was at first driven by a vertical cylinder, to which it was joined by a crank attached to the piston rod. The coupling was at the same distance from the axis of the " walking beam " as that axis was from the extremity supporting the rods. In this way the play of the piston ought to be 75 to 80 centimetres (say 29½ to 31½ inches), and mathematically equal to the maximum fall of the drill.

Play.—A play of 1 to 2 centimetres (say ·39 to ·78 inches)

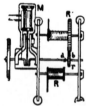

Fig. 80.—Details of Rod Windlass in the first Steam-power Drilling Plant. M, Engine; R', Cog-wheel for Hoisting Rods; R, Sand-pump Cog-wheel; r, Cog-wheel movable on its Axis, a and b, placed where the Wheel r may be capable of being geared in succession to hoist the Rods or the Sand Pump.

was allowed in fixing the crank to the " walking beam," so that the shock of the latter against the post retained all its efficacy.

Hoisting Crane.—The hoisting of the rods or the sand pumps was made by means of a steam crane driven by a small motor fed from the same boiler as that which fed the cylinder or the " walking beam ".

Almost all the borings made by steam power were wrought on this system which did not vary except in the arrangement of its different component parts.

Of such a nature was the boring—directed by Kind, a Saxon—of the artesian well of Plassy, which, we repeat, was the first boring where the principal of the free fall was applied (Fig. 81).

Combination of " Walking Beam " and Hoisting Winch.—Later on, seeing that the use of the two boring and hoisting machines required too much space and ran too dear, it was thought to combine these machines into one only.

With this end in view they grafted on to the hoisting machine a crank moved by the same motor as itself, and communicating the necessary movement to the " walking beam " by means of a coupling.

We shall now describe this machine, which, in spite of the

Fig. 81.—Plassy Boring made under Kind's directions.

addition of the drilling organisation, still retains the name of windlass or crane (German, *Bohr-Krahn*, French, *Treuil*) Fig. 82.

The engine, M, works the "walking beam" by means of the crank, B. On the axis of this crank there is a cog-wheel with eccentric teeth, R ; that is to say, it can glide along this axis and be fixed by collars in several points. These points are three in number, A *b* V, and each correspond to a different working of the machine.

At A the eccentric wheel is free, and although the axis, A, turns and the "walking beam" is in motion, the wheel does no work.

At *b* it gears with the tooth-wheeled S, and, by a movement in either the one direction, or in the other, lowers the sand pump down or raises it up.

At V it gears with the wheel, T, to effect the lowering or raising of the rods.

All these changes only require the shifting of the position of the eccentric wheel.

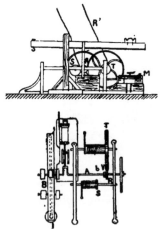

FIG. 82.—First Windlass or Machine combining in a Single Apparatus all the Necessary Parts of both Drilling and Hoisting Machinery.

"*Walking Beam*" *may be put out of Gear.*—Let it be understood, however, that when no drilling is being done the "walking beam" remains independent by lifting the movable coupling which connects it with the crank, B, and pushing it a little to one side so as to afford a free passage into the well for the sand-pump tackle.

Horse Power.—The use of this system requires 12 to 15 horse power.

Objections.—It is but little used for the following reasons :—

1. The shifting of the motions takes up too much time, and necessitates rather long stoppages of the work.

2. Machines working by gearing have the disadvantage of not being so sound as others.

3. Repairs are more costly.

4. The price of the machine itself is very high.

It, however, Drills Effectively.—As far as drilling is concerned, however, this system presents all the advantages of Fabian's system. The effects of drilling—the result of the same number of blows of the drill are even more energetic in the simple system of free-fall drilling by steam—than in its predecessor, the combined system, and that on account of the efficacy of the shock on the post absolutely necessary for the proper working of the Fabian system.

Fig. 83.—Fauck's Drilling Machine.

Fauck the Great Improver of the System.—The learned engineer, Fauck, is the apostle of this system which he has notably improved and which he has made, by his alterations, a rival of the Canadian system, and even an instrument which, from certain points of view, cannot be compared to any other.

Fauck's System.—Fauck has altogether changed the crane which we have just described and the latter tends to disappear more and more. Fauck's system on the other hand appears to get more and more into the good graces of those in search of oil and to compete with the Canadian system. It must, however, be mentioned that Fauck's machines retain a serious inconvenience of the old system which consists in the great cost of the plant.

Number at Work.—In spite of that there are at the present time in Galicia and Germany more than 200 wells being drilled by Fauck's system.

Description.—Fauck's " crane " or " rig " (Fig. 83) is like its predecessor wrought by gearing (cog-wheels) but, instead of being done slowly by hand the changing of the movements is effected by simple pressure of the levers adapted for the purpose. Fauck's crane is mounted on a base of strong woodwork on the top of which is the " walking beam "; this is not moved by coupling with a crank connected with the wheel which receives the motion . of the engine, but it receives the motion of the engine by gearing which, diminishing the number of oscillations of the " walking beam," increases their amplitude.

In this way one factor of disintegration speed is replaced by another force. The system does not lose anything thereby, and

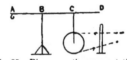

Fig. 85.—Diagrammatic representation of Fauck's " Walking Beam ".

Fig. 84.—Oscillating Head-piece of " Walking Beam ".

this arrangement considerably facilitates the movements to be executed by the foreman borer.

Modifications of " Walking Beam ".—The " walking beam " has also undergone a change in so far that the end at which the rods are attached is provided with a cast-iron headpiece (Fig. 84) oscillating freely on a hinge attached to the " walking beam " in such a manner that the rods which hang from it remain, in virtue of this oscillation, always in the axis of the well in spite of the movements of the " walking beam ".

This " walking beam " of Fauck's system is divided theoretically into three equal parts shown by the points A B C D (Fig. 85).

The extremity A is that to which the rods are attached, B is the point of suspension and oscillation of the " walking beam," C the point of attachment to the motor coupling, and finally, D the extremity which strikes against the post.

This post is retained in Fauck's system; it constitutes one of the peculiar features of that machine.

Substitute for Temper Screw.—Fauck in his apparatus has replaced the temper screw by a small hand windlass fixed on the right side of the woodwork of the "rig," which drives by means of a transmission chain a pulley fixed on the "walking beam" and round which is rolled a chain which supports the rods. This system of elongation has the advantage of being the most expeditious which is important in drilling by steam power.

Another advantage of Fauck's system is that the "walking beam" has not to be shifted during the hoisting or lowering of the rods, and during sand pumping; it suffices in fact to raise the movable head of the "walking beam" to absolutely free the mouth of the bore hole. The drilling rods, the free-fall system and the drills are the same as for all free-fall systems. The rods ought to have a width and thickness of 26 millimetres by 26 millimetres (say 1 inch), so as not to bend in the rapidity of drilling.

Fauck's System—Steam power—Practical Ideas on its Working.—The staff and boring equipment on Fauck's system consists of one foreman borer, two assistant borers, a mechanic and a stoker—in all, five persons.

During the time drilling is in progress, the foreman borer and his first assistant take their turns in succession at the boring handle; the second assistant looks after the elongation of the rods by means of the windlass, the lubrication and the driving of the engine—he will take care that the belt does not slip on the driving wheels, and in such a case, to prevent loss of energy, he will powder them with rosin. The "walking beam" oscillates normally sixteen times in a minute, and the fall of the drill varies between $\frac{3}{4}$ of a metre and 1 metre (between $2\frac{1}{2}$ feet and $3\frac{1}{4}$ feet).

The Workmen's Posts.—During the time of raising or lowering the rods, the foreman borer will remain at his levers to command the movements of the machine; the first assistant at the mouth of the well will screw on (or unscrew) the rods, which the second assistant will attach, or detach, from the hoisting cable; to do this the latter will stand on the first stage of the derrick at a height of 10 metres (say $32\frac{1}{2}$ feet) (Fig. 100, p. 362).

Technical Data as to the Working of Fauck's System.—We take from M. Fauck himself the different technical data as to a drilling period and a sand-pumping period.

The observations were made at a boring made in March, 1889, at Bilin, Bohemia, in search of mineral water. The well had a depth of 128 metres (say 420 feet), the width of the bore hole was 31 centimetres (12 inches), the number of workmen employed was four, finally, the fly-wheel of the engine made five revolutions for one oscillation of the " walking beam ".

1st Period : Drilling, 1,726 oscillations of the " walking beam " .
  Height of the fall of the drill, 90 centimetres (say 35½ inches)
  Disintegrating force of the drill, 535·59 kilos. (1,204 lb.)   .
  Drilled, 51 centimetres, say 20 inches in   .   .   .   .   100 min.
2nd Period : Raising the rods   .   .   .   .   .   .   20 ,,
3rd Period : Sand-pumping, repeated four times   .   .   .   45 ,,
4th Period : Lowering the rods .   .   .   .   .   .   15 ,,

Result of the operations, drilled 51 centimetres, say 20 inches in   3 hrs.

The effective working day being 21 hours, that gives a result of 3·57 metres (say 11·4 feet) per day. The result may be accepted as an average, says M. Fauck.

Error in Fauck's Deductions.—This would appear to us to be correct. However, M. Fauck does not take into account neither the time occupied in casing, nor the inherent difficulties of certain ground ; this is the reason why we state that—although often more than 10 metres (say 32·8 feet) have been drilled in a day by Fauck's machine, and even 6 metres (say 19½ feet) in 12 hours at Nowosielce—the average depth drilled by this system is 65 metres per month (say 213·2 feet). *A well can therefore be drilled to a depth of 300 metres (say 984 feet) in less than five months by Fauck's machine.*

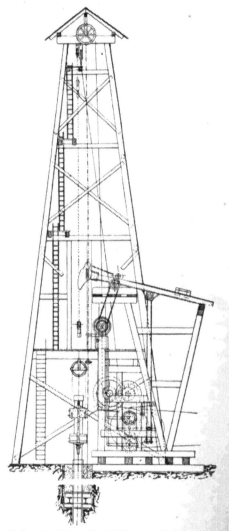

FIG. 86—Surface Installation.   Fauck's System.   Elevated " Walking E

FAUCKS' SYSTEM OF BORING. ECONOMICAL DATA.

	Francs.	£	s.	d.
CONSTRUCTION OF THE DERRICK. (*See* details, p. 312). Total	283	11	6	4·8
COST OF EQUIPPING THE WELL.				
1 shed, 20 metres × 10 metres (say 65·6 ft. × 32·8 ft.) for protecting the machines . . .	110	4	8	0
1 shed, 20 metres × 10 metres (say 65·6 ft. × 32·8 ft.), for protecting the boilers	110	4	8	0
1 portable boiler, 13 to 15 horse power, and its feed pump	3,520	140	16	0
1 horizontal engine, 13 to 15 horse power, diameter of the cylinder 24 centimetres (say 9·4 in.), range of the piston 30 centimetres (say 11·8 in.), pulley with a diameter of 1 metre (39¼ in.), belts included	2,300	92	0	0
1 drilling "rig," Fauck's system (model No. 3), including the different implements for drilling, raising, sand pumping and elongation, fall of the drill capable of varying (according to the coupling of the motion transmitting link with the "walking beam" with 2 cranks) from 1 metre (39¼ in.) to ½ metre (19⅝ in.)	4,840	193	12	0
1 pulley for the hoisting cable, 63 centimetres (say 24·7 in.) in diameter	90	3	12	0
1 pulley for sand-pump cable, 34 centimetres (say 13·4 in.) in diameter	38	1	10	4·8
1 steel cable, 50 metres (say 164 ft.) long, including the safety hook	390	15	12	0
1 sand-pumping cable, 330 metres (say 1,082 ft.) long, and 1 centimetre (say 0·40 in.) in diameter	155	6	4	0
3 bits 36 centimetres (say 14 in.) wide . . .	755	30	4	0
„ 30 „ („ 11·8 in.) wide . . .	660	26	8	0
„ 25 „ („ 9·8 „) „ . . .	555	22	4	0
„ 21 „ („ 8·2 „) „ . . .	498	19	18	4·8
„ 17 „ („ 6·7 „) „ . . .	408	16	6	4·8
„ 14 „ („ 5·5 „) „ . . .	378	15	2	4·8
2 auger stems, 5 metres (say 16¼ ft.) long, the greatest width 0·185 metre (say 7·3 in.) (Fauck's model No. 1)	1,000	40	0	0
2 auger stems, 6 metres (say 19½ ft.) long, the greatest width 60 centimetres (say 23½ in.) (Fauck's model No. 4)	552	22	1	7·2
4 free-fall instruments of different dimensions, maxima diameters, 16, 14, 12 and 8 centimetres (say 6·3, 5·5, 4·7 and 3·1 in.) . . .	1,400	56	0	0
2 adjustment pieces to hold the free-fall instrument on one end and the rods on the other . . .	72	2	17	7·2
300 metres of iron rod (say 984 ft.), 2·6 centimetres × 2·6 centimetres (say 1 inch square) . .	1,660	66	8	0
1 working lever	28	1	2	4·8
Spanners, safety forks	95	3	16	0
3 sand pumps of different dimensions :— One 3 metres (say 9¾ ft.) long × 30 centimetres (say 11·7 in.) diameter . 72·5 fr. One 4 metres (say 13 ft.) long × 19 centimetres (say 7½ in.) diameter . 65·0 fr. One 5 metres (say 16¼ ft.) long × 18 centimetres (say 5 in.) diameter . 58·2 fr.	195·7 [1]	7	16	6·72
Total . . .	19,809·7	792	7	9·12

[1] The authors make this add to 395·7, consequently the total in the original work is 200 francs more.—*Tr.*

For details of the cost of equipping the smithy, workshops,
etc., see the Fabian System, page 332.

	Francs.	£	s.	d.
Total cost of equipping forge, stores and workshops	2,412	96	9	7·2
To which we shall add, for scientific instruments and instruments of precision, which these machines involve	830	33	4	0
Leather belts and duplicates of parts for renewal .	500	20	0	0
	3,742	149	13	7·2

General total for construction and equipment of derrick, well
and smithy, 23,834·7 francs (£953 7s. 9·12d.).

We have said that it takes five men to work the well, this
staff will be doubled by a night shift.   They receive per day :—

	Francs.	£	s.	d.
2 mechanics at 4 francs (say 3s. 2·4d) each . .	8	0	6	4·8
2 stokers at 2 francs (say 1s. 7·2d) each . .	4	0	3	2·4
1 foreman borer at 5 francs (say 4s.) . .	5	0	4	0·0
1 second foreman borer at 3 francs (say 2s. 4·8d.) .	3	0	2	4·8
4 assistant borers at 1·6 francs each (say 1s. 3·36d.)	6·4	0	5	1·44
	26·4	1	1	1·44

The casing of the well, with tubing, comes to 2·4 francs (say
1s. 11·04d.) the metre (say 3¼ feet), or, at the rate of 1s. 9d. the yard.

As to the smithy and its staff (see page 336), it costs 220 francs
(say £8 16s.) per month for wages, and 300 francs (say £12) for
coal and sundry tools, say a total of 520 francs (£20 16s.) per
month.

Different repairs, and partial renewals of plant come to 200
francs (say £8) per month.

The lodging of the workmen costs 60 francs (£2 8s.).

The monthly expenses, therefore, in the case of a well sunk
under normal conditions, at the rate of 65 metres (say 213·2 feet)
per month, amount to 1,728 francs (£69 0s. 2·48d.), which is equal to
26·55 francs (say 21s. 2·88d.) per metre (3·28 feet), or say 6s. 6d.
per running foot.

*It may therefore be estimated that a well, 300 metres (say 984 feet),
may be drilled by Fauck's System (steam power) in less than five months,
and the cost will amount to 7,965 francs (£318 12s.).*

CHAPTER XXV.

DRILLING BY THE CANADIAN SYSTEM.

THE peculiarity of the Canadian system is its lightness, the lightness of all its parts, from the derrick which we have studied (chap. xxiii.) down to its "walking beam," windlasses, and its rods of oak.

Imported into Galicia in 1883 by Mr. MacGarwey, it was not long in superseding all the other systems of drilling, thanks more especially to the rapidity of its boring, which was quite a revelation to contractors and owners of wells in that country.

If the drilling effects of each of the blows are less than in the free-fall system, the Canadian plant compensates grandly for this defect by the number of blows. In fact, whilst Fauck's machine, in one minute, gives fifteen blows of the drill—a falling blow, it is true—1 metre (say 3¼ feet) in height, the Canadian machine gives sixty ; and, on account of this rapidity, it has the advantage.

Engine.—The engine required for an equipment on the Canadian system ought to have a horse power of 13 to 18.

The Rig.—The rig used in the Canadian system is the simplest, lightest, the most easily manipulated, and, what forms no objection, the cheapest of steam drilling rigs.

The dominating material of its construction is wood. The transmission wheels, the drums, the "walking beam" are made of this material. Alongside Fauck's it shows no cog-wheel gearing, with the exception of that of the rod elongation windlass.

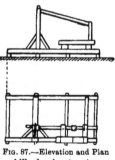

FIG. 87.—Elevation and Plan of Woodwork. American Drilling Machine.

Of simple construction (Fig. 87), easily erected, its light wood-

23

work should be built on the spot.   Every intelligent foreman
borer, with the help of a carpenter, can erect it in a few days.

It consists of a wooden " walking beam ".   The end, A (Fig. 88),
supports the rods ; the extremity, B, forms the point of attachment of the motor coupling ; the axis of the " walking beam " is at
an equal distance from the two extremities.   This " walking beam "
has a length of about 6 metres (say 19½ feet) ; its thickness and
its width are 30 centimetres by 30 centimetres (say 12 inches)

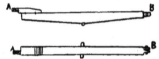

FIG. 88.—Canadian " Walking Beam ".   Profile and Plan.

at the extremities, and 45 centimetres (say 18 inches) at its
centre.

The post does not exist as there is no free fall.

The elongation windlass is fixed on the axis of the " walking
beam".   This windlass (Fig. 89) maintains the chain which sustains
the rods round its axis, by means of a spring which catches
a toothed wheel.   This spring can be held down all the

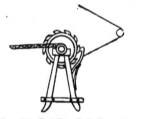

FIG. 89.—Canadian Rod Elongation   FIG. 90.—Adjustment of Auger
Windlass or Reel.   Band Wheel and Bull Wheel.

time that the foreman borer draws the rope, one end of which
is attached to it.   As long as the cog-wheel is unclutched, the
axis will turn, unrolling gradually the elongation chain, without
it being necessary to stop drilling.   When the foreman borer
releases the rope, the spring falls back of its own accord, stopping
the elongation from proceeding farther.

The end of the " walking beam " above the bore hole consists
of a cylindrical head, 50 centimetres (say 20 inches) in length by

20 centimetres (say 7·86 inches) in diameter, the circumference of
which is traversed by a groove in the form of a screw.   Through
this groove the elongation chain passes, which, although able to
glide along and elongate itself at will, remains, however, firmly
fixed to the " walking beam ".

The motor coupling is movable; it may be taken off when
other work than drilling is being done.   During boring it is put
in motion by the crank, which is itself fixed on the axis of the
wheel, Pb (auger band wheel, Figs. 90 and 91), and which is driven
from the engine by a belt which drives the wheel, Pb, fixed on
the axis of the wheel, Pa.   Above the wheel, Pb, there is another
wheel, Pc (the bull wheel), on the axis of which is the hoisting
drum.   The cable for raising the rods is rolled round this

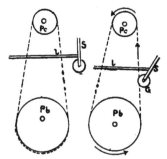

FIG. 91.—Transmission of Power for Hoisting—Canadian System.

drum; this cable, looking to the lightness in weight of the
rods and the speed of rotation of the drum, should not be made
of iron but of hemp, of a diameter of 4½ centimetres (say 1¾
inches), and a length of 50 metres (say 164 feet).

The wheel, Pc, is connected to the wheel, Pb, by a driving
belt; but this belt is not stretched, and glides or slips along
during the rotation of Pb without communicating motion to Pc.

In order to put Pc in motion, and consequently the drum, that
is to say, to lower or raise the rods, the belt must be stretched
in such a manner that it conveys the motion from the one wheel
to the other.

For this purpose, use is made of a cylinder, G (Fig. 91), 25
centimetres (say 10 inches long), which is hung from the end of

two wooden supports, S and S' (Fig. 92), themselves fixed to the
woodwork of the crane by means of hinges.

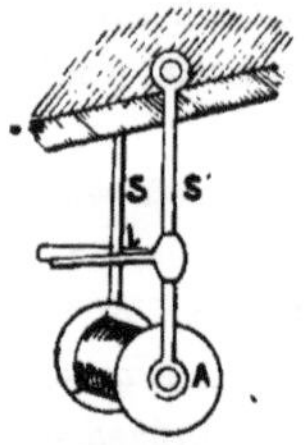

Thanks to this arrangement, the cylinder, G,
can oscillate like the pendulum of a clock.

By means of a lever, l, commanding the movement of the supports of the cylinder, the foreman
borer can apply it to the slack transmission belt,
which causes it to catch and communicate the
motion of Pb to Pc (Fig. 91).

As soon as the pression on the lever is interrupted the roll, G, regains its first position and
the rotary movement of the drum, T, is stopped.

FIG. 92.—
Oscillating
Cylinder.

In the Canadian rig, the drum used for the sand pump
cable is not in existence. Sand pumping is done by screwing the
sand pump on to the rods, and that as quickly as with special
sand-pumping tackle.

SYSTEM OF RODS, DRILLS, ETC.

The drilling tackle rods, auger stem and drill, are very light,
which explains the surprising speed of drilling on the Canadian
system.

The rods are hung to the "walking beam" by a chain 10 metres
long (32·8 feet) and 2 centimetres (say 0·786 of an inch) thick, which
rolls round the axis of the elongation windlass fixed on the "walking
beam" itself. Its rods differ altogether from those employed in the
various free-fall systems. Of the same length they are much
lighter, a fact which easily explains itself when we consider that
nothing but oak wood is used in making them.

When sunk in the water which
fills the bore hole, these rods lose
almost all their weight, which is
counterbalanced by the pressure of
the liquid displaced. It may, therefore, be said that the whole drilling
tackle plunged into a bore hole filled

FIG. 94.—Top of the Canadian
"Walking Beam".

with water weighs almost nothing. It will thus be easily understood that the 15-horse-power engine which works so light a tackle
can expend all its energy in speed, and that by that very fact there

Pl. XIV.—View of an Oil Field wrought according to the Canadian Method. Photograph communicated by M. Zipperlen (Vol. i, *to face p. 357*).

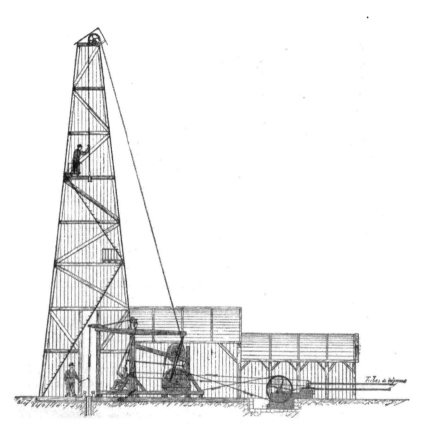

FIG. 93.—Canadian Rig.

is no need of gearing (cog-wheels) to work the rig, which thus gains in lightness, simplicity and solidity.

The wooden rods are cylindrical and 8 to 12 metres (say 26·24 feet to 39·36 feet) in length (the length usually employed being 11 metres, say 35·48 feet) ; the diameter of the body of the rod varies between 5 and 5½ centimetres (say 2 inches), the usual dimensions being 5·2 centimetres (say 2⅛ inches).

The wood of which the body of the rod is composed should be sound, light, and present great resistance to the action of moisture. That is why pine wood is used, and more especially oak.

To the two ends of the body of the rod there are firmly riveted a male and a female conical screw. In the middle is an armature of iron riveted to the body of the rod. The adjusting rods are made in the same way.

WEIGHT OF WOODEN RODS PER SQUARE CENTIMETRE OF
SECTION IN AIR AND WATER.

		Spruce Fir.	Oak.	Poplar.
		Kilog'ms.	Kilog'ms.	Kilog'ms.
Weight per metre in length and per square centimetre of section of wooden rods . .	In the atmosphere .	0·0066	0·0940	0·0041
	Weight of the surrounding air included .	0·127	0·215	0·125
	In water . .	0·007	0·115	0·005

THE " JARS ".

To the rods is screwed the instrument which in the Canadian system takes the place of the free fall. It consists essentially of two flaps pierced longitudinally by a mortice, which fit into each other strideways, and are able to glide one within the other throughout their whole length. The use of this instrument (Fig. 95), which is termed in French *glissière* (glider) (German, *Rutscheere* ; English, *jars*), is that it prevents rupture accidents, which might be produced with rigid apparatus without "jars," and whose system of wooden rods would become a fixture of the drill, and would support along with it the more or less rude shocks which it is the mission of the latter to impart.

The play of the jars varies between 30 centimetres (say 12 inches) and 60 centimetres (say 2 feet), but the fall of the drill is always augmented, by the rebounds which are given to it, by the

speed of the oscillation of the "walking beam". The rebounds, far
from impeding the work, are rather an advan-
tage thereto, by the fact that the force of the
fall of the drill is increased by the fall of the
rods regaining their first position. We must
here remark that the use of the " jars " is not
an American invention, but really the conception of a German engineer, M. Oenhausen.[1]

The *auger stem* (French, *la tige lourde ;*
German, *Schwerstange*) connects the jars with
the drill. It is the same as that used in the
free-fall system, therefore we need give no
further details regarding it.

The bit (French, *trépan ;* German, *Meisel*)
is almost the same as in the free-fall system,
but, on account of the number of blows which
are given in one of its rotations, and which insure the perfect rotundity of the bore hole, the
Canadian bit is not provided with small knife
edges, "cheeks," perpendicular to the large
one of which we have spoken. The energy
of disintegration is spread over a smaller aggregate of cutting lines, hence it increases in
intensity. The bit is made of wrought iron,
its knife edges are of steel.

Fig. 95.—
" Jars " in
Sheath.

The sand pump is generally 10 metres (say
$32\frac{1}{2}$ feet) long, and on account of the speed at
which the rods can be lowered and raised the
sand pumping is effected by screwing the sand
pump to the end of the rods.

Fig. 96.—
Canadian
Bit.

CANADIAN SYSTEM.   PRACTICAL WORKING HINTS.

Canadian drilling requires five workmen,
viz., a foreman borer, two assistant borers, one mechanic,
one stoker.

Fig. 97.—
Canadian
Drilling
Tools.

When drilling, the foreman borer watches the movements of the engine, and from time to time he takes the

[1] Boverton Redwood, *Cantor Lectures on Petroleum*, p. 22, says the "Jars" were
invented by William Morris in 1832.—*Tr.*

rods in his hand to ascertain the nature of the ground being
traversed and its resistance.   By the touch, thanks to experience,
he gains precise enough data, which he supplements by examining
the drill and the materials brought to the surface by the sand
pump.   He also sees to the elongation which he gets by means of
the rope, A, commanding the windlass " temper-screw ".   He also
by means of a rope, B, hanging at his side, regulates the entrance
of the steam into the cylinder, and obtains the effects of speed
or energy necessary for rapidity of drilling.   This rope also serves
to blow the boiler steam whistle, and enables the foreman borer to
communicate, by previously arranged signals, with the mechanic.

During drilling the assistants communicate to the drill the
necessary rotatory movement by working a mobile lever fixed on
the rods.   During the time the rods are being raised or lowered,

Fig. 98.—Foreman Borer guiding the Work—Canadian system.

the foreman borer sits on a small seat in front of the bore hole ; he
has at hand all the levers which command the different movements.

Under his left foot is a pedal, by the pressure of which he
regulates the entrance of the steam into the cylinder of the engine ;
under his right foot is the pedal which regulates the reversing of
the motion.   At his left hand is the brake lever, F, and at his
right hand, E, the lever which commands the movement of the
roller, G, and consequently the drum and the raising of the rods.

During the time the foreman borer is directing the movements
of the plant, the assistants unscrew or screw on the rods, the one
at the bore hole and the other on the first stage of the derrick.

In this way all the movements are easily executed, and no one
will be surprised at the extraordinary effects of drilling by the

Canadian system when we state that the oscillations of the beam at a normal period are from 40 to 65 per minute.

We proceed further to present to the eyes of our readers the observations which we have made during the boring of a well with the Canadian plant.

The well in question, situated at Holowecko, was 133 metres (say 426 feet) deep, when the observations were taken. The width of the bore hole was 21 centimetres (say 8¼ inches). The number

Fig. 99.—Interior View of a Canadian Installation, after a Photograph by Zipperlen.

of workmen was five; finally the engine (mark *Buffalo* of Chicago) was 15-horse-power.

	Hours.	Mins.
First period : Drilling, 4,400 oscillations of the walking beam ; height of the fall of the drill, 0·46 metre (say 18 inches).		
Drilled, 0·42 metre (say 17½ inches) in	1	20
Second period : Raising the rods	0	6
Third period : Sand-pumping (twice)	0	18
Fourth period : Lowering the rods	0	6
Result of the operations : Drilled, 0·42 metre (say 17½ inches), in	1	50

That gives 4·95 metres (say 16·3 feet) per day. Making allowance

for accidents and the time occupied, in casing we may, therefore, say that 95 metres (311·6 feet) may be bored in a month, and that in three months and one week a well 300 metres (say 984 feet) deep may be drilled. It is unnecessary to add that, under good conditions, drilling may be executed much faster. The authors have seen 15 metres (say 49·2 feet), and sometimes 20 metres (say 65·6 feet) drilled in 24 hours by the Canadian system.

Fig. 100.—Manipulating the Rods. *Rôle* of the Workmen.

The Americans claim the record for the Canadian system with the following results : A well (the Denis, No. 1.—District of Bradford) drilled from the 29th of November, 1877, to the 2nd February, 1878, say, deduction made for stoppages a depth of 1,719 English feet (523·77 metres) in 47 effective working days. The average progress made per day being 36½ feet.

The authors, as far as they are concerned, believe this record to have been beaten by the surpassing speed results obtained at Uherce, Galicia.

Here, we shall make a simple remark, without commentary. During a normal period of working with Fauck's system the total duration of the time spent in drilling occupies about 45 per cent. of the whole period of working. In the Canadian system the time occupied in drilling is almost 80 per cent. of the whole period of working. The final result is that the speed of drilling feels these differences, and that is what constitutes the superiority of the Canadian system.

In Canada, the equipment for drilling a well including derrick comes to 1,715 dollars, which is sub-divided thus :—

	Dollars.
Construction of derrick	90
Engines	800
Walking beam, etc.	97
Drilling instruments and various tools	728
	1,715

CANADIAN SYSTEM OF DRILLING.   ECONOMICAL DATA.

	Francs.	£	s.	d.
CONSTRUCTION OF DERRICK. *(See details, p. 312).*				
Total	288	11	6	4·8
COST OF EQUIPPING THE WELL.				
Sheds for engines and boilers (*see* details, Chapter XXIV., p. 351)	220	8	16	0
Boilers, engines, pulleys (*see* details, Chapter XXIV., p. 351) . . . . :	11,071 [1]	442	16	9·6
Canadian drilling crane	1,700	68	0	0
Set of bits (*see* Chapter XXIV., p. 351) . .	3,254	130	3	2·2
,, ,, auger stems (*see* Chapter XXIV., p. 351) .	1,552	62	1	7·2
1 jars, 140 millimetres (say 5·5 in.) . .	247·5	9	18	0
1 ,, 130 ,, (,, 5·07 ,,) . .	234	9	7	2·4
1 ,, 110 ,, (,, 4·3 ,,) . .	160	6	8	0
1 ,, 95 ,, (,, 3·7 ,,) . .	85	3	8	0
1 ,, 80 ,, (,, 3·14 ,,) . .	70	2	16	0
Spanners, lifting tongs for holding rods . .	95	3	16	0
300 metres (say 984 feet) of oak rods . .	1,200	48	0	0
A retaining chain, 8 metres (say 26 ft.) long .	60	2	8	0
1 aloes cable, 40 metres (say 130 ft.) long ; 22 millimetres (say ¼ in.) in diameter . . .	80	3	4	0
Set of sand pumps of the dimensions of the bits .	2,290	91	12	0
Smithy and repairing and renewal of tools .	3,500	140	0	0
Total for equipment of well and derrick . . .	26,101·5	1044	1	2·2

The expenses incurred by the Canadian system are slightly lower than those incidental to Fauck's system, the staff not being so numerous, the total monthly expenses for a well, drilled under normal conditions, of 95 metres (311 feet) per month being estimated at 1,450 francs, say £58, which is equal to 15·25 francs the metre, or 3s. 4d. per foot.

Summary.—It may therefore be estimated that a well 300 metres (984 feet) deep may be drilled in accordance with the Canadian system in about three months, and that its cost will be 4,575 francs (say, £183), or 3s. 8½d. per foot.

[1] This includes the above 283 francs for derrick, which the authors have included twice in their total.—*Tr.*

CHAPTER XXVI.

DRILLING ON THE COMBINED SYSTEM.

Definition.—Derived from the Canadian and the Fabian systems, the combined system of drilling consists in drilling by steam power, with free fall of the drill, and Canadian rig, "walking beam," etc., or *vice versâ*, with an ordinary windlass and wooden rods.

Adopted for Pseudo-Economical Reasons.—Fauck's system being the most rational system of boring with free fall and steam power, what is the utility and the advantages of the combined system of drilling? We cannot tell. It has been said that the Canadian system has the advantage of speed in ordinary ground, and that Fauck's system was the better of the two in hard or difficult ground. We are in agreement in regard to that. But, then, why do the owners of wells who use the combined system seldom or never bore drill by the Canadian system? We believe that the reason is an economical one. The owners referred to had possessed for a long time drilling equipment on Fabian's hand system, and desiring to work by steam power, recoiled before the considerable price of a Fauck drilling machine and bought a cheaper Canadian rig.

Objections to the System.—These, we may at once say, are not so fit for the work. In fact :—

1. *Deficient Unclutching.*—The shock on the post is very useful in boring by free fall. Now, the Canadian system has no post, so that a great number of oscillations of the "walking beam" are lost, the unclutching of the free-fall apparatus not having been effected.

2. *Reduced Speed.*—The speed of the oscillations of the Canadian "walking beam" is in this instance more of a disadvantage than an advantage, seeing that the foreman borer cannot impart

to the working lever 60 times in the minute the movement which gives rise to the free fall. The speed of the engine has, therefore, to be slackened, which causes a loss in the effects of the drill, the force of which is not increased, as in the Fauck system, by gearing.

3. *Less Effective Blow.*—Loss of force on account of the insufficient fall of the drill, the oscillations of the " walking beam " in the Canadian system being too small for a free-fall system.

4. *Bending or Buckling of Rods.*—Finally, the metallic rods

.FIG. 101.—Fauck's Combined Drilling Machine.

often bend on account of the speed of the oscillations, and thus cause accidents.

Speed, the Main Advantage of the Canadian System, Lost.—In a word, the advantages of the Canadian " walking beam," etc., lie in the speed which it can impart to a light tackle. If, for good reasons, and·more especially by the use of heavy boring tackle, this speed cannot be attained, the advantages of the Canadian system disappear.

Moreover, the plant in question must undergo a slight change by the addition of a drum for the sand-pump cable—a drum the

working of which is the same as that of the cable for raising the
rods.

Fauck's Light Steam Crane.—M. Fauck makes a crane (Fig.
101) lighter than that of which we have spoken (Chap. XXV.), in
which the coupling crank of the "walking beam," A, is driven
directly from the engine without any gearing, the "walking
beam" can strike on the post, B, when drilling by free fall.
When drilling, according to the Canadian system, the coupling
is sunk on to a crank shorter than that used in free-fall drilling,
the movable counterpoise is shifted, the oscillations of the beam
are shorter, and the shock on the post does not occur. This
crane possesses an equipment of geared drums for sand-pumping
and rod raising, during the use of heavy tackle in the free fall, and
a drum for raising the rods by a transmission belt, according to
the Canadian system, during the adoption of the latter method.
The different organs of the mechanism of this crane—like all the
Fauck's cranes—are made of iron, and the crane itself is mounted
on heavy woodwork. This should be the real combined crane,
that is to say, a crane uniting the two systems in one single
crane. After all it is but little used, because no one resorts to
the combined system except for economy.

Practical Hints.—The working staff is the same as in Fauck's
system, and the manipulations of the crane are those of the
Canadian rig, and that of the rods those of the Fabian
system.

Observations made at Ropa during the drilling of a well 121
metres (say 397 feet) deep.

Width of the bore hole, 23½ centimetres; five workmen employed; engine (mark *Buffalo*) 15-horse-power; nature of the
ground sandstone of average hardness.

	Hrs. Mins.
First period: Time of drilling (2,826 oscillations of the "walking beam," the height of the fall of the drills, say 2 feet), 29 centimetres (11·4 inches)	1 21
Second period: Raising the rods	0 28
Third period: Sand-pumping 4 times	0 30
Fourth period: Lowering the rods	0 21

29 centimetres (11·4 inches) in 3 hours.

That makes 2·03 metres (say 6 feet 8 inches) per day. Making
allowance for accidents and the time taken up in casing, we may

say that 57 metres (say 187 feet) may be drilled in a month ; that
is to say, a well 300 metres (say 984 feet) in less than six months
by the combined system.

Economical Data.—The economical data appertaining to the
Fabian and Canadian systems may serve for the equipment of a
boring on the combined system.

The cost of working will be the same, but with less progress.
The cost of the running metre (3·28 feet) will come to about
27 francs (say £1 1s. 7.2d.), or say 6s. 8d. the running foot.

CHAPTER XXVII.

COMPARISON BETWEEN THE COMBINED FAUCK SYSTEM AND THE CANADIAN.

In order to be quite impartial, we place before our readers the observations relative to the merits and defects of the Canadian, Fauck's and the Combined System; these observations emanate for the most part from inventors, boring engineers *(ingénieurs perforateurs)*, or manufacturers of drilling machinery. We are gratified at being able to submit them to our readers, who will thus form an opinion as to the merits of the different systems, an opinion which may, perhaps, influence their choice.

Lippmann's Objections to the Canadian System.—At Paris, in 1878, at the International Congress of Civil Engineers, M. Lippmann presented the following objections to wooden rods: "I am quite convinced that my audience is divided into two distinct schools, the one school is for wood, the other for iron, and since I have the opportunity of speaking, I shall crave your permission to give my opinion on the matter forthwith. I shall be proud if I am able at the outset to convince those who are prepared to contradict me, whilst at the same time assuring them beforehand that I have not taken the slightest means to ascertain whether the occasion of having to use wooden rods, even in isolated cases, has been demonstrated.

Legendary Origin of Wooden Drilling Rods.—"The Saxon engineer, Kind, has bequeathed a legend as to their origin: 'A carpenter dropped his three-feet rule into a bore hole almost full of water, the engineer in charge of the work was vexed at having to draw out of the bore hole what he believed to be a metallic tool. 'Don't put yourself about,' said the workman to him, 'my foot-rule is a wooden one, it will float to the surface again.' As he saw it reappear, Kind said to the engineer, 'But our rods also would come to the surface if they were wooden ones'. No sooner said than done, and wood was substituted for iron in the body of the drilling rods.

Modifications suffered by Wood under Heavy Water Pressure.— "However, experiments which every one may repeat, prove that the facts are in contradiction to the conclusion or *morale* of the legend, for a board, lowered to a depth of 100 to 150 metres (say 328 to 492 feet) in water, that is, under a pressure of 10 to 15 atmospheres, undergoes a molecular transformation which, besides the weight of the imbibed water, considerably increases its density; and I have seen such pieces of wood, after having been once submerged for a few minutes, retain, after several months' exposure to the air and to the sun, such a weight that when plunged into water they sank like a stone; a wisp of straw raised from a depth of 400 to 500 metres (say 1,312 feet to 1,640 feet), and dried for several days, sank again to the bottom of a tub of water like a lump of iron. From these facts it is easy to draw another conclusion : it is that in increasing the density, the immersed body ought to undergo a diminution in its external dimensions.

Shrinkage of Wood from Iron Joints loosens the Drilling Tackle.— "Now, whatever system of rods be adopted, the joints or connections used to join them to one another must be made of iron, and this iron must be fixed on the wood by bands, clamps, or other riveted or bolted arrangements. You then see immediately afterwards that the wood has ceased to remain in contact with the iron under the effect of pressure, and then comes bending, the play which is produced in the bolts and rivets, and causes their cutting, the dislocation of the whole assemblage of rods, etc. Besides, the fact must not be lost sight of that in all drilling operations there is often a necessity to impart considerable torsion to the drill which wood cannot resist ; finally, the considerable alteration which wood, which has been immersed in water, undergoes, more or less rapidly, when exposed to the air, must also be taken into consideration. This alteration involves a certain amount of maintenance and necessitates repairs which are not encountered in the case of iron.

Van Dijk Wood-enveloped Iron Rods.—"The only way in which wood could be utilised for drilling rods would appear to me to

24

consist in adopting the process of M. Van Dijk, chief mining engineer, a process which has been demonstrated in the Dutch Colonies, by means of a highly interesting and ingenious plant, used in the sinking of some deep borings in the Dutch Indies—a plant of which I shall have occasion to speak farther on. M. Van Dijk retains the iron rod, to which he only imparts the exact section dimensions to fit it for the traction work which it has to do, and he envelops it throughout its whole length by a polygonal prism, consisting of two envelopes of pine wood. In this way he can utilise all the advantages presented by iron, and the wooden envelope produces the double effect of completely destroying the bending and lashing of the rods, and notably diminishes the weight of the drill by increasing the volume immersed in the water."

On the 8th of September, 1888, the Congress of drilling engineers, under the presidency of M. B. Sigmondy, assembled at Vienna. After interesting communications dealing with boring, M. Fauck, a Vienna engineer, spoke as follows :—

Fauck's Objections to the Combined System.—"The merging of the Canadian rig with the free-fall system is altogether irrational and sufficiently shows that the Canadian system is incomplete when there is a desire to usefully employ a combination with the free-fall system. The combined free-fall system effected in this way cannot be utilised in great depths, seeing that the concussion on the post is awanting, the drill does not fall and nothing is changed with the Canadian. It is not hazard, therefore, that has caused my system to be almost exclusively adopted by the oldest and most experienced mining engineers."

.   .   .   .

" We have 95 wells at Kleczany, the depth of which varies from 250 to 320 metres (say from 820 to 1,050 feet). The last 50 wells have been drilled by my system. But as the district is rather poor in oil, we could not succeed there without imparting an economical and rational direction to the work, and the choice of an economical method of boring was a vital question for our undertaking. As a matter of fact, we annually bore at Kleczany 3,000 metres (say 9,840 feet), on an average, and the annual yield of oil fetches 40,000 to 50,000 florins. The total expense per metre drilled is 12 to 14 florins (say 7s. 3d. to 8s. 6d. per foot), so that under our management, and in spite of the feeble yield of the district, we are able to work at a profit. With the expenses which the Canadian system entails, according to the discourse of Director Hofer, the 3,000 metres (9,840 feet) would cost 90,000 florins, which would involve us in an annual deficit of 40,000 to 50,000 florins."

Immediately after this speech, M. Jurski, a Cracow engineer, spoke and declared that no system of boring could be compared with the Canadian.

Jurski's Points in Favour of Canadian System.—" The Canadian system of drilling is a system of drilling by concussion, to effect which use is made of ' jars ' and wooden rods.

1. *Speed.*—" The peculiarity of this system is the speed obtained in the whole of the movements. The results obtained are extraordinary, the drilling tackle, which often supports a weight of over 12 quintaux (say 24 cwt.), strikes 60 to 70 blows a minute. This speed is facilitated by the lightness of the rods, which are of oak wood and have a thickness of 5 centimetres (say 2 inches). Therefore at each oscillation all the tackle of rods and drills is projected upwards and afterwards delivers a blow in falling freely downwards. The height, or play of the oscillations, ought to be in mutual proportion to their rapidity in order to obtain the maximum effect from the work. The progress of drilling may be 22 metres (say 72·16 feet) per day as a maximum, and 10 metres (say 32·8 feet) per day on an average in rather resistant beds. A depth of 1,000 feet may be reached in 40 days. These figures speak for themselves in favour of this system.

2. *Repairs less Frequent and Rods raised more Easily than in other Systems.*—" The simplicity of the mechanism renders repairs rarely necessary, in regard to this point, and so that we may not have to allude to it again, we shall recall how often, in other systems, repairs cause loss of time. In the Canadian system the well is freed from the rods with a rapidity almost equal to that in boring with the rope. Thus it only takes 12 minutes to raise the rods from a depth of 1,000 feet.

3. *Wide Drills may be used.*—" However, the majority of engineers think that the Canadian system is not available, except for bore holes of small diameter.

" Nevertheless that is an error, because boring may be done very well with the ordinary drills. We, ourselves, having lately drilled at Wiertzno with drills of large dimensions (16 inches), and have cased the well with riveted sheet iron tubes."

At the Paris International Congress of Mines and Metallurgy, M. Léon Syroczinski expressed himself thus in regard to the Canadian system of boring :—

Wiertzno Results.—" As an example of the work we shall quote the figures of a boring of 225 metres, executed by M. Zenon Suszwycki, mining engineer of Wiertzno. Commencing with a diameter of 40 centimetres (say 15·7 inches), and finished with one of 14½ centimetres (say 5·7 inches), it was executed in 90 days, of which 70 were effective working days. The average progress was therefore 3·20 metres (say 10 feet, 6 inches) ; the maximum progress was 9·8 metres (say 32 feet) per day. The same contractor has already made a progress of 20 metres (65·6 feet) in 24 hours.

Canadian System emboldens Contractors to undertake large Contracts to be executed within Stated Time.—" The peculiar characteristic of this system of drilling is the confidence and assurance it inspires to the contractors engaged in the work. One who possesses but a single equipment of plant has no hesitation in binding himself to sink 800 to 900 metres (say, in round numbers, 2,500 to 3.000 feet) ; and a single contractor with the same plant has bored 12 bore holes in a year. The large undertaking for the working and development of petroleum in the domains of the State of M. Stanislas Szczepanowski, which employs several boring contractors, has executed more than 3,000 metres of bore holes in a single year as a proof of the good and prompt acknowledgment of a vast oil field. These figures were not obtainable five years ago when work was carried on under the old system."

The speaker at the end of his discourse declared that it was desirable that every one should recognise that the Canadian system is the cheapest, and that it is adapted for every kind of boring work as well as for petroleum.

Author's Remarks on the Discussion.—Our readers will be able to perceive from these short extracts how opinions are shared as to the merits of different drilling systems.

ARGUMENTS IN FAVOUR OF CANADIAN SYSTEM.

A Testimony from Galicia.—We have, however, an argument in favour of the Canadian : 20 to 25 different steam power systems have been used in Galicia for 20 years. The Canadian appeared in that country only 10 years ago, and at least one half of the wells which have been bored there by steam have been drilled on the Canadian system.

Italy.—In Italy, to which it went with M. Zipperlen, barely 2 years ago, it tends to dethrone all other systems.

Germany.—In Germany and at Baku it has asserted itself. In France the authors have seen it used successfully at Gabian (Hérault).

United States.—In the United States and Canada it is the only system used in difficult ground.[1]

---

[1] By difficult ground the authors apparently mean hard rock. An American oil-well machinery manufacturing firm, speaking of the Canadian System, states that "These tools are especially adapted for drilling hard rock, having water-bearing strata, as where poles will drop free in water, rope will not. When quicksands and soft formations are met the revolving process is the only method that will penetrate them." They also give the following advantages of pole tool plant over cable tools :—

1. The poles are turned continuously in a forward direction, which better insures the turning of the drill at the bottom and the making of a round hole than with the rope when it has to be turned first forward and then backward.

2. An ordinary person will know if he keeps turning the pole at the top that the drill will turn at the bottom. It requires an experienced person to know that with the cable tools.

3. If a pole break it is much easier to take hold of than a rope, and is better suited for drilling in water than the rope. The hydraulic and revolving processes work in harmony with this rig.—*Tr.*

CHAPTER XXVIII.

THE AMERICAN SYSTEM OF DRILLING WITH THE ROPE.

The Principle of the System.—The American system of drilling consists in drilling with steam power and with a rope (instead of iron or wooden rods); the main principle involved in its construction is the same as in that of Drake's old system of drilling (chap. xxiii., p. 296), and it may be said that the motor has alone been changed.

This method of drilling presents an enormous advantage, *viz.*, *speed*, and as great a disadvantage, *viz.*, *want of safety*.

Liability to Deviation.—In fact the absence of rigid rods favours the liability of the bore hole to deviate, and this very greatly increases the risk of so far blocking the work that it may be impossible to proceed with any further drilling of the bore hole.

Only Practicable in Easy Ground.—This is the reason why the system, in spite of its rapidity, is only adopted in a very restricted manner in Europe; in fact only in some undertakings involving the drilling of easily wrought ground has it been used advantageously.

THE ROPE AND DRILLING TOOLS.

Rope.—The rope which maintains the drill is a hemp cable with a diameter generally of 45 millimetres (say 1¾ inches), provided, at one of its ends, with a cylindrical piece of metal, drilled

Fig. 102.--Rope Socket.

out in the form of a screw—the "rope socket" (German, *Seil-hülse*)—which is used to hold the drilling tools. These consist of :—

Pl. XV. —An American Rig. From a Photograph communicated by MM. Fenaille and Despeaux (Vol. I., *to face p. 876*).

(a) *The bit* (French, *trépan* ; German, *Bohrmeisel*).

(b) *The auger stem* (French, *la tige lourde* ; German, *Schwer-stange*), 10 to 11 metres (say 32·8 to 36 feet) long, and often of

(c) The *jars* (French, *glissière* ; German, *Rutscheere*), the *rôle* of which is to give greater rigidity to the tackle by the fact that by its play the boring rope is always stretched.

The drilling tools are the same as those used in the Canadian system, only the jars ought to have a diameter almost as great as that of the bore hole so as to act as a guide for the drill.

Temper Screw. — The elongation of the rope is effected by means of the temper screw, which maintains it by pressure.

Combined Rope and Free-fall Systems.—We ought here to mention boring with the rope and the free-fall system.

FIG. 108.— American Bits: Narrow Section.

Experiments far from practical have been made, with this end in view, with the free-fall systems of Kind, Zobel, Sparre and Przibilla, the unclutching of which is automatic, and does not necessitate the intervention of the foreman borer, which, seeing the non-rigidity of the system, would be quite uneffectual.

FIG. 104.—Surface Arrangement of American Drilling Machinery.

The American Drilling Machinery.—Machinery for working the rope, similar to that used in drilling by the free-fall system, has been in use for a long time, but from the first years of its use this machinery was transformed in so far that the motion rod of the " walking beam " was linked to a crank hanging from the " walking

beam " itself, and was not driven directly from the cylinder. The
" walking beam " was suspended from a point at equal distance
from its two ends.

The machinery at present in use (Fig. 104) is certainly the
simplest of all.

" *Walking Beam.*"—The " walking beam " is the same as the
Canadian ; it therefore does not possess a bouncing post, and is
balanced on its middle point.

It is wrought by a coupling rod termed the pitman, which
receives its motion from a crank attached to the axis of a wooden
wheel—the band wheel—with a diameter of 2·15 metres (say 7 feet).
This wheel running on bearings resting on two uprights is made
to revolve by a belt receiving its motion from the engine. Behind
this wheel there is another of smaller dimensions, called the sand
reel, on the axis of which is the drum used to work the sand pump ;
this wheel by means of a lever can be put in contact with the face
of the band wheel and driven by the resulting friction.

The drum used for lowering and raising the drilling tackle is
fixed round the axis of a transmission wheel—" the bull wheel "—
situated apart from the general mechanism of the system and on
the opposite side of the derrick in such a manner as to be driven
from the band wheel by a belt which during the time drilling is
going on will simply be taken off.

PRACTICAL HINTS ON THE AMERICAN SYSTEM.

Staff.—The number of workmen employed in the method of
drilling is four, *viz.* :—

> 1 Foreman borer.
> 1 Assistant   ,,
> 1 Mechanic.
> 1 Stoker.

A single assistant borer is sufficient, from the fact that there
are no rods and that the drill only requires the handling of but
one man.

The foreman borer and his assistants must take care during
drilling, the one that the engine is working regularly, and the
other to communicate the rotatory motion to the drill by means
of a lever, by turning the drilling rope to the right to the extent

of one-eighth of its circumference each time; then, when the rope by its torsion renders this movement difficult, he will continue the movement, but by turning to the left and so on.

We have already stated that the American system of drilling, so extensively employed in Pennsylvania, was altogether defective in the difficult ground of Europe.

The rapidity with which the tools may be raised and lowered in this system, thereby increasing the number of drilling periods, and causing the work in consequence to progress more rapidly, is almost equalled by the Canadian system, in which the raising of 300 metres (say 984 feet) of rods only occupies 15 minutes on an average.

Slow Progress in Hard Ground.—Moreover, in hard ground progress is only made extremely slowly with the American system. If 15 to 20 metres (say 50 to 65 feet) have been drilled per day with this system, it was in exceptional ground, and the European oil fields possess but little ground of that nature.

That is the reason why the American system has not been completely acclimatised there.

Impracticable with Wide, Heavy Drills.—Yet it is much older than the Canadian. So far back as 1869, Schütter, the engineer, drilled at Mencina, in Galicia, 1 metre per hour. But the cause of the unsuccess is due to another inconvenience of rope boring, the small diameter of the bore holes, drills of great diameter being of such a weight that their use is hardly safe.

Galician Rope-bored Wells.—However, it is but just to state that some borings of this nature were brought to a successful issue at Harklowa, for example, a great number of wells were bored, and are still being bored, by this system. We have been able to obtain the following data in regard to this enterprise :—

The well, 100 metres (say 328 feet) deep, 19½ centimetres (say 7½ inches) in diameter ; 4 workmen are employed ; ground easy.

	Hrs.	Mins.
First Period : Drilling (5,000 oscillations of the "walking beam" per hour; height of the fall of the drill 30 centimetres), 45 centimetres (17¾ inches)	1	40
Second Period : Raising the drill	0	3
Third Period : Sand-pumping (thrice)	0	20
Fourth Period : Lowering the drill	0	4
Result of the operations : drilled 45 centimetres (say 17¾ inches) in	2	7

This would give 4·85 metres (say 16 feet) per day.

Making allowance for accidents, and the time occupied in casing, we shall say that, in easy ground, 100 metres (say 328 feet) may be drilled per month, and that a well of 300 metres (984 feet) may be drilled in 3 months.

In bad ground the American system is *impossible*. There is the risk in using it of not succeeding in boring a single well in a whole year, but the possibility of blocking up half a score.

ECONOMICAL DATA. AMERICAN SYSTEM.
ESTIMATE.

	Francs.	£	s.	d.
CONSTRUCTION OF DERRICK AND BUILDINGS.				
See the Canadian system, p. 363	508	20	2	4·8
Engines	5,820	232	16	0
American rig	1,440	57	12	0
Rope, 400 metres (1,312 ft.)	800	82	0	.0
Drilling tools, same as Canadian, less the rods .	7,987·5	319	10	0
Other expenses, *see* Canadian, p. 363 . . .	3,500	140	0	0
	20,050·5	802	0	4·8

These are the European prices.   In the United States the cost of installing a boring according to this system is as follows :—

	Dollars.
Derrick and buildings	350
Engines and rig	750
Rope	150
Drilling tools	1,220
Other expenses	750
Total	3,220

Say 16,100 francs, or £644.

The wages are the same as those which have been estimated for the Canadian system, *viz.*, 1,450 francs (say £58) per month. In that time 100 metres (say 328 feet) or thereabout, will have been drilled, which will give a cost of 14·5 francs (say 11s. 7·2d.) the metre, or say 3s. 7d. the foot.

CHAPTER XXIX.

HYDRAULIC BORING WITH THE DRILL BY HAND AND STEAM POWER.

Principle—Simultaneous Drilling and Hydraulic Sand-Pumping.— In the preceding chapters concerning drilling by shock we have seen that one of the causes of loss of time was the sand-pumping which, each time it required to be done, necessitated the raising of the rods and the interruption—often lasting for a long time—of the work. We have realised, in fact, in many cases, even in drilling by steam power, that sand pumping took up, in itself alone, one half of the time, occupied in sinking the well.

It was this inconvenience which inspired the idea of hydraulic drilling with simultaneous sand-pumping.

Method.—It is, in fact, by causing a current of water to circulate continually from the surface of the ground to the bottom of the well, and again from the bottom to the surface, that the mud is mechanically carried away, and by this means the time occupied in drilling is no longer subordinate to anything but the wear and tear of the bit.

The Inventor of the System, the famous Perpignan Boring.—The innovator of this principle was an Englishman, named Beart, who was the first to make a practical use of it, at Perpignan, with the machine of the French engineer, Fauvelle, in sinking the artesian well, which is to be seen in the place Saint-Dominique. The ground was propitious for this new system of drilling, and this first attempt was crowned with success. The work was started on the 1st June, 1846, and was finished on the 23rd of the same month, by striking the water-bearing bed at a depth of 170 metres (557·6 feet).

From the 23 days which the boring lasted, there must be

deducted 3 Sundays and 6 days lost, leaving, therefore, 14 effectual
working days.   In those 14 days 170 metres were drilled, or more
than 12 metres (39·36 feet) per day.   The Fauvelle system, therefore, emphasised itself victoriously.

Degoussée's Criticism.—In spite of this success, M. Degoussée,
one of the most authoritative drilling engineers of the period,
sharply criticised the Fauvelle system.

We allow him to speak : "Although the result obtained at
Perpignan is a very remarkable one, we think that the Fauvelle
system—excellent as far as the locality in which it has been
wrought is concerned—can never be generally adopted like drilling
with rigid rods, solid or hollow.

"We shall here recall the report, made in 1842 by M.
Gaymard on M. Freminville's boring who, in the department of
Isère, had started in an equally remarkable manner, and which
ended by engulfing itself in the Parisian tertiary strata.

"We have explained why these borings absorbed considerable
quantities of water, now we ask ourselves how M. Fauvelle can,
whatever quantity of water he may press through the hollow rod,
bring the water from the bottom, charged with detritus of the
boring below ground, when he has come across an ascending
non-spouting sheet of water, this sheet becoming, as a general rule,
absorbent as soon as it is charged above its own level ; we ask ourselves how he will act when he encounters subterraneous currents.

"It will happen, in our opinion, that, in this case, the detritus,
driven from the bottom of the boring, not being able to reach the
surface, will block up the bore hole, and that it will not sometimes
be possible to draw up the drill.

"We could find, in the nature of the beds to be drilled, many
other objections to the ingenious system, which has made its
début in such a brilliant manner, but we believe that we have said
enough to make people wait until the results obtained in other
localities confirm the efficacy of the method in different circumstances."

In 1878, at the Civil Engineering Congress at Paris, M.
Lippmann maintained the same thesis as Degoussée.

In spite of these opinions, Fauvelle's system has stood its
trials so well that, in 1874, M. Fauck successfully used it in seeking

for petroleum, and equipped several borings of this kind, bringing to bear upon them, it is true, several improvements in matters of detail.

FIG. 105.—Bits on the Fauvelle System.

Whilst the Fauvelle system was employed on a grand scale (in Austria, Germany and Russia), its use was neglected in France, where it made its first *début*.

We have seen that this system of boring dates from 1846. In 1847, the society of Mineral Industry of St. Etienne, one of the most influential and most competent societies of mining engineers, asked M. Nougarède, of the Bourgnies Collieries (Aveyron), who used the Fauvelle system, for information as to the working of the "new system" of drilling used by him, and explanations in regard to its construction and manipulation. The reply formed the substance of a communication to the afore-named society.

It has finally been decided to use the Fauvelle system in France and to gradually abandon the plant, as costly as it is obsolete, which found favour with French commercial men.

FIG. 106.—Rod and Auger Stem on the Fauvelle System.

Rods and Drills. — The rods of the Fauvelle system are hollow, so as to conduct to the bottom of the bore hole the column of water sand-pumping the well, there are no "jars" and the rods are screwed directly on to the auger stem. The drill, analogous to an ordinary drill, is pierced in each face by a hole through which the water conduit passes which reaches the drill through its neck.

The mouth of this conduit ought to be situated as nearly as possible to its cutting edge, so that the sand-pumping action may be more energetic (Fig. 105).

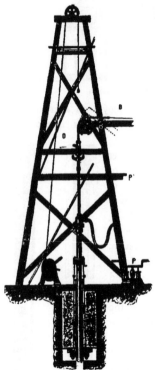

The Auger Stem.—The auger stem is also traversed throughout its whole length by the conduit of water ; it is provided with a guide (Fig. 106).

The rods consist of pipes screwing into one another; they have rather thick sides so as to sustain the weight of all the apparatus without danger of rupture. This was the essential condition to be realised in the new mode of drilling, and this was not done without numerous gropings in the dark.

Gas Piping tried and failed.— At the outset, ordinary gas-piping was used, but it was soon seen that these pipes at great depths burst, especially at the joints ; and the comparatively thin sides of such piping was continually becoming twisted and deformed.

Special Pipes.—Pipes with a thickness of side of 1½ centimetres (say 0·59 inch), and with a diameter for the water conduit of 3 centimetres (say 1·18 inch) were used. The outside diameter was thus 6 centimetres, or 2·36 inches. Finally, so as to obtain great strength at the screw joints, the water conduit was attenuated in these points, and its diameter brought to 1½ centimetres (say 0·59 inches), the exterior diameter remaining the same.

FIG. 107.—Fauck's Hydraulic Oil-Well Drilling Rig. B, "Walking Beam"; P, Pump ; P′, Derrick Stage ; D, Rod-hoisting Cable.

Steel Pipes.—But these improvements, whilst diminishing the chances of rupture, did not suppress them altogether. Pipes were next tried with brass joints without any greater success. Finally, in 1876, steel pipes made their appearance, which are still used at the present time (Fig. 106).

They have an average exterior diameter of 6 centimetres (2·36 inches) and a side diameter of 5 millimetres (say 0·2 inches). It will thus be seen how the use of steel has enabled the thickness of the sides to be reduced, and how lightness of weight in the rods has thus been gained; moreover, the retrenchments made on the water conduit have been suppressed, and the joints are only strengthened by a screwed jacket forming a slight enlargement on the exterior.

We give below a table showing the different dimensions which the hollow steel rods may attain in more or less important borings, whether by hand or steam power.

DIMENSIONS OF THE HOLLOW RODS USED IN HYDRAULIC DRILLING, FAUVELLE'S SYSTEM.

Exterior Diameter.		Diameter of the Water Conduit.		Thickness of the Sides.	
Metres.	Inches.	Metres.	Inches.	Metres.	Inches.
0·033	1·30	0·024	0·94	0·0045	0·177
0·042	1·65	0·032	1·26	0·005	0·196
0·048	1·89	0·038	1·49	0·005	0·196
0·0515	2·02	0·0415	1·62	0·005	0·196
0·050	1·97	0·049	1·92	0·005	0·196
0·076	2·99	0·068	2·67	0·005	0·196
0·089	3·39	0·079	3·00	0·005	0·196

It will be seen from this table that in all the hollow drilling rods, with slight exceptions, the sides have a thickness of 5 millimetres, and that this thickness is not proportional to the diameter of the rod.

HYDRAULIC SAND-PUMPING EQUIPMENT.

By Natural Fall of Water.—If there be in the neighbourhood of the well an abundant stream of water, we can, by means of a reservoir, or a dam, situated at a higher level than the mouth of the well, bring by means of conduits the water required for the hydraulic sand pumping which in this case operates by its **own** weight.

By Pressure Pumps.—But, under certain conditions, it would **be** too costly or too difficult to construct a dam ; a pressure pump is, in such a case, used to convey the water to the bottom of the bore hole through the system of hollow rods and drill ; **this water** rises again in the space between the casing of the well and **the** rods, and brings in its train all the disintegrated particles **of** rock.

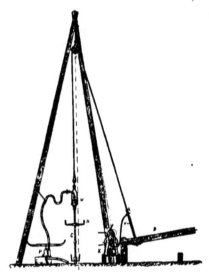

FIG. 108.—Portable Oil-Well Drilling Rig; D, " Walking Beam"; K, Hoisting Winch; N, Hook; S, Pulley; P, Pump; G, Rods.

Sand-Pumping by Reversed Current of Water.—Certain contractors have highly extolled the circulation of the water in a contrary direction to that which we have indicated ; they advise that the water should be pressed into the space between the sides of the bore hole and the rods, and to raise the sand-pumping water through the conduit of the hollow rods. As far as we are concerned we are opposed to this method of using the hydraulic system, considering that the water which has sand-pumped the well is often very muddy, especially in soft ground, and that it may block up the hollow conduit of the rods. This choking up of the rods has always an injurious effect on the work.

Drilling Crane.—The " walking beam," which will only be

FIG. 109.—Arrangement of a Boring on Fauvelle's Hand System. *A*, Support of " Walking Beam "; *G*, " Walking Beam "; *P*, Post; *S*, Adjustment of the Current of Water; *K*, Working Lever; *T*, Head of " Walking Beam ".

raised but about 1 metre (say 3·28 feet), will move as far as possible at the rate of 20 to 25 oscillations a minute, the play of the drill will be 20 to 30 centimetres (say 8 to 12 inches). Both in

FIG. 110.—Fauck's Combined Oil-Well Drilling Machine fitted up for Hydraulic Boring.

working by hand power, as well as in working by steam power, elongation is effected by an elongation windlass.

Drilling effected by Rapid Blows from Low Elevation.—When steam is the motor power used, the drilling crane employed will

25

preferably be that of Fauck, which we have described under the combined system ; only as the oscillations of the " walking beam " are 15 to 25 centimetres (say 6 to 10 inches), the crank which works the motion rod will have a smaller radius than in the combined system. This feeble play of the drill finds its explanation in this fact, that the disintegrated matter ought to consist of highly comminuted particles so as to be easily floated away by the current of water. The blows falling from a low height insure this comminution, whilst their rapid succession—50 to 70 blows per minute—is the factor which constitutes the rapidity of the perforation.

Speed of Current.—The following, according to M. Przibilla of Cologne, is the speed at which the water should be pressed into the conduits so as to cause effectual and complete sand-pumping. This speed varies with the nature of the disintegrated matter to be brought up :—

	Metres per second.	Inches per second.
Fine *débris* and sand . . .	0·02	·786
Sandstone *débris*	0·10	3·937
Rock *débris*	0·15 to 0·20	5·905 to 7·874
Débris reaching 0·05 kg. (say 1·8 oz.)	0·50 to 1·0	19·685 to 39·371
Débris reaching 0·25 kg. (say 9 oz.)	2·0	78·742

PRACTICAL HINTS ON THE FAUVELLE SYSTEM.

Fauvelle System (Hand Power) in Trial Borings.—The Fauvelle system is but seldom used with hand power in the petroleum industry, its first installation being almost as costly as a steam-power equipment on the same system. It is used, however, for small trial borings with plant of very reduced dimensions and a tripod portable derrick (Fig. 108) consisting of 3 iron tubes bolted together and capable of being planted on the ground to be lifted again in a few hours.

Its use in the Colonies in Seeking for Water.—This miniature boring, very extensively employed in seeking for water in the Colonies, only requires 1 foreman borer and 2 workmen.

Hand Power Applications to Petroleum Boring.—As to hand power applications of this system on a large scale to petroleum boring there are but few. But, to be complete, we may say that the general equipment on the surface of the ground somewhat resembles the Fabian hand system.

"*Walking Beam.*"—The "walking beam," having only to impart very feeble oscillations, and comprising beneath it all the pumping system and the adjustment of the water pipes, will have to be suspended rather high, 10 metres (say 32·8 feet), and its oscillating axis will be about its one-tenth.

The Foreman Borer's Post.—The height of the "walking beam" necessitates the foreman borer standing on the platform of the first stage of the derrick so as to attend to the elongation. The pump is on the surface of the ground. Three men are sufficient to work the lever.

Fauvelle System Steam Power.—The Fauvelle system may be employed rationally with steam as the motor power. The engine which works the "walking beam" drives the pump.

Staff.—The staff consists of five persons :—

1 Foreman borer.
2 Assistant borers.
1 Mechanic.
1 Stoker.

Division of Labour.—During drilling, the foreman borer will attend to the proper working of his engine and to the normal elongation of the rods, and to the output of his pump which he will vary according to the nature of the ground being traversed. When the drill is being raised or lowered (before or after being changed), he will stand by his levers, whilst his assistants screw or unscrew the rods. It goes without saying that during these operations the pump is not at work.

Progress.—As much as 10 metres (32·8 feet) may be drilled in 24 hours by Fauvelle's system, but this figure cannot be taken as an

FIG. 111.—Method of Adjusting the Water Conduits. S', Suspension of Rods; B', Water Exit Pipe; K, Working Lever; W, Upper End of Hollow Rods; R and S, Hermetic Passage of the Rods into the Casing; O, Water Entrance Pipe; L U H, Bolts and Crown for Rendering the Rods Independent during Raising or Lowering; B, Casing.

average, because this system, excellent in ground of slight resistance, gives very hazardous results in hard rocks such as granite.
The short play of the drill very often produces but very slight
disintegrating effects on these rocks, and it then happens that
only a few centimetres are bored in a day.

Fauvelle Drill for Soft, and Diamond Drill for Hard Ground.—It is
therefore necessary to combine this equipment and to have two
drills—a Fauvelle drill for soft ground and a diamond drill for
resistant ground; this double equipment is quite practicable and
does not involve much further expense. We shall revert to this
later on.

Fig. 112.—Hydraulic Oil-Well Drilling Rig.   Fauck's Machine, capable of being
driven indifferently by either Hand or Steam Power.

Fauvelle versus *the American System.*—The Fauvelle system
can only be compared with the American system, as, rapid as
the latter, it presents more safety in working, but like the latter
system it can only with difficulty penetrate into hard ground and
there loses the time gained in passing through the soft beds.
Moreover, its price is very high, which is often sufficient to make
oil searchers recoil from it.

Average Progress by Fauvelle's System (Steam Power).—We have
seen that with the Fauvelle system on favourable ground, 12 metres
(say 39 feet), may be bored in a day; according to the results

collected by us the average daily progress by steam power would appear to be 3·4 metres (say 11·15 feet).

We may therefore say that the average progress made by steam power, drilling on this system, is 85 metres (say 278·8 feet) per month, *so that a well 300 metres (say 984 feet) may be drilled under normal circumstances in 3½ months.*

With the hand system, the daily average according to M. Nougarède is 4 metres (say 13·12 feet), above 100 metres (328 feet), and 2·55 metres (say 8·36 feet), from 100 metres to 150 metres (328 to 492 feet). For 300 metres (984 feet) we have obtained for 7 borings representing, collectively, a drilled length of 1,853 metres (say 6,077 feet), an average of 1·52 metres (say 5 feet) per day, or 40 metres (say 131·2 feet) a month, *which makes it evident that a well of 300 metres (say 984 feet) may be drilled by hand power on the Fauvelle system in 7½ months.*

ECONOMIC DATA—FAUVELLE SYSTEM.

The following is an estimate of the cost of installing a steam power boring on Fauvelle's system:—

	Francs.	£	s.	d.
Construction of the derrick and buildings . .	503	20	2	4·8
Engines and pumps	5,980	237	4	0
Modified Canadian drilling crane	1,700	68	0	0
Hollow rods, 300 metres (984 ft.), 42 millimetres (1·65 in.)	16,500	660	0	0
Drill (bits and auger stems)	620	24	16	0
Adjustment system	165	6	12	0
Other expenses	2,450	98	0	0
	27,868	1,114	14	4·8

The cost of wages, etc., with the staff which is the same as the Canadian will be 1,450 francs (say £58) per month which gives 17 francs (say 13s. 7·2d.) the metre (say 3·28 feet) or 4s. 1·2d. per foot.

CHAPTER XXX.

ROTARY DRILLING — BITS, STEEL · CROWNED TOOLS, DIAMOND TOOLS, COMPOSITION TOOLS—HAND POWER AND STEAM POWER —HYDRAULIC SAND-PUMPING.

Principle of Rotary Drilling.—We have just passed in review the drilling processes in which the bit strikes the rock with a succession of blows, and thus gradually disintegrates it. The systems based on this principle we shall call *boring by shock.*

Starting from quite a different principle altogether, another system of boring has arisen, which consists in wearing away the rocks by the friction produced by the continuous circular motion of a piece of hard substance in the bottom of the bore hole. We shall designate this method of drilling by the general term of *rotary boring.*

This system of boring is nearly as old as boring by shock. It has progressed concurrently with it, profiting by the same discoveries, and at the present day has arrived at such a state of perfection that it may be justly preferred in certain cases.

Advantages of this System on Hard Ground.—Let us say at once rotary boring presents real advantages on hard ground in rocky beds. No rock can resist the continuous action of a diamond crown.

In such ground where the drill would hardly bore a few centimetres a day, and that only provided always that this instrument was frequently changed, the crown, thanks to its irresistible bite, to its uninterrupted action, to its composition, which renders it almost unwearable, is able to conquer the greatest resistances.

Extensively adopted in America but not in Europe.—In all rocky ground, also even in rather soft ground, the crown triumphs This is the reason why, in America—a country of rocks—rotary drilling is at the present day so generally adopted. But in Europe, where the greater part of the difficulties encountered in drilling proceed, not from the hardness of the ground, but, on the contrary, from their extreme softness and their tendency to fall in, neither rotary drilling nor drilling with the rope can succeed to their full extent, except under exceptional circumstances.

Used in Trial Borings to ascertain Structure and Nature of Rocks. — Certain trial borings have, however, been attempted on this system for particular reasons which we are about to formulate. During drilling it is necessary, especially in an experimental boring, to ascertain well the position of the beds traversed. It is often even desirable to know their geological relationship, their intimate composition. To effect this it is indispensable, therefore, to procure samples of the ground traversed (*des témoins*) according to the term employed.

Deductions from Sand-pump Débris unsafe.—The *débris* brought up by the sand-pump during drilling by shock are not very well adapted for this purpose. In fact, the brutal action of the drill, its repeated blows, its continual movement, soon result in mixing the elements forming the underground of the bore hole. When amongst the disintegrated particles of several beds not forming a homogeneous mud a sample can be collected presenting a decided character, it is difficult to fix exactly the composition of the bed of which it forms part. Hence arise uncertainties, more or less risky hypotheses, and finally errors.

The Core or Circular Section of Ground Cut Out and Brought to Surface Intact by Diamond Drill.—In rotary drilling, as it is done at the present day, the rock is only worn and cut on the circumference of the bore hole. The interior of the disintegrating apparatus of the crown is hollow, and can hold the core part of the rock which it has sawed out cylindrically in such a manner as to form a sort of cylinder which, when brought to the light of day, may serve for the profound study of the ground which it is desired to know, and may even be preserved as documentary evidence.

GENERAL PRINCIPLES OF ROTARY BORING.

The Process is Continuous.—In order that rotary drilling may be efficient, it is necessary that the work of disaggregation, by the wearing of the rocks, be continuous. This is possible, because,

as we shall presently see, the boring is no longer subordinate
to the period of time which the bit will endure, and that the
apparatus used for wearing away the rocks can hardly be worn
out.   There is, therefore, no need to interrupt the work to lift all
the drilling tools out of the bore hole, a necessity which involved
a notable loss of time in all the systems of drilling which we have
heretofore studied.   Sand-pumping, unless the work of rotary
drilling is to be made completely impracticable, ought also to
be carried out without interrupting the work of disintegration.
Moreover, it also ought to be continuous, the rock *débris* forming
a soft mass adhering to the crown, would very soon neutralise its
action, from which stoppage of the work and, perhaps, serious

Fig. 113.—Different Kinds of Augers.

accidents from torsion might arise.   The problem to be solved
was, therefore, this :—

1. The possibility of cleaning out the well without stopping
drilling.

2. To render the cleaning out process continuous, so that not
a particle of disintegrated matter can remain in the bore hole.

This problem has been solved by the application to rotary
drilling of Fauvelle's water circulating process.

THE RODS AND THE CROWN.

System Primarily used in Soft Beds : The Auger.—In the beginning
rotary drilling was primarily intended for the boring of soft

ground, and for that purpose use was made of an instrument in the form of a gouge termed an auger.

Its Metamorphoses.—Several varieties of these fell to be distinguished, amongst which the following are the principal : The open spoon, the open spoon with an auger or ribbon-shaped handle, and then cylindrical or conical augers.

Corkscrew Motion of Auger.—All these instruments were designed for almost the same purpose which was to pierce soft beds without the aid of the sand pump ; the *débris*, always compact, rose inside the spoon or in the spirals of the auger. This was sunk into the ground by the movements, after the fashion of a corkscrew, which were imparted to it by the rods.

Equalising the Sides of Bore Hole by Reamer.—Before the well could be tubed the sides of the bore hole had to be equalised by a reamer (Fig. 115). This is the name given to an iron cylinder terminating in a point acting as a pivot and provided on the outside with very sharp vertical blades, which planes and equalises the sides of the bore hole by the circular motion which is imparted to it.

FIG. 115.—
Reamer.

The Crown.—The auger was only practicable in soft ground and in shallow wells ; it was, therefore, very soon replaced by a tool offering less resistance to rotation, and which should by that very fact act with greater speed on hard rocks. The idea

FIG. 114.—
Cylinder with
Conical
Auger.

then was to make a cast-iron cylinder of a diameter sensibly equal to that of the bore hole to be pierced and provided round the whole of its circumference with movable steel teeth which could be replaced in proportion as they wore out. By analogy this wreath of steel teeth was termed the *crown* of the tool (*couronne de l'outil*), a name which it still retains.

Its Metamorphoses.—Later on, as the repair and partial removal of the teeth of the crown was rather a delicate operation, it was preferred to use crowns made in a single piece and likewise toothed. This crown (Fig. 117), which was made entirely of steel, was screwed at its lower extremity to a tube, the exterior and interior dimensions of which were sensibly equal to those of the crown. This tube served to contain the *témoin* (witness or sample).

The Diamond replaces Steel.—As steel was found not to be suffi-

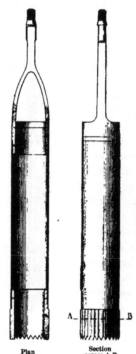

ciently resistant the idea arose of making use of the diamond. The merit of this discovery belongs to Leschot, a Swiss engineer. The crown, in this case (Fig. 118), is lined in its lower part with a layer of lead, in which are inserted some fifteen black diamonds (carbons) of

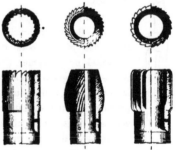

FIG. 117.—Fixed Crowns.

two carats each, or defective white jewellery diamonds (borts).

These are fixed in such a manner that each of them traces individually a different circumference, and collectively the concentric scratches which they trace make a fissure large enough to facilitate the entrance of the crown.

Plan   Section across A B

FIG. 116.—Crown Tools with steel teeth.

Artificial Substitutes for Diamonds: Iridium Crystals, Pure Alumina.—The inconvenience of the diamond crown is that it is very costly. Attempts have, therefore, been made to replace precious stones by the use of several hard bodies obtained by chemical processes, as well as by crystals of iridium and pure alumina, which were successfully employed in the Böhmisch Brod boring in Bohemia.

We shall mention in this connection the crowns of Olaf Terp, the Danish engineer, whose results obtained in Germany would appear to be most conclusive.

The " Carottier ".—Above the crown is screwed a hollow cast-iron tube of a diameter equal to that of the crown itself, and of a length which may vary between 4 and 8 metres (say 13 and 26 feet). Whilst adding, by its weight, its complement of adherence or continuity of contact with the rock to the crown, it receives in its empty interior the column of rock, the circumference of which has been sawn by the diamonds. This instrument is known in France as the *carottier* (player for low stakes).[1]

Neuburger's Improvements.—The rods are screwed on to the top of the *carottier*. These, like those of Fauvelle's system, are hollow,

FIG. 118.—Diamond Crown.

FIG. 119—Working Levers.

and as their principal merit ought to be to resist torsion we recommend Henry Neuburger's hollow tube with side linings and safety sheath.

This latter apparatus consists of a sheath, free to slip on each rod, and the interior of which is hollowed with slightly helicoidal grooves, which correspond with an equal number of relief surfaces on the gudgeon of rod. When the two rods are brought together, their grooves correspond, in such a manner, that the safety sheath can be fitted on, thus covering a part of each rod and rendering the whole unscrewable from whatever side the rod may be turned, we may add that the gudgeon of the lower rod ought to be provided with a slight enlargement or cushion which will hinder the safety sheath from slipping down lower. In rotary drilling it is often desirable to turn the rods to the right or to the

[1] *Cp.* the American "Core barrel".—*Tr.*

left, and we think that our safety sheath constitutes the most practical and cheapest method of preventing accidents in unscrewing and even of rupturing the apparatus when it is being raised.

Knuil.[1]—The last rod, that which rises out of the orifice of the bore hole, is provided with a pinion to which it acts as an axis, and which receiving the rotary movement from the drilling machine communicates it to that rod, and consequently to the others. This pinion is one of the principal pieces of the whole system.

Fig. 120.—Rotary Oil-well Drilling Rig for Shallow Depths.

It is known by the term *knuil.*[1] In addition to transmitting the circular motion to the rods, it allows them to descend, freely turning in their two bearings, one of which is in the upper part of the structure of the machine, and the other follows its ascending and descending movements.

Working by hand, the circular motion is obtained by manipulating levers (Fig. 119).

[1] ? Swivel.

The elongation of the rods does not require any particular apparatus. In fact, far from being suspended, their extremity always rests on the bottom of the bore hole, and their upper part can emerge from the orifice without inconvenience as far as may be desired. Of course, the pinion driven from the drilling crane is movable on the upper rod, and can without difficulty gear itself in any point whatever of the rod as the progress of the drilling may necessitate.

Rotary Oil-Well Drilling Machinery.

The machines employed in circular drilling are very numerous; they vary from the simple wheel with rungs to the most perfectly

Fig. 121.—Rotary Oil-Well Drilling Machine.

developed steam rig. In certain undertakings a unique rig is used (Fig. 121), which by itself performs the different operations of drilling (drilling, pumping, and pressing the water into the bore hole, and raising the drilling tools), in other cases it is preferred to have a different rig for each kind of work. We may at once say that the unique rig system answers best with mechanical motors.

Motive Power.—Putting human energy on one side, the motors may be steam and water, and even the engines may be replaced by one or two horses moving a pole.[1] Their uninterrupted movement answers very well for circular boring. The machine

---

[1] Gasoline engines are extensively used in the United States for this purpose.—*Tr.*

communicates the movement to the pinions, which commands the rods and which turns in a horizontal plane by means of an angle wheel which turns in a vertical plane and hangs directly from the machine.

CIRCULATION OF THE WATER.

The application of the principle of water circulation is the same as in the Fauvelle system (see chap. xxix., page 381 *et seq*).

CASING.

The crowns used in circular drilling, wearing very slowly and being very costly, it is important to be able to case the well without retrenching its diameter; to effect this use is made of cast-iron tubes for large diameters and 9-inch iron for the smallest. So as to impart more rigidity to these tubes they are screwed the one to the other, and are lowered at will by screwing each new portion of tube to the upper part of the casing, and allowing the whole to descend.

The following are the ordinary dimensions and the prices of these tubes :—

Kind of Tube.	Exterior Diameter.		Interior Diameter.		Unions.				Weight in kilos. per running metre.	Weight in lb. per running foot.	Cost per kilogramme.	Cost per lb.
					Interior Diameter.		Exterior Diameter.					
	M'trs.	Ins.	M'trs.	Ins.	M'trs.	Ins.	M'trs.	Ins.				
Cast Iron	0·300	11·8	0·256	10	—	—	—	—	—	—	—	—
Iron	0·229	9·0	0·216	8½	0·209	8·21	0·219	8·61	38·17	25·5	1·5 frs.	6¼d.
,,	0·203	8·0	0·190	7½	0·184	7·23	0·194	7·62	31·82	20·94	,,	,,
,,	0·178	7·0	0·165	6½	0·159	6·15	0·168	6·54	25·65	17·20	,,	,,
,,	0·102	4·0	0·089	3½	0·089	3·50	0·092	3·52	19·21	12·88	1·6 frs.	7d.

The crown, so as to pass across these tubes whose interior diameter is less than that of the bore-hole, ought to be slightly eccentric. So that, as is indicated in Fig. 122, whilst not having a greater diameter than 120 millimetres, the diamond farthest distant from the point of rotation will consequently be 70 millimetres distant from it, and can describe a circle of 140 millimetres in diameter; this diameter, which will be that of the bore hole, is 10 millimetres greater than that of the casing, and 20 millimetres greater than that of the crown.

CIRCULAR DRILLING WITH THE DIAMOND.
ECONOMICAL DATA.

Boring with the diamond by steam power is executed by an engine of 15 to 18 horse-power. The staff employed in the boring properly, so called, consists of 5 persons, *viz* :—

Working Staff.—1 Foreman borer.

        2 Assistant borers.

        1 Mechanic.

        1 Stoker.

Duties of the Foreman.—During the boring the foreman looks after the engine and satisfies himself as to the regularity of its movements by the means already mentioned. The rotation of the rods ought to be continuous, and without jerks ; its speed 5 to 6 revolutions per second.

FIG. 122.—Function of Eccentric Shape of Crown.

Jerking.—When jerking is produced it is a sign that the bottom surface is uneven. When this happens, the foreman borer lifts his drill a few centimetres, and turns on the current of water used for sand-pumping.

Raising the Carottier.—Each time that the boring has been sunk the length of the carottier or " core barrel," it will be necessary to raise the tackle and empty the former. This is done as follows :—

1. The pumps which deliver the water into the bore hole are stopped.

2. The circular movement of the drill is stopped.

3. The rods are separated by detaching the flexible tubes which join them to the pumps, and taking away all the parts of the *knuil*.

4. The upper extremity of the rods is suspended to the raising key and hoisting is proceeded with.

At this point the foreman borer remains at his levers, the assistants unscrew the rods; it is the same when lowering.

Emptying the Carottier.—When the crown and the carottier have been brought up, the foreman borer will proceed in the following manner to empty the carottier and take out the "witness" without breaking it. He turns the whole apparatus of the crown and carottier in the axis of the rods as quickly as possible. The abrupt displacement of these tools induces the exit of the "witness" or cover, which then only requires to be received with caution. It is a difficult operation to obtain a complete "witness," the successes being about 1 in 10.

Examination of Crown.—The foreman borer then examines the crown to see that all the diamonds are there. In case one or two have been disembedded, they must be sought for in the bottom of the bore hole. This is done with an old crown, which is coated with pitch; and attempts are made to recover the disembedded diamond by adherence. The raising of the tools and the extraction of the "witness" is only done but once a day, because it is rare to be able to drill more than 8 metres (26 feet) in 12 hours. However, at Wallaf, in Sweden, 33 metres (say 112½ feet), were bored in a single day.

Speed Obtained.—One of the most important diamond drilling companies, the Continental Diamond Rock-boring Company, has furnished us with a series of observations made in the borings executed by them. We extract from these the following data :—

CONTINENTAL DIAMOND ROCK-BORING COMPANY.

TIME OCCUPIED IN DIAMOND DRILLING : AVERAGE AND MAXIMUM SPEEDS.

Name of Boring.	Depth expressed both in metres and in feet.		Time occupied in drilling.	Average progress per drilling day expressed in metres and in feet.		Average progress per day expressed in metres and in feet.		Maximum daily progress expressed in metres and in feet.	
	M'trs.	Feet.	Months	M'trs	Feet.	M'trs.	Feet.	M'trs.	Feet.
Bethleem, near Liebau .	500	1,640	4	7·4	24·27	5·00	16·4	18·1	59·36
Rheinfelden . .	443	1,453	2	13·0	42·64	7·90	25·91	22·0	72·16
Villefranche d'Allier .	741	2,430	13	5·1	16·72	2·25	7·380	23·8	78·02
Ascherleben, No. 5 . .	902	2,958	9½	9·0	29·52	3·00	9·84	30·0	98·40
Ascherleben, No. 7 . .	361	1,182	2	6·0	19·68	3·1	10·168	9·2	30·27

The extreme sensibility or delicacy of circular drilling render it subject to accidents, especially in soft ground. To satisfy the reader in regard to this, it will be sufficient to quote the observations, taken during a rotary boring, made by the above-mentioned company at Villefranche d'Allier (France).

The boring, 741 metres (2,420 feet) in depth, lasted from the 28th of November, 1875, to the 5th of January, 1877, or over 13 months. During this time only two-thirds of the amount of working days were occupied in boring, one-third being devoted to repairing accidents.

The month during which these accidents occurred with the least frequency was December, 1875, when 139·28 metres (456·8 feet) were drilled with only 3 broken days due to accidental causes.

The month during which these accidents occurred with the greatest frequency was that of February, 1876, when only 33·96 metres (111·4 feet) were drilled, with 22 broken days for accidental causes.

In this same boring, the duration of the crown was studied. This question was a vital question on account of the high price of these instruments. The whole boring of the Villefranche Well of 741 metres (2,420 feet) necessitated 3 crowns; they lasted as follows :—

Surface well 10·3 metres (say 33·78 feet).

	From	To
First crown 279·186 metres (say 915·7 feet) . . .	28·1·75	16·3·76
Second crown 223·996 metres (say 734·707 feet) . .	16·3·76	12·8·76
Third crown 227·256 metres (say 745·4 feet) . . .	12·8·76	5·1·77

From the whole of these series of observations, we conclude that, in average ground, with normal working, in diamond boring by steam power, 2½ metres (say 8·2 feet) can be drilled per day. From which it would appear that with this system 63 metres (say 206·64 feet) should be drilled per month, after allowing 2 days for casing. To sum up, in less than 5 months a well may be drilled 300 metres (984 feet) deep by rotary drilling.

CHAPTER XXXI.

IMPROVEMENTS AND DIFFERENT SYSTEMS.

WE have just passed in review, in the preceding chapters, the different systems of boring. We have, in each instance, described a plant whose general adoption has seemed to constitute a type— a link of the system. Other different descriptions of plant now remain to be mentioned, and more especially some improvements which, in certain cases, may be adopted.

Peculiar Features in the Boring Systems of Different Countries.— This leads us to say a few words regarding the object kept in view by the manufacturers of boring machinery in the different countries in which it is constructed.

Peculiarities of French Boring Machinery.—In France drilling has been applied to the search for other deposits than those of petro- leum ; the old ideal would appear to have been to execute borings of large section, and the typical plant, with the exception of slight improvements, remains what it was during the drilling of the artesian wells of Grenelle and Passy. The Frenchman likes the task well finished—work which lasts ; but, on the other hand, he rebels against bold innovations, such as those, which, in the person of Degoussée, made him hurl anathemas on Fauvelle the inventor.

Further, with the exception of the invention of the latter, we only see in France minute improvements of such and such an organ of the drilling machine, which by dint of improvement in details has ended in becoming very delicate and terribly heavy. The tendency here is to the solidity of the whole and improvement in detail. The French manufacturer pays no heed to the speed of drilling, the greater part of the time he is an artist, not a practical man. He pays too little heed to the progress realised abroad, and all his connections with this subject denote an absolute indifference and less than even superficial thought ; moreover, he does not export (at least for the petroleum industry the principal branch of boring).

Complicated, Costly and Highly Adjusted.—The French machine is therefore almost always highly complicated, accurately adjusted, very complete, and very dear.   It suits perfectly for borings of great diameter.

Simple Colonial Machinery.—During the last few years however, colonial necessities, the search for springs of water in the desert, have led to the manufacture in France of a *Colonial plant*, more rustic, lighter, even portable.   These equipments are suited for drilling by hand power such as is adapted for the sandy beds of Algeria.[1]

American Machinery.—America would appear to have gone quite contrary to French methods of construction.   In that country, they concerned themselves but very little, in the early days of boring, with the method of sinking Artesian wells in use in the old continent; they wanted to work quickly and cheaply.   Risk was little thought off; and the Chinese system of boring was borrowed.

In order to work still more rapidly the steam engine was introduced, but without adding to it either heavy drilling cranes or gearing machines; belts, driving wooden wheels fixed to the old drums used in drilling by hand power, did the work.   It was very light, cost but little, was a little rusty, but, they contented themselves by promising in their own minds to arrange all that better on another well.   Better arrangements were made, and the improved American system which we have described was the result; then came the Canadian system.

Characteristic Detail.—Whilst in France they strive to substitute iron for wood, in the complement of tools, the American everywhere forswears iron for wood as lighter and more easily handled.

We have thought it would prove interesting to contrast, with an analogous Austrian system, the estimate of a boring of 300 metres (984 feet) by a French constructor—the cost of erecting the buildings and the derrick, as well as that of the steam engines and the equipment of the smithy, being considered identical.

---

[1] Nothing could exceed the variety and portability of the American Rigs for this purpose as used on ranches in Texas, etc.—*Tr.*

ESTIMATE.

	Francs.	£	s.	d.
Derricks and buildings	508	20	2	4·8
DRILLING TOOLS (LIPPMANN'S SYSTEM).				
1 bore hole head (No. 3)	30	1	4	0
1 removal key (No. 1, 2, 3)	117	4	13	7·2
1 retention key (No. 1, 2, 3)	70	2	16	0
6 wrenches	112	4	9	7·2
1 large working handle	63	2	10	4·8
1 small ,, ,, ...	39	1	11	2·4
1 hasp for chain	35	1	8	0
1 chain cable, 26 metres (say 85·28 ft.) . .	340	18	12	0
1 pulley and its axis	138	5	10	4·8
1 crane (No. 1, complete), driven at will either by hand or steam power, with renewal pieces .	2,800	112	0	0
1 crane for free fall, to be driven by steam engine with its motion rod	1,245	49	16	0
1 free-fall sheath (openable)	580	23	4	0
1 ,, suspension tool	400	16	0	0
1 Poupée (No. 1) with lever irons . . .	510	20	8	0
1 master rod (No. 1-2), 4 metres (say 13 ft.) .	250	10	0	0
2 rods (No. 1), 6 metres (say 19½ ft.) .	240	9	12	0
1 joining-rod (No. 1-2), 6 metres (say 19½ ft.) .	118	4	14	4·8
1 union (No. 1-2) 1 metre (say 3¼ ft.) . .	63	2	10	4·8
15 rods (No. 2), 6 metres (say 19¼ ft.) . .	1,440	57	12	0
1 joining-rod (No. 2-3) 6 metres (say 19½ ft.) .	94	3	15	2·4
1 union (No. 2-3), 1 metre (say 3¼ ft.) .	48	1	18	4·8
30 rods (No. 3), 6 metres (say 19¼ ft.) . .	2,160	86	8	0
1 lengthening rod (No. 3), 4 metres (say 13 ft.) .	66	2	12	9·6
1 ,, ,, (,,), 3 ,, (,, 9¾ ,,)	62	2	9	7·2
1 ,, ,, (,,), 2 ,, (,, 6½ ,,)	52	2	1	7·2
1 ,, ,, (,,), 1 metre (say 3¼ ft.) .	42	1	13	7·2
2 drills (No. 0)—one flat, one with gudgeon .	780	31	4	0
2 ,, ,, ,, ,, ,, ,, ,,	585	23	8	0
Total . . .	12,982	519	5	7·2

"SAND-PUMPING" WITH THE ROPE.

	Francs.	£	s.	d.
1 sand reel with its renewal pieces, to be wrought by steam power	1,680	67	4	0
1 pulley and its axis for wire cable . . .	140	5	12	0
1 wire cable, 450 metres (1,462½ ft.) . . .	630	25	4	0
1 bore hole head, No. 2, with immovable eye .	24	0	19	2·4
1 hasp for sand reel	15	0	12	0
1 harpoon	84	3	7	2·4
Total . . .	2,573	102	18	4·8
Boiler and engine	5,820	232	16	0
Smithy and sundry accessories . . .	3,742	149	13	7·2
General total . .	25,117	1,004	13	7·2

It will be observed from this detailed statement that a French drilling equipment on the free-fall system and wrought by steam power, making allowance for the discount on the account, comes absolutely to the same cost as an Austrian equipment on Fauck's system. Say, with the discount, 22,600 francs (£904), and 24,000 francs (£960) for the Austrian.

Britain—Diamond Boring a Feature.—Britain not being a petroliferous country, follows the beaten path pursued on the old continent; but boring with the diamond is much resorted to.

Germany.—Germany, of late years, has made enormous progress in the manufacture of drilling machines. Thanks especially to the experiments of Messrs. De Tecklemburg & Przibilla, it has remarkably improved the system of Fauvelle, the French engineer. Rotatory drilling has found there an ardent adept and a bold innovator in Olaf Terp, the Danish engineer.

The drilling plant made in Germany, whether according to the German principle of the free fall, or after the French principles of Fauvelle, is always rather heavy, but substantial enough to stand any test. Soundness and simplicity would appear to be the unique end of the improvements brought to bear on the plant by the constructors.

Austria—The Scene of Struggle between the Free-fall and the Canadian System.—Austria is the field on which the battle is taking place between the German free-fall system and the Canadian. There are convinced partisans on both sides. The Canadian would appear, however, to attract numerous proselytes. The mistrust at the outset, the theoretical doubts have been dissipated by a series of convincing proofs.

Austrians Constructing on American Models.—Many Austrian builders of plant have forsaken every other form of construction to devote themselves solely to the production of American models. On the other hand, the free-fall system has found a partisan whose innovations and ingenious improvements have enabled it to compete, sometimes victoriously, with the Canadian. We refer to Fauck, the constructor, one of the masters of boring, whose machine, moreover, we have taken as a type.

Principle of Austrian Machines.—Contrary to the French, the Austrian constructors are engaged in combining in one single

machine, easily looked after, and commanded by levers grouped
within reach of the hands of the foreman borer. They are, therefore, more or less inspired by American machines or Fauck's
cranes. Let us add that Austria exports its boring plant, even
to France and its colonies.

Russia.—Russia is a land also conquered by America. With
slight modifications, American plant dominates at Baku. A large
number of installations, however, whilst preserving windlasses,
etc., constructed on the American plan, use rigid rods in drilling,
after the old French plan. Let us hasten to add that these rods
are not made in France, where so little is done in exportation.

Having made these observations, we shall go on to describe
the different methods adopted in boring.

Rigid Rod System—Different Models of Installation.

Origin.—The principle of the rigid rod system is French.

Artesian Wells.—As their name indicates, the first artesian
wells were bored at Artois, in France, in the seventeenth
century, with a rigid rod system.

The following are, in chronological order, the principal borings of this kind undertaken before the commencement of this
century :—

1600-1700.—Artesian wells of the seventeenth and eighteenth
centuries.

1777.—Cannstadt (Würtemberg).

1781.—Sheerness (England), depth 330 feet.

1784.—Grenelle (France), commenced in 1784, and bored to
a depth of 560 feet.

1794.—London (England).

1833.—The Grenelle boring was resumed by Mulot.

1841.—Mulot reached the water-bearing bed at a depth of
547 metres (1,794 feet).

The house or firm of Mulot passed from the son of this latter
into the hands of Léon Dru ; M. Arrault succeeded the latter.

Mulot's Successors and their Machinery.—The principles of boring
on which Mulot's successors work proceed from the method of
working adopted by him. The equipment—a very important one
—includes a " walking beam " driven directly by the piston rod at

a point twice as far from the axis of this "walking beam" as is the point from which the rods are suspended. The shock of the "walking beam" is obtained by a smaller "walking beam" at a lower level, joined to the other "walking beam" by an articulated or hinged rod.

The raising of the rods and sand pumping are done by independent windlasses. Messrs. Mulot & Dru have acknowledged the principle of the free fall, and have improved its tools. We shall speak of these improvements at the proper time and in the right place. The machines sent out by this French firm are of very high quality, as much in regard to their soundness as their finish. Unfortunately, they are hardly transportable and too delicate.

Degoussée's Machinery of a different Type.—The French engineer Degoussée started upon another principle; he imparted motion to the "walking beam" by means of a vertical motion rod coupled to the crank of a transmission wheel. In the installation model of 1855, the rather long "walking beam" was moved in the way we have just explained by the motion rod of a wheel which was driven directly by the movement of an oscillating cylinder placed against the structure of the crane; in front of the latter, between the point of attachment of the "walking beam" and its point of suspension, was the hoisting windlass driven by gearing from the driving wheel of the "walking beam". The motion of the hoisting windlass was transmitted by a movable belt to the "sand reel," which was placed in the rear, and independent moreover of the "walking beam".

The different organs of this complicated machine were rendered independent by means of cog-wheels, free to move on their axis and capable of being fixed on any point whatever of that axis.

Laurent's Improvements on Degoussée's System.—Laurent, in 1872, improved Degoussee's system by simplifying it. He made the "walking beam" of larger dimensions; the eccentric wheels were replaced by removable belts, and the oscillating cylinder was transformed into a horizontal one. The formidable plant, comprising a rather complicated assemblage of gearing, consisted, as in Degoussée's system, of two windlasses :—

1. The drilling windlass under the "walking beam".

2. The hoisting windlass behind the latter.

The rather costly machine took up much space; it, however, for a long time constituted the model of free-fall drilling, and it remains to the present time a model for drilling wells of wide diameter. M. Fauck, we believe, was inspired by it in the construction of his first drilling crane.

Lavé's Simultaneous Sand-pumping and Drilling Plant.—Lavé's drilling plant showed rather a curious, if not very practicable,

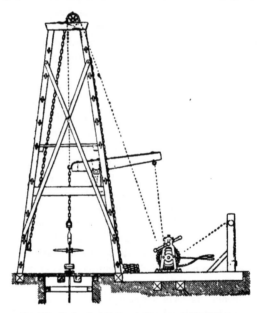

Fig. 123.—Boring Installation on Lippmann's Hand System.

innovation, simultaneous drilling and sand-pumping by means of a tube placed directly above the bit and provided in its lower part with a system of valves.

Machines on Mulot's and Free-fall Principle.—Arrault of Paris, J. Winter de Carnen (Germany), and Koller (Germany) have constructed or invented beautiful strong machines on Mulot's principles, which they have improved for free-fall drilling.

Machines on Degoussée's System.—On Degoussée's system, his suc-

cessor Lippman of Paris, Wolf of Buckau (Germany) and Mather
& Platt of Manchester have equipped numerous borings.

Modified Laurent System adopted in the Caspian.—One of the
systems most generally adopted in the Caspian basin, and, especially, by the Schibaeff Co. at Baku, is directly derived from
Laurent's system.   On one side of the bore hole is placed the
horizontal engine which drives by a belt the double windlass,
placed in front of it, for hoisting and sand-pumping.   On the
other side of the bore hole is placed the drilling crane, with gearing, which is driven by a belt, which joins it with the double
windlass, on the other side of the bore hole.   This movable belt
renders the two windlasses independent.

Fig. 124.—Boring Installation at Baku after a Model deposited at the *Conservatoire
des Arts et Métiers.*

Modified American System at Baku.—On the other hand, a very
simple equipment, derived from the American rope-boring system,
is much employed at Baku.   Only a longer " walking beam," the
adjunction of a sand reel and the use of rigid rods cause it to differ
from the latter.

In it there is no gearing, simply a wooden bull wheel and
band wheel.   All its component parts are driven by belts.   The
" walking beam," however, merits a special description: rather
long, and working with speed, it requires a certain stability.
This stability is obtained, not after the French style, but rather

after the American, by an innovation which costs but little. The " walking beam " consists of two parallel beams firmly tied together (tie beams) and a sort of ladder laid flat down on its support, thus occupying a greater base without loading it with useless weight. This rather cheap " walking beam " is provided at its extremity, opposite to where the rods are attached, with a box into which stones may be placed as a counterpoise.

DIFFERENT MODELS OF RODS.

Léon Dru's and Van Dijk's Modifications. — M. Léon Dru adopted rods which exhibit an ingenious method of screwing. The gudgeons of the rods are cylindrical, with round threads ; they are joined together by an independent sheath or case, which clasps itself on one extremity of the bar. This method of coupling together does not necessitate the consignment of the sheath to the scrap-iron heap when it is worn. When a sheath has become no longer serviceable, it is unclasped, and by turning it upside down an altogether new screw furrow is obtained once it is tightened at the bottom on the rod. If, on the other hand, it be the thread of the upper part of the rod that is deteriorated, the dismounted sheath is replaced on this thread, and thus a new drilling tool may be obtained by also turning the rod upside down.

Let us likewise quote, as a reminder, the wooden rods provided with a central square iron rod, used by Van Dijk, the Dutch engineer.

DIFFERENT MODELS OF FREE-FALL INSTRUMENTS.

Fabian's the Model and Typical Instrument.—In the detailed review of drilling tools we have quoted Fabian's apparatus as the model free-fall instrument. The reason for this choice was the fact that this instrument is still, almost universally, used at the present day. Moreover, as we shall see, it is from Fabian's apparatus or Kind's instrument that almost all the other free-fall instruments have sprung.

Origin.—The free-fall instrument dates from 1845. Its inventor was Kind, the Saxon engineer, who directed the works of the artesian well of Passy. Irritated at the difficulties of the work, he invented the first free-fall instrument (Fig. 125).

First Free-fall Instrument—Description.—This consists of a flat piece, A, contained by and capable of gliding between two plates bolted to another upper flat plate, B. The latter serves to connect the apparatus with the upper drilling rods by a square rod ending in a nut. To prevent the piece, A, from leaving the plates, the latter are joined together in their lower part by a solidly-bolted iron ring. The lower piece, A, is provided with small lugs, c, and terminated above the lugs by a square rod, ending in a triangular elongation, a, which is seized, on raising the rod, by the pincers SS'. These pincers consist of two branches, terminated by cams moving on bolts fixed to the two plates, and connected on the other side by hinges to a double vertical rod, which glides inside the plates and the piece B. The two branches of this double rod are connected with a disc, R, formed of superimposed rounds of leather, with a diameter almost equal to that of the bore hole, and flexible enough to be bent by the resistance of the water, and pressed by bolts and nuts between two sheet-iron discs of much smaller diameter. The disc, R, is pierced and can move freely on the rod, F, from the two plates as far as the enlargement of the rod which limits its ascension to the upper part.

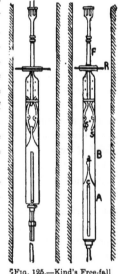

Fig. 125.—Kind's Free-fall Instrument.

Manner in which it acts.—When the descending movement of the rods commences, the disc, R, pressed from below upwards by the water of the bore hole, tends to rise, and bringing with it the double rod, which commands the branches of the pincers SS', forces the overture of the latter, and thus determines the free fall of the lower piece, A, which sustains the drilling tools. At the bottom of the run, the pincers, which the descending movement has left open, close, by the first ascending shock, on the projection, a, of the lower piece, and draw it after them in the ascending movement. During all this latter movement the pincers, commanded by the disc, R, remain closed, the latter receiving a pressure from above downwards.

Acts automatically.—It will be seen that Kind's apparatus was automatic. It will find its utility later in its improved form, in borings of great diameter, or in very deep ones where another tool would rebel against the hand of the foreman borer.

Rost's Free-fall Instrument.—In 1847 Rost's instrument appeared. It presented the same principal parts as Kind's, but instead of plates an upper piece of tubular shape was used.

Fabian's Free-fall Instrument.—Fabian's instrument (1848) was based on an altogether different principle. We refer the reader to our detailed description, chap. xxiii., pp. 318-320.

Klecka's Free-fall Instrument.—Klecka's instrument (1851) is on the Fabian principle, with two grooves and two shoulders.

Mulot's Free-fall Instrument.—Mulot's instrument (1851) presents some originality. It is a tube, to which the lower part is screwed ; it hangs altogether from the upper part by a rod, along which it can glide and fall freely down a certain space. The lower tubular part is reseized by a circular movement of the rods, imparted by a kind of caracole, which guides itself naturally, on a projection *ad hoc* of the lower piece of the free-fall instrument. The instrument would appear to us to be uncertain in its action.

Gault's Free-fall Instrument.—Gault's instrument (1854) is the first instrument which acts from a point of support. The unclutching is produced, at the moment of the introduction, from the upper part of two vertical hooks, in a kind of ring enveloping the sliding part. It is maintained, at the height from which the desired fall is to be given, by a rod resting on the bottom of the boring.

Werner's Free-fall Instrument.—Werner's instrument (1856) is a Fabian instrument, actuated by a Kind's disc.

Wlach's Free-fall Instrument.—Wlach's instrument resembles Klecka's ; it is complicated by two springs, intended to hold collars on the shoulders, during the ascending movement. The upper part is not cylindrical ; it consists of two plates which may be separated for repairs. The apparatus is in fact delicate.

Zobel's Free-fall Instrument.—Zobel's instrument (1859) is a Fabian instrument, actuated by a hood, a sort of cone, with sides

inclined towards the bottom at an angle of 45 degrees, and intended
to replace Kind's disc. Zobel's instrument is an excellent automatic
apparatus; it is preferable to Kind's in consequence of the sup-
pression of the pincers, which eventually got worn out, and
whose play, at the end of a certain time, became very irregular.
The unclutching is actuated by means of a hinged piece of
mechanism.

Seckendorff's Free-fall Instrument.—Seckendorff's instrument
(1860) is an improved Kind.

Degoussées Free-fall Instrument.—Degoussée's instrument (1861)
has also sprung from Kind's; it has pincers, actuated by a pin,
which raises a disc.

Esche's Free-fall Instrument.—Esche's instrument resembles
Zobel's; the shoulders of the upper piece, supporting the collars of
the upper piece during raising, move on hinges actuated by a
Zobel hood.

Greifenhagen's Free-fall Instrument.—Greifenhagen's instrument
(1866) is on the same principle as that of Kind and Zobel.

Ermeling's Free-fall Instrument.—Ermeling's instrument (1875)
is an improved Kind.

Wilke's Free-fall Instrument.—Wilke's instrument (1875) resembles Zobel's in many points.

Maldener's Free-fall Instrument.—Maldener's instrument (1877)
is an amalgamation of Zobel's and Degoussée's.

Léon Dru's Free-fall Instrument—Description.—Léon Dru's instruments merit special mention. This free-fall system (Fig. 126) has
no resemblance to any of the instruments which we have enumerated. It is based upon the principle of the resistance of water,
and the inventor has termed it a water-pressure groove (*coulisse à
pression d'eau*). The tool consists of a cylinder, *a*, *b*, with two
diameters joined by a conical union : it is closed at the bottom and
open on the top. Inside the cylinder is placed an obturator, *r*,
pierced by the air-holes which are covered by a plate; the rod of
the disc is adjusted to a fork led on the outside to the cylinder to
which the drill is screwed.

The cylinder is also fixed to a fork constituting a part of the
drill.

Function and Working of Dru's Instrument.—There are, therefore,

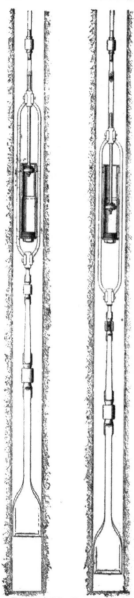

two arrangements with very distinct functions entering into the construction of the drill :—

1. The cylinder from which the drill is suspended.

2. The bit fixed to the compressor piston.

The tool, manipulated as has been stated, is let down to the bottom of the bore hole; the bit carries the first, and the cylinder continues its course until the piston occupies the top of the cylinder.

The "walking beam" then raises the whole system very rapidly; the water compressed in the cylinder by the weight of the piston to which the drill is fixed escapes, slowly, in contact with the small cylinder; but, when the disc reaches the large cylinder, the resistance is destroyed and the tool falls freely. In the descent through the small cylinder the tool descends but slowly, it is therefore necessary to provide for this loss of play, and to supplement it by increasing the amplitude of the "walking beam".

At each oscillation of the drilling tools, the piston is always raised to the top and removed to effect the free fall, which is accelerated on passing to the enlarged part. This groove does away with the clutch, axis and catch indispensable to the other instruments; it therefore constitutes a new principle of free-fall very different from those hitherto used in drilling.

FIG. 126.—Léon Dru's Water-pressure Free-fall Instrument

Automatic Free-fall Instruments on the Bayonet Principle.—The automatic instruments of Laurent (1872), Dru, Lippmann and Fauck are likewise based on a different principle. These instruments are constructed on what is known as the bayonet principle. By bayonet is meant a rather heavy metal rod parallel with the system of drilling rods and independent of it, to which it is attached by rings in which it can glide freely, the lower extremity of the bayonet always resting on the bottom of the bore hole, notwithstanding the movement of the rods. It will be seen that these instruments are improvements on Gault's apparatus.

Laurent's Free-fall System.—In Laurent's system the pincers retaining the lower part of the freefall instrument were actuated by one of the immovable rings of the bayonet in the upward movement ; it encountered this ring always in the same point of its ascension. A second lower ring acted as guide to the auger stem and to the drill.

Free-fall Instruments at the Paris Exhibition of 1878.—The Paris exhibition of 1878 brought numerous free-fall instruments to light. The most original innovation has to be credited to M. Saint-Just Dru, but

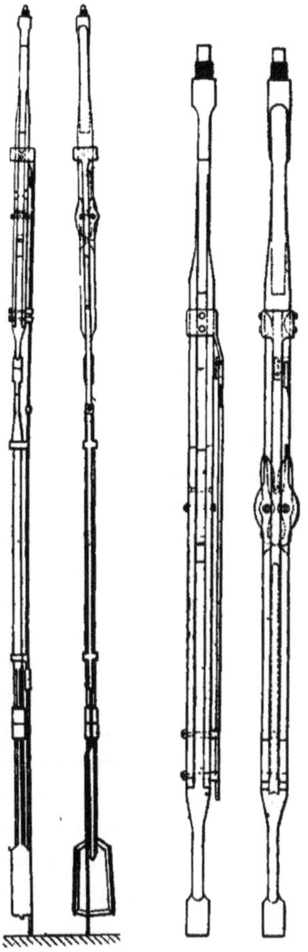

FIG. 127.—Laurent's (Bayonet System) Free-fall Instrument.

on this point we shall let M. Lippmann speak.

Lippmann on Free-fall Instruments. Principle of Kind's Instrument.—" Whilst observing Kind's system at work, one was struck with this fact, that instead of imparting to the suspension lever of the drill an oscillation movement, pure and simple, as the theory of the resistance of the water for the function of the disc would indicate, it causes the tail part of the lever to strike with violence against an abutment at the top of its course. We must therefore conclude that the action of water in an irregular hole having sometimes a much greater diameter than that of the disc was not sufficient to conquer the adherence of the top part of the drill to the catches of the hooks, and that the object of this shock was to impart such a sharp movement to the mobile parts of the apparatus—that is to say, to the top part of the drill which was raised simultaneously with the disc—as would produce with the same blow the opening of the hooks and the fall of the drill.

Kind's Idea improved by Dru.—" This idea was immediately seized by Saint-Just Dru who has constructed the first free-fall groove which really and truly works by shock. The two hooks cross each other in this case as in a pair of scissors ; they have a unique axis of rotation, the two extremities of which are borne by the two cheeks of the upper part of the "jars" into two openings having almost the form of ellipses with great vertical axes. The "jars" carries on the upper part two inclined faces in the form of a very hollow V. During the time the drill is being raised, seized by the catches of the two hooks, their axis lies in the heart of the ellipse and each of the hooks is in contact with one of the branches of the V. The suspension lever, arrived at the summit of its course, striking against an abutment, the axis leaps in its ellipses, the upper part of the hooks slip on the oblique faces, separating in so doing to liberate the drill.

Fig. 128.— Free-fall Instrument on Dru's Principle.

Dehulster's Application of Kind's Idea.—" We again find the application of shock to the production of the unclutching in M. Dehulster's apparatus, who makes use of a stirrup, on a bascule and counterpoise principle, to obtain the hooking and the fall of the drill by means of an angle block fixed on the right angle of this stirrup. This stirrup rotates around an axis, eccentric in relation to that of the drill ; during the descent of the drill it occupies by its shape an inclined position a little below the horizontal ; the angle block is shifted by the head of the drill, which gears under its catch when the stirrup falls back very soon afterwards into its first position, and the drill, just seized, reascends with the drilling rods until the moment when the shock bascules the stirrup and causes the drill to fall to the bottom.

Van Dijk's Free-fall Instrument.—" A free-fall apparatus, worthy of engaging our attention, very specially, because it works really and in a unique manner by the effect of the resistance of water on a disc, is represented only by plans in the exhibition of the Dutch Indies ; it has been imagined by M. Van Dijk whom I have already named. I have had the opportunity of seeing it work, in the workshops where it has been constructed, according to the plans and instructions furnished by this eminent engineer, and I am able to affirm that it is impossible to look upon a more ingenious combination. Unfortunately, the number of the parts which compose it is too great—although forming an aggregate working with the greatest simplicity—for it to be possible to give a comprehensible description without accompanying it by an illustration, or by following it in the plan itself. I will content myself with saying that M. Van Dijk, having also concluded that in Kind's free-fall system the action of the water on the disc was almost unable to overcome the adherence of the hooks to the catches on the head of the drill, only causes his disc to play the *rôle* of a gentle-slipping guide, which, during the ascension of the drill, holds a unique hook in the vertical position which it ought to have for raising the drill, and as soon as the latter begins to re-descend it lets go the hook, which bascules of its own accord to effect the fall of the drill.

Sigmondi's Free-fall Instrument.—M. Sigmondi's (of Buda-Pesth) " . . . a free-fall apparatus also working by the resistance of water on a disc which works a little like the preceding ; by means of an elliptical bolt which it carries, it maintains the head of the bit a prisoner in a notch during the ascension of

27

the drill, and at the moment of descent the bolt opens for the tool to fall."

François' Free-fall Instrument.—M. Charles François (Lens [1] Mines), a modification of Oenhausen's sheath as a frée-fall instrument, by an arrangement almost similar to Fabian's instrument.

Lippmann's Free-fall Instrument.—Lippmann's instrument is an improvement of Laurent's apparatus.

Fauck's Free-fall Instrument. — In regard to Fauck's, it consists of two points of support and an upper piece, which hangs from them. The object of this upper piece, the lower part of which is bevelled, is to displace the collar of a Fabian apparatus, and to make it quit the shoulder when this collar reaches as high as the bevel.

DIFFERENT MODELS OF "JARS".

Origin.—The "jars" was invented in 1834 by Oenhausen.[2] This very simple apparatus has hardly tempted innovators. M. Fauck, however, has used a "jars" consisting of a cylindrical body, (lower part), and gliding after the manner of the Fabian instrument in and out of a sheath likewise cylindrical.

Attempts have also been made (we do not know why) to suppress the "jars," and to replace it by a series of tubes containing india-rubber plugs to deaden shocks. The use of this instrument has not become general.

FIG. 129—Fauck's Free-fall Instrument (Bayonet Principle).

ROPE DRILLING—VARIOUS MODIFICATIONS.

Round Ropes versus Flat Ropes.—Amongst different contractors a controversy is going on, on the question of the rope—a round rope or a flat rope. Opinions are divided.

Rope Rods.—On the other hand M. Fauck has made rope rods,

[1] (Pas de Calais).   [2] See note page 359.

that is to say, portions of rope 10 metres (32·8 feet) in length, provided with drilled sheaths and screws on their ends. We do not very well see the utility of this modification.

Rope Substitutes.—Attempts have also been made to replace the rope by chains and hinged rods. All such attempts have been without success.

FREE FALL APPLICABLE TO BORING WITH THE ROPE.

The first to adopt the innovation of the free fall to rope boring, was M. Gaiski in 1868. This engineer used a system of pincers, guided between two parallel rods, joined together in their lower part by a ring, which allowed the auger stem to pass and guided it.

Let us also mention in this line of ideas the instruments of Sontag, Kleritj (pincers actuated by a Zobel hood), Straka, Sparre, Noth, Sisperle and Benda.

DRILLING BY CIRCULATION OF WATER. DIFFERENT MODIFICATIONS.

Neuburger's Improvements.—We shall only recall, in the system of drilling rods, the rods with supports resisting torsion and the safety sheaths rendering the drilling system unscrewable of Henry Neuburger (see p. 395).

Free Fall in Hydraulic Boring.—The Germans are much occupied in improvements to be brought to bear on Fauvelle's system. They have sought to solve the problem—a very difficult one, in fact—of a free fall in a rod system, which is at the same time used as a water conduit.

Introduced by Köbrich in 1877.—Köbrich, in 1877, first introduced a free-fall instrument on Fabian's principle, but without exterior communication. The whole of the instrument was absolutely closed and impermeable. The point where the slipping of the lower part—that is to say, the bottom of the upper sheath—was effected, was terminated by a stuffing-box, likewise impermeable. The sheath, naturally hollow, and likewise the lower piece, served as

FIG. 130.—Hollow Auger Stem and Bit.

a water conduit, without there being much loss in the force of
the liquid projected into the bore hole.

Various Other Attempts.—After this first attempt other instruments were constructed. The inventors of them are Messieurs
de Tecklenburg, Schumacher, Przibilla and Fauck.

Fauck's Method.—The latter does not recognise hollow free-fall
instruments. Screwed to the hollow ends, which
have the orifice of their conduit above it, the
whole of the rather slender instrument is enveloped in a sheath.

It is in the length of the sheath
that the apparatus works. Sometimes
it only includes the free-fall instrument,
and joining the auger stem below it in a
hermetic fashion, allowing the slipping,
but not the escape of water, it is closed,
and the water, pressed into the conduits,
runs out through the openings of the
hollow auger stem so as to reach the
bit through this channel.

More Simple Apparatus.—In other
and more simple apparatus the sheath
descends, comprising all the lower part
of the drill, right to the bottom, where
the water is projected, without having
to use a hollow auger stem and bit (Fig.
130).

FIG. 132.
—Fauck's
Hydraulic
" Jars ".

*Total Substitution of Sheath for Hollow
Rods.*—Attempts have been made to
generalise this system of the sheath, by starting
it from the top of the bore hole, so as to do away
with the hollow rods and thus to obtain perfect
circulation. We do not believe that this system has a very great
future before it (Fig. 131).

FIG. 181.--Com-
pletely Sheathed
System of Rods.

Hydraulic " Jars ".—M. Fauck has introduced a " jars," with
water circulation, in the form of a cylinder, in which a piston
with a hollow rod moves. It has the advantage of being absolutely hermetical (Fig. 132).

Hydraulic Boring by Hand.—We are indebted to the same con-structor for a hand installation for the water circulation system, which, from the point of view of lightness, simplicity and the ingenuity of its details, is a veritable model.

Prolongation of Crane substituted for Derrick.—In this transport-able equipment the derrick is done away with; it is replaced by a prolongation of the structure of the crane, the two vertical branches of which, in the form of a guillotine, rise 6 metres (say 19·68 feet) above the level of the ground. These two branches support the part which serves for hoisting.

The plant, which is light, has but a slight base, does not com-prise a " walking beam," like the old French system, the bell-ringing movement being obtained by cranks with arms. But there

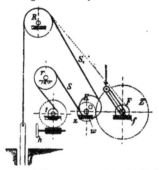

FIG. 133.—Diagrammatic Representation of Fauck's Hydraulic System.

is the real improvement. In order to impart this bell-ringing movement to the drill, the handles of the winch must be animated by an inverse movement, according to the rise and fall of the drill. This rather fatiguing change of movement could only be done very slowly.

Crank Motion.—In the system now being dealt with the motion of the crank is mechanical, uniform and in the same direction.

Manner in which the Apparatus Works.—The manner in which the apparatus works may be ascertained from Fig. 133. The end of the cable, SS_1, which supports the drill, is attached to the drum, t, which serves for elongation by means of the screw, h. Fixed to the drum, t, the cable, SS_1, runs in the groove of the wheel, r, then on a wheel, K, geared eccentrically to the axis, w, of the gearing, *z*, which itself receives the movement of *Z*, from which it is sent on to the pulley, *R*, situated at the top of the building, and in the axis of the rods.

It will be remarked that the parts, S and S_1, of the drilling cable are almost parallel, but the part, S_1, being fixed by the drum, *t*, will not budge during the movements of the eccentric, and thus the part of the cable, S_1, and consequently the rods, are animated by an oscillation of 8 centimetres (say 3 inches), that is to say, exactly double the excess of the diameter of the eccentric on the axis, *w*.

Raising of the Rods.—The raising of the rods is done on the drum, *F*, by means of cog-wheels and by simply removing the crank, *f*, to bring it to *w*.

With this apparatus, the number of oscillations, per minute, may be 80 to 260. The casing is from 89 to 68 millimetres interior diameter ($2\frac{1}{2}$ to $3\frac{1}{2}$ inches).

Speed of Boring.—With this portable equipment (hand power), during trials in favourable ground, 17 minutes sufficed to bore 1 metre (3 feet 3 inches).

Cost. — The whole cost of such an equipment, without buildings, casing, or forge, may come to 4,900 francs (say £196).

The Raky Rapid System of Hydraulic Boring.—" Much attention is now being given," says M. Lippmann, " to a system in which the principal agent is, once more, water, injected under pressure, to keep the bottom of the bore hole continually free. Here, again, it is a question of the hollow drill; but what is peculiar about this method of boring, patented under the name of the Raky system, is the method adopted to work the drill at a speed hitherto unadopted. Steam is the motor power; the motion is imparted by a pulley which actuates by a belt a plate crank, and the motion rod, imparting a to-and-fro movement to the horizontal ' walking beam,' to which the drill is suspended ; the adherence of the belt is maintained by a roller mounted on a counterpoise bascule ; one of the arms of the lever of the latter has its extremity in contact with the pulley of the plate crank, which carries, on the rim of its felly, a slight enlargement which has the effect of a small cam, in removing the roller screw at the moment the drill approaches the bottom of the bore hole. The concussion of the belt produces the fall of the drill, which is instantly raised by the plate-crank, the roller immediately resuming its effectual position. The axis of the 'walking beam' is supported by strong springs, which, by their elasticity, deaden the shocks, which results in imparting to the drill, and to the whole of the mechanism, the desired irregularity of movement.

Number of Blows of Raky Drill per Minute.—" By this means 80 to 120 blows of the drill may be given in a minute, and as the attack is made on a perfectly free and clear bottom, the rate at which deepening is effected is extraordinarily rapid. The suspension of the hollow rod to the lever is effected by a special method, which permits of regulating, during the shower of blows, the gradual lowering of the drill. The rotation of the drill is done by hand, as in drilling with solid rods ; the injection of water is effected by a pressure pump as in drilling by Fauvelle's system.

Rapidity of Boring.—" Borings have been instanced of 100 metres (say 328 feet) in depth, which have been finished in 7 or 8 days' drilling ; a boring of 395 metres (1295·6 feet) was made with an average rate of progress of 11 metres (35·48 feet) per day ; another of 500 metres (say 1,640 feet), with an average daily rate of progress of 7 metres (say 23 feet), but which in the 200 and the 300 metres reached depths of 18, 14 and 11 metres (say 59·04, 45·92 and 35·48 feet) per day.

" The above method constitutes a very ingenious idea of the application of a very simple mechanism to the production of a rapid succession of blows, which adds to the indisputable advantages of the Fauvelle system, by acting on an unencumbered bottom and in economising the manipulations incidental to the sand pumping of the well."

Raky's system has commenced to be very much employed in Galicia.

ROTARY BORING.   DIFFERENT MODIFICATIONS.

Augers.—The augers have been very much improved in France ; they constitute part of the Colonial equipment of which we have spoken p. 403. As they are but seldom used in petroleum boring, we shall pass over these improvements in silence ; moreover they are in points of detail (improvements in the steels, slight changes in shape).

Diamonds.—The diamond crowns, at the hands of inventors, have had to submit to changes of quite a chemical nature relative to the drilling substance used.

FIG. 134.—The Combined Diamond and Fauck System of Boring.

Borts, etc., Artificial Diamonds.—The *borts* and the *carbons* are very dear; moreover, they only adhere but slightly to the lead of the crown. The theoretical researches of the celebrated French savant, M. Moissan, on the artificial diamond have led to the realisation of real progress in actual practice by the foreign engineers, MM. Olaf Terp and Przibilla (the latter makes use of iridium crystals).   Let us also note the existence of carborundum.

Carborundum.—This substance is a crystallised carbide of silicium.   Discovered in 1892 by F. G. Acheson of Monongahela (Pennsylvania, United States of America), carborundum was patented in February, 1893, and the moment draws near when factories will be established in Germany, England and France. Carborundum would appear to replace emery and even diamond powder, in virtue of the three following properties : incombustibility, infusibility, extreme hardness.

COMBINED SYSTEMS.

According to the arrangement or lie, and the nature of the ground to be drilled, combinations of several systems have been often used and applied in the same well, according to circumstances. Let us enumerate amongst these combinations :—

1. The free-fall system with the auger system, used in the French Colonial equipment.

2. The Fauvelle system, combined with the free-fall system, with circulation of water.   In this case use is always made of hollow rods.

3. The Fauvelle system, combined with diamond drilling.

4. The diamond system, combined with the Fauck system (Fig. 134), using hollow rods for both systems.   It was with an equipment of this nature that Verbunt, the German engineer, drilled 58 metres (190 feet) in 24 hours at Beuthen, Silesia.

Record Speed.—This is, we believe, the record speed obtained in drilling.   The strata traversed consisted of coarse-grained sandstones and beds of gypsum.

PART IV.

ACCIDENTS.

CHAPTER XXXII.

BORING ACCIDENTS—METHODS OF PREVENTING THEM—METHODS OF REMEDYING THEM.

Risk involved in Boring.—The owner of a petroleum installation is exposed, during the boring of a well, to accidents which may, according to their seriousness, stop the work for a certain time, and even render the continuation of it impossible. It is easy to understand that no industrial undertaking is subject to more risk than that which consists in prying into the bowels of the earth, in perforating a succession of beds of unknown material, in directing a tool—the drill—sometimes several hundreds of metres from the hand of the operator; in a word, in executing a work of precision blindfolded.

Different Causes of Accidents.—The accidents which the manager of a boring has to fear are of several kinds. The causes of these accidents are multiple, and may be classified as follows :—

1. DEVIATION OF DRILL.

2. LANDSLIPS.

3. FRACTURES OF RODS AND DRILL, AND DEFORMATION OF CASING.

4. FALL OF FOREIGN OBJECTS INTO THE BORE HOLE.

5. FLOODING.

6. UNFORESEEN SPOUTING.

We shall, forthwith, make a special study of each of these cases, and, as we pass them in review, we shall indicate the preventive measures and the line of conduct to pursue, if, in spite of all precautions, they happen to occur.

1. Deviation of Drill.

Partial and Total Deviation.—Deviation is one of the most frequent eventualities amongst these accidents.   This deviation may be *partial* or *total*, according to whether the bore hole has gone out of the vertical by a length shorter or greater than that of the drill.

Partial Deviation, Causes—Hard Highly Inclined Strata.—The causes of deviation may proceed from two sorts of circumstances : the first—a quite natural one—is the direct consequence of the structure and arrangement of the stratifications of the bottom of the bore hole.   In petroliferous ground the stratifications often assume, as we have already seen, an almost vertical inclination. In these highly inclined strata, if the drill, coming out of a relatively soft bed, attacks a hard bed of this nature, and if its cutting edge is parallel to the plane of the latter (*i.e.*, parallel to the plane of bedding), there will be many opportunities for this cutting edge, instead of cutting the hard rock, of gliding over its surface (which, as we have seen, is highly inclined), and thus deviate from the vertical gradually as the work of drilling is prolonged.   If the hard rock, even highly inclined, had been attacked by a drill, the position of which had been perpendicular to its surface, the risks of deviation would have been much less. In fact, in that case, the cutting edge of the drill would have attacked the rock *in a corner*.   This rock would have been cut away, or at least a violent shock, making itself felt throughout the whole of the system of rods, would have warned the foreman borer of the presence, in the bottom of the bore hole, of a hard point, on which he would have felt the necessity of dwelling upon with his drill.

In that way, deviation would not have occurred.

How Partial Deviation may be Avoided.—The deviation, produced by the conformation of the beds to be traversed, may be avoided by taking particular care, in drilling, to keep the bottom of the bore hole perfectly uniform.   The foreman borer should bring all his attention to bear upon it, and with this end in view he should constantly verify the condition of his underground by imparting a continuous movement of rotation to the drill by means of the working lever. This rotation will be effected by more decided turns as the ground gets more difficult, and will only be suspended when the foreman borer feels, from the vibrations of his tackle, the necessity of dwelling upon a resistant point.

Total Deviation. — The second cause of deviation arises generally from quite a fortuitous cause, such as preliminary deviation, for some accidental reason, of the last section of casing of the bottom part of the bore hole. In this case the system of the auger stem and drill is guided falsely, and the boring is continued in the axis of the deviated tube. This deviation is much more serious from the fact that it has always assumed the form of *complete deviation* before it has been perceived.

It can only be avoided by a perfect system of guiding the rods, by the addition of a second guide placed at a distance of some 10 metres (say 32·8 feet) above the first.

Remedies. — In both kinds of deviation the remedy is the same. The crushed debris of hard rocks, agglomerated with clay, or even cement, is run into the bore hole, till it fills it up to the height of the original deviation, taking care to raise the tube if this deviation commenced in a tubed part. It is on this fictitious bottom that the fresh drilling of the deviated part is recommenced.

2. LANDSLIPS.

General Prevalence.—As a general rule, there is no boring in which landslips do not occasion mishaps.

The Cause of Various Accidents.—Landslips are the direct cause of several varieties of accidents, such as : the adherence and even the engulfing of that part of the drilling tackle which is working in the non-tubed portion of the bore hole. In the tubed part the displacement movement of the shifting ground may shift, deform and even rupture the casing.

Precautions, Use of Water.—The precautions to be taken against landslips in the lower part of the bore hole consist in the care which should be taken to maintain there a column of water of a sufficient height to counterbalance by its pressure the tendency to slip of the shifting beds.

Strong Casing Pipes, Danger of " Lost" Pipes.—As regards the accidents produced by landslips, and as far as the tubing is concerned, there is only one preventative : to use very strong tubes and never to use " lost " tubes.

Forewarnings.—The foreman borer is warned that a landslip has occurred in the bore hole by the adherence of the drill at the bottom of the boring.⸱ This adherence manifests itself by the difficulty which is experienced in lifting the tackle by the " walking beam ".

Measures to Adopt.—In view of these symptoms of landslips the foreman borer stops drilling.   By means of the temper screw he raises the drill until it is free from the *débris* brought down by the landslip.   This result will be attained when the adherence of the drill ceases, or at least no longer presents any danger of rupturing the rods.   If the landslipping continues, the foreman borer raises his drill to the surface, and by means of the sand pump ascertains the nature of the beds which have fallen in.   When these prove to be very mobile he proceeds to completely remove them by repeated sand-pumping.   When, on the contrary, the ground which has fallen in consists of rock *débris* it is necessary to recommence drilling.

Landslips may render Daily Progress nil or even Negative.—Under such circumstances the work is most deceiving.   Very often the daily progress made in drilling has been *nil*, or even negative, from the fact that the foreman borer has had to raise his drill, on account of adherence, further up than the latter has descended during the whole working day.   We have seen a boring not advance one centimetre (0˙393 inch) during a period of twenty days in spite of incessant work, taken up in drilling masses of landslips, which appeared to be inexhaustible.

Exasperation Aggravates Disaster.—The exasperating effects of such results on the managers of borings are the cause of almost irreparable accidents and even the loss of the well by its getting deadlocked.

In fact, when the drill sticks the foreman borer feels an extreme repugnance to sacrifice the ground already acquired by raising his drill.

The adherence, however, becomes greater and greater, then

comes a moment when the effort of the " walking beam " becomes insufficient to wrench up the drill—the latter has become engulfed.

Necessity for Prompt Action.—This situation is disastrous ; it necessitates immediate energetic measures, because each minute aggravates the situation.

The delicate manœuvre of wrenching is proceeded with.

Wrenching.—Wrenching, the word says so, is the operation which consists in raising by force the boring tackle which is engulfed under the agglomeration of the landslipped masses. This operation, which it would be much better to avoid, is one of the most dangerous, as it may complicate the engulfment by

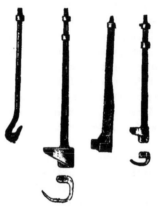

Fig. 135.—" Fishing " Tools (Wrenches).

rupturing the rods.   It would be well, therefore, to use force measures with the greatest precaution.

Use of Temper Screw in Freeing the Drill, A Safe Operation.—If the boring tackle includes a temper screw, by raising the screw by hand the embedded drill is disengaged centimetre by centimetre.

This first manœuvre, on account of the division of the effort, presents few dangers of rupturing the rods.

Wrenching Proper.—The 65 centimetres (say 25½ inches), the length of the temper screw, having been gained, operations are conducted by aid of an instrument *ad hoc* (Fig. 135), which consists of a rod of about 1½ metres in length, whose upper extremity is screwed like that of an ordinary rod, and the lower extremity of which presents at a right angle the form of a horse shoe or two parallel branches of 6 centimetres (say 2·34 inches) in length, leaving a groove between them rather wider than that of the drilling rods, and destined to hold these after the manner of a raising key.

The instrument is termed a *caracole* or tool wrench.

It is fixed by its upper extremity to the hoisting cable, and brought towards the rods in such a manner that the claw of the hook of the *caracole* comes under the safety bulge of one of them.

In this way the raising cable supports the system of rods and drill. The temper screw may now be unscrewed, and it is then attempted, without insisting, however, for fear of ruptures, to wrench the drill up by means of the raising windlass. If this tentative measure does not succeed, the caracole is screwed to the temper screw, the screw of which has previously been completely unscrewed, and the first operation of raising the drilling tackle 65 centimetres ($25\frac{1}{2}$ inches) is recommenced.

Thus very slowly, and unless the landslipping continues and triumphs over the efforts of the men, this delicate operation is performed satisfactorily.

If the equipment does not include a temper screw, the assemblage of rods is hooked on to the hoisting cable by means of the *clef de relevée*. In this case the wrenching manœuvre commences by a bell-ringing motion, which is intensified more and more before attempting the great effort.

For forced measures of this nature it is well to use a rather worn hoisting cable. The chances of breaking this cable, less strong than the rods, ensure the safety of the latter.

Raising an Embedded Sand-Pump.—If, instead of a drill, a sand-pump has stuck, the wrenching up of the latter can only be done by aid of the rods. A hook (Fig. 135) or a caracole at the end of the latter seizes the handle of the sand pump. It would be a measure of prudence, in ground very liable to landslips, to always use the rods for sand pumping.

Precaution.—In this description of ground it is more than ever necessary to adhere to the general rule to never leave a tool, in the untubed portion of the bore hole, during the suspension of work.

3. DAMAGED RODS AND CASING.

The accidents which may happen to the tubing or casing in consequence of landslips are, as we have said, of two kinds :—

(1) DEFORMATION, (2) FRACTURE.

1. *Deformation.*—Deformation necessitates the passage of an expander of an oval form (Fig. 136), called a *pear*. This "pear," ballasted by the auger stem, is lowered at the end of the rods into the well. By its passage it rounds the deformed portions of the casing. If this expedient be not successful in the rather frequent instance in which the rivets of the tubes have been sprung, the wrenching up of the whole or part of the damaged tube must be proceeded with. This wrenching manœuvre is done with the rods, working in the manner previously indicated.

FIG. 136.—" Pear " Pipe Expander.   FIG. 137.—Tap Grab.

In order to seize the tube, which, as we have said, is pierced by some holes intended for the ends of the hooks, the rods are provided with a hook, and they are manipulated by groping in such a way as to reach one of the hooking-on holes. A tap borer may also be used for raising tubes (Fig. 137).

Tools for Cutting Casing Pipes.—For cutting casing pipes use is made of Fauck's instrument (Fig. 138), designed for this purpose. This instrument consists of an extensible ring, *s*, furnished with steel rings, *k*, on its cutting circumference. The extensible ring, *s*, may be increased in diameter by the more or less accentuated introduction of an extension triangle ; this triangle regulates the total diameter of the instrument in such a manner that the cutting wheels are pressed against the side of the tube. A circular movement of the apparatus, and the simultaneous introduction of the extension triangle in the ring produces the division of the tube.

Dru's instrument (Fig. 139) is a cylinder fixed eccentrically to the rods.   Another manipulating rod works the knife, which is pressed against the tube with the whole of the weight of the rods, which hang vertically.

Of course, in ground liable to landslips—shifting beds—more

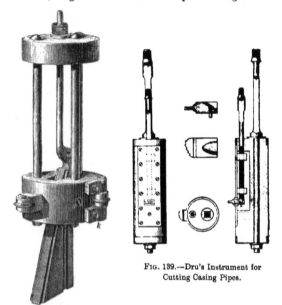

FIG. 139.—Dru's Instrument for
Cutting Casing Pipes.

FIG. 138.—Fauck's Instrument for
Cutting Casing Pipes.

especially than in any other ground, casing which has been raised ought to be at once replaced.

(2) FRACTURE.

Accidents due to fracture are of comparatively infrequent occurrence with good plant.

Fractures generally happen either to the neck of the drill (result of a shock) or to the rods.

Broken Drills.—A rupture of the neck of the drill is a serious

28

matter, especially if the instrument, losing its vertical position, falls flat on the bottom of the bore hole. In this latter case, the accident is almost irreparable. No attempt should be made to perforate the mass of steel forming the drill. All efforts should therefore be limited to shifting this mass.

If the drill remains in an almost vertical position, and if it be not entirely smooth, if it has any projections, such as the cheeks or orifices of the water conduit, as is the case with the drill used in Fauvelle's system, attempts are made in the first instance to seize it between the legs of a toothed-spring tongs (Fig. 144), or, in the second place, to hook it on to the end of a simple hook.

Fractured Rods.—The fracture of the rods, if it be not complicated by landslips or deformation of the casing, is a matter of a much less serious nature. A simple tool wrench (Fig. 135) can easily seize the rods by one of their shoulders.

Unscrewed Tackle.—The unscrewing of the tackle in the bore hole may be dealt with as a case of fracture.

Unscrewed Drill.—The unscrewed drill may be easily seized by means of the tapping cone, which is capped with its screw gudgeon. The tapping cone works either to the right or to the left, it is turned like a draw plate on the object which it is desired to seize. It would, perhaps, be more expeditious in bringing the drill to the surface if a tool wrench were used, but this method is less certain from the fact that the drill runs the risk of falling whilst being raised.

Unscrewed Rods.—The unscrewing of the rods necessitates the same remedial measures as that of the drill, tool wrench and tapping cone.

4. FALL OF FOREIGN OBJECTS INTO THE BORE HOLE.

The " Fishing" Instruments Numerous and Varied.—Foreign objects fallen into the well will, according to their nature, either be crushed during the drilling process, or seized hold of and brought up. The form, nature and position of these will determine the nature of the hooking instruments to be used. These instruments and their form are extremely various; the details of their construction are dictated by circumstances. Often in the case of

.a single accident, a great number may be tried, and often also it may be necessary to modify them.

Principal Types.—All these instruments, however, approach a few principal types :—

These are hooks, toothed or otherwise (Fig. 135), anchors, hooked keys (Fig. 135).

Tongued tool wrenches and spring tool wrenches.

Tools on the pincers principle, ordinary pincers, angle pincers (Fig. 144), which are manipulated by turning the rods which regulate the movement of a pressure screw pushing an angle on an inclined plane at the top of the two handles of the pincers

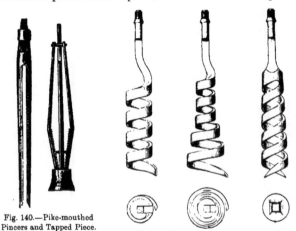

Fig. 140.—Pike-mouthed Pincers and Tapped Piece.

Fig. 141.—Worm Grabs (" Ram's Horns ").

which force the extremities of the latter to approach the objects to be seized, clicking pincers, pike-mouthed pincers (Fig. 140).

The tap grabs (Fig. 137) which are used to seize the tubes or objects tapped.

The tapped cones, the use of which has been already indicated.

The single or double-branched worm grabs (Fig. 141).

The drags.

The claw-spring and tongue-shaped cylinders or sockets. The latter instruments are used to hook rods without the projections required for the use of tool wrenches, or the screwed parts indispensable to the use of the tap grabs or the tap cones.

Tooth springs destined to act by pressure on the sides of the tubes to be seized.

The different kinds of bells—clapper and roller.

5. FLOODING.

Preventive Measures.—The only way to prevent flooding is to use hermetically sealed casing. This tubing consists of an assemblage of cast-iron tubes screwed together every 5 metres (say 16½ feet).

Hermetically Sealed Casing.—In addition to the protection

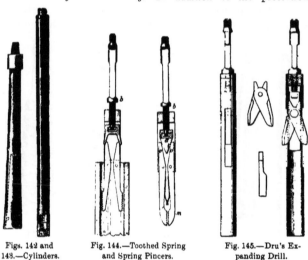

| Figs. 142 and 143.—Cylinders. | Fig. 144.—Toothed Spring and Spring Pincers. | Fig. 145.—Dru's Expanding Drill. |

afforded against flooding, the hermetically sealed casing has the advantage of being so solid as to be proof against all landslip attacks.   But its inconvenience, besides its great cost, is its great thickness which does not allow telescoping casing.   Thus the bore hole, instead of having pipes of several diameters increasing upwards to the top, has only a single metallic interior without any reduction whatever.   The portions of casing to be added, instead of being sunk into the well, are screwed on above, and the whole casing descends correspondingly.   The following are the prices of hermetic tubes or pipes according to their dimensions :—

DIMENSIONS OF HERMETICALLY-SEALED PIPES.

Interior Diameter.		Thickness of the Sides.		Cost in Francs per Running Metre.	Cost per Running Foot in Shillings.
Millimetres.	Inches.	Millimetres.	Inches.		
180	7·186	4½	·176	32	7·3
150	5·9	4½	·176	21·5	5·3
108	4·25	4¼	·167	14·5	3·7
90	3·5	4	·157	10·0	2·5
62	2·4	3½	·137	7·25	1·9

Ratio of Diameter of Bore Hole to that of Casing.—It will be understood that the hole bored by the drill ought necessarily to have a larger diameter than the interior diameter of the casing. At first sight this would appear to be an absolute impossibility, it being impossible for the drill to have a larger diameter than the casing.

Dru's Enlarging Scissors.[1]—In order to obviate this difficulty use is made of an instrument called an enlarging scissors (*ciseaux élargisseurs*). These scissors are intended to develop a larger section than that of the pipe, whilst at the same time they pass through its interior diameter. For this purpose two small drills—mounted on hinges after the manner of the blades of a scissors—are used, which during their passage through the tubes are shut, and enter as if into a sheath into grooves cut out of a cylindrical body, *ad hoc*, which is screwed immediately above the drill. When they get to the bottom the first shock of the instrument causes a knife to advance which opens the blades, and the latter, pushed forward by springs, project from the circumference of the cylindrical body. By imparting a percussion movement to the tool the sides are enlarged, and the tubing can be sunk into this enlargement. Dru's enlarging scissors are wrought by a screw, the pressure of which determines the extent to which the blades open.

Fauck's Enlarging Scissors.[1]—Fauck's enlarging scissors (Fig. 146) are kept in a downward position during their passage through the tube by an iron wire which care has been taken to attach round the apparatus in such a manner that whilst holding the scissors in position it passes below the cutting blade of the drill. At the first blow of the latter the iron wire is cut and the scissors pushed by springs assume the horizontal position.

[1] Expanding Drill.

The Old-fashioned Eccentric Drill.—At the present day, enlargement scissors are everywhere employed ; we shall only advert
here, therefore, as a matter of history, to the eccentric drill which
was formerly destined to fulfil the same office. As its name
indicates, the prolongation of the rods do not form the axis of this
drill, which, for example, having on one side of the axis of the
rods 10 centimetres (say 4 inches) of cutting width, had on the
other side only 7 centimetres (say $2\frac{3}{4}$ inches). The total section
of this drill was only 17 centimetres (say $6\frac{3}{4}$ inches) and yet the
diameter of the bore hole drilled by it was 20 centimetres (say
8 inches).

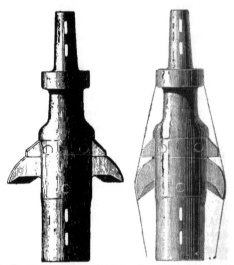

Fig. 146.—Fauck's Expanding Drill (*Cf.* American Paddy and Expansion Drills).

Its use now Casual.—The eccentric drill is now no longer used
except under altogether transient circumstances, and in borings in
which the casing not being hermetically sealed, they do not
possess enlarging scissors.

Use of Freezing Machines to Prevent Flooding.—We shall mention
equally as a matter of history, and as a simple matter of curiosity,
the process experimented with by the Anzin Mining Co. (Anzin,
a coal mining district in the Department du Nord, 2 kilometres

(1¼ miles) from Valenciennes).   In two borings, which had to pass through water-bearing beds 90 metres (say 295·2 feet) in thickness, the dangers of flooding were fought against by the help of two powerful freezing machines.   The flooded ground was strongly frozen in such a way that the work of drilling was done inside an immense solid block of ice.

6. Unforeseen Spouting.

Regarded as Accident.—Unforeseen spouting, although, on the whole, a rather fortunate occurrence, amidst the incidents of drilling, ought, nevertheless, to be regarded as a case of accident, as much from the point of view of the loss sustained as the material damage it may cause to the plant.

Unforeseen Spouting Rare : Previous Signs and Subsequent Manifestations.—It is, however, a rare case in which spouting has not been anticipated, unless it be the case of striking, unexpectedly, a pocket of oil or gas.   Petroleum itself announces its presence. The signs of that presence are multitudinous.   They are to be seen in the rock *débris* brought up by the sand-pump ; this *débris*, impregnated with oil, assumes a more and more deep greenish colour ; its mass is agitated by a peculiar boiling due to the presence of a large proportion of gas.   This gas becoming stronger and stronger impedes the sand-pumping by keeping the valves of the sand-pump open by its pressure.   Very soon it rushes out of the orifice of the well with a peculiar whistling sound, which the oilmen term the sighs of the earth—*les soupirs de la terre.*   These symptoms increase, gradually the ascent of gas changes into a clamorous eruption, into a spout of deleterious emanations, which renders the margins of the bore hole untenable.   This column of gas becomes more and more charged with petroliferous mud, then, suddenly, a black column shoots forth in a powerful jet of the same diameter as that of the bore hole, annihilating everything in its passage,—a jet of petroleum mixed with sand, stones, and often even of wrenched-up tubing ; it is a real volcano.

Intervals between Signs and Actual Spouting.—The order of sequence of these symptoms is almost always the same.   However, the interval which elapses between the ascension of the gas and the spouting of the petroleum is very variable, depending on the nature of the ground in which the well is bored.

Sometimes the evolution of gas manifests itself several weeks —perhaps even several months—before the spouting, which may only occur several hundred metres lower down than the point at which signs of petroleum began to appear. In such a case there will, always, be ample time to make the necessary preparations, and take all precautions for "capping" the well as soon as spouting occurs. However, one cause—which we shall qualify as one which is more or less unthought of—militates against these precautionary methods being adopted. This cause is superstition. It is feared that, by providing against spouting, one would discount the probabilities too soon. One does not wish to count his chickens before they are hatched (on ne veut pas vendre " la peau de l'ours ").

At other times spouting is totally unexpected.

The first appearance of symptoms precede but by a few hours, indeed, but by a few minutes, the outburst of petroleum.

The Potok Wielki Spouting Well of Galicia.—We may quote as an example of this latter circumstance the sudden spouting of the great well of Potok Wielki (Galicia). The gas rushed forth, unexpectedly, with such a force as to almost totally wrench asunder the covered-in portion of the derrick. It carried in its train a waterspout of sand and stones, which very soon covered the surrounding ground. The workmen ran to save their lives as quickly as possible to a respectable distance. A few minutes had scarcely elapsed before the gas was replaced by a column of petroleum, which, turning round and round upon itself in the air, and rising in a majestic shower to a height of some 40 metres (130 feet), continued the work of destruction commenced by the outburst of gas. All attempts to approach the well were made in vain.

The spouting occurred at nine o'clock in the morning; at three o'clock in the afternoon it diminished in intensity. At that hour they were able to commence the operations of "capping the well," which were finished at five o'clock that evening. Ten minutes after the well had been *put under lock and key* the petroleum suddenly ceased flowing. Only gas continued to issue abundantly.

Duration of Spouting and Amount of Petroleum Wasted.—As will have just been seen, it was impossible to collect a single drop of the petroleum of the great well of Potok Wielki. All was lost, scattered at the mercy of the caprice of the earth. The spouting lasted eight hours. According to the force of the jet the quantity of petroleum lost may be estimated at 12,000 barrels, representing at the time (1890) the respectable sum of 140,000 francs (£5,600).

Total Damage.—As far as the undertaking itself was concerned, this memorable day closed its account in addition to the dead loss

Fig. 147—A Spouting Well (*un puits emballé*).

of 12,000 barrels by damages amounting to 22,700 francs, caused by the impetuosity of the spouting.

This day may be described as a dreadful day, and it will be readily understood from this example why we have classified unforeseen spouting amongst boring accidents.

Precautions and Procedure to Stop Unforeseen Spouting.—When the manager of the works finds himself in presence of analogous circumstances, his first care should be to extinguish all fire in the

neighbourhood.   The boiler fires should be drowned out to prevent
all causes of fire.

Capping the Well.—The "capping" apparatus the (*kalpak*)
should be held in readiness, and the progress of the spouting
attentively followed; if it relaxes, which may occur when the

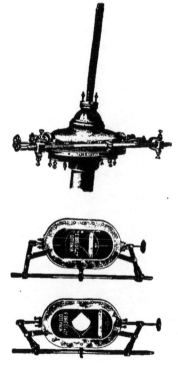

FIG. 148.—Bredt's Patent Oil-well Cap.

projected *débris* obstructs the bore hole, an attempt should be
made to cap the well.

Method of Procedure.—This operation, although a difficult one,
may be successfully performed if the orifice of the well be free;
but if, unfortunately, the drilling tackle is in the well, which is
frequently the case, the rods projecting out of the orifice will form
an almost insurmountable object to screwing the tubulure intended

for the cap on to the well.   The hoisting up of the drilling tackle takes time, and moreover, the raising of the rods would displace and destroy the providential obstruction which would have enabled the well to be capped.   It will be seen that this operation may be classified as one of the most delicate which may be met with in the development of the petroleum industry.

Bredt's Patent Well Cap.—M. Bredt of Ottynia, Galicia, has patented a capping apparatus which would appear to meet all exigencies ; Bredt's apparatus is so much the more perfect by the fact that it allows of the well being capped without withdrawing the drilling tackle from the well.   This apparatus consists of two parts, A and B, which screw into each other, and thus form a sort of box with its lid on.   In each of these parts there is a double register, which leaves in its middle an opening of a quadrangular form of variable size, capable of being closed completely by a screw, or only in such a manner as to allow a drilling rod to pass through.   The first part, A, of this apparatus is screwed on the bore hole, then the second part, B, is applied against the first, to which it is adjusted by a screwing system.   By tightening the screw gradually the closing of the bore hole is effected, not abruptly, but progressively, and it is thus without any great difficulty capped (Fig. 148).

Telescopic Method of Capping.—A telescopic method has also been adopted for shutting the mouth of the bore hole, consisting of cylindrical bodies of gradually diminishing diameter, which are screwed on the top of the bore hole.   This method of capping is much more delicate than that of Bredt.

CHAPTER XXXIII.

I. The "Torpedo."

When Used.—The "torpedo" is not resorted to during boring. However, boring having been finished, if the yield begins to diminish, two alternatives present themselves: either of boring farther down to reach a fresh petroliferous bed, or, if the well, already very deep, is judged to have reached the lowest petroliferous strata, other means are taken to cause it to furnish its yield.

Causes which Tend to Stop the Yield of a Well.—The decrease in the yield of a well may be due to two causes. On the one hand, the petroleum does not flow because the ground between the bore hole and the neighbouring petroliferous vein (perhaps lateral and which cannot be got at by boring) is choked up, or, it may be, because a fissure, dependent on this vein, but without direct communication with it, has been exhausted by pumping. In these two instances it will be advantageous to create new fissures artificially, bringing the bottom of the bore hole into direct communication with the petroleum-bearing bed. Resort is then had to the "torpedo," or *levigation.* The "torpedo" is the most energetic method; dynamite is its impetuous agent.

Chemical Properties of Dynamite.—Dynamite, says M. Louis Figuier, is a solid body, oily and pasty, formed by mixing different substances with a highly explosive oil—nitro-glycerine.

$$C_6H_{10}N_6O_{18} = 6CO_2 + 5H_2O + 3N_2 + O$$

Nitro-glycerine.		Carbonic Acid.	Water.	Nitrogen.	Oxygen.
454		264	90	84	16
454			454		

Nitro-glycerine Unsafe and its Use Discontinued in Europe.—In the beginning this oil was used by itself alone, but its use was inconvenient and dangerous. In consequence of numerous accidents which occurred in transit, it was recognised in Europe that nitro-glycerine could not be utilised except by being prepared on the spot, *in situ*, and immediately used.

Same Effects Produced by More Safe Dynamite.—The invention of dynamite, a mixture of nitro-glycerine and an inert substance, afforded the means of producing, without danger, almost the same effects as nitro-glycerine.

Varieties of Dynamite.—Dynamite is divided into two categories according to the nature of the substance, which is associated with the nitro-glycerine.

Dynamite with an Inert Base.—In the first—

$$C_6H_{10}N_6O_{18} + \text{Silica} = 6CO_2 + 5H_2O + 3N_3 + O + \text{Silica.}$$

		Carbonic Acid.	Water.	Nitrogen.	Oxygen.	
454	151 grammes	264	90	84	16	151
605			605			

the absorbent matter is *inert*, and, consequently, only serves as a vehicle for the explosive oil. For a dynamite of this nature to be of good quality, it is necessary not only that there should be admixture between the solid body and the nitro-glycerine, but also real absorption, in such a manner that, under the influence of atmospheric variations, or shaking induced by prolonged transport, no separation may take place between the liquid and the solid. It is the judicious choice of this absorbent which has assured the success of Nobel's dynamite No. 1.

Moreover, although the presence of the sand, called *infusorial earth* (*kieselguhr*), which constitutes about 25 per cent. of the weight of the dynamite, has the effect of diminishing the force of the explosive oil, its advantages are such that the use of pure nitro-glycerine has been, and ought to be, totally abandoned.

Dynamites with an Active Base.—The dynamites of the second category, called, in contradistinction, dynamites with an *active* base, are mixtures of different explosive substances with nitro-glycerine. The most widespread type is the dynamite No. 3 of the Paulilles factory of the *Société générale de dynamite*, which, not so strong as No. 1, is, however, sufficient for the greater number of operations ;

besides, it is cheaper and presents very great guarantees of security, because it is but slightly sensible to shock, and only inflames with difficulty.

Properties of Good Quality Dynamite.—All good quality dynamites possess the following properties :—

1. *Appearance.*—They present themselves in the form of a more or less greasy plastic mass of great density (about 1·6).

2. *Behaviour on Ignition.*—They burn simply by contact with a flame or an ignited body, and burn tranquilly without exploding.

In order to cause them to detonate, the detonating priming capsule must be used, the use of which will be indicated farther on. Nevertheless, it is always prudent to keep dynamite far away from fire. If a certain quantity of dynamite may burn with impunity, it is to be feared that it would not be the same with a large mass.

3. *Congealing Point.*—All dynamites congeal and lose their plasticity at a rather high temperature, at 7° or 8° C. To use it with good effect it must be thawed and brought back to the soft condition.

Comparative Sensibility to Shock of Frozen Dynamite and Soft Dynamite.—It is an error to suppose that frozen dynamite is more sensible to shock than soft dynamite; but it may detonate by the shock of a metallic body if this shock be violent enough.

Dynamite Cartridges.—Dynamite is always sold in cartridges made of *parchment paper ;* a very sound paper which does not tear and is not traversed by the explosive oil.

All dynamite enclosed in a rigid resistant envelope or one which allows itself to be traversed by the nitro-glycerine ought to be rejected in trade.

Dynamite powders are put on the market in slightly plastic cylinders called cartridges, the diameter of which is generally from 2 to 2½ centimetres (say ¾ to 1 inch) by 2 centimetres (¾ inch) in length. They are sold in cases containing 20 or 25 kilogrammes of cartridges.

Blasting Gelatine.—In 1878 Nobel produced a new explosive

$$C_{24}H_{31}(NO_2)_9O_{20} \quad + \quad 25\cdot5(C_6H_{10}N_9O_{18}) \quad = \quad 177CO_2 \quad + \quad 143H_2O \quad + \quad 81N_2$$

Collodion.	Nitro Glycerine.	Carbonic Acid.	Water.	Nitrogen.
1053	11,577	17,788	2574	2268

| | 12,630 | | 12,630 | |

(dynamite gum). This product is a combination of 8·3 per cent. of soluble collodion with 91·7 per cent. of nitro-glycerine.

This gum has the appearance of jujube paste.

Properties of Blasting Gelatine.—Theoretically it should be a little more powerful than pure nitro-glycerine, not that the 7 or 8 per cent. of gun-cotton is of greater energy than the quantity of nitro-glycerine which it replaces, but because each equivalent of nitro-glycerine, by its perfect combustion, leaves an unused equivalent of oxygen, whilst the small fraction of gun-cotton which is added, in virtue of its carbon and hydrogen, comes as a suitable combustible to use up that oxygen. Hence more heat and greater energy is produced.

A more powerful body than nitro-glycerine is thus obtained, the latter not being available for easy use in the liquid state on account of the danger attending its use.

Inferiority of Dynamite to Nitro-glycerine greater than that Deduced from Ratio of Nitro-glycerine in the Dynamite to that of the Inert Body therein.—It is advantageous to know that dynamite is comparatively very much inferior in theoretical force to nitro-glycerine, and this inferiority is a point which, perhaps, has not received from engineers the attention which it deserves.

One would naturally imagine because dynamite contains 75 per cent. of explosive oil, combined with 25 per cent. of silica, that it ought to possess about three-quarters of the force of nitro-glycerine. As a matter of fact it does nothing of the sort.

Berthelot, in fact, has told us that the absolute effect of an explosive is generally proportional :—

1. To the quantity of gases developed.

2. To the quantity of heat which these gases may receive.

Now, dynamite, containing only three-quarters of nitro-glycerine, will produce, in the first place, three-quarters of the quantity of gases which an equal weight of pure nitro-glycerine would produce. Again, it would produce three-quarters of the heat which nitro-glycerine would produce. This heat, being applied to the heating of 250 grammes of silica, on the one hand, and 750 grammes of gases, on the other hand, it will be seen that the gases resulting from the combustion of a kilogramme of dynamite will only receive, supposing that the co-efficient of heat be the

same, the three-quarters of the three-quarters of the heat which
the gases resulting from the combustion of a kilogramme of pure
nitro-glycerine would assimilate to themselves.

The effect produced by dynamite will therefore be $\frac{3}{4} \times \frac{3}{4} \times \frac{3}{4}$ of
that of nitro-glycerine, and as the cube of 3 divided by the cube
of 4 yields less than half, it follows that a kilogramme of dynamite
is less strong theoretically than less than half a kilogramme of
nitro-glycerine.

Blasting Gelatine : Theoretical Explosive Force.—Blasting gelatine
ought therefore, theoretically, to be more than twice as strong
as dynamite; in actual practice its explosive power is, compared
with dynamite, in the ratio of 100 to 74, which is a superiority in
force of 26 per cent.

These explosives detonate like nitro-glycerine under the influence of a capsule of fulminate of mercury. This capsule cannot
be exploded in an oil well except through the agency of electricity.

Pictet's Fulgurite.—In 1893 M. Raoul Pictet invented a new
explosive fulgurite which would appear to possess all the remarkable properties which an ideal explosive should, according to M.
Pictet himself, present.

Properties of an Ideal Explosive.—These properties are as
follows :—

1. *Safety in Manufacture.*—The manufacture of the explosive
should be perfectly safe.

2. *Bear Handling in Transit.*—It ought to bear transportation
without danger, without having to dread an explosion resulting
from handling, concussions or accidental falls.

3. *Unaffected by Weather.*—It should not change its physical
state under the influence of variations of temperature or hygro-
metric conditions; it should neither be deliquescent, congealable
nor liquefiable.

4. *Non-volatile and Non-pulverent on Storage—Non-poisonous.*—It
should not evaporate nor crumble to powder by prolonged storage
in depôts; it should not be poisonous neither in itself nor in virtue
of the gases resulting from its explosion.

5. *Easily made from Abundantly Occurring Raw Materials.*—It
ought to be easily made from raw materials which exist abundantly and are easily obtained.

Now, Pictet's fulgurite presents all those different advantages.

Pictet has compounded three different fulgurites which differ simply from one another in the proportion of the ingredients.

His No. 3, the action of which is the weakest, is destined, in the opinion of the inventor, to replace gunpowder in the loading of war arms; it is a slow progressive explosive, which presents itself in liquid form.

Nos. 1 and 2, which are solid bodies, are violent explosives, the effects of which are analogous to those of dynamite.

But these three explosives possess the common property of only detonating when they are brought to a temperature higher than 800° C. (1,472° F.), either by the inflammation of a capsule of fulminate of mercury, or by the passage of an electric current through a metallic wire in contact with the explosive.

The following table, moreover, shows the comparative properties of Pictet's three fulgurites, and those of the principal explosives in use :—

COMPARATIVE EXPLOSIVE PROPERTIES OF PICTET'S FULGURITES AND THE PRINCIPAL EXPLOSIVES IN USE.

Denomination.	Temperature of the Explosion.	Volume of the Gases produced brought to 0° C. (32° F.).
		Litres.
Fulgurite (No. 1)	1,575° C.	761
,, (No. 2)	3,822° ,,	817
,, (No. 3)	1,900° ,,	841
Fulminate of mercury	4,000° ,,	314
Nitro-glycerine	6,980° ,,	713
Dynamite	5,378° ,,	535
Gunpowder	3,514° ,,	300

Fulgurite, notwithstanding the fact that it is composed of very simple ingredients, cannot be manufactured except with quite special and costly plant. In virtue of this circumstance, the question arises whether advantage should not be taken of Pictet's discovery to suppress dynamite, and to replace it by a substance which excludes all danger of unforeseen explosion, and above all, of clandestine manufacture.

Other Explosives.—There are yet other explosives used in blasting than those just mentioned.

29

Ammonium Nitrate.—In the first place there is nitrate of ammonia,

$$2NH_4NO_3 = 4H_2O + 2N_2 + O_3.$$

Nitrate of Ammonia.   Water. Nitrogen. Oxygen.

$$80 \times 2 = 160 \qquad 72 \quad 56 \quad 32$$

$$160$$

or its compounds, of which one of the most frequently used is a mixture of this product (80 per cent.) with dynamite (20 per cent).

$$C_8H_{10}N_{16}O_{18} + Silica + 29\cdot2NH_4NO_3$$

Nitro-glycerine.   Ammonium Nitrate.

$$454 \qquad 151 \quad + 29\cdot2 \times 80 = 2,336 \quad = 2,941$$

$$605$$

$$6CO_2 + 63\cdot4\,H_2O + 30\cdot2O + 64\cdot4N + Silica.$$

Carbonic Acid. Water.   Oxygen. Nitrogen.

$$264 \qquad 1,141\cdot2 \quad 488\cdot2 \quad 901\cdot6 \qquad 151$$

$$2,941$$

Chlorate and Picrate Powders.—We need only mention chlorate and picrate powders as a matter of reference. Their use is delicate and hardly safe.

EXPLOSIVE FORCE, VOLUME AND WEIGHT OF GASES FROM DIFFERENT EXPLOSIVES, CALCULATED ON 1 KILOGRAMME.

Designation of Explosive.	Explosive Force.		Weight of the Gases per Kilogramme of Second Order.	Reduced Volume of the Gases per Kilogramme of Second Order.	Heat Disengaged.	
	2nd Order.	1st Order.			2nd Order.	1st Order.
					Cal.	Cal.
Fulminate of Mercury	—	9·28	—	—	—	758
Nitro-glycerine . .	4·80	10·13	0·800	—	1,720	1,777
Dynamite No. 1 (75 per cent.) . .	—	—	0·600	455	1,290	—
Pyroxylin . . .	3·00	6·43	0·850	700	1,056	1,060
Picric acid . . .	2·04	5·50	0·892	—	828	868
Picrate of potash .	1·82	5·31	0·740	576	787	852
Mixture of 55 per cent. picrate of potash and 45 per cent. saltpetre . . .	—	—	0·485	334	916	—
Mixture of equal weights of chlorate and picrate of potash	—	—	0·466	329	1,180	—
Picrate of baryta .	1·71	5.50	0·719	—	671	705
,, ,, strontia .	1·35	4·51	0·624	—	637	745
,, ,, lead .	1·55	5·94	0·668	—	555	668

It will be seen from these figures that the force of the *simple explosion* of a substance is proportional to the product of the weight of the gas which it yields, multiplied by the heat it disengages.

EXPLOSIVE FORCE OF VARIOUS EXPLOSIVES, MEASURED BY THE LEAD CUBE.

Designation of Explosive.	Explosive Force.	Remarks.
Dynamite (Nobel's No. 1)	147·50	St. Etienne trials. Holes not rammed.
„ (Vongue's „ 1) . . .	100·00	
„ („ „ 2) . . .	46·73	
„ („ „ 3) . . .	21·30	
„ (Paulille „ 3) . . .	70·59	
Blasting gelatine (*Société générale*) . .	100·00	Anzin trials, 1888. Holes rammed for 50 millimetres. Total length of holes, 140 millimetres. 10 grammes of explosive.
10 per cent. nitro-benzol, 90 per cent. ammonium nitrate	62·29	
30 per cent. dynamite, 70 per cent. ammonium nitrate	54·84	
12 per cent. gun-cotton, 85 per cent. ammonium nitrate	71·09	
Dynamite No. 1 (*Société générale*) . . .	80·81	

DETONATION TEMPERATURES (CALCULATED); DENSITY AND EXPLOSIVE FORCE IN PRACTICE OF SOME SAFETY EXPLOSIVES (ANZIN TRIALS, 1889).

Nature of the Explosive.	Temperature of the Detonations.	Density of the Explosives.	Practical Explosive Force of Equal Weights.
10 Binitro-benzol . . . . . 90 Ammonium nitrate	1,820° C.	0·93	100
10 Gun-cotton octonitric . . . . 90 Ammonium nitrate	1,562° „	0·84	100
Grisoutine B 12 Nitro-glycerine (gelatinised) . . 88 Ammonium nitrate . . .	1,494° „	1·04	111
Grisoutine F 20 Nitro-glycerine (gelatinised) . . 80 Ammonium nitrate . . .	1,638° „	1·17	118
Grisoutine Gum 30 Nitro-glycerine (gelatinised) . . 70 Ammonium nitrate . . .	1,871° „	1·22	122

Measured according to this table, the practical explosive force of blasting gelatine would be 130, and that of Nobel's No. 1 dynamite 100.

Firing by Electricity.—Whatever be the nature of the explosive used—whether dynamite or any other—the tools required for firing it by electricity consist of (1) the exploder, (2) the conducting wires, (3) the fuse or detonators.

The exploder generally employed is imported from Germany by M. Viau, one of the directors of the *Société générale de dynamite.*

It consists of two plates of ebonite or hardened caoutchouc, to which a rapid movement of rotation is imparted by means of exterior gearing driven by a handle.

These plates turn between rubbers lined with catskin. The electricity with which they are charged by rubbing is collected by combs, protected by an insulating disc of ebonite, and put in communication with the exterior armatures of a Leyden jar.

An independent screen also of ebonite is placed in front of the exterior armatures of the jars, between these armatures and the discs, with the view of preventing loss of electrical fluid.

The electricity produced by the small exterior Leyden jar is led to the conductors. The apparatus is enclosed in a rectangular wooden box, 38 centimetres high, with a base of 55 by 27 centimetres divided into two compartments by a partition. One of these, always closed, contains the generators of electricity, the other, containing the exterior armatures and the button, is furnished with a lid, which may be thrown back on the top of the box.

The following is the method adopted in firing a mine by this apparatus: the handle is seized by the right hand and turned at a moderate speed (2 turns per second). During the eleventh turn the button, which puts the electric current in contact with the exterior wire, is pressed, and the pressure kept up for a second.

If after eleven turns sparks are not produced, it is necessary to recommence by doubling the number of turns of the handle; then they are tripled, and so on, etc., until the spark is produced.

Particular care must be bestowed in laying the electrical conductors. Their *rôle* is to lead all the electricity which the apparatus may produce, to the explosive, of which the charge consists. They consist, therefore, of an uninterrupted continuation of insulated wire—a good conductor of electricity. In each conduit two principal parts are to be distinguished :—

(1) The principal conductor ; (2) The return wire. When complete success is desired to be ensured, insulated conductors must be used. In such a case the circuit must never be closed by running the wire to "earth" to save the cost of a return conducting wire, because to produce the spark it then requires a greater number of turns of the handle, and it is sufficient for one of the connecting wires to touch the ground for the current to return by accidental contact.

Conducting wires in dry tunnels may be made of annealed iron, but in petroleum wells, where moisture would soon oxidise the iron, well-annealed brass wire should be used of half a millimetre, or copper wire covered with gutta-percha. It is well known that the electric conductivity of copper is six times greater than that of iron.

The principal conducting wires are stretched on insulators, which themselves lie upon goblets or on pieces of wood arranged for the purpose.

Insulators.—The insulators commonly used are made of glass, porcelain or vulcanised caoutchouc ; they have the form of a bell, which guarantees insulation whatever may be the state of the weather. Thus, in spite of the utmost downpour of rain, the under part of the bell always remains perfectly dry.

The handles of the insulators are made of horn, and are furnished with two grooves or gutters for laying the conducting wires. If the insulators are exposed, care must be taken that the wire does not touch any surrounding objects, and that it is freely suspended through space ; it is only under these conditions that perfect isolation is obtained.

However, if the use of these conductors is not to be of long duration, it suffices to lay the wire freely on the ground.

The conducting wires rolled on bobbins are unwound in proportion to the descent down the well, and kept at 20 metres (say 65 feet) from the bottom to avoid projected rock *débris*.

They are in addition joined outside the well to the poles of the machine.

The conductors, which reach as far as 250 metres (820 feet), have a thickness of 0·002 metres; they weigh 36 grammes and cost 0·6 francs the running metre (say 1½d. per foot).

Those which reach a length of 300 metres have a diameter of 0·0023; they weigh 70 grammes and cost 0·75 francs per metre (say 2¼d. the foot).

Priming.—The third important factor in firing by electricity is the priming, which causes the production of the electric spark destined to set fire to the capsule.

The priming consists of a small cylinder of an insulating mastic, which keeps the two extremities of the two copper wires which form the conductor apart. A small paper cartridge containing an explosive substance is attached to the cylinder, and in which the two ends of the conductor dip, at a distance of about a quarter of a millimetre apart from one another.

Fulminating Capsule.—On the top is placed a capsule containing fulminate of mercury. The whole is covered with a coating of pitch.

It will, therefore, be seen that as soon as the spark transmitted by the electrical apparatus is given off it inflames the explosive material, and consequently detonates the capsule.

When the "torpedo" has to be submerged, the small paper cartridge is replaced by a metallic one.

Certain engineers prefer a simple battery to the apparatus just described.[1]

With the latter it is possible even during manipulation to verify constantly by means of a very feeble battery of common salt whether or not the current passes through the whole circuit. Moreover, the insulation of the principal conductors is not an absolute necessity.

When, on the other hand, use is made of an apparatus which gives off an electrical spark, it is quite necessary that the machine, the conductors, and the priming are in perfect communication.[2]

Method of using Explosives in Petroleum Borings.—Now that we have passed in review the explosives and the apparatus used in the "torpedoing" of oil wells, we proceed to indicate the method of working so as to accomplish this operation with success.

[1] In American *electric* blasting outfits a magneto battery is used.—*Tr.*
[2] Louis Figuier, *Les Poudres de Guerre* (Supplement).

Precautions.—In order to avoid injuring the casing by the effects of the explosion, it is hoisted up some 30 metres (say 100 feet). Care is also taken that the bore hole is completely filled with water, the *rôle* of the liquid being to serve as a wad to the charge and to act as an obstacle to the projection of rock *débris* into the upper part of the bore hole. These precautions having been taken, the " torpedo " containing the charge of explosives is lowered down the well. When the " torpedo " touches the bottom, the agent in charge of the " torpedoing " introduces the ends of the principal conductors into the holes of the receivers of the electrical generator. He fixes the handle to the apparatus, charges the latter by giving it 20 to 30 turns of the handle, and presses the button which determines the inflammation of the charge. On the surface nothing is heard but a dull noise made by the water in the bore hole, but within a certain radius the ground trembles and oscillates as if there had been an earthquake.

Operation done by Contract.—The use of " torpedoes," as we have said, is comparatively rare, for few establishments possess " torpedo " apparatus. This operation is generally undertaken by contract by companies who in the oil fields lay themselves out for this especial line of business.

The " Torpedo " Monopoly in America.—In America, the country in which the " torpedo " originated, they, even up to now, persist in the use of nitro-glycerine, a dangerous substance, it is true, but the use of which is not so complicated in virtue of the fact that it does not necessitate the use of batteries or exploders. A violent shock to the " torpedo " produces the explosion, and, by this fact, time is economised. The Americans are the " torpedo " apostles, and this process is in general use in the oil fields of the United States and Canada.

Colonel Roberts' Patent.—It was an old army colonel—Colonel Roberts—who, in 1861, first used nitro-glycerine in the petroleum oil wells. A patent was taken out for the invention, and at the present day a company has acquired the monopoly of the " torpedo " throughout the whole of the States of the Union.

First Use.—The first use of the " torpedo " was at Marysville (Ohio), and spouting immediately followed the explosion.

The Phenomenal Results at Thorn Creek.—One of the most beautiful results obtained by this process was observed on the 27th October, 1884, at Thorn Creek, in the well belonging to Messrs. Semple, Bayd & Armstrong.

"It was perceived that the 'torpedo' had exploded," relates M. W. de Fonvielle, "because a column of water was seen to issue from the well and rise to a height of more than 30 metres (say 100 feet). After this effort came a moment of repose, then a current of blackish water was seen to ascend, carrying in its train the residual products of the combustion of the dynamite, sand and mud. Little by little the current changed its colour and became yellow, then all at once came a rush of gas with a frightful crash. For a few moments the derrick was hidden from the view of the workmen who, after having dropped the piece of iron,[1] had taken refuge in a neighbouring house situated some distance off. The cloud dispersed itself, and there was seen to surge in its place a column of about 50 metres in height, tinged with a golden yellow by the sand in its train. It was the oil which was spouting. Moses' rod could hardly have produced more marvellous results. Very soon the whole of the neighbourhood was inundated by a sheet of petroleum which at last succeeded in finding an exit by precipitating itself into Thorn Creek. The little river swelled to such an extent that the frightened riverside inhabitants saved themselves by flying to the neighbouring hills, carrying with them their most precious valuables. In vain was it attempted to keep back the river by dams, the oil carried away all the obstacles which could be placed in its path. It was estimated that in this firstday 1,400,000 litres (say 300,000 gallons) were lost. But the spring, soon losing some of its impetuosity, was not long in being capped."

"*Torpedoing*" *an American Oil Well before Delegation of the French Society of Civil Engineers.*—During the excursion which the delegation of the Society of French Civil Engineers made to the United States, they had the opportunity of being present at the "torpedoing" of a well.

The President's Description of the Process.—The following are the impressions which this spectacle left on the mind of M. Rey, the president of the delegation : "When a boring yields petroleum, but in insufficient quantity, this may be due to the fact that the

---

[1] The "Devil" used to cause the explosion by concussion. *See* **next page.**

impregnated rock is too compact and does not present a sufficient number of fissures for an abundant flow of liquid petroleum or gas.

"The 'torpedoing' of the well is then proceeded with. This operation, the object of which is to dislodge the rocks and increase the fissures, does not always succeed, but is at times tried.

"The Edward Macdonald well No. 12, 2,300 feet deep, being in this condition, its owner decided to 'torpedo' it, and at the request of our American colleagues, he consented to postpone the operation until the day of our visit.

"We were therefore present at this really original and interesting spectacle, which I shall describe to you in a few words.

"Forty lb. of nitro-glycerine are poured into a white-iron cartridge of a diameter slightly inferior to that of the well, and this cartridge is cautiously lowered to the bottom of the well.

"A hollow cast-iron cylinder is then released, weighted in its lower part by a wad which is called the 'devil'; when it strikes the nitro-glycerine the concussion produced inflames this latter and detonation occurs.

"The owner of the well asked the president of your delegation to let go the 'devil,' which I did, hoping that the result would be favourable.

"After having withdrawn to a distance from the well and sheltered from the liquid and solid projections which might issue forth from one moment to another, we had the satisfaction of seeing a magnificent jet of petroleum spout upwards at the first bound to a height of 80 feet (25 metres), and at the second 120 feet (36½ metres). A slight wind spread out the jet in the form of a plume, and thus enhanced the beauty of the spectacle " (*Mémoires de la Société des Ingénieurs civils*, October, 1893).

II. LEVIGATION.

Levigation is another kind of operation which ought—when there is reason to believe that the sides of the well are choked up, and that this choking up hinders the exudation of the petroleum—to lead to the same end as "torpedoing". Levigation is carried out by means of a steam jet projected into the previously dried well.

In installations like Fauvelle's, which includes a hollow drill, the interior conduit of the rods is used for projecting the steam. In ordinary borings an india-rubber hose may be successfully employed.

CHAPTER XXXIV.

WHEN the owner of a well has struck oil, his *rôle* as a technological expert diverges into another *rôle* almost as difficult to maintain.

The deposit brought to light ought to be wrought and developed in a rational manner—the whole yield utilised to the best advantage. Waste would be a grave error. The earth is miserly of its treasures, and providential good luck does not renew itself on short maturity. For one productive well, time and money has often to be spent in unfruitful researches.

Storing Oil from Spouting or Flowing Wells.—If the well in question be a *spouting* well, or a *flowing* well, its working is an easy matter.

The tubulure of the *kalpak* or cap is prolonged by tubes which connect it with the reservoir. The petroleum, by its natural upward force, in the case of a spouting well, or, in the case of a flowing well, by the pressure of the gas, can be conducted into the reservoirs.

Oil from Non-flowing Wells: Pumping-Wells, Draw-Wells.— In non-flowing wells, the extraction of the oil is effected in two different ways—by drawing or pumping.

Draw-Wells. — Drawing is done by means of a large-sized sand-pump with ball valve. This rather inexpensive method is admirably adapted for wells with a moderate production.

Pumping-Wells: Capacity of Pumps.—Pumping necessitates a more complicated, and, consequently, a more costly equipment. One of the principal points to be observed, and one which the managing director ought never to lose sight of, is that the pump which he buys ought to correspond in output with the maximum yield of the well. No consideration of economy ought to stop him on this question.

Necessity for Thorough and Exhaustive Pumping.—A well whose oil is not exhausted daily will soon be seen to diminish its proportions in real earnest.

On the other hand, extreme pumping has often had the best results on the yield of a petroleum bearing vein, which, at the outset, has hardly been regarded as productive.

Beneficial Effects of Extreme Pumping.—Upon this point we may quote this curious fact, which the authors had the oppor-

Fig. 149.—Oil Pumps at Balakhany.

tunity of observing: At Ze Gleboka (Galicia) a well sunk to the depth of 244 metres (800 feet) gave, in the beginning, a daily yield of 42 barrels.

During eight months the well was pumped with the greatest regularity. The yield kept up, during the eighth month, a daily output of 28 barrels. At this time, a grave accident happened to the pumping system. The repairs, of rather a delicate nature, required the presence at Za Gleboka of specialist workmen, foreign to the establishment.

The repairs lasted ten days. According to the normal yield

of the last days, before the breakdown of the pumping system, the total oil accumulated in the well should have been about 280 barrels. Instead of the looked-for harvest, 45 barrels were pumped.

This sudden diminution of the yield of the well could only be attributed to the temporary stoppage of the pumping. The fact was, moreover, explained in that way in the note, inscribed in regard thereto, in the boring journal.

In fact, the sequel confirmed this hypothesis. The day following the resumption of work 17 barrels only were pumped; this figure was 11 barrels below the average yield. Extreme pumping was resorted to, emptying the well daily of its last drop of water.

This manœuvre fully succeeded. The sixth day the well had regained its normal yield. Nevertheless, through curiosity, the experiment was continued, and at the end of 15 days the following surprising result was obtained: a daily yield of 45 barrels, *superior to the initial yield of the well.* Thanks to extreme pumping, this important yield was maintained for several weeks.

The following are the types of pumps in current use:—

HAND PUMPS.

Width of Cylinder.		Output per Minute in—	
Centimetres.	Inches.	Litres.	Gallons.
1	$\frac{1}{16}$	50	11·0
$1\frac{1}{4}$	$\frac{1}{2}$	90	19·8

STEAM PUMPS.

Width of Cylinder.		Output per Minute in—	
Centimetres.	Inches (approximately).	Litres.	Gallons.
7	$2\frac{3}{4}$	120	26·4
8	$3\frac{1}{8}$	180	39·6
11	$4\frac{1}{3}$	280	61·6
12	$4\frac{3}{4}$	350	77·0
13	$5\frac{1}{8}$	430	94·6
14	$5\frac{1}{2}$	500	110·0
15	6	750	165·0

Wooden, Metallic and Natural Reservoirs.—As it issues from the bore hole the crude oil is led into reservoirs. These reservoirs, according to the magnitude of the yield, and also the structure of the ground in the neighbourhood of the works, may be made of metal or wood, or they may be natural, that is to say, consisting simply of excavations made in the ground.

Natural Reservoirs.—These latter are, more especially, adopted in the Caspian oil fields. There, the abundance of petroleum, and its consequent depreciation, allow, perhaps, of its being treated with less care than elsewhere. The nature even of the mineral oil of Baku, very often mixed with sand, renders its sojourn in a reservoir, which there is no risk of choking up, advantageous. This reservoir consists of a natural depression of the ground, or a hole dug around the borders of the well. A reservoir of this kind always absorbs a certain amount of oil ; when it has become sufficiently saturated it is almost impermeable. At Baku the oil is left to sojourn provisionally in these ponds; it clarifies and disembarrasses itself of the water and sand which it holds in suspension ; it can then, without fear of blocking them up, be run through the metal pipes into the iron reservoirs.

Wooden Reservoirs.—Wooden reservoirs are not utilisable, except for rather minimum quantities of liquid, 400 barrels at the most. They ought to be made of pine wood boards and encircled with iron bands. In spite of the economy which might appear to result from this method of storage it presents many inconveniences, the principal of which is leakage. The reservoir is besides subject to the destructive action alike of the solar rays and rain. On that account it would be very imprudent to leave it empty for any length of time ; in that case the planks exposed to the variations of temperature would soon become detached from one another.

Wrought-iron Reservoirs.—A riveted, wrought-iron reservoir is from many points of view that which ought to be preferred. It is perfectly staunch and tight, its solidity is proof against the weather, and its metallic structure guarantees it against all risk of fire. Unfortunately the price of these reservoirs is high ; only a great yield justifies their purchase.

Construction.—The erection of a wrought-iron reservoir necessi-

tates both earthwork and masonry, which according to the nature
of the ground in the locality of the site, may increase its cost to
a notable extent.

Site.—It will be advantageous to locate the reservoir, not in
the immediate neighbourhood of the works, but as near as possible
alongside the means of communication and transport roads, canals
or railways.   If these points are situated at lower levels than that
of the well, this difference in level will enable the petroleum to
flow naturally as far as the reservoir.   When the neighbouring
point of the means of transport is at a much higher level than that
of the well, the storage of petroleum becomes more costly.   It is
for the director of the works to decide whether the purchase and
maintenance of a pumping engine to pump the oil into the pipes
under pressure would be less costly than tentative measures to
facilitate other methods of conveyance.

Conveyance.—These methods of conveyance differ in form.
Petroleum is conveyed either in bulk, or in barrels.

Bulk Transport.—The first method is always to be preferred by
the owner ; it presents the same advantages as storage in metallic
reservoirs ; it reduces leakage almost to *nil*.   On the other hand,
it ought also to be preferred on account of the decrease in dead
weight.   Nevertheless transport in bulk has not yet been utilised,
except on waterways or railways ; if the owner of the installation
has not a production which warrants him in laying down a pipe
line as far as the nearest line of railway, transport in barrels
can alone be adopted.

Barrels.—The barrels used in conveying petroleum are of three
kinds :—

The *American barrel*, containing 42 American gallons (say 160
litres) (say 35·2 English gallons).

The *English barrel*, the contents of which are 163½ litres.

The Galician barrel, containing about 200 litres.

Tank Waggons.—For the transportation of petroleum in bulk by
railway, the companies, as well as the great petroleum firms, have
built special vehicles which go under the name of tank cars, or
tank waggons (*wagons citernes*).   These waggons are built on the
French principle, according to Lepage's system, as practised and
constructed by M. de Bonnefond, and contain about 10,000 litres

Pl. XVII.—Filling Petroleum Barrels at Baku.   Photograph communicated by M. André (Vol. 1., *to face p. 462*).

Pl. XVIII.— Loading Petroleum Tank Waggons at a Railway Station in United States Oil Fields. Photograph communicated by MM. Fenaille and Despeaux
(Vol. i., to face p. 468).

of liquid. The cylindrical reservoir is made of boiler-plate iron, and is firmly fixed in the axis of a covered waggon.

The petroleum may be transferred to it by a rotary pump if it comes from different sources and in different receptacles. Arrangements are made to avoid leakage, either by default in filling at the point of departure, or in consequence of evaporation in transit. With the view of preserving the liquid from the action of the exterior temperature, the case consists of a double flooring, a double roof or ceiling, and on the contour of a double panelling applied horizontally in the interior and vertically on the exterior. The case is mounted on a framework made wholly of iron. It is

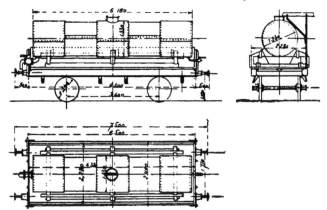

Fig. 150.—French tank waggon ; State Railway model.

on this model, but without the covered waggon, that the Russian tank waggons of Nobel, and the Austrian tank waggons of Mr. MacGarvey are built.

American Tank Cars.—Some firms and some companies prefer the American type of tank car. In these, the platform supports a cylindrical reservoir with a capacity of 2,500 gallons (9,500 litres). A small cupola rises up in the centre and thus enables the petroleum to dilate under the action of heat. These waggons do not carry feed pumps ; they are filled directly from the reservoirs built along the railway lines which cross the oil fields.

Tank-car Charging Stations.—The stations where the waggons are charged are arranged in a peculiar way, along the length of permanent way on which the tank-car trains are marshalled and over the whole length of the train is a plank platform, supported by metal uprights. Along this kind of quay, and on a level with the orifice of the tank waggons, runs the pipe which hangs from the reservoirs. This pipe is furnished with elbow-pieces, every 9 metres (say 30 feet), which are themselves prolonged by a caoutchouc tube, and closed by a tap. These are the pipes which are used to charge the waggons. Thanks to this arrangement, the loading of a train is effected very rapidly and almost without manipulation.

This work cannot, let it be well understood, be done by night on account of the risk of fire incidental to lights.

Tank Steamers.—Mineral oil may likewise be transported in bulk by waterways. Nobel Brothers were the first to initiate this method of conveyance.

Nobel Brothers' Innovations.—The following are the facts with which the firm of Nobel Brothers have been good enough to supply the authors as to the innovations made by the founders of that firm.

The End in View.—The end which they had in view was to remedy as much as possible the defects of the then existing method of working, and to reduce the cost of conveyance in such a manner as to give Russian petroleum the supremacy, at least on all the markets of the Russian empire.

The Method Adopted.—The essential point was to connect, by means of pipes, the naphtha wells near Balakhany, at a distance of about 15 kilometres from Baku, with the shores of the Caspian Sea, where they had installed their refineries. The most improved methods were adopted in the installations. The refined petroleum was charged into barrels no longer, but into tank steamers, which were filled by means of pumps from a tubular conduit stretching from the factory to the point of embarkation.

Tank Barges.—Moreover, large tank steamers were used to convey the oil across the Caspian Sea. These, owing to the considerable amount of water which they drew, were not adapted for going up the rivers. It was, therefore, necessary to have steamers and lighters to receive the oil at Astrakhan, and to

Pl. XIX.—Crude Oil Receiving Tanks.—Nobels' Works at Baku (Vol. i., *to face p. 464*).

convey it up the Volga to Tzarytzin, where the ramifications of the Russian railways commenced. Immense depôts were established at Tzarytzin, whence the petroleum was distributed all over Russia. Barrels were also suppressed in the conveyance by railway, and the oil was despatched to its destination in tank waggons. This organisation was completed by the creation of depôts and reservoirs at the principal railway stations.

Conveyance in Bulk from Baku to St. Petersburg.—Thanks to this method, the mineral oil was conveyed directly from the place of extraction to St. Petersburg, to Warsaw, and even to the frontiers of Austria-Hungary and Germany without having been put in barrels, and with no other method of charging than the steam pump.

Per mare et per terram.—The most important and the most remarkable innovation was the conveyance of oil in tank steamers, without which there would have been no need for tank waggons. This was the method of conveyance by land and sea, *per mare et per terram*, without the use of barrels inaugurated by Ludwig Nobel, who thus imparted a special feature to the Russian petroleum industry.

Effect : Widespread Use of Russian Petroleum.—By the adoption of these improved methods of transport, and of charging and discharging, the Russian mineral oil of the Caucasus has penetrated not only over the whole of the Russian empire, but also over the whole of the States of Western Europe, Africa and Asia, where it competes with American petroleum.

Petroleum Reduced in Price.—Moreover, illumination, which was, in fact, very costly in Russia before the radical reform of the industry, is at the present day extraordinarily cheap.

Illustration.—Thus a pood, 16·4 kilogrammes (say 36 lb.) of naphtha cost *at Baku* in 1873 about 45 copecks ; at the present rate of exchange of 2·7 francs for a rouble = 1·21 francs (say 11½d.), *and at Nijni-Novgorod* 6 roubles [at 2·7 francs = 16·2 francs (say 13s.)]. At the present time, it is sold, at Baku, at from 1 to 2 copecks [0·27 to 0·54 francs (2½d. to 5d.)], and at Nijni-Novgorod at 35 copecks [0·94 francs (say 9d.) the pood, or 5¾ francs the 100 kilogrammes (say 2s. 4d. per cwt.)], excluding the octroi duties, which amount to 60 copecks per pood on refined illuminating oil.

Thus the realisation of the programme elaborated by Messrs. Nobel created an enterprise of unexpected dimensions.

Initial Difficulties.—In the beginning, however, funds were wanting to carry this project to a successful issue. In order to confront this difficulty Ludwig Nobel applied, so far back as 1876, to the principal owners of the Baku oil wells asking them to associate themselves with him for the construction of pipe lines to connect the wells with the refineries.

Railway and Steamship Opposition.—He proposed to the steamship companies of the Caspian Sea and of the Volga *Caucasus and Mercury*, and the Railway Company of *Griazi Tzarytzin* (these companies were subventioned by the State) to construct, in conjunction with them, or on their own account, several *tank steamers* and *tank waggons* for the conveyance of naphtha. The two companies thought the project of the dauntless innovator chimerical and irrealisable ; they scorned Ludwig Nobel's proposals, and his plans were rejected.

Determination to carry out Proposals Personally.—In consequence of this fact the two brothers, instead of being discouraged, and confining themselves to their single line of business, decided to take into their own hands the working and the sale of petroleum products throughout the whole Russian Empire. Their enterprise embraced a vast extent.

Ultimate Universal Adoption.—Matters changed rapidly. When it was well demonstrated that pipe lines worked admirably and very economically, that tank steamers, tank waggons and reservoirs were perfectly adapted for their purposes, even the most infatuated of the old school comprehended the immense advantage of the innovation. In a short time, the environs of Balakhany-Sabountchi were covered with pipe lines, and nobody dreamt of transporting the crude products from the wells to the refineries in barrels any longer. Tank steamers followed each other in file on the Caspian Sea and on the Volga.

Railways add Tank Waggons to Rolling Stock.—The railway companies were thus compelled, in their own interests, to adopt the innovation conceived by Ludwig Nobel, and to add tank waggons to their rolling stock.

Present Strength of Tank Waggons, Rolling Stock and Fleet of Tank

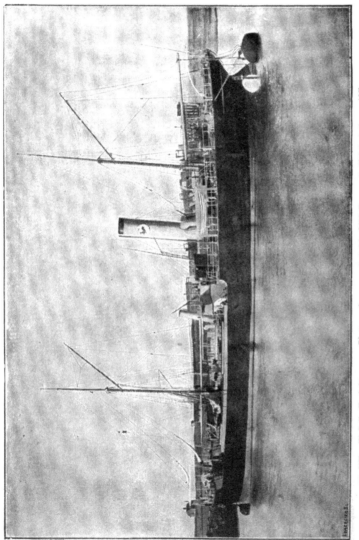

Pl. XX.—*El Gallo*, Tank Steamer. Photograph communicated by MM. Desmarais (Vol. i., *to face p. 466*).

Pl. XXI.—Group of Petroleum Reservoirs at Batoum. Photograph communicated by M. André (Vol. i., to face p. 467).

Steamers and Lighters.—There are now on the Caspian Sea and the Volga several hundreds of tank steamers and tank lighters.

On the Black Sea thirty tank steamers, specially constructed for the conveyance of petroleum, pass to and fro continually, and the different railway companies of Russia have more than fifteen thousand tank waggons for the conveyance of mineral oil.

Routes.—The oil is conveyed in two directions :—

1. By way of the Caspian Sea and the Volga, and

2. By the Trans-Caucasian Railway, *viá Batoum*, to the south-west of Russia and abroad.

Caspian Sea and Volga Route.—In the first of these directions the oil is charged into large tank steamers of a capacity of 750 to 900 tons. There are a dozen of them engaged in conveying the petroleum from the quays of the Baku factory to the *nine-feet* road situated 150 kilometres from Astrakhan, which is separated from the Caspian Sea by channels which the Volga has dug in bygone centuries. Arrived at the *nine-feet* road, the oil is transferred from the tank steamers, partly into river steamers (of which there are eight), and partly into tank lighters (of which there are eighty-nine). These vessels convey the oil up the Volga to Tzarytzin, Saratoff, and Nijni-Novgorod, the termini of the railway lines.

The Siberian Route.—The petroleum intended for Siberia is embarked on the river Kama, a tributary of the Volga, for Perm, whence it is despatched to the depôts of the company at Omsk, Tomsk, and other central markets.

German and Scandinavian Route.—The conveyance of mineral oils from St. Petersburg and Libau for Scandinavia and the north-west of Germany is effected by a tank steamer, built specially for the Baltic Sea, and which bears the name of *Ludwig Nobel.*

Trans-Caspian Route viá *Batoum.*—For exportation into the countries of Western Europe the company chiefly use the Trans-Caucasian line of railway, and, to facilitate the conveyance of its products from Baku to Batoum before the construction of the Suram tunnel, it had laid a pipe line of four inches in diameter connecting the stations of Michailoff and Quirilli, 65 kilometres apart.

Batoum Reservoirs.—For storing the oil at Batoum the company constructed reservoirs with a capacity of 1,300,000 poods (22,000 tons), in addition to other facilities for transport. From Batoum the products are conveyed to the West of Europe and to the South of Russia generally in tank steamers, some of which are the property of the foreign agents of the company, or are chartered by them. So as to ensure regularity in the service of its tank waggons, the company organised a special railway service, and despatches its own equipped workmen to make necessary repairs.

Docks and Workshops.—The company constructed two docks at Astrakhan, and workshops for repairs of steamers at Tzarytzin, Astrakhan, and Baku for the use of the line of tank steamers on the Caspian Sea and the Volga.

Tank Waggons versus *Barrels.*—The construction of tank waggons is of especial interest, from a technical point of view, seeing that in them the problem of transporting a liquid freight is solved with complete success.

Rapid Transport.—The advantage of this system is beyond doubt, especially for short distances. Charging and discharging, in this instance, is done in the most simple manner by steam pumps, and the voyage of the steamers is accomplished with rapidity.

Such a rate of speed was not dreamt of with the old barrel method, which involved many cares, and the discharging of which was accomplished with such a great loss of time.

Leakage Reduced to a Minimum.—Moreover, the considerable leakage, incidental to the use of barrels, is reduced to a minimum by the tank system.

Charging and Discharging Tank Steamers.—A tank steamer, with a capacity of 800 tons, is charged at Baku in $4\frac{1}{2}$ hours, and can be discharged in an equal period of time.

Fuel: Mazout.—These tank steamers, which at present ply on the Caspian Sea, the Volga and its tributaries, do not use any other fuel than naphtha residuum, or *mazout*, as it is commonly called in Russia. This liquid fuel, independently of its cheapness, has the great advantage of its calorific intensity, or heating power, being double that of coal. It may be added that it may be charged on board by pumps as rapidly as the cargo. Moreover, it only requires but a minimum amount of manual labour to supervise

Pl. XXII.—Batoum Petroleum Depôt. Photograph communicated by M. André (Vol. I., *to face p. 468*).

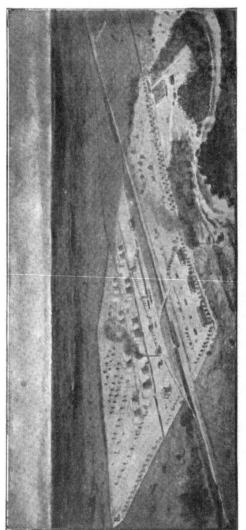

Pl. XXIII.—Russian Petroleum Depôt at Domnino.   Photograph communicated by M. André (Vol. i., *to face p.* 469).

the boilers, as the stoker only requires to manipulate a tap to regulate the consumption of the liquid.

Coming back once more to the enterprise of Nobel Brothers, the following facts may be added to those already mentioned :—

Tzarytzin Mechanical Coopérage.—In laying out the plan of the enterprise, it was decided that the town of Tzarytzin should be the centre of operations, and up to then, as far as the Russian markets were concerned, it had maintained that character.   To meet the requirements of the sales along the Volga, and the transport of benzene and lubricating oil to other markets, the firm erected a mechanical cooperage at Tzarytzin, even although the transport of these is principally accomplished in tank waggons, constructed by the company at its own expense.

Other Depôts on the Volga.—It was finally necessary to establish other depôts along the Volga—at Saratoff, Batraki, Nijni-Novgorod and Perm (on the Kama).   The object of all these depôts was to facilitate the sale of naphtha residuals and lubricating oil.   Before the opening of the Trans-Caucasian Railway, the greater portion of these was sent abroad by way of St. Petersburg.

From Tzarytzin, the tank waggons (the firm then possessed, approximately 1,633) conveyed the petroleum to the west, to the north-east and to the south of Russia, in a word, towards all the points where railways existed, and beyond the frontiers of Germany and Austria-Hungary.   So as to marshal and regulate the working of these tank waggons, the firm built at Domnino, near to Orel, a central depôt capable of containing 4,000,000 poods = 400,000 barrels of petroleum.

Storage Capacity of Depôts.—The distribution of these mineral products across Russia is effected, therefore, by depôts established in every centre, provided with reservoirs of suitable capacity, corresponding to the general demand of the district.   Thus, the depôts for the local sale or home consumption of St. Petersburg, Moscow and Warsaw, have a capacity of 50,000 to 85,000 barrels, whilst at other points, depôts of a few hundred barrels suffice.

Capital of Company.—The assets of the company, movable and immovable, amount at the present moment to 26,000,000 roubles.

Total Transport and Storage Capacity.—The transport services, comprising sea-going steamers, river steamers, tank barges and tank waggons, have a total capacity of 6,400,000 poods, about 106,000 tons, whilst the reservoirs and other depôts can contain 56,000,000 poods (930,000 tons) of crude petroleum and all the products which may be extracted from it.

Tank Steamers: Fireproof Compartments.—At the present day, both in Europe, and to an equal extent in America, 1,000 *tank steamers* are engaged in the trade.   To annihilate the risk of fire, these ships are built in a peculiar way ; they are divided into several compartments, each of which is completely independent of all the others, and separated therefrom by absolutely incombustible and perfectly tight partitions.   If a fire burst out in one of the compartments of the ship, there would be nothing to fear in regard to the remainder of the freight ; on the other hand, each of the compartments of the ship being hermetically sealed, and capable of being easily drenched by means of pipes, the fire, however violent, would soon be extinguished for want of air.

Typical Observation.—Amongst over 900 tank steamers, American, Russian, British and Belgian, we recognise four of them as flying the French flag.   The *Ville de Dieppe*, belonging to M. Robbe fils, and built at Southampton in 1888 ; the *Ville de Douai*, belonging to MM. Pax & Cie., built at Newcastle in 1890 ; the *Lion*, belonging to MM. Deutsch & Cie., built at Havre in 1893 ; the *Madeleine*, belonging to the French South-American Petroleum Company.

That is, indeed, an eloquent record !   We shall not detract from it by any commentary !

Pipe Lines.—The transport of petroleum in bulk may also be done by *pipe lines.*

Pipe lines are an extension of the system of pipes used to convey the petroleum from the wells to the reservoir.   It is the same pipes, of a dimension in accordance with the transit, which are used to convey the oil often to very considerable distances, either by means of a pumping engine or by its own weight.

American Origin : Circumstances which led to their Adoption.— This system was inaugurated in America.   In that country, in fact, the wells are often situated in districts absolutely isolated from all centres of population, and deprived of all means of communication, especially in Canada.   A consular report of the

Pl. XXIV.—Tank Steamer moored before the Offices of the Nobel Administration at Baku. Photograph communicated by M. André (Vol. I, *to face p. 470*).

period when the American oil fields began to extend gives us precious information on this point.

French Consul's Gloomy Picture of Results of want of Communication in Early Days of Petroleum.—The picture painted by M. Gauldree - Boileau in his report as French Consul is not a brilliant one.

He shows an industry which, in a manner, sprung out of the ground in a few months. The products, for want of roads, accumulated and blocked the way, were lost, or, at all events, " paid in cost of freight sums which exceeded their selling price ten times over. The only outlet possible for the whole of this indusdustrial zone was the Great-Western line, using Wyoming station, at a distance of 27 kilometres (16¾ miles) from the centre of these undertakings. All the barrels of petroleum were brought with greater or less difficulty to Oil Spring, whence a wooden tramway plank road facilitated their conveyance as far as Wyoming. From there the barrels of petroleum were loaded into special trains which came to a standstill at Port Sarnia, on the shore of the Gulf of St. Lawrence, where merchant steamers took them aboard to convey them to Liverpool. These different transit operations involved various expenses, which were distributed over a barrel of petroleum of 158 kilogrammes (3⅛ cwt.) as follows :—

COST OF DELIVERING A BARREL OF PETROLEUM (F.O.B.),
PORT SARNIA.

Items of Expense.	Francs.	s.	d.
Cost of the barrel	2·0	1	7·2
Cost of extracting oil	0 07	0	0·672
Conveyance from wells to Wyoming	0·85	0	8·360
Expenses at Wyoming	0·05	0	0·480
Freight from Wyoming to Sarnia	1·48	1	2·208
Various expenses at Sarnia	0·04	0	0·384
Total . . .	3·99	3	2·304

" We may further remark that these figures are taken under conditions where conveyance by plank road is made during the fine season, because, during autumn and spring, the cost of conveyance from the wells to Wyoming may vary in the proportion of one to five.

Total Cost of Delivery in Liverpool.—" By continuing to reason in the most favourable sense in regard to conveyance, the figure of 14·79 francs is arrived at as the cost per barrel of 158 kilogrammes, containing 182 litres of mineral oil, delivered from the wells from which it was extracted at Port Sarnia. From Port Sarnia to Liverpool 28·01 francs (say 22s. 6d), are reckoned as the cost of Custom House dues, warehouse and commission charges, etc. It results, therefore, that a barrel of 182 litres (40 English gallons) delivered at Liverpool costs 42·8 francs (34s. 3d.), or a barrel of 1,365 litres (say 300 gallons), weighing a ton, costs 318 francs (say £12 14s. 4·8d.). These prices allow of a certain profit being made by the Enniskillen proprietors of the oil wells; but when the cost of conveyance during the bad season reaches higher limits than the preceding, the owners would have to deliver their oil at prices which would hardly cover the cost of production, seeing that petroleum is sold in Liverpool at the rate of 330 francs per ton (£13 4s.)."

Lack of Practical Methods of Conveyance.—The lack of practical methods of conveyance was also acutely felt in the United States. Pittsburg was the first centre of the petroleum refining industry; it was, therefore, towards this town that from all parts of Pennsylvania there came waggons loaded with barrels. This method of conveyance was not carried on without great expense, nor without great loss by leakage. The roads were bad and scarcely kept up in this new country, and the cost of production of the petroleum, in the same way as in Canada, was at the mercy of the bad or goodwill of the carting contractors. The works, which were situated near a navigable stream of water, made use of barges (flat boats), which, in the rapid-running torrents of the oil region, often ran the greatest danger.

Moreover, at the time of the first flowing wells, barrels, either through real scarcity or on account of the jealousy of rival proprietors, were often not obtainable by the owners of wells of great yield. The Americans, under these circumstances, took a heroic departure; they converted their petroleum barges into tank barges. Long before Nobel they created the tank steamer. But when they encountered the great Russian industrialist, they had not improved this method of transport, which, in the case of steamships, which were not intended for this purpose, became not only a doubtful but a dangerous expedient.

Conflagrations in Transit. — Conflagrations were frequent in these masses of oil carried in the open air without any precaution. On the 12th of May, 1863, a flotilla of tank barges took fire. The banks of the Alleghany were ravaged by the fire which this infernal wave carried everywhere throughout its passage.

They then made rafts of barrels, which they let go with the current of water, but not without numerous losses and wrecks. The rivers were infected.

First Pipe Line.—-In 1863 an inhabitant of Oil Creek (Mr. Hutchinson) laid down the first pipe line which joined the oil wells with the Alleghany River. The aggregate length of this line of metal pipes, laid on the surface of the ground, was about 5 kilometres.

The Laughing-stock of the Oilmen: Final Success. — Unfortunately, this enterprise did not prosper. The pipes leaked to such an extent that the conduit became the laughing-stock of the oilmen of the district. Hutchinson did not get discouraged ; he searched for and found joints which would hold oil. The pipe lines were created ; their number increased in an extraordinary manner. Numerous companies were founded to develop this new method of conveyance.

Glut of Pipe Line Companies with Insufficient Capital.—However, as these hastily founded companies had, in the greater number of cases, insufficient capital and defective plant, and as competition obliged them to adopt excessively diminished tariffs, Hutchinson's innovation did not appear to be able to hold its own against the multiplication of railway lines, whose strong organisation presented a greater guarantee when, in 1876, the *United Pipe Lines Company* was founded. This company, furnished with extensive capital, absorbed the small companies, which it grouped in a network, covering the whole region.

The Standard Oil Company.—At the present time two powerful *pipe line* organisations divide the United States between them ; these are the *Standard Oil Company* and the *Tide Water Company*.

The Standard Oil Company, by far the most powerful, serves the whole of Pennsylvania.

Extent of Pipe Lines and Dimensions of Piping.—It possesses more than—2,000 kilometres of pipes of 150 millimetres in diameter, (say) 1,240 miles of pipes of 6 inches in diameter ; 700 kilometres of pipes of 100 to 125 millimetres in diameter, (say) 434 miles of pipes of 4 to 5 inches in diameter; and nearly 10,000 kilometres (say) 6,200 miles of pipes of a diameter under 100 millimetres (4 inches), descending as far as 50 millimetres (2 inches) diameter, the latter bringing the oil from the wells to the main conduits.

The principal lines of the *Standard Oil Company* are—

1. The New York Line, 714 kilometres (say 443 miles) in length.

2. The Philadelphia Line, 372 kilometres (say 231 miles).

3. The Ohio Line, 160 kilometres (say 100 miles).

4. The Baltimore Line, 106 kilometres (say 66 miles).

5. The Buffalo Line, 101 kilometres (say 63 miles).

Dimensions, Capacity, Manufacture and Testing of Pipes for Mains.—These principal conduits are 150 millimetres (say 6 inches) in diameter ; they are calculated in such a manner as to afford passage to 25,000 barrels of petroleum per day. The tubes of which they consist are of wrought iron, rounded, whilst hot, on an iron mandril, and welded, likewise in the hot state, by the pressure of a series of rollers ; they are tested to withstand an inside pressure of 600 kilogrammes (say 1,320 lb.) per square inch.

Laying Down a Pipe Line.—The laying down of these pipe lines requires as delicate planning as that of a railway. The slopes of the ground ought to be utilised, always bearing in mind the adherence of the mineral oil to the sides of the conduit. So long as the slopes have a sufficient inclination, the flow of the material being transported may go on naturally, and even do so on inclined planes in virtue of the law of equilibrium.

Interval Between Pumping Stations.—The pumping establishments may therefore be far apart if the line be well laid. This interval is, on an average, 35 kilometres (22 miles) on the *Standard Oil Company's* system.

Receiving Stations.—The receiving stations are organised in the same manner as the Russian petroleum depôts. They consist of wrought-iron reservoirs of formidable dimensions. These depôts are almost all situated on the coast in the neighbourhood of large

Pl. XXV.—Pumping Station of the "United Pipe Line" at Clarion. Photograph communicated by MM. Fenaille and Despeaux (Vol. I., *to face p. 474*).

Pl. XXVI.—Nobel Brothers' General Pumping Station at Baku.   Photograph communicated by M. André (Vol. i., *to face p.* 474).

ports.   The principal are at New York, Philadelphia and Baltimore. The importance of these petroleum depôts will be understood when we have shown that the companies often hold consider-able stocks there.

According to Riche and Halphen, the following are the quantities for Pennsylvania and Ohio.

Stocks of Pennsylvanian and Ohio Petroleum in Depôts, 1885-1896.

		Barrels.
·On the 1st January,	1885	37,366,126
,,	1886	34,428,841
,,	1887	34,156,605
,,	1888	28,006,211
,,	1889	18,995,814
,,	1890	11,562,593
,,	1891	9,443,744
,,	1892	15,354,233
,,	1893	17,386,389
,,	1894	12,116,183
,,	1895	6,336,777
,,	1896	5,161,904

The Standard Oil Company's Monopoly.—*The United Pipe Lines Company*, afterwards the *Standard Oil Company*, is the work of two worthy Yankees, a labourer, Samuel Andrew, and a commercial employé, Rockefeller.   How did they find the money to start with ? How did they come to think of engaging in a business which, in 1862, might have been regarded as risky?   That is a matter which we Frenchmen, in our country of France, cannot conceive. . . . In America, the country of free initiative backed by genius, it is possible.   However that may be, the fact remains that Rockefeller was the first of the *oil kings*, that his phenomenal fortune enabled him not only to get the conveyance of petroleum into his own hands, but also to swallow up the refineries in addition.

The Russian and American Fight for European Trade.—At the present day, the *Standard Oil Company* is no longer content with America, it is fighting for Europe.   It is open warfare, the pacific but desperate struggle between the Nobels and the Rockefellers, between American and Russian petroleum.

The yield is greater at the present day at Baku, but on the

other hand, the wiles of the *Standard Oil Company* are proverbial, even in America. It may therefore happen that the *Standard Oil Company* will monopolise the refineries of Central Europe, the best customers of Russia, in the same way as it did the American refineries. . . . But these are trade secrets, and this digression carries us beyond the limits which we have fixed for the contents of the present volume. The *Standard* always remains the *Standard*, although to-day, swelled by several other companies, it calls itself the *National Transit Company.*

Tide Water Company.—The other American pipe line company is the *Tide Water Pipe Company*, which may at the present time be regarded, speaking commercially, as affiliated to the *Standard Oil Company.* It consists of: 275 kilometres of pipes of 150 millimetres in diameter, (say) 178½ miles of pipes of 6 inches in diameter; 755 kilometres, (say) 468 miles of pipes of less dimensions.

The main line connects the district of Bradford to the State of New Jersey through Williamsport.

Pipe Line Companies' Method of Transacting Business. — The following is the manner in which the pipe line companies, with the view of abridging formalities, deal with the producers. As soon as those have, in their own reservoirs (which are always connected with the net-work of the company) a sufficient quantity of oil which they judge should be sold, they make a declaration to the company's agent, who simply determines the amount of oil which flows from this particular reservoir into the conduit of the pipe line, the orifice of which he has opened. Only the agent possesses the key which regulates the entrance of the oil into the system, and he withdraws after having closed the branch pipe and handed to the producer, in exchange for his oil, not coin but a voucher, an *acceptance* of so many barrels, a sort of bank cheque negotiable at the current price of oil fixed by the company without distinction of quality.

European Ground Unsuited for Pipe Lines. — In Europe, the situation of the most important oil fields lends itself but very little to the construction of pipe lines. At Baku, in spite of the fact that the wells are concentrated within, comparatively, a very small area, we have seen that Messrs. Nobel have laid down

there, as well as in the Caucasus, the first pipe line connecting the
wells with the distilleries of the *Black Town* and with the shipping
port.   Since that time, the pipe lines have multiplied in the
Apshéron territory.   At the present time they have developed
to the extent of about 150 kilometres.   The principal are the
conduits of :—

Mirzoez	.	. 11	kilometres (say 6·82 miles) in length.				
Lianosow	.	. 10	,,	,,	6·2	,, ,, ,,	
Caspi	.	. 12	,,	,,	7·4	,, ,, ,,	
Artichew	.	. 12	,,	,,	7·4	,, ,, ,,	

Pipe Lines in Galicia.—In Galicia the first pipe lines were

Fig. 151.—On the Pipe Line.

private ones.   They were those of MacGarvey, the American,
who connected the oil wells of Wiertzno with the railway station
at Krosno over a length of 14 kilometres, then afterwards those
of the English company of Wankowa, the branches of which
connect the wells of this company in the Ustrzyki oil fields with
the Uherce station.   This pipe line has been developed to the
total extent of 25 kilometres.

First Galician Pipe Line.—The first *pipe lines* established in

Galicia were dependent on particular installations. At the present time a movement analogous to that which took place in Pennsylvania tends to group, in a syndicate, the principal producers of this country, so as to create pipe lines with acceptances similar to those of America. Already, at Schodnica, important works in which the principal petroleum companies of the country are co-operating are being undertaken, and pushed vigorously forward for the extension of the pipe line which joins this deposit with the station of Boryslaw.

Java and Peru.—The great oil fields of Java and Peru likewise possess a fairly complete network of *pipe lines.*

Specification for Pipes for Pipe Lines.—The following is a model specification for the purchase of pipe lines of a certain magnitude for any particular installation. This model specification is the common market one; it will be of use to a certain extent to know its terms.

Art. 1.—The contracting party agrees to supply to the company . . . such and such a quantity of cast-iron pipes of . . . millimetres in diameter.

Art. 2.—The pipes are to be with ball and socket joints, or with two sockets or two collars, as may be necessary for connections with the reservoirs, as well as bent pipes and pipes with branch tubes. The straight ball and socket pipes are to be . . . in length ; they must weigh . . . kilogrammes, with allowance of one-twentieth at least. Each ball and socket pipe will count its actual weight, but if the weight exceeds . . . kilogrammes the excess will not be taken into account. The union pieces alone will always be taken into account at their actual weight.

Art. 3. — The cast iron is to be of the best quality, not brittle, homogeneous, capable of being turned on the lathe without flying off or breaking at the edges. All the pipes are to be turned at the ends ; the moulding must be done with care, so that there is no seam inside the socket nor outside the male end nor on the collars. Every seam will therefore be carefully reamed at the expense of the seller; the insides of the pipes are to be smooth and perfectly free from sand.

Art. 4.—Each pipe to bear the name of the maker.

Art. 5.—At a distance of 1 centimetre from where they commence, the ball joints of the pipes are to be hollowed along an annular surface of 6 millimetres in diameter.

ART. 6.—The party supplying will be liable to submit, at the factory, to whatever verifications the company may judge right to prescribe, to satisfy themselves as to the quality of the cast iron, and that all proper precautions are taken to guarantee good execution, not only in regard to the perfect dressing of the models, but also in regard to the perfect adjustment of the frame, and to see that due care is taken in setting and piercing.

ART. 7.—Deliveries are to be made at the station, and despatched at intervals as desired.

ART. 8.—The pipes will be examined by an agent appointed by the company's engineer. Every defective piece, or every piece not being of the desired shape, will be refused. The expense of collection of the rejected pieces will be defrayed by the manufacturer. The pipes will be tested separately, and subjected to a charge of water of 100 metres (328 feet) in height, by means of a hydraulic press. When exudation occurs with bubbling, however feeble it may be, or small jets of water, the pipe will be rejected. A report will be made of this operation.

ART. 9.—All the castings will be paid for at the rate of . . . per kilogramme.

PETROLEUM DEPÔTS.

Galician Petroleum Depôts.—The principal petroleum depôts in Galicia are : Kolomea, Drohobycz, Uherce, Krosno, and Gorlice. It is from these depôts, provided with pumping stations, that the greater part of Galician petroleum is despatched.

Roumanian Petroleum Depôts.—In Roumania the forwarding stations are situated along the railway from Gernowicz to Bucharest. These are :—

Bacau, Romniculu-Saratu, Buzau and Breza.

Russian Petroleum Depôts.—In Russia the railway depôt of the northern oil field of the Caucasian Isthmus is Wladikawkas. The southern oil field is served by the Trans-Caucasian stations of Koutais, Tiflis and Elisabethpol, as well as the ports of Noworossirsk, Poti and Batoum on the Black Sea.

As we have seen, Russian petroleum is conveyed to Western

Europe by two routes: the Trans-Caucasian Railway takes part
of it, which is conveyed over the Black Sea towards Batoum. In
this port the petroleum is loaded into ships to convey it to the
ports of Fiume, Gênes, Brindisi, Marseilles, Barcelona, Bordeaux,
Havre and Liverpool.

Tzarytzin.—Another portion of the petroleum of Baku goes by
the way of the Caspian Sea and the Volga as far as Tsarytzin, the
first railway station on the river. On the arrival of the tank
steamers at Tzarytzin, the oil is pumped into the reservoirs of the
immense depôts of Messrs. Nobel Brothers. The total storage
capacity of the reservoirs at this point exceeds a million barrels.
From Tzarytzin petroleum trains convey the petroleum to Orel.

Orel.—Orel is the point of concentration of Russian petroleum ;
it is from there that the greater part of the production of Baku is
despatched to the different parts of Russia. Reservoirs have been
constructed there capable of holding 1,500,000 barrels of petroleum, thereby rendering this depôt the most important one in all
Europe.

There have, moreover, been established in Russia two depôts
possessing reservoirs with a capacity of more than 500,000 barrels
of petroleum : these are Moscow and St. Petersburg. We may
also mention amongst Russian petroleum depôts possessing reservoirs ranging from 100,000 to 500,000 barrels : Warsaw, Saratow,
Astrakhan, Nijni-Novgorod, Kharkow, Kiew, and Minsk. The
following are Russian reservoirs, with a capacity of under 100,000
barrels : Kazan, Voroneje, Koslow, Riazan, Smolensk, Toula,
Ekaterrinoslaw, Krementschong, Tchernigow, Koronew Polotsk,
Dunaborg and Riga.

German Petroleum Depôts.—Germany gets its Russian petroleum
from the frontier points of Alexandrovno and Granitza ; the petroleum market, or exchange, of Germany is in Bremen. The
principal German depôts are at the latter town, Hamburg, Berlin,
Hanover and Cologne.

French Petroleum Depôts.—In France, where petroleum only
comes by sea, the principal depôts, in addition to the ports of
arrival, are : Paris (Pantin, Clichy, Colombes), Rouen, Nantes,
Nancy, Lille, Bordeaux, Châlon, Dijon, Roanne, Lyons, Besançon,
Charleville, Troyes, Saint Etienne, Valence, Limoges, Périgueux,

Angoulême, Poitiers, Tours, Agen and Toulouse. These depôts, in the greater number of instances, have been established by the firm of Deutsch of Paris.

Spanish Petroleum Depôts.—This same firm has installed in Spain the depôts of Santander, Seville, Alicante, Madrid, Bilbao, Saragossa, Gigon, La Corogne, Vigo, Cadix, Malaga, Almeiria, Garrucha, Carthagena and Valence.

American Petroleum Depôts.—There are depôts for American petroleum at Port Sarnia, Quebec and Gaspe, for Canada; and at Rixford, Bradford, Washington, Macdonald, Milton, Pittsburg, Buffalo, Franklin, Parker (Pennsylvania), Cleveland (Ohio), Whiting (Indiana), New York, Philadelphia, Boston and Baltimore, for the United States.

CHAPTER XXXV.

GENERAL ADVICE.—PROSPECTING, MANAGEMENT AND CARRYING ON OF PETROLEUM BORING OPERATIONS.

Prospecting : Necessity for Rational and Methodical Work.—The work of the mining prospector, or the searcher after petroleum, can only be accomplished satisfactorily on condition that the party in question conducts his researches, and carries on his work, in a methodical and rational manner.

Necessity for Technical Knowledge.—It is therefore a matter of prime importance that the owner of the right of searching for petroleum and developing the same should have at least a general idea of all the work involved in this search and development.

Capitalists Deceived by Speculators.—Capitalists have too often been seen to scatter their fortune to the winds in barren work and vain research. Deceived by the plausible talk and glib tongues of speculators, swindled by unprincipled contractors interested in the job, not only unprofitable, but even ruinous to them who gave the contract, these capitalists have purchased mediocre ground in which they had been made to see inexhaustible lakes of petroleum. Like real fools they have not even taken the trouble to study the nature of the ground ; they have trusted to experts, and biassed prospections or surveys. One well, two, . . . ten are undertaken at places, dictated by caprice, the anxiety of preserving the beauty of the landscape, or even by superstition. In such rich ground, they have thought any place will do, and people have become accustomed to regard the earth as a lottery, where at each pull they gain ; . . . they expect to gain the grand prize, *the spouting well*.

This way of working is more general than may be imagined ; in it is hidden the secret of so many barren undertakings (44 per cent.).

This number, let us repeat it, let us shout it to the uninitiated who consider the profession of_an *oilman* as a game of chance, a lottery where blind fortune turns the wheel.

Chance or good luck has no more to do with petroleum undertakings than it has with any other. The chances of success, like in everything else, are determined by constant study and rational work.

Preliminary Prospecting Work a Necessity.—Before taking a boring in hand, the price of which always amounts to some 30,000 francs (and that under the most favourable conditions), a searching examination of the ground to be wrought must be undertaken.

This examination is termed *prospection*.

Too much Importance not to be Attached to Surface Signs.—The prospector will only attach but a very relative importance to surface signs ; that is, to petroleum exudations, even on the surface. These exudations certainly indicate the *presence* of oil, but not the *position* of the springs from which it has been derived—often very far distant—nor the depth at which they exist. They do not indicate the ground to be traversed before reaching this depth, nor the disposition of the beds which cover the producing vein.

Let us at once say that embryo prospectors are always greatly impressed by surface signs. Unscrupulous speculators know this little failing, and take advantage of it, faking up, as it were, the ground to be sold on purpose for the occasion, artfully laying down fictitious exudations which they produce by means of a few litres of crude petroleum, the origin of which is a matter of little or no consequence as far as they are concerned.

One can imagine the effect of these exudations ; the future purchaser is enraptured, he takes a sample of this petroleum which announces its presence in person, so as to bind him as it were not to. let this chance of making his fortune escape him ; . . . a sample of the oil coming from these exudations is taken, it is analysed "Excellent Oil ". . . (they all are) . . . the prospector is in raptures. He then and there sinks his fortune, unless, having read our treatise, he has some doubts as to the origin—perhaps artificial—of the petroleum signs which might have convinced him.

We have said above that the prospector's examination should bear upon four principal points ; these are :—

1. THE PRESENCE OF THE OIL.

2. ITS APPROXIMATE DEPTH.

3. THE GEOLOGICAL ARRANGEMENT OF THE STRATA.

4. THE NATURE OF THE STRATA.

1. *The Presence of the Oil.*—The presence of oil is determined in two ways : either by (*a*) deductive reasoning if the ground be situated in territory already wrought and developed ; or (*b*) by surface signs of petroleum.

(*a*) *By Deductive Reasoning.*—The first of these methods consists in taking—as a point of departure and comparison—the nature of the beds traversed by the borings between which the ground to be prospected lies. By a careful examination, often over a rather large distance, of the beds exposed by surface denudation, land-slips or the disintegrating action of water, or, in default of these kinds of beds, by a succession of small surface borings made by means of a cane drill, capable of boring in a few minutes from 1 to 5 metres (say 3¼ to 16 feet), one is easily able to find in a precise enough manner the passage of the petroliferous line of strike.

(*b*) *Surface Signs.*—The surface signs consist either of exudations or of outcrops of strata essentially petroliferous (petroliferous eocene sandstone, shales of the same or of miocene formation, etc.). Exudations are, as already said, always to be taken with great caution. As a general rule, an exudation ought always to be confirmed by the presence, in its immediate neighbourhood, of a bed, or at least a small vein of an essentially petroliferous nature. It is for the prospector, by clearing away the surface, to proceed to lay bare the bed from which the petroleum directly flows, or the sands, impregnated with naphtha, which has served as a vehicle for it.

The conscientious prospector leaves no stone unturned, disregards no sign to determine definitely and precisely the place where the petroleum originates. This search is often very minute and circumstantial ; a stone, rolled by the waters of a stream, may afford a precious indication ; the stream then becomes the thread of *Arianne*, which guides the prospector towards the bed from which it has been wrenched.

2. *Approximate Depth of Petroleum-bearing Bed.*—The depth of the deposit can only be determined in an approximate manner,

and only by comparison with neighbouring borings. It is from the examination on the surface of the out-cropping bed that the position of this same bed above the petroleum-bearing bed in the substructure of neighbouring borings is deducted. For this purpose it will be useful to well examine this bed, to establish in a certain manner its identity (geological nature, chemical composition, dip, line of strike).

3. *The Geological Structure and Arrangement of the Strata.* —The geological structure and arrangement of the strata, both below and above the prospected bed, will indicate, by the directions of its layers, the bearings of the future workings, by their dip, the anticlines, the *saddlebacks*, on the slopes of which the wells will be grouped.

4. *The Nature of the Strata.*—The nature of the beds to be traversed can only be exactly determined after the first boring has been made. We can, however, always draw inferences from, and benefit by, the experience presented by neighbouring borings if there be any in existence.

Trial Borings.—The ground having been prospected, the site of the first well or wells having been chosen, the preliminary work incidental to boring executed, the study of this ground, which is only known in a superficial manner, must be continued. The first three wells in a real business undertaking will be regarded as *trial wells.* They will not be located in a straight line, but rather in such a fashion that the imaginary line which joins them forms a triangle. In this way, by quite a geometrical method, the direction and inclination of the beds of which the underground strata of the borings consist is deduced in the most definite and complete manner. All precautions having been taken to secure the most complete data, for the study of each well, the foreman borer is entrusted with supplying all such relating to his boring; he is responsible for their correctness. For the agreement of these data custom has established invariable formulæ which, at the present time, are in use in all countries.

The following is the way in which our honourable predecessor, M. Degoussée, delineated the duties of the foreman borer in regard to this matter, according to a practice which has, moreover, become universal in all borings.

Collections of Samples of Strata being Bored.—" Before commencing a boring the man in charge takes the precaution to provide himself with a set of pigeon holes in the form of a draught board, and places on each of the holes a consecutive number corresponding with the journal, the model of which is about to be given. Each time the drill is raised he examines the earth brought up by the bottom of the tool, and every time it varies he places one or two samples in the pigeon holes, in numerical order, so as to show at once, and without difficult examination, the succession of the strata traversed. When the beds are rather thin, he only preserves a sample of each ; when they are thick, he takes a sample every metre. Each sample ought to consist of a small prism of paste taken from the bit and a few fragments of the crushed rock. By taking the trouble to lift the mud brought up by the valve or by the drill, enough can in every instance be collected. In some cases the solid fragments of the rocks may suffice ; but often also the paste is required for the purpose of testing it with acid. The box of samples formed, in proportion as the work progresses, is used to control the journal, and at the end of the boring to establish a complete geological section.

Boring Journals or Diaries.—" We append below (table, p. 491) the model of the journal to be kept at a boring ; its minute posting up each day is indispensable, if it is desired to keep an account of what has been done. Well-kept boring journals are quite a necessity for a contractor, so as to be able to avoid—after what may be sometimes a very long lapse of time—having to relearn what may have been a very costly lesson ; when he does not know the nature of the ground to be bored, they yet serve to fix in his mind in a positive manner the time and the expense of the borings to be undertaken, and consequently enable him to make rational contracts.

Different Entries in Journal.—The man in charge ought to mention the different water levels ; the height of the boring above the river near to which it is situated, as well as that above the sea ought to be noted at the top of the journal.

The least change in the ground ought to be indicated ; the points where the tools stop more often than in others being observed, he knows what difficulties he will have to encounter in casing the well, and what points the reamer ought to scrape longest.

Registration of Accidents and Dimensions of Drilling Tools.—" The chapter of accidents ought especially to be very detailed ; it ought to describe the result of each tool-searcher or hook, so as to be able to appropriate the one which has been used again to the situation of the hole and to that of the fragments of iron sought for ; without precautions, without the exact measurement of the rods, a tool might be sought for, indefinitely, in the bore hole, without being sure of touching it, and the evil would be aggravated instead of being remedied. Each rod ought to be inscribed in its order of descent, and a table placed in the ' remarks ' column of the journal ought to indicate not only the length of each of them, but also the thickness of the iron."

The foreman borer is responsible for the regularity of the work in his well. It rests with him to avoid all causes of accidents, and to avoid them by constant precautions. When the drilling tackle is being brought to the surface, he ought to verify minutely the state of his tools, to classify them and make an inventory of them, and, if need be, to withdraw each piece. He ought always to take care that the rods are used in the same numerical order (according to the number on them). The length of each ought to be taken as well as the diameter of the iron, so that without losing time in measuring, he knows positively the depth of the boring ; if a rupture has happened, he at once ascertains the spot where the portion of the tool remains in the hole ; his journal tells him the nature of the ground in which the " fishing " tool will have to work, and without any hesitation he lowers the appropriate instrument. The rupture of a tool or a rod is nothing so long as the borer knows his equipment and bore hole thoroughly ; he repairs the accident at the first go off, but if he hesitates and gropes about he aggravates it.

Tools to be Repaired after use and put back into Stock in Good Condition.—The foreman borer, besides keeping his journal regularly posted up and attending to his box of samples, ought to look after his tackle being kept in good working order. When a series of tools are of no more use for the moment, because the diameter has been narrowed by casing the well, he ought to cause them to be subjected to all the repairs necessary for their being put back again in good condition, also to be separated the one from the other, the screws and nuts oiled, and the threads of the screws wrapped up in such a manner as to preserve their edges.

Inspection and Lubrication of Machinery.—He ought to inspect his chain cable, from time to time, to see that no link has deteriorated, take care to keep it oiled for fear of it rusting, inspect and lubricate the gearing and shafts of his cranes, etc., remedy any play which the different pieces may have made whilst at work, especially after an exceptional effort has been made.

Engine Driver Subordinate to Foreman.—The foreman borer, during steam power drilling, will have the engine driver under his orders, the latter will obey him in everything which concerns the work in his well.

Relations with other Foremen on the Establishment.—He ought to be on friendly terms with the foremen borers of other wells on the establishment, follow their boring journals and prove the information contained therein by reference to, and comparative notes from, his own journal. Friendly relations between the different borings of the same establishment are indispensable, the one profits by the experience of the other, and this experience may afford valuable information in regard, for example, to the thickness of a bed or the necessity for casing the well.

Information from Wells outside Establishment to be Regarded with Distrust.—However, information received from neighbouring— *nearly always rival*—wells ought always to be regarded with suspicion, because it must be remarked that this rivalry not only exists in the oil fields between neighbouring establishments, but also between the foremen borers, who do not hesitate to deceive their opponent when they think it is to their interest to do so. There, the misfortune of one is the joy of the other.

Moreover, the ambition of a foreman borer, and even of a company, causes them nearly always to conceal their accidents so as to exalt their success. Almost all the information is vague or decidedly falsified.

The boring journals alone can be consulted with profit; they are deposited, as well as the series of samples, with the management, who keep a general journal of the different borings.

Boring Journal.—Herewith is inserted a specimen of this boring journal. (Unfold).

Table of Result of Comparison of Samples.—A table of the agreement of samples is also useful in the study of the beds, see p. 492.

The examination of this table—given as an example—shows that, after the first few metres drilled in the three different wells the nature of the ground traversed in Nos. 1 and 2 appears identical, only their situation differs in this fact that No. 2 would appear to be situated nearer the anticline of the beds, and therefore nearer the petroleum ; in fact, according to the table, it strikes the bed No. 3 common to the two wells at the depth of only 26 metres, whilst well No. 1 strikes it at 37 metres. According to this table, well No. 3 would appear to be drilled in ground of a different nature from that of the others.

In this table, entirely a geological comparison, no remarks relative to dates or working should be made.

Additional Documents.—These, the chief, documents may be completed by graphical representations of the progress of the work and of their cost, as well as by geological sections and groups of sections drawn from the data afforded by the samples and the positions in the boring journal.

Routine of Day and Night Shifts.—A final word to finish this chapter as to the custom of boring workmen.

If the day be a twenty-four hours' one, the Sunday rest is twenty-four hours. Work ceases at midnight, to be resumed at midnight. Of the two gangs engaged (which we shall call 1 and 2), No. 1, which leaves off work at midnight, and which has consequently had during the week which has just come to an end, the afternoon shift, resumes boring at midnight, so that during the following week it may be at work in the morning from midnight to noon. In this manner gang No. 2 will have in its turn the afternoon shift, and the two gangs will have every fifteen days turn about, the one twenty-four hours and the other forty-eight hours' rest.

Care must be taken not to clash with the superstitious ideas of the boring labourers. A joke, in regard to these, often demoralises them, and takes away all their courage on the occasion of an accident.

This superstition is so strong that a whole staff of workmen have been known to abandon a well which they believed to be bewitched.

A Few of the Customs Connected with this Subject.—In all petroleum districts the bidding "good luck" is a necessary wish to express, before any other formula of politeness when the buildings of a boring are entered; woe to him who forgets it, whatever be his rank! He will have to pay the penalty and comply with the rites intended to conjure the evil one.

In certain countries it is forbidden to whistle at the mouth of a bore hole; that would provoke the greatest misfortune, and all the exorcisms, all the magic formulæ of the foreman borer, will be powerless to prevent a probable blocking up of the well which is forthwith predicted.

Finally, we have seen Canadian foreman borers, before taking their seat for the first time, propitiate the genius of the well, or the spirit of the earth, by throwing him a coin. Often small treasures are buried in this way at the bottom of the bore hole; there are in all cases but few instances in which it has not received an offering of this kind.

The earth is a sacred divinity; it must not be traduced, least of all by the borer.

"The borer is the wooer of the earth, her secrets and riches are his."

GENERAL BORING JOURNAL OF THE DIFFERENT WELLS OF THE ESTABLISHMENT.

Date.	Well No. 1 — Foreman Borers . X and Y.						Well No. 2 — Foreman Borers . Z and K.						Well No. 3 — Foremen Borers . N and W.					
	Beds Traversed.	Sample from No. 1.	Sample from No. 2.	Observation.	Progress.	Depth of the Well.	Beds Traversed.	Sample from No. 1.	Sample from No. 2.	Observation.	Progress.	Depth of the Well.	Beds Traversed.	Sample from No. 1.	Sample from No. 2.	Observation.	Progress.	Depth of the Well.
1st May	Green shale (29)	—		Landslips	8 m.	301	Limestone (11) Limestone and sand (12) Brown shales (13)	17 18 21	— — —	Gas	16	217	Grey clays	3	7	Landslips	0	96
2nd May	Green shale (29)	15	—	Landslips	4 m.	305	Brown shale (14) Green shale (15)	28 29	— —	The bed of green shale (15) would appear to be the same as the bed 29 of well No. 1 met with at a depth of 298 metres in the latter	—	228	Grey clays	3	5	Landslips Casing Boring	1	97

TABLE OF RESULTS OF COMPARISON OF SAMPLES.

Depth. Metres.	Well No. 1.		Well No. 2.		Well No. 3.	
	Consecutive No.	Nature of Sample.	Consecutive No.	Nature of Sample.	Consecutive No.	Nature of Sample.
1	1	Grey clays and sand.	1	Grey clays and sand.	1	Grey clay and sand.
2		,,		,,		,,
3		,,		,,		,,
4		,,		,,		,,
5		,,		,,	2	Shaly limestone.
6		,,		,,		,,
7		,,		,,		,,
8		,,		,,		,,
9		,,		,,		,,
10		,,	2	Green shale.		,,
11		,,		,,		,,
12		,,		,,		,,
13		,,		,,		,,
14		,,		,,		,,
15		,,		,,		,,
16	2	Green shale.		,,		,,
17		,,		,,		,,
18		,,		,,		,,
19		,,		,,		,,
20		,,		,,		,,
21		,,		,,	3	Brown limestone.
22		,,		,,		,,
23		,,		,,		,,
24		,,		,,		,,
25		,,		,,		,,
26		,,	3	Brown limestone.		,,
27		,,		,,		,,
28		,,		,,		,,
29		,,		,,		,,
30		,,	4	Petroleum impregnated sand.		,,
31		,,		,,		,,
32		,,		,,		,,
33		,,		,,		,,
34		,,		,,		,,
35		,,		,,	4	Shaly limestone.
36		,,		,,		,,
37	3	Brown limestone.	5	Green shale.		,,
38	4	Petroleum impregnated sand.		,,		,,
39		,,		,,		,,
40		,,		,,		,,
41		,,		,,	5	Green shale.
42	5	Green shale.		,,		,,
43		,,		,,		,,
etc.		,,		,,		,,

PART V.

GENERAL DATA.—CUSTOMARY FORMULÆ.

MEMORANDA.

PRACTICAL PART.

GENERAL DATA BEARING ON PETROLEUM.

TABLE OF THE SPECIFIC GRAVITY OF LIQUIDS HEAVIER THAN WATER.[1]

Degrees Baumé.	Corresponding Density.	Degrees Baumé.	Corresponding Density.	Degrees Baumé.	Corresponding Density.	Degrees Baumé.	Corresponding Density.
0	1·0000	19	1·1516	38	1·3574	57	1·6529
1	1·0069	20	1·1608	39	1·3703	58	1·6720
2	1·0140	21	1·1702	40	1·3884	59	1·6916
3	1·0212	22	1·1798	41	1·3968	60	1·7116
4	1·0285	23	1·1896	42	1·4105	61	1·7322
5	1·0358	24	1·1994	43	1·4244	62	1·7532
6	1·0434	25	1·2095	44	1·4386	63	1·7748
7	1·0509	26	1·2197	45	1·4581	64	1·7969
8	1·0587	27	1·2301	46	1·4678	65	1·8195
9	1·0665	28	1·2407	47	1·4828	66	1·8428
10	1·0744	29	1·2515	48	1·4984	67	1·859
11	1·0825	30	1·2624	49	1·5141	68	1·864
12	1·0907	31	1·2736	50	1·5301	69	1·885
13	1·0990	32	1·2849	51	1·5466	70	1·909
14	1·1074	33	1·2965	52	1·5633	71	1·935
15	1·1160	34	1·3082	53	1·5804	72	1·961
16	1·1247	35	1·3204	54	1·5978		
17	1·1335	36	1·3324	55	1·6158		
18	1·1425	37	1·3447	56	1·6342		

[1] These densities are given according to Baumé's " Rational Scale," temperature 12·5° C. V. 1·073596 $d = \dfrac{145 - 88}{145 - 88 - N}$.

TABLE OF SPECIFIC GRAVITIES OF LIQUIDS LIGHTER THAN
WATER (PETROLEUM).

Degrees Baumé.	Correspond- ing Densities.	Degrees Baumé.	Correspond- ing Densities.	Degrees Baumé.	Correspond- ing Densities.	Degrees Baumé.	Correspond- ing Densities.
10	1·0000	23	0·9183	36	0·8488	49	0·7892
11	0·9982	24	0·9125	37	0·8439	50	0·7849
12	0·9865	25	0·9068	38	0·8391	51	0·7807
13	0·9799	26	0·9012	39	0·8343	52	0·7766
14	0·9733	27	0·8957	40	0·8295	53	0·7725
15	0·9669	28	0·8902	41	0·8249	54	0·7684
16	0·9605	29	0·8848	42	0·8202	55	0·7643
17	0·9542	30	0·8795	43	0·8156	56	0·7604
18	0·9480	31	0·8742	44	0·8111	57	0·7566
19	0·9420	32	0·8690	45	0·8066	58	0·7526
20	0·9359	33	0·8639	46	0·8022	59	0·7487
21	0·9300	34	0·8588	47	0·7978	60	0·7449
22	0·9241	35	0·8538	48	0·7935	61	0·7411

TABLE OF CORRECTIONS FOR BRINGING THE DENSITIES OF RUSSIAN
PETROLEUM—TAKEN AT VARIOUS TEMPERATURES—TO 15° C.
(*According to Riche and Halphen.*)

Temperature in De- grees Centigrade.	DENSITY OBSERVED.													
	750	760	770	780	790	800	810	820	830	840	850	860	870	880	890
	TO DEDUCT.													
0·	11·7	11·5	11·3	11·1	10·9	10·7	10·5	10·4	10·2	10·1	10·0	9·9	9·8	9·7	9·6
1·25	10·7	10·5	10·3	10·1	10·0	9·8	9·6	9·5	9·4	9·3	9·1	9·0	8·9	8·9	8·8
2·50	9·7	9·5	9·4	9·2	9·0	8·9	8·8	8·6	8·5	8·4	8·3	8·2	8·1	8·1	8·0
3·75	8·7	8·6	8·4	8·3	8·1	8·0	7·9	7·8	7·7	7·6	7·5	7·4	7·3	7·2	7·2
5·00	7·7	7·6	7·5	7·3	7·2	7·1	7·0	6·9	6·8	6·7	6·6	6·6	6·5	6·4	6·4
6·25	6·8	6·6	6·5	6·4	6·3	6·2	6·1	6·0	5·9	5·9	5·8	5·7	5·7	5·6	5·6
7·50	5·8	5·7	5·6	5·5	5·4	5·3	5·2	5·2	5·1	5·0	5·0	4·9	4·9	4·8	4·8
8·75	4·8	4·7	4·6	4·6	4·5	4·4	4·3	4·3	4·2	4·2	4·1	4·1	4·0	4·0	4·0
10·00	3·8	3·8	3·7	3·6	3·6	3·5	3·5	3·4	3·4	3·3	3·3	3·3	3·2	3·2	3·2
11·25	2·9	2·8	2·8	2·7	2·7	2·6	2·6	2·6	2·5	2·5	2·5	2·4	2·4	2·4	2·4
12·50	1·9	1·9	1·9	1·8	1·8	1·8	1·7	1·7	1·7	1·7	1·7	1·6	1·6	1·6	1·6
13·75	1·0	0·9	0·9	0·9	0·9	0·9	0·9	0·9	0·8	0·8	0·8	0·8	0·8	0·8	0·8
15·00	0·0	0·0	0·0	0·0	0·0	0·0	0·0	0·0	0·0	0·0	0·0	0·0	0·0	0·0	0·0
	TO ADD.													
16·25	0·9	0·9	0·9	0·9	0·9	0·9	0·9	0·8	0·8	0·8	0·8	0·8	0·8	0·8	0·8
17·50	1·9	1·9	1·8	1·8	1·8	1·7	1·7	1·7	1·7	1·7	1·6	1·6	1·6	1·6	1·6
18·75	2·8	2·8	2·7	2·7	2·7	2·6	2·6	2·5	2·5	2·5	2·5	2·4	2·4	2·4	2·4
20·00	3·8	3·7	3·7	3·6	3·5	3·5	3·4	3·4	3·4	3·3	3·3	3·2	3·2	3·2	3·2
21·25	4·7	4·6	4·6	4·5	4·4	4·4	4·3	4·2	4·2	4·1	4·1	4·1	4·0	4·0	4·0
22·50	5·7	5·6	5·5	5·4	5·3	5·2	5·2	5·1	5·0	5·0	4·9	4·8	4·8	4·8	4·8
23·75	6·6	6·5	6·4	6·3	6·2	6·1	6·0	5·9	5·8	5·8	5·7	5·6	5·6	5·6	5·5
25·00	7·5	7·4	7·3	7·2	7·1	7·0	6·9	6·8	6·7	6·6	6·5	6·4	6·4	6·4	6·3
26·25	8·5	8·3	8·2	8·1	7·9	7·8	7·7	7·6	7·5	7·4	7·4	7·3	7·2	7·2	7·1
27·50	9·4	9·2	9·1	8·9	8·8	8·7	8·6	8·4	8·3	8·2	8·2	8·0	8·0	8·0	7·9

TABLE SHOWING THE GENERAL PROPERTIES OF CRUDE MINERAL OIL FROM DIFFERENT COUNTRIES.

Oil Field.	Annual Production per Oil Field in Barrels.	Where Extracted.	Number of Borings in Progress.	Density of Petroleum.	Colour.	Remarks.	French Establishments.
Galicia	2,500,000	Kleczany	22	0·779	Reddish yellow	Very rich in paraffin	
		Ropa	2	0·800	Brown red	Rich in paraffin	
		Ropa (Fedorowitch)	1	0·780	Yellow		
		Siari	24	0·835	Dark brown		
		Sekowa	8	0·887	Greenish black		
		Kryg	6	0·898	Brown	Viscous	
		Ropica Ruska	3	0·808	Red brown		
		Mecina	—	0·869	Greenish black		
		Wojtkowa	4	0·820	Dark green		
		Libusza	6	0·842	"		
		Lipinki	4	0·849	Brown		
		Wiertzno	25	0·872	Greenish brown		
		Iwonycz	10	0·864	Brown		
		Wankowa	3	0·835	"		
		Wankowa (2)	2	0·844	"		1
		Holowecko	2	0·861	"		
		Holowecko (Zrubicz)	4	0·851	Dark green		1
		Strelbycz	8	0·898	Black	Bituminous	1
		Pagorzin	2	0·847	Dark green		
		Starunia	1	0·845	Brown		
		Sloboda	138	0·883	Dark brown	Lightest Sloboda petroleum	2
				0·868	"	Heaviest Sloboda petroleum	
Roumania	275,000	Bacau	49	0·890	Dark brown		
		Dragonese		0·902	Black	Thick	
		Ploesti		0·776	Brown	Very fluid	

TABLE SHOWING THE GENERAL PROPERTIES OF CRUDE MINERAL OIL FROM DIFFERENT COUNTRIES—(Contd.).

Oil Field.	Annual Production per Oil Field in Barrels.	Where Extracted.	Number of Borings in Progress.	Density of Petroleum.	Colour.	Remarks.	French Establishments.
Caucasus	1,700,000	Tamansk		0·765	Bright green	Almost translucent, illuminating oil	1
		Noworosirsk		0·985	Black	Viscous	
		Batoum		0·877	"	(According to Riche and Halphen)	
		Glinoi Balka	74	0·970	"	"	
		Koudako		0·815	Greenish black	"	
		Grosnaia		0·892	"		
		Grosnaia (2)		0·873	"		
Apshéron	55,008,200	Baku		0·886	Black	Rich in heavy oil	7
		Baku (2)		0·884	"	"	
		Baku (3)		0·988	"	Bituminous	
		Baku (Zoubaloff spouting well)		0·872	"	(According to Riche and Halphen)	
		Bibi Eibat	34	0·859		Rich in heavy oil	
		Sabounlschany	198	0·810	"	"	
		Surachany	49	0·770	Yellowish white	Almost translucent, volatile, illumin- [ating	
		Surachany (2)		0·760	"	"	
		Balakhany	224	0·905	Black	"	
		Balakhany (2)		0·882	"	Rich in heavy oil	
		Balakhany (3)		0·910	"	(According to St. Claire Deville)	
		Balakhany (4)		0·871	"	Rich in heavy oil	
		Binagadine	13	0·892	"	(According to Riche and Halphen)	
Northern Germany	65,000	Œdesse		0·849	Dark brown	"	1
		Œdesse (2)		0·892	"	Viscous	
		Wenigsen		0·850	"	(According to St. Claire Deville)	
		Sehnde	111	0·865	Black	"	
		Oelheim		0·908	"	Bituminous	
		Weetzen		0·965	Very deep brown	Viscous	
		Oberg		0·944	Black	"	

Country	Production	Locality	Wells	Sp. gr.	Colour	Remarks 2
Alsace	35,000	Pechelbron (1)		0·912	Black	
		" (2)		0·968	"	
		" (3)	14	0·892	"	
		Pechelbron p. 116		0·906	"	
		" p. 277		0·885	Deep brown	
		Olhungen	1	0·878	"	
		Woerth		0·885	"	
		Schwabwiller (1)	6	0·861		
		" (2)		0·829		
Bavaria	Prod. insig.	Tegernsee		0·811	Light brown	
Italy	24,000	Salo		0·787	Light brown	
		Parma oil field	22	0·828	Brown red	
		Parma oil field (2)		0·786		
		Miano	1	0·906		(According to St. Claire Deville) Disagreeable sulphur smell
		Piedmont		0·919	Yellowish white	
		Toco		0·942	Black	
France	Prod. insig.	Gabian	abandoned	0·894	Greenish brown	Viscous
		Limagne	commen'g.		Black	Bituminous
		Credo		0·975	"	"
		Le Gus	abandoned	0·782	Light brown	Very fluid
		Chatillon				
Spain	Prod. insig.	Huidobro	no regular	0·921	Red brown	Sulphuretted smell
		Conil	working	0·837	"	
Zante	Prod. insig.	Kieri (old well)	"	1·017	Black	Mixes with water. Bituminous
		Kieri (boring)		0·952	"	Would appear to be the real type
England	Prod. insig.	Ashwick Court	"	0·816	-	
Sweden	Prod. insig.	Nullaberg	"		Black	Bituminous
Trans-Caspian	78,000	Tcheleken	12	0·912	Black	
		Tcheleken (2)		0·988	"	Pasty
		Mikhailowsk		0·946	"	Viscous

TABLE SHOWING THE GENERAL PROPERTIES OF CRUDE MINERAL OIL FROM DIFFERENT COUNTRIES—(Contd.).

Oil Field.	Annual Production per Oil Field in Barrels.	Where Extracted.	Number of Borings in Progress.	Density of Petroleum.	Colour.	Remarks.	French Establishments.
Persia	4,400	Kashavashirin		0·864	Greenish	Volatile, almost illuminating	
		Chouster		0·773	Light yellow		
		Haf-Cheid	1	0·927	Greenish	Bituminous. Solid at + 5° C.	
		Daliki		1·016	Black		
		Kism		0·897	Greenish		
Beluchistan	7,500	Kahatan	2		Greenish brown	Rich in paraffin	
		Magalkot	1	0·819	Deep yellow		
British East Indies	12,500	Gunda (Pengale)	1	0·907	Deep brown	Traces of ozokerit	
		Digboi (Assam)	3	0·835	„		
		Digboi (Assam) (2)		0·844	Red brown	Very viscous	
		Makun	7	0·944	„		
Burmah	710,000	Létaung	3	0·826	„	Very fluid	
		Likman	1	0·831	„	„	
		Kiang Phyn		0·818	„	„	
		Cheduba		0·824	„	„	
		Baronga (East)	2	0·843 to 0·885	Black	Viscous	
		Baronga (West)	2	0·888	„		
		Arakan Coast		0·821	Brown	Viscous; solidifiable at + 14° C.	
		Tongune		0·877 to 0·814	„	„ „ + 22° C.	
		Yenangyaung	16	0·860	„	„	
		Yenangyat	8	0·956 to 0·828	„	„	
		Minbu		0·866	„	„	

Country	Production	Locality	No.	Sp. gr.	Colour	Remarks
China	?	Foo Choo-koo		0·860	Slightly coloured	Very fluid: solidifiable at 0° C.
		Tai-Li-Chen		0·881	Greenish	Viscous; rich in paraffin.
Japan	146,000	District of Echigo		0·831	Brown	
		District of Schinano		0·839	Greenish black	
		Orguni	56	0·840	Greenish brown	Viscous
		District of Tosan		0·882	Dark brown	
		Sumatra		0·771 to 0·789	Light brown	According to Boverton Redwood
		"	27	0·857	Red brown	" [wood]
		Langkat		0·765		Surf. oil (according to Boverton Red-
		Sungie Rebah		0·945		According to Boverton Redwood
		"		0·948		"
		Sungie Sichino		0·940		"
		"		0·897		"
		Sungie Penanti		0·848		"
		"		0·800		"
Dutch East Indies	468,000	Sungie Penanti (So.)		0·798		
		"		0·777		
		Beloc Telang	2	0·769 to 0·771	Light brown	
		"		0·789	Red brown	
		Java		0·885	"	Surface Petroleum
		"		0·868		
		Dandang-Ho	9	0·876 to 0·898		
		Djaba Kota		0·923	Red brown	According to St. Claire Deville
		Cheribon		0·878	Dark brown	Rich in paraffin
		Timor		0·823		According to St. Claire Deville
				0·825		
New Zealand	?	Sugar Loaves (surf.)	2	0·840	Light Brown	According to Boverton Redwood
		Sugar Loaves (So.)		0·971	Dark Brown	Rich in Paraffin
		" " (So.)	3	0·966	Greenish	
		Poverty Bay		0·864	Dark green	
		Manutahi		0·878 to 0·829	Brown	

Oil Field	Annual Production per Oil Field in Barrels	Where Extracted.	Number of Borings in Progress.	Density of Petroleum.	Colour.	Remarks.	French Establishments.
Philippine Isles	?	Cebu [Creek]	1	0·809	Brown, fluorescent	Rich in paraffin	
		Pennsylvania (Oil		0·816		(According to St. Claire Deville)	
		„ (Franklin)		0·886		(According to Riche and Halphen)	
		„ (average)		0·7904		,,	
		„ (Washington)		0·7916		,,	
		„ (Foxburgh)		0·797		,,	
		„ (Washington)		0·7912		,,	
		„ „		0·7926		,,	
		„ (Macdonald)		0·7884		,,	
		„ „		0·798		,,	
		„ (Grovetown)		0·787		,,	
		„ (Taylorstone)		0·7954		,,	
		Pennsylvania (Evans City, pump)		0·785		,,	
		Pennsylvania (Zelienople, pump)		0·7894		,,	
		Pennsylvania (Foxburgh, pump)		0·7861		,,	
		Pennsylvania (Tionsbil Stoneham)		0·797		,,	
		Pennsylvania (Harmony, pump)		0·785		,,	
		Pennsylvania Coraopolis		0·7886		,,	
		Pennsylvania (Glade Run, pump)		0·7931		,,	
		Pennsylvania Mannington		0·7694		,,	
		Pennsylvania (Elk District)		0·7924		,,	
		Pennsylvania (Warren County)		0·7906		,,	
				0·7916		,,	
				0·7988		,,	

Country	Production	Source	Specific gravity	Colour	Remarks
United States	52,999,526	Pennsylvania (Bradford County)	0·8016		(According to Riche and Halphen)
		Pennsylvania (Foster Creek, pump)	0·8062		„
		Pennsylvania (Rixford, pump)	0·8066		„
		Pennsylvania (Bolivar)	0·817		„
		Pennsylvania (Allentown)	0·8208		„
		Pennsylvania (Standard Oil Co.)	0·8226	Blackish	
		New York (Richburg)	0·807 to 0·833		
		Pennsylvania (Bradford)	0·769 to 0·817		
		Pennsylvania (Warren County)	0·798 to 0·601		
		Pennsylvania (Crawford County)	0·808		
		Pennsylvania (Oil Creek)	0·818		
		Pennsylvania (Franklin)	0·885 to 0·898		
		Pennsylvania (Clarion County)	0·904	Amber	Highly esteemed Illuminating without treatment
		Pennsylvania (Smith's Ferry)	0·778		(According to St. Claire Deville)
		Virginia (White Oak)	0·873		„
		Virginia (Burning Spring)	0·8412		„
		Virginia (Roger's Gulch)	0·897		
		Virginia (average)	0·868		
		Ohio (Mecca)	0·888		Viscous
		Lima (light oil)	0·790		
		Illinois	0·835 to 0·850	Dark green	Disagreeable sulphuretted smell
		Colorado	0·924	Brownish green	Fluid
		California (los Angelos)	0·874 0·832 to 0·924	Dark brown	Bituminous

TABLE SHOWING THE GENERAL PROPERTIES OF CRUDE MINERAL OIL FROM DIFFERENT COUNTRIES—(Contd.).

Oil Field.	Annual Production per Oil Field in Barrels.	Where Extracted.	Number of Borings in Progress.	Density of Petroleum.	Colour.	Remarks.	French Establishments.
Canada	689,000	Big Fork		0·871	Dark Brown	According to Boverton Redwood	
		Cap au goudron		0·989	,,	,,	
		Patawagia Brook		0·949	,,	,,	
		Silver Stream		0·894	,,	,,	
		Galls Brook		0·921		,,	
		Gaspé Wells		0·877 to 0·847	Greenish	,,	
		Enniskillen district		0·828 to 0·857	,,	According to St. Claire Deville	
West Indies	?	Cuba		0·892		Crude extract of chapapote	
		Barbadoes		0·952			
		Trinidad		0·978		Extracted from Trinidad pitch	
		Trinidad (Brea)		0·882		Sulphuretted	
		Trinidad (Aripero)		0·971			
				0·938			
		Columbia (Medina)		0·926	Brown	Surface oil	
		Venezuela		0·947	Dark brown	Bituminous	
		Ecuador		0·971 to 0·984		Surface oil	
		Ecuador San Raymondo		0·983	Greenish		
South America	200,000	Peru (Zorritos)		0·810 to 0·840	Dark Brown	According to Weinstein	
		,, (Tucillal)		0·940	,,	According to Boverton Redwood	
		,, (Quenada)		0·859	,,	,,	
		,, (Siches)		0·920			
		,, (Talara)		0·884	Brown fluorescent		
		,, (Puerto Grau)		0·848 to			
		Brazil		0·888		,,	
		Argentine Republic		0·935		,,	
		Ain Zeft		0·924	Dark brown	Sulphuretted	
		Wagap Oil Field		0·980	,,	Rich in Paraffin	

MEASURES USED IN PETROLEUM COUNTRIES.

GERMANY.—Metric system.

GALICIA
{
Joch = 57·598 ares.
Klafter = 1·93 metres.
Pouce = 0·0223.
Garnietz = 1·59 litre.
Koretz = 51·137 litres.
}
Concurrently with the Metric system.

ITALY.—Metric system.

ROUMANIA.—Metric system.

RUSSIA.—*Measures of Weight.*—In Russia the chief unit of weight is the *fount* (pound) for retail trade, and the *poud* for wholesale transactions. The following are the whole of the different weights used with their equivalents in French weights :—

Zole	equal to	0·044 grammes.
Zolotnik	,, ...	4·266 ,,
Loth	,, ...	12·797 ,,
Fount	,, ...	409 ,,
Poud	,, ...	16·375 kilogrammes.
Berkovetz	,, ...	163·720 ,,

Measures of Length.—The principal unit is the *sagène* or brasse ; the itinerary unit is the *verst*, which is equal to 500 sagènes.

Sagène	equivalent to	2·133 metres.
Archine	,, ...	0·711 ,,
Verchok	,, ...	0·044 ,,
Verst	,, ...	1067·000 ,,

Superficial Measures.

Square sagène	equivalent to	4·5 square metres.
Square archine	,, ...	0·5 ,,
Deciatine	,, .	1 hectare, ·09 ares, 25 centiares.

Measures of Capacity for Liquids.

Chtof	equivalent to	1·54 litres.
Vedro	,, ...	1·229 decalitres.
Botchka	,, ...	4·916 hectolitres.

Dry Measure.

Garnetz	equivalent to	3·277 litres.
Tchetverik	,, ...	2·621 decalitres.
Osmine	,, ...	1·05 hectolitres.
Tchetverte	,, ...	2·10 ,,
Last	,, ...	33·55 ,,

GREAT BRITAIN.

Length.—Inch (pouce)	. . . . .	= 0·0254 metre.
Foot (pied)	. . . . .	= 0·3048 ,,
Yard	. . . . .	= 0·91438 ,,
Fathom .	. . . .	= 2 yards.
Pole or perch .	. . . .	= 5½ yards.

Superficial.—Mile	. . . . .	= 1609·31 metres.
Acre	. . . . .	= 404·67 ares.

Capacity.—Gallon	. . . . .	= 4·543 litres.
Pint = ⅛ gallon	. . .	= 0·57 litre.
Quart = ¼ gallon	. . . .	= 1·138 litres.
Peck = 2 gallons	. . . .	= 9·087 litres.
Bushel .	. . . .	= 8 gallons.
Sack	. . . . .	= 3 bushels.
Quarter .	. . . . .	= 8 bushels.
Chaldron	. . . .	= 12 sacks.

Weight.—Dram = $\frac{1}{16}$ of an ounce .	. .	= 1·77 grammes.
Ounce = $\frac{1}{16}$ of a pound (oz.) .	. .	= 28·35 ,,
Pound avoirdupois (lb.) .	. .	= 453·59 ,,
Hundredweight (cwt.) = 112 lb. (cwt.)		= 50·802 kilogrs.
Ton = 20 cwt.	. . . .	= 1016 ,,

UNITED STATES.

Length.—Yard	. . . . .	= 0·914 metre.
Foot	. . . . .	= 0·304 ,,
Inch	. . . . .	= 0·025 ,,

Superficial.—Square foot	. . . .	= 0·0929 sq. metre.
Square yard	. . . .	= 0·8360 ,,

Capacity.—Gallon	. . . . .	= 3·785 litres.
Barrel	. . . . .	= 158·970 litres.

JAPAN.

Catty	. . . . .	= 604·53 grammes.

UNIT OF MEASURE ADOPTED IN BORING.—The unit of measure adopted in boring is universally, in addition to the metric measures, the *British* inch of 0·0254 metres.

Very often this measure is alone adopted, in order to simplify calculation. We append the following reducing table :—

Table showing the number of millimetres corresponding to English inches in round numbers from 1 to 30 inclusive, and in the parallel columns for each individual inch the number of millimetres corresponding to that inch plus the following fractions of an inch respectively, *viz.*: $\frac{1}{8}$, $\frac{1}{4}$, $\frac{3}{8}$, $\frac{1}{2}$, $\frac{5}{8}$, $\frac{3}{4}$, $\frac{7}{8}$:—

English Inches.	Even Inches.	⅛ Inch Extra.	¼ Inch Extra.	⅜ Inch Extra.	½ Inch Extra.	⅝ Inch Extra.	¾ Inch Extra.	⅞ Inch Extra.	English Inches.
	Milli-metres.	Milli-metres.	Milli-metres.	Milli-metres.	Milli-metres.	Milli-metres.	Milli-metres.	Milli-metres.	
		3·2	6·4	9·5	12·7	15·9	19	22·2	
1	25·4	29	32	35	38	41	44	48	1
2	51	54	57	60	63	67	70	73	2
3	76	79	83	86	89	92	95	98	3
4	102	105	108	111	114	117	121	124	4
5	127	130	133	137	140	143	146	149	5
6	152	156	159	162	165	168	171	175	6
7	178	181	184	187	190	194	197	200	7
8	203	206	210	213	216	219	222	225	8
9	229	232	235	238	241	244	248	251	9
10	254	257	260	264	267	270	273	276	10
11	279	283	286	289	292	295	298	302	11
12	305	308	311	314	317	321	324	327	12
13	330	333	337	340	343	346	349	352	13
14	356	359	362	365	368	371	375	378	14
15	381	384	387	391	394	397	400	403	15
16	406	410	413	416	419	422	425	429	16
17	432	435	438	441	444	448	451	454	17
18	457	460	464	467	470	473	476	479	18
19	483	486	489	492	495	498	502	505	19
20	508	511	514	518	521	524	527	530	20
21	533	537	540	543	546	549	552	556	21
22	559	562	565	568	571	575	578	581	22
23	584	587	591	594	597	600	603	606	23
24	610	613	616	619	622	625	629	632	24
25	635	638	631	645	648	651	654	657	25
26	660	664	667	670	673	676	679	683	26
27	686	689	692	695	698	702	705	708	27
28	711	714	718	721	724	728	730	733	28
29	737	740	743	746	749	752	756	759	29
30	762	765	768	772	775	778	781	784	30

TABLE FOR THE REDUCTION OF FOREIGN MONEY INTO FRANCS.

Great Britain.		Russia.		Austria.	
Pound Sterling, Shillings, Pence. 1	Francs and Centimes. 2	Roubles and Kopecks. 3	At the rate, 3½ Francs the Rouble. 4	Florins and Kreutzer. 5	Francs and Centimes. 6
£ s. d.		Rbls. Kp.	Fr. C.	Fl. Kr.	Fr. C.
0 0 1	0·10½	0 1	0 03½	0 1	0·2½
0 0 2	0·21	0 2	0 07	0 2	0·5
0 0 3	0·31½	0 3	0 10½	0 3	0·7½
0 0 4	0·42	0 4	0 14	0 4	0·10
0 0 5	0·52	0 5	0 17½	0 5	0·12½
0 0 6	0·62	0 6	0 21	0 6	0·15
0 0 7	0·73	0 7	0 24½	0 7	0·17½
0 0 8	0·83	0 8	0 28	0 8	0·20
0 0 9	0·94	0 9	0 31½	0 9	0·22½
0 0 10	1·04	0 10	0 35	0 10	0·25
0 0 11	1·14½	1 0	3 50	1 0	2·50
0 1 0	1·25	2 0	7 0	2 0	5·00
0 2 0	2·50	3 0	10 5	3 0	7·50
0 3 0	3·75	4 0	14 0	4 0	10·00
0 4 0	5	5 0	17 50	5 0	12·50
0 5 0	6·25	6 0	21 0	6 0	15·00
0 6 0	7·50	7 0	24 50	7 0	17·50
0 7 0	8·75	8 0	28 0	8 0	20·00
0 8 0	10·00	9 0	31 50	9 0	22·50
0 9 0	11·25	10 0	35 0	10 0	25·0
0 10 0	12·50	100 0	350 0	20 0	50·0
1 0 0	25			30 0	75·0
2 0 0	50			40 0	100·0
3 0 0	75			50 0	125·0
4 0 0	100			60 0	150·0
5 0 0	125			70 0	175·0
6 0 0	150			80 0	200·0
7 0 0	175			90 0	225·0
8 0 0	200			100 0	250·0
9 0 0	225				
10 0 0	250				
100 0 0	2,500				

PRELIMINARY WORK.

Earthworks.—The weight of a cubic metre of :—

Mould or soil is	. . . . .	1,200 to 1,400 Kilogrammes
Fine dry sand is	. . . . .	1,400 ,,
Wet sand is	. . . . .	1,800 [1] ,,
Clay soil is	. . . . .	1,600 ,,
Potter's earth is	. . . . .	1,900 ,,
Moorland is	. . . . .	650 ,,
Marl	. . . . .	1,600 ,,

Time occupied in excavating a cubic metre of :—

	H. M.
Ordinary soil	54
Running sand	57
Clay or potter's earth	1 45
Gravel, very compact	1 57

[1] This is possibly a misprint for 1,500. It is against all reason that wet sand should weigh less than dry sand.—*Tr.*

Forty minutes are required in which to load a metre of soil by barrow.

Transports.—The cost of wheeling away a cubic metre of rubbish to the distance d, p being the labourers' day's wages is—

$$\frac{2\,pd}{1000}.$$

DATA FOR THE CONSTRUCTION OF DERRICKS AND BUILDINGS.

Resistance of materials which may be used in construction :—

Resistance to Traction Parallel to the Fibre.			
Per Square Centimetre.		Per Square Millimetre.	
Oak	600 to 800 kilogrs.	Wrought iron	36 kilogrs.
Pine	800 ,, 900 ,,	Sheet iron	33 ,,
Ash	1,200 ,,	Cast iron	12 ,,
Elm	1,040 ,,	Steel	70 ,,
Beech	800 ,,	Copper	34 ,,
Teak	1,100 ,,	Lead	2·85 ,,
Boxwood	1,400 ,,	Zinc	4 ,,

Resistance to Crushing.			
Per Square Centimetre.		Per Square Millimetre.	
French Oak	380 to 700 kilogrs.	Wrought iron	40 kilogrs.
Pine	400 to 538 ,,	Sheet iron	25 ,,
Red Pine	379 to 528 ,,	Cast iron	60 to 70 ,,
Beech	540 to 650 ,,	Steel	70 ,,
Elm	725 ,,		
Poplar	220 to 360 ,,		

The load per square millimetre which may be laid on an oak or pine post decreases as follows, as the ratio of the height to the thickness of the piece of timber increases :—

	Kilogrammes.			Kilogrammes.
From 1 to 8	0·40	From 1 to 48		0·07
,, ,, 12	0·33	,, ,, 60		0·07
,, ,, 24	0·20	,, ,, 72		0·02
,, ,, 36	0·11			

In the case of timber, in buildings intended to last some time, the load should not exceed $\frac{1}{10}$ of the breaking strain.

RESISTANCE OF THE DERRICK TO THE WIND.

Pressure of the wind per square metre :—

	Metres per Second.	Kilogrammes per Square Metre.
Fresh wind	7	6
Very strong wind	15	30
Tempest	24	78
Great hurricane	45	265

RESISTANCE OF ROPES AND CABLES.

Rope Cables of Vegetable Fibre.—The *breaking strain* Q which a hemp or aloes rope may bear before breaking is 300 kilogrammes per circular centimetre.

The *practical load* Q' which it can be made to carry in normal working is $\frac{1}{5}$ of the breaking strain.

The weight *per running metre p* is about—

> 80 grammes per circular centimetre for hemp.
> 75   ,,   ,,   ,,   aloes.

both being tarred.

Let *d* be the diameter of the cable expressed in centimetres, then we have the following formulæ :—

Hemp.	Aloes.
Q = 300 d^2.	Q = 300 d^2.
p = 0·08 d^2.	p = 0·0775 d^2.
Q = $\frac{100}{0·08}$ p = 3,750 p.	Q = $\frac{300}{0·075}$ p = 4,000 p.
Q' = $\frac{1}{5}$ Q = 60 d^2 = 750 p.	Q' = $\frac{1}{5}$ Q = 60 d^2 = 800 p.

If *a* be the area of the section of the cable in square centimetres we then have $a = \dfrac{\pi d^2}{4}$ or $d^2 = \dfrac{4a}{\pi}$, and therefore

$$Q' = 60 d^2 = 76a.$$

To sum up, a rope cable of vegetable fibre may be loaded with 60 kilogrammes per circular centimetre or 76 kilogrammes per square centimetre or with a charge equal to 750 to 800 times its own weight per running metre.

The dimensions of the cables most generally used in actual practice, their weight per metre and the load which they can bear are given in the following table :—

ROUND HEMP CABLES.

Untarred Hemp.			Untarred Hemp.		
Diameter in Millimetres.	Weight per Metre in Kilo-grammes.	Load which may be carried in Kilo-grammes.	Diameter in Millimetres.	Weight per Metre in Kilo-grammes.	Load which may be carried in Kilo-grammes.
16	0·21	200	46	1·65	2,250
20	0·32	300	52	2·13	3,000
23	0·37	400	59	2·67	3,600
26	0·53	500	65	3·70	4,500
29	0·64	750	72	4·00	5,000
33	0·80	900	78	4·80	6,200
36	0·96	1,000	85	5·60	7,500
39	1·06	1,250	92	6·40	8,700
46	1·55	1,500	98	7·46	10,000
52	2·03	2,000	105	8·53	12,000

The composition of a round cable is defined by stating that it consists of so many strands of so many ply of such and such a number, and that of a flat cable by saying that it consists of so many small ropes of so many ply per small rope.

Taking 55 kilogrammes per square centimetre as the average resistance to fracture, the numbers 12 to 17 used in their manufacture give the following results :—

Number of the Ply.	Diameter in Tenths of Milli-metres.	Section in Square Millimetres.	Weight per Metre in Grammes.	Breaking Load of the Ply in Kilo-grammes.
12	18	2,545	19·84	140
13	20	3,142	24·48	173
14	22	3,801	29·64	209
15	24	4,524	35·28	249
16	27	5,725	44·63	315
17	30	7,068	55·13	389

Starting from the above data, the cables are manufactured in accordance with the strain which they are to bear.

In practice the following formula is generally adopted for iron wire cables :—

$$Q = 7,000\, p \qquad Q' = 1,000\, p$$

p being the actual weight of the tarred cable per current metre.

This is equivalent to the statement that an iron wire cable will support a load equal to 1,000 times its weight per current metre.

The following formula may be adopted for steel wire cables, the quality of which may be relied upon : $Q' = 1,500\, p$.

In the following tables there will be found the dimensions of the round and flat cables most generally used in practice, their weight per metre, and the load which they carry.

DIMENSIONS OF ROUND IRON OR STEEL WIRE CABLES.

Diameter in Metres.	Weight per current Metre in Kilogrammes.	Load that can be lifted from a depth of 400 Metres in Kilogrammes.	Steel Wire Cables.
0·038	3·25	3,000	The steel wire cables capable of raising the same load may weigh ½ less per running metre than the iron wire cables.
0·028	2·50	2,500	
0·025	1·90	2,000	
0·021	1·50	1,000	
0·018	1·30	1,000	
0·016	1·00	750	
0·015	0·75	500	
0·013	0·50	200	

DIMENSIONS OF FLAT IRON, WIRE OR STEEL CABLES.

Number of Strands.	Dimensions.		Weight per current Metre in Kilogrammes.	Load that can be raised from a depth of 400 Metres in Kilogrammes.	Steel Wire Cables.
	Width (Metres).	Thickness (Metres).			
8	0·130	0·022	800	5,000	The steel wire cables capable of raising the same load may weigh ½ less per running metre than the iron wire cables.
8	0·120	0·020	650	4,500	
6	0·100	0·021	600	4,000	
8	0·110	0·017	550	3,500	
6	0·090	0·020	500	3,000	
6	0·080	0·017	450	2,500	
6	0·080	0·016	400	2,000	
6	0·070	0·015	350	1,800	
6	0·060	0·014	800	1,500	

DRILLING WITH THE AID OF A CIRCULATING CURRENT OF WATER.

I.—When instead of a pump a natural fall of water is utilised, the following is the formula generally used to estimate its strength :—

Neglecting the actual force of the current, which should be reduced as far as practicable in a good hydraulic installation, the force of a fall in kilogrammetres is expressed by the product of the flow or output P in litres per second by the height H of the fall expressed in metres :—

$$T = PH.$$

PRACTICAL PART. 511

By dividing this number by seventy-five the horse-power is obtained; that is, the theoretical work of which the fall is capable.

II.—In order to effect complete " sand pumping " of the well, the speed of the current of water is varied as follows, for :—

	Metres per Second.
Fine débris and sand	0·02
Sandstone débris	0·10
Rock débris	0·15 to 0·20
Débris, weighing 0·05 kg.	0·50 to 1·00
„ 0·25 kg.	2

CALCULATIONS USED IN ESTIMATING THE OUTPUT FLOW OF A CONDUIT.

If we call the radius of the pipe r, and represent the loss of charge per running metre, that is to say, the quotient of the length of the pipe by the height which separates the orifice of the pipe from the free surface of the reservoir, by j, the rate of flow output per second by q, the average velocity by u, that is, the quotient of the amount of flow by the section, the fundamental equation established by Darcy as to the flow of liquids in tubes is written thus :—

$$(1) \qquad rj = b_1 u^2.$$

When dealing with pipes which have been a long time in use, and which are the only ones which can be considered in actual practice, the following figures are taken as the coefficient, b :—

Radius in Metres.		Radius in Metres.	
0·01	0·0023	0·05	0·0013
0·02	0·0017	0·07	0·0012
0·03	0·0014	0·10	0·0011

and for radii above 0·10 metre the coefficient is 0·0011.

Equation (1) combined with the rate of flow equation :—

$$(2) \qquad q = \pi . r^2 u,$$

enables us to solve all equations relating to the flow of pipes.

EXAMPLE I.—*Two reservoirs the levels of which differ by 10 metres are connected by a pipe 1,000 metres long. What ought to be the diameter of this pipe so that it should deliver 0·5 cubic metres per second ?*

Eliminating u from equations (1) and (2) we have—

$$r^5 = \frac{b_1 \, q^2}{\pi^2 \, j}$$

the loss of charge j per current metre is equal to the difference of level 10 metres. Taking 0·0011 divided by the length 1,000 metres, say 0.01 metres, as the value of b_1 we get $r = 0·31$ metres. It therefore requires a pipe of 0·62 metres in diameter.

EXAMPLE II.—*What is the output of a pipe of 0·62 metres in diameter, 1,000 metres long, connecting two reservoirs the free level of which differs by 10 metres ?*

$$q^2 = \frac{\pi^2 j \, r^5}{b_1}$$

$r = 0·31$ $j = 0·01$ $b_1 = 0.0011$, which gives $q = 0·5$ cubic metre.[1]

VISCOSITY OF PETROLEUM AND HEAVY OILS AT DIFFERENT TEMPERATURES.

(According to Wischin.)

Temperature. (Degrees C.)	Pressure in Kilogrammes.	Rectified Petroleum D = 0·823.	Heavy Oil D = 0·836.	Crude Petroleum D = 0885.	Lubricating Oil D = 0912.	Mazout D = 0·907.	Water taken as Unity.
10	0·70	6·56	12·00	18·50	126·5	186·5	5·25
	1·40	4·50	7·25	10·25	85·0	94·5	8·75
	2·11	3·75	5·00	7·85	56·5	65·0	8·25
	2·81	3·25	4·25	6·25	48·0	45·2	2·75
25	0·70	6·00	8·25	11·00	61·0	72·5	5·10
	1·40	4·25	5·00	6·25	44·5	36·5	8·75
	2·11	3·50	3·90	4·50	21·0	20·7	8·00
	2·81	3·10	3·36	3·75	17·7	17·2	2·69
40	0·70	5·90	7·25	7·50	26·2	22·0	5·00
	1·40	4·25	4·75	5·00	14·2	11·7	8·50
	2·11	3·50	3·75	3·75	10·0	8·0	3·00
	2·81	3·00	3·00	3·25	8·0	6·2	2·66

[1] As Example II. seems to be the exact converse of Example I., the translator has altered the value of q to 0·5 cubic metre per second, the value of 0·15 given by authors being evidently a misprint.

WEIGHT OF RODS.

It is sometimes useful to know exactly the weight of the drilling tackle. Nothing could be done more easily than by consulting the following tables :—

WEIGHT PER LINEAL METRE OF SQUARE SECTION AND ROUND SECTION IRON RODS BOTH IN THE AIR AND IN WATER.

Thickness or Diameter in Millimetres.	Iron Square Section in the Atmosphere	Iron Square Section in Water.	Iron Circular Section in the Atmosphere.	Iron Circular Section in Water.	Thickness or Diameter in Millimetres.
5	0·195	0·170	0·153	0·133	5
6	0·280	0·244	0·220	0·191	6
7	0·381	0·332	0·299	0·260	7
8	0·498	0·434	0·391	0·340	8
9	0·630	0·549	0·495	0·429	9
10	0·778	0·678	0·611	0·532	10
11	0·941	0·820	0·739	0·643	11
12	1·120	0·976	0·880	0·766	12
13	1·185	0·996	1·033	0·900	13
14	1·525	1·329	1·198	1·044	14
15	1·751	1·526	1·375	1·198	15
16	1·992	1·736	1·564	1·362	16
17	2·248	1·959	1·756	1·539	17
18	2·521	2·197	1·980	1·725	18
19	2·809	2·448	2·206	1·922	19
20	3·112	2·712	2·444	2·129	20
21	3·422	2·981	2·695	2·384	21
22	3·726	3·242	2·957	2·576	22
23	4·116	3·587	3·232	2·816	23
24	4·481	3·905	3·520	3·067	24
25	4·863	4·238	3·819	3·328	25
26	5·259	4·583	4·131	3·600	26
27	5·672	4·943	4·455	3·882	27
28	6·100	5·316	4·791	4·176	28
29	6·543	5·702	5·139	4·478	29
30	7·002	6·102	5·499	4·792	30
31	7·477	6·516	5·872	5·117	31
32	7·967	6·952	6·257	5·452	32
33	8·382	7·298	6·654	5·798	33
34 ·	8·994	7·838	7·064	6·093	34
35	9·531	8·316	7·485	6·522	35
36	10·080	8·784	7·919	6·901	36
37	10·650	9·281	8·365	7·289	37
38	11·230	9·786	8·823	7·689	38
39	11·830	10·309	9·294	8·099	39
40	12·450	10·850	9·776	8·519	40
41	13·060	11·399	10·270	8·849	41
42	13·690	11·926	10·780	9·394	42
43	14·390	12·541	11·300	9·847	43
44	14·900	12·964	11·830	10·039	44
45	15·750	13·725	12·370	10·779	45
46	16·460	14·344	12·930	11·268	46
47	17·190	14·981	13·500	11·765	47
48	17·930	15·626	14·080	12·270	48
49	18·680	16·276	14·670	12·784	49
50	19·450	16·950	15·280	13·316	50

33

WEIGHT PER LINEAL METRE OF SQUARE SECTION AND ROUND
SECTION IRON RODS BOTH IN THE AIR AND IN WATER—(*Contd.*).

Thickness or Diameter in Millimetres.	Iron Square Section in the Atmosphere.	Iron Square Section in Water.	Iron Circular Section in the Atmosphere.	Iron Circular Section in Water.	Thickness or Diameter in Millimetres.
52	21·040	18·336	16·520	14·396	52
54	22·690	19·774	17·820	15·529	54
56	24·400	21·264	19·160	16·696	56
58	26·170	22·806	20·560	17·917	58
60	28·010	24·410	22·000	19·172	60
62	29·910	24·066	23·490	20·470	62
64	31·870	25·774	25·080	21·813	64
66	33·580	28·974	26·620	23·198	66
68	35·980	31·356	28·260	24·628	68
70	38·120	33·220	·29·940	26·091	70
72	40·320	35·136	31·680	27·608	72
74	42·600	37·124	33·400	29·159	74
76	44·920	39·144	35·290	30·753	76
78	47·320	41·236	37·180	32·401	78
80	49·790	43·390	39·110	34·083	80
85	56·210	48·985	44·150	38·475	85
90	63·020	54·920	49·490	43·128	90
95	70·210	61·185	55·150	48·061	95
100	77·800	67·800	61·160	53·206	100
105	85·550	77·525	67·370	59·710	105
110	94·140	82·040	73·940	64·436	110
115	102·900	89·675	80·810	70·423	115
120	112·000	97·600	88·000	76·690	120
125	121·600	105·975	95·480	83·209	125
130	131·500	114·600	103·800	90·027	130
135	141·800	123·575	111·400	97·087	135
140	152·500	132·900	119·800	104·407	140
145	163·600	142·575	128·500	111·987	145
150	175·100	152·600	137·500	119·929	150
155	186·900	162·875	146·800	127·931	155
160	199·200	173·600	156·400	136·294	160
165	209·600	182·375	166·600	145·218	165
170	224·800	195·900	176·100	153·402	170
175	238·300	207·675	187·400	163·384	175
180	252·100	219·700	198·000	172·554	180
185	266·300	232·075	209·100	182·220	185
190	280·900	244·800	220·600	191·348	190
195	295·900	257·875	232·300	202·436	195
200	311·200	281·200	244·300	212·884	200
205	327·000	284·975	256·700	223·694	205
210	343·100	299·000	269·300	234·664	210
215	359·600	313·375	282·300	245·995	215
220	376·600	328·200	295·600	257·587	220
225	393·900	343·275	309·200	269·540	225
230	411·600	358·700	323·100	281·553	230
235	429·700	374·475	337·100	293·727	235
240	448·100	390·500	351·800	306·561	240
245	467·000	406·975	366·600	319·457	245
250	486·800	423·800	381·700	332·613	250
255	605·900	439·875	397·100	346·030	255

N.B.—If the millimetres in the above table be multiplied by ·00393, the result
will be the diameter of the rods in inches and decimal inches. If the kilogrammes
be multiplied by 2·47 and divided by 3·28 the quotient will be the weight in lb. per
running foot.—*Tr.*

TABLE FOR GAUGING BARRELS.

Interior Length in Centimetres.	Height at the Bung in Centimetres.	Bottom Diameter in Centimetres.	Contents in Litres.	Interior Length in Centimetres.	Height at the Bung in Centimetres.	Bottom Diameter in Centimetres.	Contents in Litres.	Interior Length in Centimetres.	Height at the Bung in Centimetres.	Bottom Diameter in Centimetres.	Contents in Litres.
0·75	0·56	46	164·7	0·75	0·61	51	197·2	0·76	0·59	38	167·3
"	"	47	166·5	"	"	52	199·2	"	"	39	168·8
"	"	48	168·4	"	"	53	201·3	"	"	40	170·4
"	"	49	170·3	0·75	0·62	37	177·8	"	"	41	172·0
0·75	0·57	47	170·9	"	"	38	179·3	"	"	42	173·6
"	"	48	172·8	"	"	39	180·8	"	"	43	175·3
"	"	49	174·7	"	"	40	182·4	"	"	44	177·0
"	"	50	176·7	"	"	41	184·0	"	"	45	178·8
0·75	0·58	43	168·4	"	"	42	185·6	"	"	46	180·6
"	"	44	170·1	"	"	43	187·3	"	"	47	182·5
"	"	45	171·9	"	"	44	189·0	"	"	48	184·4
"	"	46	173·7	"	"	45	190·7	"	"	49	186·3
"	"	47	175·7	"	"	46	192·5	"	"	50	188·3
"	"	48	177·3	"	"	47	194·3	"	"	51	190·3
"	"	49	179·2	"	"	48	196·2	"	"	52	192·3
"	"	50	181·2	"	"	49	198·1	0·76	0·60	37	170·5
"	"	51	183·2	"	"	50	200·0	"	"	38	172·0
0·75	0·59	41	169·7	"	"	51	202·0	"	"	39	173·5
"	"	42	171·3	"	"	52	204·0	"	"	40	175·1
"	"	43	173·0	"	"	53	206·1	"	"	41	176 7
"	"	44	174·7	"	"	54	208·2	"	"	42	178·4
"	"	45	176·5	0·75	0·63	37	182·7	"	"	43	180·0
"	"	46	178·2	"	"	38	184·2	"	"	44	181·0
"	"	47	180·1	"	"	39	185·7	"	"	45	183·5
"	"	48	181·9	"	"	40	187·3	"	"	46	185·4
"	"	49	183·8	"	"	41	188·9	"	"	47	187·2
"	"	50	185·8	"	"	42	190·5	"	"	48	189·1
"	"	51	187·8	"	"	43	192·2	"	"	49	191·0
"	"	52	189·8	"	"	44	193·9	"	"	50	193·0
0·75	0·60	39	171·2	"	"	45	195·6	"	"	51	195·0
"	"	40	172·8	"	"	46	197·4	"	"	52	197·1
"	"	41	174·4	"	"	47	199·2	"	"	53	199·1
"	"	42	176·0	"	"	48	201·1	0·76	0·61	37	175·2
"	"	43	177·7	"	"	49	203·0	"	"	38	176·8
"	"	44	179·4	"	"	50	204·9	"	"	39	178·3
"	"	45	181·1	"	"	51	206·9	"	"	40	179·9
"	"	46	182·9	"	"	52	209·0	"	"	41	181·5
"	"	47	184·7	"	"	53	211·0	"	"	42	183·2
"	"	48	186·6	"	"	54	213·1	"	"	43	184·9
"	"	49	188·5	"	"	55	215·3	"	"	44	186·6
"	"	50	190·5	0·76	0·56	47	168·7	"	"	45	188·4
"	"	51	192·4	"	"	48	170·6	"	"	46	190·2
"	"	52	194·5	"	"	49	172·6	"	"	47	192·0
"	"	53	196·5	0·76	0·57	45	169·6	"	"	48	193·9
0·75	0·61	36	171·6	"	"	46	171·4	"	"	49	195·8
"	"	37	173·0	"	"	47	173·1	"	"	50	197·8
"	"	38	174·5	"	"	48	175·1	"	"	51	199·8
"	"	39	176·0	"	"	49	177·1	"	"	52	201·9
"	"	40	177·5	"	"	50	179·0	"	"	53	204·0
"	"	41	179·1	0·76	0·58	42	169·0	0·76	0 62	37	180·2
"	"	42	180·8	"	"	43	170·7	"	"	38	181·7
"	"	43	182·4	"	"	44	172·4	"	"	39	183·2
"	"	44	184·1	"	"	45	174·2	"	"	40	184·8
"	"	45	185·9	"	"	46	176·0	"	"	41	186·4
"	"	46	187·7	"	"	47	177·8	"	"	42	188·1
"	"	47	189·5	"	"	48	179·7	"	"	43	189·8
"	"	48	191·4	"	"	49	181·6	"	"	44	191·5
"	"	49	193·3	"	"	50	183·6	"	"	45	193·3
"	"	50	195·2	"	"	51	185·6	"	"	46	195·1

TABLE FOR GAUGING BARRELS—(Contd.).

Interior Length in Centimetres.	Height at the Bung in Centimetres.	Bottom Diameter in Centimetres.	Contents in Litres.	Interior Length in Centimetres.	Height at the Bung in Centimetres.	Bottom Diameter in Centimetres.	Contents in Litres.	Interior Length in Centimetres.	Height at the Bung in Centimetres.	Bottom Diameter in Centimetres.	Contents in Litres.
0·76	0·62	47	196·0	0·77	0·59	44	179·6	0·77	0·62	52	209·5
,,	,,	48	198·8	,,	,,	45	181·2	,,	,,	53	211·6
,,	,,	49	200·7	,,	,,	46	183·0	,,	,,	54	213·8
,,	,,	50	202·7	,,	,,	47	184·9	0·77	0·63	37	187·6
,,	,,	51	204·7	,,	,,	48	186·8	,,	,,	38	189·1
,,	,,	52	206·8	,,	,,	49	188·7	,,	,,	39	190·7
,,	,,	53	208·9	,,	,,	50	190·7	,,	,,	40	192·3
,,	,,	54	211·0	,,	,,	51	192·8	,,	,,	41	193·9
0·76	0·63	38	186·7	,,	,,	52	194·9	,,	,,	42	195·6
,,	,,	39	188·2	0·77	0·60	36	171·3	,,	,,	43	197·3
,,	,,	40	189·8	,,	,,	37	172·7	,,	,,	44	199·0
,,	,,	41	191·4	,,	,,	38	174·3	,,	,,	45	200·8
,,	,,	42	193·0	,,	,,	39	175·8	,,	,,	46	202·7
,,	,,	43	194·7	,,	,,	40	177·4	,,	,,	47	204·5
,,	,,	44	196·5	,,	,,	41	179·0	,,	,,	48	206·5
,,	,,	45	198·2	,,	,,	42	180·7	,,	,,	49	208·4
,,	,,	46	200·0	,,	,,	43	182·4	,,	,,	50	210·4
,,	,,	47	201·9	,,	,,	44	184·2	,,	,,	51	212·5
,,	,,	48	203·8	,,	,,	45	186·0	,,	,,	52	214·5
,,	,,	49	205·7	,,	,,	46	187·8	,,	,,	53	216·6
,,	,,	50	207·7	,,	,,	47	189·7	,,	,,	54	218·8
,,	,,	51	209·7	,,	,,	48	191·6	,,	,,	55	221·0
,,	,,	52	211·7	,,	,,	49	193·5	0·78	0·56	44	167·6
,,	,,	53	213·8	,,	,,	50	195·5	,,	,,	45	169·4
,,	,,	54	216·0	,,	,,	51	197·6	,,	,,	46	171·3
,,	,,	55	218·1	,,	,,	52	199·7	,,	,,	47	173·2
0·77	0·56	45	167·3	,,	,,	53	201·8	,,	,,	48	175·1
,,	,,	46	169·1	0·77	0·61	37	177·6	,,	,,	49	177·1
,,	,,	47	171·0	,,	,,	38	179·1	0·78	0·57	40	165·4
,,	,,	48	172·9	,,	,,	39	180·7	,,	,,	41	167·0
,,	,,	49	174·8	,,	,,	40	182·3	,,	,,	42	168·7
0·77	0·57	43	168·3	,,	,,	41	183·9	,,	,,	43	170·4
,,	,,	44	170·0	,,	,,	42	185·6	,,	,,	44	172·2
,,	,,	45	171·8	,,	,,	43	187·3	,,	,,	45	174·0
,,	,,	46	173·6	,,	,,	44	189·0	,,	,,	46	175·9
,,	,,	47	175·4	,,	,,	45	190·8	,,	,,	47	177·7
,,	,,	48	177·4	,,	,,	46	192·7	,,	,,	48	179·9
,,	,,	49	179·4	,,	,,	47	194·6	,,	,,	49	181·7
,,	,,	50	181·4	,,	,,	48	196·5	,,	,,	50	183·7
0·77	0·58	40	167·9	,,	,,	49	198·4	0·78	0·58	40	170·1
,,	,,	41	169·5	,,	,,	50	200·4	,,	,,	41	171·7
,,	,,	42	171·2	,,	,,	51	202·5	,,	,,	42	173·4
,,	,,	43	172·9	,,	,,	52	204·5	,,	,,	43	175·1
,,	,,	44	174·7	,,	,,	53	206·6	,,	,,	44	176·9
,,	,,	45	176·4	0·77	0·62	37	182·6	,,	,,	45	178·7
,,	,,	46	178·3	,,	,,	38	184·1	,,	,,	46	180·6
,,	,,	47	180·2	,,	,,	39	185·6	,,	,,	47	182·5
,,	,,	48	182·1	,,	,,	40	187·2	,,	,,	48	184·4
,,	,,	49	184·0	,,	,,	41	188·9	,,	,,	49	186·4
,,	,,	50	186·0	,,	,,	42	191·5	,,	,,	50	188·4
,,	,,	51	188·1	,,	,,	43	192·8	,,	,,	51	190·5
0·77	0·59	36	166·5	,,	,,	44	194·0	0·78	0·59	36	168·6
,,	,,	37	167·9	,,	,,	45	195·8	,,	,,	37	170·1
,,	,,	38	169·5	,,	,,	46	197·6	,,	,,	38	171·7
,,	,,	39	171·0	,,	,,	47	199·5	,,	,,	39	173·2
,,	,,	40	172·6	,,	,,	48	201·4	,,	,,	40	174·8
,,	,,	41	174·2	,,	,,	49	203·4	,,	,,	41	176·5
,,	,,	42	175·9	,,	,,	50	205·4	,,	,,	42	178·2
,,	,,	43	177·6	,,	,,	51	207·4	,,	,,	43	179·9

TABLE FOR GAUGING BARRELS—(Contd.).

Interior length in Centimetres.	Height at the Bung in Centimetres.	Bottom Diameter in Centimetres.	Contents in Litres.	Interior Length in Centimetres.	Height at the Bung in Centimetres.	Bottom Diameter in Centimetres.	Contents in Litres.	Interior Length in Centimetres.	Height at the Bung in Centimetres.	Bottom Diameter in Centimetres.	Contents in Litres.
0·78	0·59	44	181·7	0·78	0·62	51	210·1	0·79	0·56	40	162·8
,,	,,	45	183·5	,,	,,	52	212·2	,,	,,	41	164·5
,,	,,	46	185·4	,,	,,	53	214·4	,,	,,	42	166·2
,,	,,	47	187·3	,,	,,	54	216·5	,,	,,	43	168·8
,,	,,	48	189·2	0·78	0·63	37	190·1	,,	,,	44	169·8
,,	,,	49	191·2	,,	,,	38	191·6	,,	,,	45	171·6
,,	,,	50	193·2	,,	,,	39	193·2	,,	,,	46	173·5
,,	,,	51	195·3	,,	,,	40	194·8	,,	,,	47	175·4
,,	,,	52	197·4	,,	,,	41	196·4	,,	,,	48	177·4
0·78	0·60	36	173·5	,,	,,	42	198·1	,,	,,	49	179·4
,,	,,	37	175·0	,,	,,	43	199·9	0·79	0·57	40	167·5
,,	,,	38	176·5	,,	,,	44	201·6	,,	,,	41	169·2
,,	,,	39	178·1	,,	,,	45	203·4	,,	,,	42	170·9
,,	,,	40	179·7	,,	,,	46	205·3	,,	,,	43	172·6
,,	,,	41	181·4	,,	,,	47	207·3	,,	,,	44	174·4
,,	,,	42	183·0	,,	,,	48	209·1	,,	,,	45	176·3
,,	,,	43	184·8	,,	,,	49	211·1	,,	,,	46	178·2
,,	,,	44	186·6	,,	,,	50	213·1	,,	,,	47	180·0
,,	,,	45	188·4	0·78	0·64	37	195·2	,,	,,	48	182·0
,,	,,	46	190·2	,,	,,	38	196·8	,,	,,	49	184·1
,,	,,	47	192·1	,,	,,	39	198·3	,,	,,	50	186·1
,,	,,	48	194·1	,,	,,	40	200·0	0·79	0·58	36	166·0
,,	,,	49	196·1	,,	,,	41	201·6	,,	,,	37	167·5
,,	,,	50	198·1	,,	,,	42	203·3	,,	,,	38	169·0
,,	,,	51	200·1	,,	,,	43	205·0	,,	,,	39	170·6
,,	,,	52	202·2	,,	,,	44	206·8	,,	,,	40	172·2
,,	,,	53	204·4	,,	,,	45	208·6	,,	,,	41	173·9
0·78	0·61	36	178·4	,,	,,	46	210·5	,,	,,	42	175·6
,,	,,	37	179·9	,,	,,	47	212·4	,,	,,	43	177·4
,,	,,	38	181·5	,,	,,	48	214·3	,,	,,	44	179·2
,,	,,	39	183·0	,,	,,	49	216·3	,,	,,	45	181·0
,,	,,	40	184·6	,,	,,	50	218·3	,,	,,	46	182·9
,,	,,	41	186·3	,,	,,	51	220·3	,,	,,	47	184·8
,,	,,	42	188·0	,,	,,	52	222·5	,,	,,	48	186·8
,,	,,	43	189·7	,,	,,	53	224·6	,,	,,	49	188·8
,,	,,	44	191·5	,,	,,	54	226·8	,,	,,	50	190·9
,,	,,	45	193·3	,,	,,	55	229·1	,,	,,	51	192·9
,,	,,	46	195·2	,,	,,	56	231·8	0·79	0·59	36	170·8
,,	,,	47	197·1	0·78	0·65	38	202·0	,,	,,	37	172·3
,,	,,	48	199·9	,,	,,	39	203·8	,,	,,	38	173·9
,,	,,	49	201·0	,,	,,	40	205·2	,,	,,	39	175·4
,,	,,	50	203·0	,,	,,	41	206·9	,,	,,	40	177·1
,,	,,	51	205·1	,,	,,	42	208·6	,,	,,	41	178·8
,,	,,	52	207·2	,,	,,	43	210·3	,,	,,	42	180·5
,,	,,	53	209·3	,,	,,	44	212·1	,,	,,	43	182·2
0·78	0·62	37	184·9	,,	,,	45	213·9	,,	,,	44	184·0
,,	,,	38	186·5	,,	,,	46	215·8	,,	,,	45	185·9
,,	,,	39	188·1	,,	,,	47	217·7	,,	,,	46	187·8
,,	,,	40	189·7	,,	,,	48	219·6	,,	,,	47	189·7
,,	,,	41	191·3	,,	,,	49	222·2	,,	,,	48	191·6
,,	,,	42	193·0	,,	,,	50	223·6	,,	,,	49	193·6
,,	,,	43	194·7	,,	,,	51	225·7	,,	,,	50	195·7
,,	,,	44	196·5	,,	,,	52	227·8	,,	,,	51	197·8
,,	,,	45	198·3	,,	,,	53	229·9	,,	,,	52	199·9
,,	,,	46	200·2	,,	,,	54	232·1	0·79	0·60	37	177·2
,,	,,	47	202·1	0·79	0·55	45	167·0	,,	,,	38	178·8
,,	,,	48	204·0	,,	,,	46	168·9	,,	,,	39	180·4
,,	,,	49	206·0	,,	,,	47	170·8	,,	,,	40	182·0
,,	,,	50	208·0	,,	,,	48	172·8	,,	,,	41	183·7

TABLE FOR GAUGING BARRELS—(Contd.).

Interior Length in Centimetres.	Height at the Bung in Centimetres.	Bottom Diameter in Centimetres.	Contents in Litres.	Interior Length in Centimetres.	Height at the Bung in Centimetres.	Bottom Diameter in Centimetres.	Contents in Litres.	Interior Length in Centimetres.	Height at the Bung in Centimetres.	Bottom Diameter in Centimetres.	Contents in Litres.
0·79	0·60	42	185·4	0·80	0·56	49	181·6	0·80	0·60	49	201·1
,,	,,	43	187·2	0·80	0·57	36	163·2	,,	,,	50	203·2
,,	,,	44	189·0	,,	,,	37	164·8	,,	,,	51	205·2
,,	,,	45	190·8	,,	,,	38	166·3	,,	,,	52	207·4
,,	,,	46	192·7	,,	,,	39	167·9	0·80	0·61	38	186·1
,,	,,	47	194·6	,,	,,	40	169·6	,,	,,	39	187·7
,,	,,	48	196·6	,,	,,	41	171·3	,,	,,	40	189·4
,,	,,	49	198·6	,,	,,	42	173·0	,,	,,	41	191·1
,,	,,	50	200·6	,,	,,	43	174·8	,,	,,	42	192·8
,,	,,	51	202·6	,,	,,	44	176·6	,,	,,	43	194·6
,,	,,	52	204·8	,,	,,	45	178·5	,,	,,	44	196·4
0·79	0·61	37	182·2	,,	,,	46	180·4	,,	,,	45	198·3
,,	,,	38	183·8	,,	,,	47	182·3	,,	,,	46	200·2
,,	,,	39	185·4	,,	,,	48	184·3	,,	,,	47	202·1
,,	,,	40	187·0	,,	,,	49	186·4	,,	,,	48	204·1
,,	,,	41	188·7	,,	,,	50	188·5	,,	,,	49	206·2
,,	,,	42	190·4	0·80	0·58	37	169·6	,,	,,	50	208·2
,,	,,	43	192·2	,,	,,	38	171·2	,,	,,	51	210·3
,,	,,	44	193·9	,,	,,	39	172·8	,,	,,	52	212·5
,,	,,	45	195·8	,,	,,	40	174·4	,,	,,	53	214·7
,,	,,	46	197·7	,,	,,	41	176·1	0·80	0·62	38	191·3
,,	,,	47	199·6	,,	,,	42	177·9	,,	,,	39	192·9
,,	,,	48	201·6	,,	,,	43	179·6	,,	,,	40	194·5
,,	,,	49	203·6	,,	,,	44	181·5	,,	,,	41	196·2
,,	,,	50	205·6	,,	,,	45	183·3	,,	,,	42	198·0
,,	,,	51	207·7	,,	,,	46	185·2	,,	,,	43	199·7
,,	,,	52	209·8	,,	,,	47	187·2	,,	,,	44	201·6
,,	,,	53	212·0	,,	,,	48	189·2	,,	,,	45	203·4
0·79	0·62	37	187·3	,,	,,	49	191·2	,,	,,	46	205·3
,,	,,	38	188·9	,,	,,	50	193·3	,,	,,	47	207·3
,,	,,	39	190·5	,,	,,	51	195·4	,,	,,	48	209·3
,,	,,	40	192·1	0·80	0·59	37	174·5	,,	,,	49	211·3
,,	,,	41	194·6	,,	,,	38	176·1	,,	,,	50	213·4
,,	,,	42	195·5	,,	,,	39	177·7	,,	,,	51	215·5
,,	,,	43	197·2	,,	,,	40	179·8	,,	,,	52	217·7
,,	,,	44	199·0	,,	,,	41	181·0	,,	,,	53	219·8
,,	,,	45	200·9	,,	,,	42	182·8	,,	,,	54	222·1
,,	,,	46	202·8	,,	,,	43	184·5	0·80	0·63	38	196·5
,,	,,	47	204·7	,,	,,	44	186·4	,,	,,	39	198·1
,,	,,	48	206·7	,,	,,	45	188·2	,,	,,	40	199·8
,,	,,	49	208·6	,,	,,	46	190·1	,,	,,	41	201·5
,,	,,	50	210·7	,,	,,	47	192·1	,,	,,	42	203·2
,,	,,	51	212·8	,,	,,	48	194·1	,,	,,	43	205·0
,,	,,	52	214·9	,,	,,	49	196·1	,,	,,	44	206·8
,,	,,	53	217·1	,,	,,	50	198·2	,,	,,	45	208·6
,,	,,	54	219·3	,,	,,	51	200·3	,,	,,	46	210·6
0·80	0·55	45	169·1	,,	,,	52	202·4	,,	,,	47	212·5
,,	,,	46	171·0	0·80	0·60	37	179·5	,,	,,	48	214·5
,,	,,	47	173·0	,,	,,	38	181·0	,,	,,	49	216·5
,,	,,	48	175·0	,,	,,	39	182·7	,,	,,	50	218·5
0·80	0·56	40	164·9	,,	,,	40	184·3	0·80	0·64	39	203·4
,,	,,	41	166·6	,,	,,	41	186·0	,,	,,	40	205·1
,,	,,	42	168·3	,,	,,	42	187·7	,,	,,	41	206·8
,,	,,	43	170·1	,,	,,	43	189·5	,,	,,	42	208·5
,,	,,	44	171·9	,,	,,	44	191·3	,,	,,	43	210·3
,,	,,	45	173·8	,,	,,	45	193·2	,,	,,	44	212·1
,,	,,	46	175·7	,,	,,	46	195·1	,,	,,	45	214·0
,,	,,	47	177·6	,,	,,	47	197·1	,,	,,	46	215·9
,,	,,	48	179·6	,,	,,	48	199·1	,,	,,	47	217·8

TABLE FOR GAUGING BARRELS—(Contd.).

Interior Length in Centimetres.	Height at the Bung in Centimetres.	Bottom Diameter in Centimetres.	Contents in Litres.	Interior Length in Centimetres.	Height at the Bung in Centimetres.	Bottom Diameter in Centimetres.	Contents in Litres.	Interior Length in Centimetres.	Height at the Bung in Centimetres.	Bottom Diameter in Centimetres.	Contents in Litres.
0·80	0·64	48	219·8	0·81	0·58	44	183·7	0·81	0·62	43	202·2
,,	,,	49	221·9	,,	,,	45	185·6	,,	,,	44	204·1
,,	,,	50	223·9	,,	,,	46	187·5	,,	,,	45	206·0
0·80	0·65	39	208·8	,,	,,	47	189·5	,,	,,	46	207·9
,,	,,	40	210·5	,,	,,	48	191·5	,,	,,	47	209·9
,,	,,	41	212·2	,,	,,	49	193·6	,,	,,	48	211·9
,,	,,	42	213·9	,,	,,	50	195·7	,,	,,	49	213·9
,,	,,	43	215·7	,,	,,	51	197·8	,,	,,	50	216·0
,,	,,	44	217·5	0·81	0·59	38	178·3	0·81	0·63	39	200·6
,,	,,	45	219·4	,,	,,	39	179·9	,,	,,	40	202·8
,,	,,	46	221·3	,,	,,	40	181·6	,,	,,	41	204·0
,,	,,	47	223·2	,,	,,	41	183·8	,,	,,	42	205·7
,,	,,	48	225·2	,,	,,	42	185·0	,,	,,	43	207·5
,,	,,	49	227·3	,,	,,	43	186·8	,,	,,	44	209·4
,,	,,	50	229·3	,,	,,	44	188·7	,,	,,	45	211·3
0·80	0·66	39	214·3	,,	,,	45	190·6	,,	,,	46	213·2
,,	,,	40	216·0	,,	,,	46	192·5	,,	,,	47	215·2
,,	,,	41	217·7	,,	,,	47	194·5	,,	,,	48	217·2
,,	,,	42	219·4	,,	,,	48	196·5	,,	,,	49	219·2
0·81	0·55	40	162·2	,,	,,	49	198·5	,,	,,	50	221·3
,,	,,	41	163·9	,,	,,	50	200·6	0·81	0·64	39	206·0
,,	,,	42	165·7	,,	,,	51	202·8	,,	,,	40	207·6
,,	,,	43	167·5	,,	,,	52	205·0	,,	,,	41	209·4
,,	,,	44	169·3	0·81	0·60	38	183·3	,,	,,	42	211·1
,,	,,	45	171·2	,,	,,	39	184·9	,,	,,	43	212·9
,,	,,	46	173·2	,,	,,	40	186·6	,,	,,	44	214·8
,,	,,	47	175·1	,,	,,	41	188·3	,,	,,	45	215·7
,,	,,	48	177·2	,,	,,	42	190·1	0·81	0·65	40	213·1
0·81	0·56	40	166·9	,,	,,	43	191·9	,,	,,	41	214·8
,,	,,	41	168·6	,,	,,	44	193·7	,,	,,	42	216·6
,,	,,	42	170·4	,,	,,	45	195·6	,,	,,	43	218·4
,,	,,	43	172·2	,,	,,	46	197·6	,,	,,	44	220·2
,,	,,	44	174·1	,,	,,	47	199·5	,,	,,	45	222·1
,,	,,	45	175·9	,,	,,	48	201·5	0·82	0·55	43	169·6
,,	,,	46	177·9	,,	,,	49	203·6	,,	,,	44	171·4
,,	,,	47	179·8	,,	,,	50	205·7	,,	,,	45	173·4
,,	,,	48	181·9	,,	,,	51	207·8	,,	,,	46	175·3
,,	,,	49	188·9	,,	,,	52	210·0	,,	,,	47	177·3
0·81	0·57	37	166·8	,,	,,	53	212·2	,,	,,	48	179·3
,,	,,	38	168·4	0·81	0·61	38	188·4	0·82	0·56	40	169·0
,,	,,	39	170·0	,,	,,	39	190·1	,,	,,	41	170·7
,,	,,	40	171·7	,,	,,	40	191·7	,,	,,	42	172·5
,,	,,	41	173·4	,,	,,	41	192·5	,,	,,	43	174·3
,,	,,	42	175·2	,,	,,	42	195·2	,,	,,	44	176·2
,,	,,	43	177·0	,,	,,	43	197·0	,,	,,	45	178·1
,,	,,	44	178·8	,,	,,	44	198·9	,,	,,	46	180·1
,,	,,	45	180·7	,,	,,	45	200·8	,,	,,	47	182·1
,,	,,	46	182·7	,,	,,	46	202·7	,,	,,	48	184·1
,,	,,	47	184·5	,,	,,	47	204·7	,,	,,	49	186·2
,,	,,	48	186·7	,,	,,	48	206·7	0·82	0·57	38	170·5
,,	,,	49	188·7	,,	,,	49	208·7	,,	,,	39	172·1
,,	,,	50	190·8	,,	,,	50	210·7	,,	,,	40	173·8
0·81	0·58	37	171·7	,,	,,	51	213·0	,,	,,	41	175·6
,,	,,	38	173·3	,,	,,	52	215·2	,,	,,	42	177·4
,,	,,	39	174·9	,,	,,	53	217·4	,,	,,	43	179·2
,,	,,	40	176·6	0·81	0·62	39	195·3	,,	,,	44	181·1
,,	,,	41	178·3	,,	,,	40	197·0	,,	,,	45	183·0
,,	,,	42	180·1	,,	,,	41	198·7	,,	,,	46	184·9
,,	,,	43	181·9	,,	,,	42	200·4	,,	,,	47	186·8

TABLE FOR GAUGING BARRELS—(Contd.).

Interior Length in Centimetres.	Height at the Bung in Centimetres.	Bottom Diameter in Centimetres.	Contents in Litres.	Interior Length in Centimetres.	Height at the Bung in Centimetres.	Bottom Diameter in Centimetres.	Contents in Litres.	Interior Length in Centimetres.	Height at the Bung in Centimetres.	Bottom Diameter in Centimetres.	Contents in Litres.
0·82	0·57	48	189·0	0·82	0·62	40	199·4	0·88	0·57	44	183·3
,,	,,	49	191·0	,,	,,	41	201·1	,,	,,	45	185·2
,,	,,	50	193·1	,,	,,	42	202·9	,,	,,	46	187·2
0·82	0·58	38	175·4	,,	,,	43	204·7	,,	,,	47	189·1
,,	,,	39	177·1	,,	,,	44	206·6	,,	,,	48	191·3
,,	,,	40	178·8	,,	,,	45	208·5	,,	,,	49	193·4
,,	,,	41	180·5	,,	,,	46	210·5	,,	,,	50	195·5
,,	,,	42	182·3	,,	,,	47	212·5	0·83	0·58	39	179·2
,,	,,	43	184·1	,,	,,	48	214·5	,,	,,	40	181·0
,,	,,	44	186·0	,,	,,	49	216·6	,,	,,	41	182·7
,,	,,	45	187·9	0·82	0·63	40	204·8	,,	,,	42	184·5
,,	,,	46	189·9	,,	,,	41	206·5	,,	,,	43	186·4
,,	,,	47	191·9	,,	,,	42	208·3	,,	,,	44	188·3
,,	,,	48	193·9	,,	,,	43	210·1	,,	,,	45	190·2
,,	,,	49	196·0	,,	,,	44	212·0	,,	,,	46	192·1
,,	,,	50	198·1	,,	,,	45	213·9	,,	,,	47	194·2
,,	,,	51	200·3	0·82	0·64	40	210·2	,,	,,	48	196·3
0·82	0·59	38	180·5	,,	,,	41	211·9	,,	,,	49	198·4
,,	,,	39	182·1	,,	,,	42	213·7	,,	,,	50	200·5
,,	,,	40	183·8	,,	,,	43	215·6	,,	,,	51	202·7
,,	,,	41	185·5	,,	,,	44	217·4	0·83	0·59	39	184·3
,,	,,	42	187·3	,,	,,	45	219·3	,,	,,	40	186·0
,,	,,	43	189·2	0·83	0·54	40	161·5	,,	,,	41	187·8
,,	,,	44	191·0	,,	,,	41	163·3	,,	,,	42	189·6
,,	,,	45	192·9	,,	,,	42	164·1	,,	,,	43	191·4
,,	,,	46	194·9	,,	,,	43	166·9	,,	,,	44	193·3
,,	,,	47	196·9	,,	,,	44	168·8	,,	,,	45	195·3
,,	,,	48	198·9	,,	,,	45	170·7	,,	,,	46	197·3
,,	,,	49	201·0	,,	,,	46	172·7	,,	,,	47	199·3
,,	,,	50	203·1	,,	,,	47	174·7	,,	,,	48	201·3
,,	,,	51	205·3	0·83	0·55	38	162·8	,,	,,	49	203·5
0·82	0·60	39	187·2	,,	,,	39	164·5	,,	,,	50	205·6
,,	,,	40	188·9	,,	,,	40	166·2	,,	,,	51	207·8
,,	,,	41	190·7	,,	,,	41	168·0	,,	,,	52	210·0
,,	,,	42	192·4	,,	,,	42	169·8	0·83	0·60	39	189·5
,,	,,	43	194·3	,,	,,	43	171·6	,,	,,	40	191·2
,,	,,	44	196·1	,,	,,	44	173·5	,,	,,	41	192·8
,,	,,	45	198·0	,,	,,	45	175·5	,,	,,	42	194·8
,,	,,	46	200·0	,,	,,	46	177·4	,,	,,	43	196·6
,,	,,	47	202·0	,,	,,	47	179·5	,,	,,	44	198·5
,,	,,	48	204·0	,,	,,	48	181·5	,,	,,	45	200·5
,,	,,	49	206·1	0·83	0·56	38	167·7	0·83	0·61	40	196·5
,,	,,	50	208·4	,,	,,	39	169·3	,,	,,	41	198·2
,,	,,	51	210·4	,,	,,	40	171·1	,,	,,	42	200·0
,,	,,	52	212·6	,,	,,	41	172·8	,,	,,	43	201·9
,,	,,	53	214·9	,,	,,	42	174·6	,,	,,	44	203·8
0·82	0·61	39	192·4	,,	,,	43	176·5	,,	,,	45	205·7
,,	,,	40	194·1	,,	,,	44	178·4	,,	,,	46	207·7
,,	,,	41	195·8	,,	,,	45	180·3	,,	,,	47	209·7
,,	,,	42	197·6	,,	,,	46	182·3	,,	,,	48	211·8
,,	,,	43	199·5	,,	,,	47	184·3	,,	,,	49	213·9
,,	,,	44	201·3	,,	,,	48	186·4	,,	,,	50	216·0
,,	,,	45	203·2	,,	,,	49	188·5	0·88	0·62	40	201·8
,,	,,	46	205·2	0·88	0·57	38	172·6	,,	,,	41	203·6
,,	,,	47	207·2	,,	,,	39	174·2	,,	,,	42	205·4
,,	,,	48	209·2	,,	,,	40	176·0	,,	,,	43	207·2
,,	,,	49	211·3	,,	,,	41	177·7	,,	,,	44	209·1
,,	,,	50	213·4	,,	,,	42	179·5	,,	,,	45	211·1
0·82	0·62	39	197·7	,,	,,	43	181·4	,,	,,	46	213·0

TABLE FOR GAUGING BARRELS—(*Contd.*).

Interior Length in Centimetres.	Height at the Bung in Centimetres.	Bottom Diameter in Centimetres.	Contents in Litres.	Interior Length in Centimetres.	Height at the Bung in Centimetres.	Bottom Diameter in Centimetres.	Contents in Litres.	Interior Length in Centimetres.	Height at the Bung in Centimetres.	Bottom Diameter in Centimetres.	Contents in Litres.
0·83	0·62	47	215·1	0·83	0·63	41	209·0	0·84	0·47	36	125·7
,,	,,	48	217·1	,,	,,	42	210·8	,,	,,	37	127·3
,,	,,	49	219·2	,,	,,	43	212·7	,,	,,	38	128·9
,,	,,	50	221·4	,,	,,	44	214·6	,,	,,	39	130·6
0·83	0·63	40	207·3	,,	,,	45	216·5	,,	,,	40	132·5

N.B.—To bring centimetres to inches multiply by 0·393 ; to bring litres to English gallons divide by 4·543 ; and to bring litres to American gallons divide by 3·635.—*Tr.*

CONTINENTAL REGULATIONS UNDER WHICH MINERAL OILS ARE ACCEPTED FOR TRANSPORT.

We append the clauses relating to the conveyance of petroleum —in the arrangement agreed to on 16th July, 1895, for the regulation of this traffic between Switzerland, Germany, Belgium, France, Italy, Luxembourg and Russia, and which have been in force on the continent since 1st July, 1896 :—

XX.

Crude and rectified petroleum, if it have at least a density of 0·780 at 17·5° of the centigrade thermometer (Celsius), 63·5 Fahrenheit, or if it do not give off inflammable vapours at a temperature of at least 21° of the centigrade thermometer (Celsius), 69·8° Fahrenheit, with Abel's apparatus, and at a pressure of the barometer of 760 millimetres reduced to the level of the sea.

Oils prepared from lignite tar, if they have at least the above specific gravity. Solar oils, photogen, etc.

Oils prepared from coal-tar (benzol, toluol, xylol, cumol, etc.), as well as essence of mirbane (nitro-benzole), are subject to the following regulations :—

1. These substances, unless conveyed in waggons specially constructed for the purpose, can only be transported :—

    (*a*) In particularly good sound casks.

    (*b*) Or in tight resistent metallic vessels.

    (*c*) Or in glass or earthenware vessels, provided always that the prescriptions indicated below be observed :—

        (*aa*) When several vessels are packed together in one package, they should be packed firmly in strong wooden cases, lined with straw, hay, chaff, sawdust, infusorial earth (kieselguhr), or other mobile substances.

        (*bb*) When the vessels are packed singly, they may be sent in crates, or sound vats, provided with well-fitting covers and handles, and lined with a sufficient quantity of packing materials. The cover consisting of straw, rushes, reeds, or analogous material, ought to be impregnated with milk of lime ; or a cream of clay or other equivalent substance mixed with soluble glass. The gross weight of the single package should not exceed 60 kilogrammes (132 lb.) in the case of glass vessels, nor 75 kilogrammes (165 lb.) in the case of earthenware vessels.

2. Vessels which have become damaged in transit will be immediately unloaded and sold, with the remaining contents, to the best advantage of the sender.

3. Conveyance is only made in open waggons. If the operations of passing through the Customs necessitates waggons provided with lead-sealed sheets, the goods will not be accepted for transport.

.4. The regulations in No. 3 preceding are also applicable to the casks and other vessels in which these substances have been transported. (*Returned empties.*) They should always be declared as such.

5. In regard to packing with other goods, see No. XXXV.

6. It ought to be declared on the consignment note that the substances indicated under paragraphs 1 and 2 of this section have a density of at least 0·780, or that the petroleum is of the quality indicated in the first paragraph of this Section (XX.), in regard to flash-point. When such a declaration is not on the consignment note, the goods become subject to the conditions of conveyance prescribed in No. XXII.

XXI.

Crude and Rectified Petroleum, Petroleum-naphtha, and Products of the Distillation of Petroleum and Petroleum-naphtha.—When these substances have a specific gravity of less than 0·780 and more than 0·680, at a temperature of 17·5° of the centigrade thermometer (63·5° Fahrenheit), (benzine, ligroin, and spirits for cleaning), they are subject to the following regulations :—

Paragraphs 1, 2, 3, 4, and 5 of XX. (which see) apply to goods coming under this section, but gross weight of vessels packed singly must not exceed 40 kilogrammes (88 lb.).

6. In loading or unloading, the crates or tubs containing glass carboys (*ballons en verre*) should not be carried on trucks, nor carried on the shoulders, nor on the back, but simply by the handles.

7. The crates and tubs ought to be solidly packed or wedged together, and attached to the sides of the waggon. The packages should not be loaded one above the other, but side by side, and without superposition.

8. Each individual package ought to bear, on a prominent label, the word "inflammable," printed on a red ground. The crates or vats containing glass or earthenware vessels, ought, moreover, to be labelled " To be carried by hand " (*à porter à la main*). The waggons should carry a red label bearing the inscription, "To be handled with caution " (*à manœuvrer avec precaution*).

9. It ought to be declared on the consignment note that the substances named in the first paragraph of the present section have a specific gravity of less than 0·780 and more than 0·680 at a temperature of 17·5° centigrade (63·5° Fahrenheit). When that declaration is not borne on the consignment note the goods become subject to the conditions of conveyance prescribed in No. XXII. regarding petroleum spirit, etc.

XXII.

Petroleum spirit (gasoline, neoline) and other easily inflammable products prepared with petroleum-naphtha, or with lignite tar, when these substances have a specific gravity of 0·680, at least, at a temperature of 17·5° centigrade (63·5° Fahrenheit) are subject to the following conditions :—

1. These substances can only be transported—

    (a) In tight resistent metallic vessels.

    (b) Or in glass or earthenware vessels in the manner described in (*aa*) and (*bb*), as in XX. The gross weight of single packages must not exceed 40 kilogrammes (88 lb.).

    (c) In hermetically sealed tank-waggons (perfectly tight tank-waggons). Pars. 2, 3, 4, 5 of XX. and 6, 7, and 8 of XXI. also apply (which see in each instance).

MEMORANDA.

THEORETICAL PART.

USEFUL FORMULÆ.

FACTORS COMMONLY USED IN CALCULATIONS.

Various factors of π ratio of the circumference to the diameter.

$$\pi = 3\cdot1415926 = \tfrac{2\cdot2}{7} \text{ approximately,}$$
$$\pi^2 = 9\cdot869604,$$
$$\frac{1}{\pi} = 0\cdot318310,$$
$$\sqrt{\pi} = 1\cdot772453.$$

Various factors of g acceleration of gravity at Paris.

$$g = 9^{m}\cdot80896,$$
$$\sqrt{g} = 3\cdot13,$$
$$\sqrt{2g} = 4\cdot429,$$
$$\sqrt{\frac{1}{2g}} = 0\cdot226,$$
$$\frac{1}{\sqrt{g}} = 0\cdot320.$$

The value of g decreases by $0\cdot00020$ every hundred metres ascended; it increases by $0\cdot00088$ as the latitude increases 1°.

Square roots :—

$$\sqrt{2} = 1\cdot4142,$$
$$\sqrt{3} = 1\cdot7321,$$
$$\sqrt{5} = 2\cdot2361,$$
$$\sqrt{6} = 2\cdot24480,$$
$$\sqrt{10} = 3\cdot1623,$$
$$\sqrt{\frac{1}{2}} = 0\cdot707,$$
$$\sqrt{\frac{1}{3}} = 0\cdot577,$$
$$\sqrt{\frac{1}{5}} = 0\cdot447.$$

Cube roots :—

$$\sqrt[3]{2} = 1\cdot2599,$$
$$\sqrt[3]{3} = 1\cdot4422.$$

ALGEBRA.

Equation of the second degree :—

$$ax^2 + bx + c = 0.$$
$$x = \frac{-b \pm \sqrt{b^2 - 4ac}}{2a},$$
$$x' + x'' = \frac{-b}{a},$$
$$x'x'' = \frac{c}{a}.$$

PROGRESSIONS.

a, 1st term ; r, ratio of an arithmetic progression ; q, ratio of a geometrical progression ; n, number of terms ; l, the principal term ; S, sum of the terms from a to l.

ARITHMETICAL PROGRESSIONS.

$$l = a + (n - 1)r,$$
$$S = (a + l)\frac{n}{2}.$$

GEOMETRICAL PROGRESSIONS.

$$l = aq^{n-1},$$
$$S = \frac{lq - a}{q - 1} = a\frac{q^n - 1}{q - 1}.$$

DEPTH OF A WELL.

t, the time elapsed from the time a stone commenced to fall freely until the moment the sound is heard ; g, acceleration of gravity $= 9^m\cdot80896$; a, velocity of sound $= 333^m$; the depth $x = at + \dfrac{a^2}{g} - \dfrac{a}{g}\sqrt{a^2 - 2agt}$, say for $t = 1''$, $2''$, $3''$, $4''$, $5''$, about $4^m\cdot9$, $19^m\cdot6$, $44^m\cdot0$, $78^m\cdot5$, $122^m\cdot6$.

GEOMETRY.

Lines.—Circumference of radius $= 2\pi r$.

Length of arc of 1° in the circumference of radius :—

$$1 = 0\cdot017453293 ; 1 \text{ grade} = 0\cdot015707963.$$

Angle at the centre corresponding to the arc of length S, in the circumference of the radius :—

$$r, \frac{S}{r} \cdot 57^\circ, \quad 29578 = \frac{S}{r} \cdot 63^\circ, 6620.$$

Area of Plane Surfaces.—Triangles ; h being the height corresponding to the side a, and $2p$ the perimeter, the area.

$$S = \tfrac{1}{2}ah, \text{ or } S = \sqrt{p(p-a)\,(p-b)\,(p-c)}.$$

Curved Surfaces.—Surface of cylinder or prism with parallel bases = length of axis multiplied by the perimeter or circumference of the base.

Surface of truncated cylinder (or prism), if it be a right figure, = the product of the perimeter of the lower base by the distance from the centre of gravity of the area of the bases, if the figure be not a right figure this product must be multiplied by the sine of the slope of the axis to the base.

Surface of right cone = $\pi r l$; l axis ; r radius of the base.

Right truncated cone = $\pi l(r + r')$; r and r' radii of the bases.

Sphere = $4\pi r^2$.

Zone of a sphere = $2\pi r h$; h, the distance of the two planes limiting the zone.

Spherical spindle $\dfrac{\pi r^2 a}{90}$ a number of degrees of the spindle.

Spherical triangle $S = \pi r^2 \dfrac{S - 180}{180}$; S is the sum of the three angles of the triangle in degrees.

Surface of revolution = $2\pi r l$; l, the length of the axis ; r, the distance from the centre of gravity of this line to the axis of rotation.

Volume.—Prism or cylinder. Area of base multiplied by the height, if the bases be parallel ; if not, the area of the base multiplied by the distance from the centre of gravity of the other section to the base.

Truncated triangular prism $v = B\left(\dfrac{h + h' + h''}{3}\right)$; B being the area of the base ; h, h', h'', perpendiculars from the three summits to the base.

Any Truncated Prism.—The area of the base by the distance from the centre of gravity of the upper section to the plane of the base.

Pyramidal Cone.—The area of the base by $\tfrac{1}{3}$ of the height. Truncated pyramid with parallel bases :—

$$v = \frac{h}{3}(B + b + \sqrt{Bb}) ; \, h, \text{ height} ; B, b, \text{ bases.}$$

Heaps of stones, excavated matter, rubbish with rectangular

bases between parallel planes at a distance h have as sides a, b, $a'b'$ respectively parallel $v = \dfrac{h}{6} [(2ab + a'b') + ab + ba']$.

A solid in the form of a roof (b' is *nil*) $v = \dfrac{hb}{6} (2a + a')$.

Truncated cone with parallel bases with radius r and R :—

$$v = \tfrac{\pi h}{3}(R^2 + r^2 + Rr).$$

Sphere from diameter and radius r. $v = \dfrac{4}{3}\pi r^3 = \dfrac{1}{6}\pi d^3$.

Segment of a sphere with parallel bases $v = h\left(\dfrac{B + b}{2} + \dfrac{\pi h^2}{6}\right)$; B and b, bases ; h, distances of bases.

Spherical cap $v = h\dfrac{B}{2} + \dfrac{\pi h^2}{6}$.

Spherical sector engendered by a circular sector revolving on an axis. The third of the product of the radius by the surface of the zone.

Ellipsoid $v = \dfrac{4}{3}\pi abc$; $2a$, $2b$, $2c$, length of the three axes.

Revolving Solid.—$v = 2\pi rS$; S the revolving area ; r, the distance of the centre of gravity of this area from the axis of rotation.

Timber Measuring.—c, average circumference of radius r ; the average section is $\pi r^2 = \dfrac{c^2}{4\pi} = 0\cdot0796c^2$; the average square piece that can be got out of this tree, or theoretical purchasing section, would be $a^2 = 0\cdot05066c^2$.

In practice timber is bought reduced to the $\frac{1}{5}$, $\frac{1}{4}$, or $\frac{1}{3}$, that is to say, in order to obtain the side of the square to pay on there is deducted from the circumference $\frac{1}{5}$, $\frac{1}{4}$, or $\frac{1}{3}$, and the $\frac{1}{4}$ of the remainder is taken.

Gauging of Barrels.—l, inside length ; D, diameter at the bulge ; d, at the cross grooves ; $v = \dfrac{\pi l}{15}\left(2D^2 + Dd + \dfrac{3}{4}d^2\right)$, or approximately $\dfrac{\pi l}{36}(d + 2D)^2$.

To gauge a barrel in the process of being emptied, insert a graduated rule by the bung and read the great diameter D and the moistened height h ; the relation of $\dfrac{v}{V}$ is given below in reference to the different values of $\dfrac{h}{D}$.

$\frac{h}{D}$: 1	0·9	0·8	0·7	0·6	0·5	0·4	0·3	0·2	0·1
$\frac{v}{V}$: 1	0·95	0·86	0·75	0·63	0·50	0·37	0·25	0·14	0·05

CENTRE OF GRAVITY OF LINES.

Perimeter of a Triangle.—At the centre of the circle inscribed in the triangle which joins the middle of the sides.

Arcs of a Circle. — Subtending the centre of the arc at a distance from the centre equal to $\frac{rc}{l}$ (c, cord ; l, length of the arc ; r, radius ;) or about $\frac{1}{3}$ of the arrow starting from the summit.

CENTRE OF GRAVITY OF SURFACES.

Area of a Triangle.—At the third part of the line starting from the base which connects the middle of that base with the apex.

Area of a Parallelogram.—At the intersection of the diagonals.

Area of a Trapezium—On the straight line joining the middle of the bases at a distance from the large base B equal to $\frac{h}{3}\left(\frac{B+2b}{B+b}\right)$, approximately : $h\left(\frac{1}{3}+\frac{1b}{6B}\right)$; B and b are the bases, h the heights. Graphically, prolong each of the bases by a length equal to that of the other base, and in an inverse sense ; join the two points thus obtained by a straight line, the point of intersection of which with the straight line joining the bases gives the centre of gravity.

Area of a Quadrilateral.—Mark the middle I of one of the diagonals, and draw DO' to the other = AO. The centre of gravity is on IO' in G at $\frac{1}{3}$ of its length starting from I.

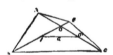

FIG. 152.—Area of a quadrilateral.

Area of any Polygon.—Decompose it into two systems of two quadrilaterals, the centres of gravity of which are determined. The centre of gravity of the polygon in question is at the point of intersection of the lines which join the centres of gravity of the polygons of the two systems.

Area of Half a Circle.—Distance from the centre equal to $\frac{4r}{3\pi}$ = 0·424 (r, radius).

Area of a Circular Sector.—Distance from the centre equal to $\frac{2rc}{3l}$; c, cord ; l, length of the arc.

Area of a Semi-ellipse bounded by One Diameter. — On the conjugated diameter at a distance from the centre equal to $\frac{4b}{3\pi}$.

Area of a Segment or Zone of a Sphere.—At the middle of the height.

VOLUMES (SOLIDS).

Prism or Cylinder.—At the middle of the straight line which joins the centre of gravity of the two bases.

Pyramid or Cone.—At the fourth part starting from the base of the straight line which joins the centre of gravity of the base with the vertex.

Frustum of a Pyramid with Parallel Bases.—On the straight line joining the centre of gravity of the base, at a distance from the greater base B, equal to $\dfrac{h}{4} \times \dfrac{3b + B + 2\sqrt{Bb}}{B + b + \sqrt{Bb}}$.

Frustum of a Cone with Parallel Bases.—On the axis, at a distance from the great base equal to :—

$$\frac{h}{4} \times \frac{(R + r)^2 + 2r^2}{(R + r)^2 - Rr}$$

Sector of a Sphere.—Distance from the centre :—

$$\frac{3}{4}\left(r - \frac{f}{2}\right); f, \text{ arrow of the arc.}$$

Segment of a Sphere.—Distance from the centre equal to :—

$$\frac{\left(r - \frac{f}{2}\right)^2}{r\frac{f}{3}}$$

MECHANICAL DATA AND FORMULÆ.—GENERAL FORMULÆ AND DEFINITIONS.

Fall of Bodies.—h, space traversed in the time t; v, velocity at the end of the time ; g, acceleration of gravity (double the space traversed in 1″) :—

$$h = \frac{1}{2} gt^2 \qquad v = \sqrt{2gh}.$$

Simple Pendulum.—T, duration of an oscillation (a simple passage from arc to arc[1]); l, length of the pendulum; $T = \pi\sqrt{\dfrac{l}{g}}$. The length of the simple pendulum beating the minute at the Paris observatory is : $l = 0·9939$ metre at the level of the sea ; at the latitude λ, $l = 0·991026 + 0·005072 \sin^2\lambda$.

Conical Pendulum.—$T = 2\pi\sqrt{\dfrac{h}{g}}$, h being the projection on the vertical of the rod of the pendulum.

Kilogrammetre.—Unit of mechanical work corresponding to the

---

[1] *De l'arc de l'arc.*

elevation of 1 kilogramme 1 metre in height in 1 second; abridged form *kgm*.

Tonnemetre.—Elevation of 1,000 kilogrammes 1 metre in height in 1 second; abridged form *tm*.

Mass of a Solid Body P ; $M = \dfrac{P}{g}$.

Motive Force.—F, intensity of the force; M, mass of a body; J, acceleration due to the force $F = MJ$.

Work done by a Force.—Product of that force by the distance traversed by its point of application in the direction of the force.

Momentum or Quantity of Motion $= MV$ (V, velocity).

Vis Viva $= MV^2$.

Velocity.—Product of the intensity of the force by the duration of its action, $V = Ft$.

DIFFERENT METHODS OF CARTAGE.

A man, working piece work, lifts with a shovel, and loads into a barrow, 12 to 16 cubic metres of earth, in 10 hours.

When the earth is thrown, horizontally, from 2 to 4 metres, or raised 1·60 metres, or loaded into a cart, this cube is only 10 cubic metres. Reduce these numbers by $\frac{1}{2}$ for day work.

Length of relay of barrow: horizontal 30 metres; on a plank 1 in 12, 20 metres.

Number n of relays of barrow between 2 points. D being the horizontal projection of the run, h, the difference of level between the two horizontal points, or in descending slope: $h \bar{\angle} o, n = \dfrac{D}{30}$.

On ascending plane greater than $\dfrac{1}{12} : 12h \angle D, n = \dfrac{D}{20} = 0\!\cdot\!6\,h = \dfrac{h}{1\cdot67}$;

on a less highly inclined plane than $\dfrac{1}{12} : 12h \angle D, n = \dfrac{D + 6h}{30} = \dfrac{D}{30} + \dfrac{4}{5}$.

Capacity of a barrow: 0·0333 cubic metre $= \dfrac{30}{1^{\,m.\,c.}}$.[1] Weight of a barrowful of earth: 85 to 90 kilogrammes.

A horse in harness or loaded can only do from 48 to 56 kilometres per day.

TRACTION OF VEHICLES.

Experimental Data.—The friction of paved or macadamised roads opposed to the movement of vehicles and reduced to the axis of the axle, is directly proportional to the pressure, and inversely to the radius of the wheels, independent of the number of the

---

[1] This formula is apparently reversed; it should read $\dfrac{1^{\,m.\,c.}}{30}$. The authors evidently mean to say that there are 30 barrow loads in a cubic metre.—*Tr.*

wheels and almost of the width of the rims. On compressible ground, it decreases as the width of the rim increases. On soft compressible ground, at ordinary speeds, it is independent of that speed.

On macadamised or paved roads it increases with the speed and almost proportionally thereto, starting from 1 metre per second; it is so much less in proportion as the vehicle is well suspended and the road uniform.

At a walking pace, on a compact uniform pavement, it is only ¾ of that of the best macadamised roads; at the trot this still holds good with well-suspended vehicles, and greater on a mediocre pavement than on macadamised road.

The inclination of the draught, corresponding to the maximum useful effect ought in general to increase with the resistance of the ground, and to be greater the smaller the radius of the fore wheels.

The most favourable varies, for different speeds, from 6° 15′ to 7° 10′ on sand and gravel ground; it is about 1° on macadamised roads. In practice the former numbers are aimed at because it is in bad ground that it is necessary to make the most of the horses' strength.

The action of several horses harnessed to a vehicle is lower than the sum of their individual efforts owing to their not all pulling together and the arrangement of the traces. For 4, 6, or 8 horses harnessed in pairs the work of each is respectively proportional to the numbers 8, 7, 6.

Do not exact from a horse in harness working continuously 8 to 10 hours per day a greater effort than 40 kilometres.

RATIO OF THE TRACTION TO THE LOAD DRAWN: VEHICLE IN-
CLUDED.

	Nature of the Supposed Horizontal Road.	Ratio of the Traction to the Load Drawn.
Ground	Natural, not trodden, clayey, but dry . .	0·250
	Natural, untrodden, sandy and chalky . .	0·165
	Firm, trodden, very uniform . . .	0·040
Causeway or Highway	Sand or pebbles, newly laid . . .	0·125
	Macadamised { in the ordinary condition	0·080
	{ perfectly maintained, easy	0·033
	Paved in the ordinary manner { walking pace .	0·030
	and the vehicle suspended . { quick trot .	0·070
	Paved with sandstone blocks, { walking pace .	0·025
	well kept . . . { quick trot .	0·060
	Unplaned oak blocks	0·022
Tramways	With flat cast-iron ruts, or very hard uniform stone slabs	0·010
	With projecting track in good condition . .	0·007
	With projecting track kept in perfect condition, and the axles always oiled	0·005

The weight of the vehicle generally varies between $\frac{1}{3}$ and $\frac{1}{4}$ of the total load.

The number of horses required to draw horizontally, on a road of any given nature, a vehicle the weight of which, including the load, is known, as well as the distance to be traversed, may be determined from the indications of the table and the preceding data.

Say a load of 3,000 kilogrammes on a vehicle of 1,000 kilogrammes; the traction effort on firm trodden very uniform ground would be $4000 \times 0\cdot040 = 160$ kilogrammes.

A horse, at a walking pace, exerting an effort of 70 kilogrammes, it requires $\frac{160}{70}$, that is to say 2·3, say 3 horses.

With 3 horses, the daily work will be $3 \times 2\cdot168\cdot000 = 6\cdot504\cdot000$, which corresponds to a journey of $\frac{6\cdot504\cdot000}{160} = 40\cdot650$ metres, about 40·500 kilogrammes.[1]

RESISTANCE OF SOLIDS.

Resistance of Ropes.—The load P in kilogrammes which cannot be exceeded with a hemp rope of circumference C in centimetres is $P = 20C^2$.

If the rope has to be moistened take

$$P = \frac{2}{3}\,20C^2.$$

HORIZONTAL PIECE MORTICED AT ONE OF ITS EXTREMITIES.

Charged with a weight $p \times X$ spread uniformly and urged at its free extremity by a weight P.

$$R\frac{I}{n} = \frac{pX^2}{2} + PX,$$
$$f = \frac{I}{EI}\left(\frac{pX^4}{8} + \frac{PX^3}{3}\right).$$

R, permanent resistance to extension to be attributed to the material of the prism per cubic metre; I, moment of inertia of the transverse section of the prism with reference to the neutral fibre;

---

[1] The translator does not follow the reasoning in this paragraph. According to the authors' own data, the three horses, instead of exerting a collective effective force of 210 kilogrammes, exert a traction force of 160. There is thus, supposing each horse to exert an equal effective force, a loss of $\frac{1}{7}$ of the total force to be exerted. Again, 40 kilometres is the distance the authors allow to be covered in a day; the work done in a day by the three horses would therefore appear to be a traction force of 160 kilogrammes exerted over a space of 40 kilometres. Again, the authors make no allowance for the nature of the vehicle and the size and breed of the horses. Two horses in the kind of carts known in London as a railway van, but which is in reality a two-horse spring cart, can easily traverse the streets of London at the trot with a load of 4 to $4\frac{1}{2}$ tons over above that of the vehicle.

n, distance from the surface of the neutral fibres of the point of section furthest distant from this surface ; E, modulus of elasticity of the material of the prism ; X, length of the unmorticed portion of the piece ; p, intensity per unit of length of the weight spread uniformly over the length X ; f, the supporting post (*flèche de la piece*).

HORIZONTAL PIECE RESTING ON TWO POINTS OF SUPPORT.

Loaded with a weight $p \times X$ spread uniformly on the length X separating the points of support and acted on in its middle part by a weight P.

$$R \frac{I}{n} = \frac{pX^2}{8} + \frac{PX}{4},$$

$$f = \frac{1}{48EI}\left(\frac{5pX^4}{8} + PX^3\right).$$

Horizontal piece morticed at its two extremities and loaded as above.

$$R \frac{I}{n} = \frac{pX^2}{8} + \frac{PX}{4} - \frac{py^2}{2} - \frac{py}{2},$$

$$f = 192EI + \left(\frac{pX^4}{2} + PX^3\right).$$

FIG. 158.—Resistance of solids.

y, distance of the two points of flexure which the piece presents to the vertical plane of section passing through its middle.

$$y = \frac{1}{2}\left(-\frac{P}{p} + \sqrt{\frac{P^2}{p^2} + \frac{PX}{p} + \frac{X^2}{3}}\right).$$

for

P = 0 $y = \dfrac{X}{\sqrt{12}} = 0{\cdot}288X$,

P = 0·2pX $y = 0{\cdot}283X$,

P = 0·5pX $y = 0{\cdot}270X$,

P = pX $y = 0{\cdot}264X$,

P = 2pX $y = 0{\cdot}252X$,

P = 10pX $y = 9{\cdot}252X$,

P = pX∞ or $p = 0$ $y = 0{\cdot}25X$,

For P = 0, $R\dfrac{I}{n} = \dfrac{pX^2}{12}$, and for 0, $R\dfrac{I}{n} = \dfrac{PX}{8}$.

The formula

$$R \frac{I}{n} = \frac{pX^2}{12}\frac{PX}{8},$$

is sometimes used to calculate the balancing of the formulæ, which gives too low a result. The error committed, in the second member, is lower than the product of the multiplication of the two weights, P or pX, which is found to be the greatest by the factor $\dfrac{X}{10{\cdot}000}$.

GLOSSARY OF TECHNICAL TERMS

(ENGLISH AND FOREIGN)

USED IN THE PETROLEUM INDUSTRY.

Acre.—A measure of superfices employed in America, and about equal to 40 ares (exactly 40·4671 ares).

Auger Stem.—French, *Tige lourde*; German, *Schwer-Stange.*

Avancement (*Fr.*).—Progress made in boring during a fixed time.

Ball-wheel.—German, *Förderwelle.* In rope boring, the wheel on the axis of the windlass used for hoisting or lowering the drilling tools, and which is driven from the band-wheel by the bull-rope.

Band-wheel.—German, *Gross reimschiebe.* In rope boring the wheel —the circumference of which is hollowed into a groove in which the band from the engine runs—which drives the walking beam, etc.

Bassin (*Fr.*).—English, *Oil field.* The petroleum deposits belonging to any single petroliferous district taken collectively.

Barils (*Fr.*).—English, *Barrels.* There are three kinds of petroleum barrels: (1) The American barrel of 160 litres, of about 8 barrels to the ton; (2) the Galician American barrel of 200 litres, of about 6 barrels to the ton; (3) the large barrel containing 4 hectolitres, 3 of which make a ton.

Bergoel (*Ger.*).—French. *Petrole brut;* English, *Crude oil, sp. "Rock" Oil.*

Bit.—The cutting part of the drill. French, *Trepan;* German, *Meisel.*

Bohrloch (*Ger.*).—French, *Trou de sonde;* English, *Bore hole.*

Bohrmeister (*Ger.*).—French, *Maitre-sondeur;* English, *Foreman borer.*

Borts.—White diamonds, defective as jewels and utilised in boring.

Bulk (Transport).—French, " *Transporter en vrague* ".

Cal (*Slav*).—Crude petroleum mixed with sand.

Carbons.—French, *Diamant noirs* (black diamonds).

Casing.—French, *Tubage;* German, *Bohrröhren.*

Centre-iron.—(Walking beam and its support). German, *Schwen gel-lager.*

Clef (mettre sous) (*Fr.*).—To " cap " a spouting well.

Derrick.—Tower or spire-shaped scaffolding used in drilling. German, *Bohrthurm.*

Dessiatine.—Measure of superfices used in Russia; its value is 1·03 hectares.

Dollar.—American coin worth about 5·1 francs, say 4s. 2d.

Dreifuss (Ger.).—French, *Derricks à trois pieds;* English, *Tripod (three-legged) derrick.*

Drilling hook.—French, *Attache du balancier;* German, *Schwen-gelhaken.*

Enclouer (*Fr.*).—To render a well unworkable by accident. To render the further continuation of work impossible.

Erdoel (*Ger.*).—Petroleum.

Flowing Well.—A spouting well the eruption of which does not rise higher than the orifice of the casing.

Frei fall (*Ger.*).—French, *Chute libre;* English, *Free fall.*

Gallon.—English measure of capacity equal to 4·543358 litres. The American gallon is equal to 3·785 litres.

Garnetz (*Russ.*).—Measure of capacity; the Russian garnetz is equal to 3·277 litres, say 5·6 English pints; the Polish garnetz to 4 litres, say 7·04 English pints.

Hurd (*Slav*).—A gang of working miners working piece work.

"Jars."—French, *Glissières;* German, *Rutscheere.*

"Joch" (*Polish*).—Measure of superfices used in Galicia; a square the sides of which are 80 metres, say 262·4 feet in length.

Kalpack (*Russian*).—Oil well "cap".

Klafter (*Ger.*).—Old measure still used in Galicia = 1·93 metres, say 6·23 feet.

Knuil[1] (*English*).—Piece attached to the pinion of the rods in rotatory drilling.

Krahn (*Ger.*).—French, *Treuil;* English, *Crane, windlass, winch.* (French engineers apparently also apply the term *treuil* synonymously with "rig" to the whole aggregate of drilling machinery.)

Land Interest.—Proprietor's ground rent over a petroleum concession or undertaking.

Link.—French, *Alésoir.*

Loffel (*Ger.*).—French, *Curette;* English, *Sand-pump.*

Matka (*Polish*).—Sudden eruption of ozokerit.

Meisel (*Ger.*).—French, *Trepan;* English, *Bit.*

Nachlass-Schraube (*Ger.*).—French, *Vis de réglage;* English, *Temper Screw.*

Oil field.—French, *Bassin pétrolifère.*

Oil lease.—French, *Concession du droit de forage.*

[1] ? Swivel.

Oil man.—A man engaged in searching for, or in the extraction of, petroleum.

Pipe line.—French, *Ligne de tuyaux.*

Poche (*Fr.*).—English, *Pocket ;* subterraneous cavern filled with petroleum.

Poud (*Russ.*).—Russian measure of weight = 16·372 kilo.

Pumping-well.—A non-flowing well from which the petroleum is extracted by a pump.

Rig.—The entire boring equipment.

Ropa (*Polish*).—Crude petroleum. *Crude Oil.*

Rope Socket.—Piece of screwed metal ending the cable for raising and lowering the rods. German, *Seilhülse.*

Rutscheere (*Ger.*).—French, *Glissière ;* English, " *Jars* ".

Sagene (*Russ.*).—Measure of superfices = 2·13 square metres.

Sand-pump.—French, *Curette ;* German, *Löffel.*

Sand-pump wheel (*sand-reel*).—French, *Treuil de curage ;* German, *Löffel-Seiltrommel.*

Sandstone.—French, *Grès friable à gros grains ;* German, *Sandstein.*

Schact (*Ger.*).—Well dug by pick and shovel.

Schwengellager (*Ger.*).—The walking beam and its support ; English, *Centre iron.*

Schwerstange (*Ger.*).—French, *Tige lourde ;* English, *Auger stem.*

Seed bag.—Sack of linseed used to cap a well.

Sinker bar.—French, *Tige lourde supérieure.*

Spouting well.—French, *Puits jaillisant.*

Stange (*Ger.*).—French, *Tige de sonde ;* English, *Drilling rod.*

" **Strike oil.**"—French, *Arriver à la couche pétrolifère.*

Tank.—French, *Réservoir.*

Tank car.—French, *Wagon-citerne.*

Temper screw.—French, *Vis de réglage ;* German, *Nach lass schraube.*

Tool wrenches.—French, *Clefs de retenue ;* German, *Schlüssel.*

Trocken loch (*Ger.*).—French, *Sondage infructueux ;* English, " *Dry hole* ".

Tubing.—French, *Tuyaux de pompe* (? hose).

Zoll (*Ger.*).—Venetian inch 0·026 metre (1·0218 English inches).

	Cost of a Well of 300 Metres (984 ft.), including 5 per cent. interest on Price of Plant.				Cost of Plant and Equipment.				Land for which each Boring is Adapted.
d.	Francs.	£	s.	d.	Francs.	£	s.	d.	
7·2	32,200	1288	0	0	1,892·8	75	14	2·88	Soft ground, no gas nor water.
·6	12,912	516	9	7·2	10,288·0	411	10	4·8	Suits all grounds that are not liable to shift or slip.
·8	9,136	365	8	9·6	23,834·7	953	7	2·4	Suits all grounds that are not liable to shift or slip.
)	5,880	235	4	0	26,101	1044	0	9·6	Suits all grounds that are not liable to shift or slip.
)	9,060	366	8	0	19,200	768	0	0	Suits all grounds that are not liable to shift or slip.
)	5,390	215	12	0	20,800	832	0	0	Strata of little inclination; accidents frequent in upheaved strata.
)	6,083	403	6	4·8	27,868	1114	14	4·8	Soft ground (very slow progress in hard rocks).
·6	9,500	380	0	0	15,900	636	0	0	
)	9,275	371	0	0	34,000	1360	0	0	Hard ground (frequent accidents in soft ground).
)	7,090	283	12	0	39,500	1580	0	0	Suits all ground.

Oilfield of X...
Well No.
Names of

Height above sea level + 618 metres.
Height above the river level + 24 metres.
Method of Boring adopted : Canadian.

Number of Sample.	Dat		Total Depth of the Well.		Diameter of the Drill.
	Boring.	bservations.	In Metres.	In Feet.	
1	1st May	nced by excavating with the pick ell (boarded up); the width of the y 2 ft. 2 in.). This well being d. Four men are sufficient.	2·20	7·2	
2	2nd May		3·70	12·1	
	4th ,,		4·50	14·8	
	5th ,,	(A workman added to staff).	6·50	21·3	
	6th ,,		8·0	26·2	
		e rods are of oak of a uniform y 36·08 ft.), except the adjusting . All the screws are of conical			
	8th May	s the ground is soft, it is possible auger stem, thus avoiding the e surface well deeper.	11·80	38·7	0·305 (say 13 in.)
3 4, 5 6	9th May	s possible to add an auger stem [t.) long. Boring is proceeding iable beds. The wear of the drill uent sand-pumping is necessary.	19·10	62·6	

INDEX.

A.

Abandoned oil wells, law as to filling up, 245.
Abandonment—
    Of ancient American oil pits, 183. *See* Mound Builders.
    Of brine wells on account of petroleum, 187.
Abatement of salt manufacture on account of contamination with petroleum in
      U.S.A., 187.
Abbruzzi oil field of Italy, 118, 119, 120, 127, 277.
    Authors' no faith in, 119.
    E. Fairman on, 118, 120.
Abdoul Djalil Outchieff, 79.
Abibekow and Lianosow No. 1 spouting well, 95.
Abich, origin of petroleum, 18-19.
Abolition of Russian petroleum monopoly, 108.
Aborigines. *See* Native.
Abortive attempts to find petroleum in France, 127-146.
Absence of petroleum from—
    Carboniferous strata of Europe, 3.
    Plutonic rocks, 22.
Acceleration of gravity formulæ, 528.
Acetylene—
    Addition products of, 16, 17.
    Condensation products of, 16, 17.
    Synthetical production of, 17-19.
Acetylide, monosodic, 16.
Acetylides—
    Alkaline, 16.
    Metallic, 19.
Accidental deaths at Boryslaw. *See* Boryslaw.
Accident fund, master contributes to, in Galicia, 54.
Accidents—
    All, to be reported to Galician mining officials, 54.
    Deformation of casing, 426.
    Deviation of drill, 426-428.
    Fall of foreign objects into bore hole, 426, 434-436.
    Flooding, 426, 436-439.
    Fracture of rods and drill, 426, 433-434.
    Landslips, caving in of bore hole, 426, 428-432.
    Preventing and remedying, 426-448.
    Unforeseen spouting, 88-94, 439-443.
Accumulation, aggregate daily, of oil less than that removed by pumping, 455.
Accuracy, drilling oil wells a work of (done blindfold), 426.
Achagian, Ecuador, 251.
Acheson's carborundum, 425.
Achievements in oil-well drilling. *See* Records.
Acre (0·40467 hectare), 504, 533.
Acrolein, 14.
Action of water on metallic acetylides and other metallic carbides, 20.
Active base, dynamite with, 445.
Addition products of acetylene, 16.
Adepts differ as to origin of petroleum, 10.
Aderbighian, The, 168.
Adipocere and the origin of petroleum, 15.

Adjacent oil wells, depth at which oil is struck varies in, 84.
Adjustment pieces of rod and drill, 335.
Administration, mining. *See* Mining Administration.
Admixture of Galician petroleum and sand. *See Cal.*
—— of dried Russian petroleum and sand. *See Kir.*
Adoration of sun and stars. *See* Fire Worship.
Adriatic coast. *See* Abruzzi.
Adulterant, petroleum as an, 1.
Adulteration, petroleum as an agent of, 1.
Advance, comparative daily, made in oil-well drilling by different systems. *See*
        Table, p. 425a ; also Oil-well drilling, speed of.
Advanced post of European civilisation (Baku), 81.
Advantages and disadvantages of oil-well drilling by different systems. *See* Ad-
        vocates, etc., and oil-well drilling by each individual system.
Adventurers, etc., flock to Pennsylvania during "oil fever," 195-196.
Adversaries—
        French *versus* British in China, 162.
        Neighbouring well owners as, 214.
        —— foreman borers as, 488.
Advertisement, broadcast, of medicinal petroleum, 189.
Advice, general, 482-490.
        on purchasing land—
            In Galicia, 50-52.
            In Russia, 105.
Advocates or adopters of different oil-well drilling systems—
        Armitage (Canadian), 266.
        Fauck (Fauck's system), 370.
        Jurski (Canadian system), 371.
        Leon Syroczinski (Canadian), 372.
        Lippmann (Raky system), 422.
        —— (Van Dyk system), 417.
        McGarwey (Canadian), 26.
        McIntosh (Canadian), 46.
        Zipperlen (Canadian), 373.
Aeration of excavated (hand-dug) wells in—
        Burmah, 156-157.
        Galicia, 282, 284, 285.
        Japan, 164-165.
Affections, chest, petroleum a remedy for, 71.
Affidavit. *See* Declaration.
Affinity of petroleum for stratified rocks, 22.
Africa, French. *See* Algeria, Madagascar and Tunis.
Age. *See* Epoch, Era.
Agen, dépôt, France, 481.
Agent, pipe-line, 476.
Aggravation of disaster, 471.
Aggregate petroleum production. *See* under different countries and producing
        centres.
Agriculture, natural gas injurious to, 144.
Agrigentum (Girgenti), Italy, 118.
Ailments, petroleum a sovereign remedy for all, 184, 189-190.
Aimkechlach, Turkestan, 149.
Ain, France, 139.
Aïn-Dellah, Algeria, 264, 265, 267.
Aïn-el-Kellah, Algeria, 262, 264.
Aïn-Louise, Algeria, 264, 265, 267.
Aïn Zeft, 260, 261, 262, 264, 265, 267, 268, 269, 270, 276.
Air, comparative weight of iron rods in, and in water, 513-14.
—— wooden rods in air and in water, 358.
Air propeller, 164, 165, 282-285. *See* Aeration.
Aix gypsum, Baku petroliferous strata geologically contemporaneous with, 85.
Akhverdoff (Lianbimoff, Kokoreff and Akhverdoff)—
        Borings south of Petrovsk, 79.
Alarmists, coal, 4.
Alarum bell of hand-dug wells, 285.
        Code of signals, 285.
Alaska, discovery of vast petroleum lake in (?), 241.
Albany, U.S.A., oil fields near to, 203, 207, 279.
Albatagio, Italy, 277.
Alberta, Canada, 239, 278.

Alberti's improvements in rope boring, 303.
Alertness of Red Indians' faculties, 183.
Alexander the Great, 67.
Alexandrovno, Russia, 480.
Algae (fossil), 279.
    Of corniferous limestone of Canada, 289.
Algebraical formula, equations of the second degree, 524.
Alger, department of, Algeria, 261, 272.
Algeria (French Africa), 239-272, 276, 277.
    Alger (Algiers), department of, no oil in, 261, 272.
    Ammi Moussa, 261.
    Armitage uses Canadian system in first boring, 266.
    —— fractional distillation products of petroleum of, 270.
    Baille's report on Aïn Zeft, 265-266.
    Ben Lakdar and the goat, 259-260.
    Bitumen, 272.
    Black Spring, the, 259-260.
    Canadian system in, 266.
    Constantine, department of, 261, 272.
        Deposits of—
            Beni Siar, 272.
            Chebka des Sallouah, 272.
            Ferdjiouah, 272.
    Depth at which oil is struck, 261.
    D'hara, 270.
    Djidjelli, 272.
    Dip of strata, 266.
    Distillation products of oil, 270.
    Evrard's report on geology of the Aïn Zeft oil fields, 265.
    Gas, 268, 272.
    Guelma, 272.
    Gypsum, prevalence of in, 263, 264.
    Hilil, 271.
    Little Kabylie Mountain, 272.
    Medgerda Mountains, 272.
    Milah, 272.
    Mostaganem, 273.
    Oran oil field, 261-272, 276, 277.
        Principal deposits in—
        First line of strike—
            Aïn-Dellah, 264, 265, 267.
            Aïn-el-Kellah, 262, 264.
            Aïn-Louise, 264, 265, 267.
            Aïn Zeft, 264-270, 276.
            Beni Chougran, 262.
            Beni Zenthis, 264.
            Mascaza, 261.
             Mazouna, 265.
            Palikao, 261.
            Port aux Poules, 262, 265.
            Sidi Brahim, 262, 264, 265.
            Tarria, 265, 270.
            Vieux Jardin, 263.
        Second line of strike—
            Oued Ourizaine, Lainé's Well at, 271.
            Relizane, 271.
            Tiliouanet, 271.
    Petroliferous strata, geology of—
        Apennine sub-marls, 270.
        Astien marls, 270.
        Cenomanian shales, 272.
        Cretaceous, 269.
        Lines of strike of, 262.
        Pliocene, 264.
        Sahalien, 264-266.
        Selenitic marl, 267, 269.
        Sidi Slimann, Aïn Zeft, geological section, 269.
    Pirates of, caulked their ships with petroleum from Black Spring, 260.
    Spaniards and petroleum of Black Spring, 260.
    Spouting wells, 268.

Algeria (French Africa) (*continued*)—
    Sulphur, deposits of, 263, 270.
    Water potable, 270.
    Wincquz, J. D., 265.
Alicante, depôt, Spain, 481.
Alienate underground mines in Galicia, legal right of women to own land, 52.
Aliens cannot own land in Roumania, 61.
Aligned streets of Baku, 81.
Alkaline acetylides. *See* Acetylides, Alkaline.
—— water, association of petroleum with, 77.
Alkali, volatile, in fossil shells of Jura, 10.
Allegation (former) against production of Virginian oil wells, 220.
Alleghany coal measures, 205.
—— mountains, 202.
—— pool, 217, 279.
—— river, 204, 213, 473.
—— town, 217.
Allen, Ohio, U.S.A., 223.
Allen spouting well, 235.
Allentown, Pennsylvania, 501.
Alliance of French and Senecas, petroleum bonfire to celebrate, 185.
Allier, department of (France). *See* Limagne.
—— river (France), 136.
—— Ville franche d' (France), results of diamond boring at, 400.
Allocation of work in oil-well drilling. *See* Oil-well Drilling.
Almeiria, a Spanish depôt of petroleum, 481.
Aloes rope, 508. *See* Cable, Aloes.
Alps, French, 127, 146.
Alsace, oil fields of, 4, 6, 7, 116, 118, 277.
    Antoine le Bel on petroleum of, 6, 116.
    Asphaltum of Pechelbronn, 117.
    Degousée signalised petroleum at Schwabwiller, 116.
    Density, etc., of the different petroleums of, 118.
    Geological horizon (tertiary), 117.
    History of petroleum in, 4, 6, 116-118.
    Middle Ages, Walsbronn fountain famous in, 117.
    Pechelbronn, view of petroleum works in, 117.
    Producing centres and properties of the respective oils—
        Lobsann, 116, 277.
        Olhungen, 116, 118.
        Pechelbronn, 115, 118.
        Schwabwiller, 116, 118.
        Walsbronn, 117.
        Wœrth, 116, 118.
        Other centres, 277.
Alteration of density with temperature, table showing, 494.
—— of viscosity with temperature, table showing, 512.
Altitude of Turkestan oil wells, 149.
Aluminium carbide—
    Synthetical production, 20.
    Yields methane (marsh gas) with water, 40.
Ambaratobe, Madagascar, 273.
Ambohitsalika, Madagascar, 273.
America, United States of, oil fields of—
    Abandoned wells have to be filled up in, 245.
    Abandonment of oil pits by mound builders, 183.
    Age of Indians. *See* Epoch.
    —— of mound builders of. *See* Epoch.
    —— of salt makers of. *See* Epoch.
    —— of trappers of. *See* Epoch.
    Agents, pipe line, of, 476.
    Alaska, alleged discovery of petroleum in, 241.
    Alliance between French and Red (Seneca) Indians sealed by light of **inflamed** petroleum, 185.
    Anglo-Saxons and the history of petroleum in, 186.
    Animals of the forest and history of petroleum in, 183, 186.
    Apothecaries of, vend petroleum as a universal panacea, 189-190.
    Artificial gems and metals, use of, as substitutes for diamond in rock boring in, 394.
    Asphaltum of California, 227-228.

America, United States of, oil fields of (*continued*)—

Barrel, American. capacity of compared with English and Galician, 57.
Basins of. *See* Geology.
Beasts of the forests of, and the history of petroleum in, 183, 186.
Carboniferous strata of. *See* Geology.
Carpenters, derrick, of, paid by day, 242.
Cask. *See* Barrel.
Casual hands, derrick carpenters are, 242.
Cattle, etc., of, flock to salt lakes, 185, 186.
Cave dwellers of, 294.
Chronology, French, of petroleum of, ceases with British conquest of Canada, 186.
Civilised peoples of, ancient, 183.
Colonel Drake and casing oil wells, 193-197.
—— Roberts and torpedoing of oil wells, 455.
Combat, petroleum rendered Indians of, invulnerable in, 184.
Comparison between barrels of, and English and Galician, 57.
—— —— geology of oil fields of, and European, 32, 202.
—— —— wages of, and those paid at Baku, 242.
Competition, European, between petroleum of, and Russia, 475.
Contents of petroleum barrels of, 57.
Conveyance of petroleum. *See* Pipe Lines.
Cretaceous strata of. *See* Geology of.
Crude oil, exports of, 246.
Cyclopean tombs of mound builders of, 182.
Decadence of Pennsylvanian production, 218-220, 231.
Defunct race, the mound builders of, a, 183.
Delaroche, the French missionary to, and petroleum of, 185.
Delegation of French institute of civil engineers to, and the underground blasting of oil wells, 456-457.
Depôts, petroleum, of, 481.
Depth to be drilled before striking oil in, 198, 199, 201, 209, 225, 244.
Derricks of, 309, 310. *See* Oil-well Drilling.
Despatches from Montcalm to France relative to petroleum of, 185.
Devonian strata of. *See* Geology of.
Dip of petroliferous strata, 202, 207.
Divide or watershed of the " Rockies " of. *See* Canada.
Drake, Colonel, 193-197.
Drastic purgative, petroleum as a, in, 184.
Dr. Brewer, 191, 192.
—— Hildreth, 188, 190.
Drilling machinery of, oil well, expensive in Japan, 165.
—— of oil wells in. *See* Oil-well Drilling.
Drug stores of, sale of petroleum in, 187.
Dry holes, ratio of, to productive wells in Pennsylvania, 219.
Dynamite, extensive use of, in the underground blasting of oil wells, 455.
—— how exploded, 455-457.
Embrocation, petroleum as an, in, 186.
Emoluments of Colonel Drake, 196.
Epoch of Cooper's heroes of, Indians (Senecas) of, 186.
—— of mound builders of, 182, 183.
—— of saltmakers of, 186, 187.
—— of trappers of, 186, 187.
Era. *See* Epoch.
Exports of crude oil, 246.
Faculties of observation of Indians of, 183.
Festal day of French and Red Indians (Senecas) of 185.
" Fever oil " of, 195.
Fierce competition in Europe between petroleum of, and Russia, 475.
Flannel used in early days to filter petroleum skimmings in, 199.
Forerunner in use of petroleum as an illuminant in Dr. Hildreth, 182-185.
Forest, tracts of, burned by inflamed petroleum during wars of Indians of, 184.
Geology of : basins of—
    Eastern (Ohio and Mississippi) oil fields—
        Bradford, 201, 204, 205, 207-208.
        Connecticut, 32.
        Indiana, 202, 224.
        Kentucky, 202.
        Lower Field, 205, 218-216.
        Middle Field, 205, 210-213.
        New York, 202, 207, 208.

America, United States of, oil fields of (*continued*)—
  Geology of : basins of (*continued*)—
    Eastern (Ohio and Mississippi) oil fields (*continued*)—
      Ohio, 222, 224.
      Pennsylvania, 202-220, 279.
      Tennessee, 32, 202, 225.
      Washington District, 205, 216-218.
      Western Virginia, 32, 202, 220-221.
    Northern and Central (Missouri and Upper Mississippi) oil fields of—
      Carolina, N., 32.
      Dakota, 202, 226.
      Illinois, 202, 226.
      Kansas, 202, 227.
      Minnesota, 202, 225.
      Missouri, 202, 227.
      Texas, 202, 227.
      Wisconsin, 202, 225.
    Southern Basin (rivers of the Mexican Gulf) oil fields—
      Alabama, 202, 226.
      Louisiana, 202, 226.
    Western Basin (Rocky Mountains)—
      California, 202, 228.
      Colorado, 32, 202, 226.
      Utah, 32, 202, 226.
      Wyoming, 202, 226.
    Petroliferous strata, classification of—
      Carboniferous—
        Crops up in Ohio and Kansas, 203.
        Dips towards the south, 203.
        Found in south of Pennsylvania, 203.
        Least rich in oil, 203.
        Occupies irregular zone in environs of Pittsburg, 203.
        Oil often met with in, of Virginia, 32.
          Subdivisions of—
            Anthracite beds, Mauch Chunk, Pocono and Butler sandstones, 205.
            Barren beds, 205.
            Butler sandstone, 213.
            Lower coal measures, Alleghany district, 205.
            Mahoning sandstone, 205, 217.
            Pittsburg coal, 205.
            Pottsville conglomerates, 205, 213, 217.
            Section of, encountered at Pittsburg, 205, 206.
      Cretaceous, 204.
        California and Utah oil fields belong to, 204.
      Devonian, 203, 205.
        Oil fields of New York and Pennsylvania—
          Alleghany, 203.
          Armstrong, 203, 213.
          Beaver. 203, 213, 215, 216, 222.
          Butler, 203, 213, 215.
          Clarion, 203, 206, 213, 215, 219.
          Lawrence, 203, 213.
          MacKean, 203, 205, 208-219.
          Macksburg, 222-223.
          Nobel (Ohio), 203.
          Venango, 203, 213, 214, 217.
          Warren, 203-205, 212, 219.
          Western Virginia (Washington), 203, 215.
        Classification of strata—
          Catskill red sandstone, including Venango sands, 205.
          Chemung series (comprising MacKean sandstones and Warren sands). 205.
          Section of Devonian strata encountered at Pittsburg, 205.
      Jurassic, 227, 276.
        Oil fields of—
          Colorado, 227, 276.
      Permian : ideal section, 205.
      Silurian, 203, 207, 223, 227.
        Crops out towards the north, 203.

America, United States of, oil fields of (*continued*)—
    Geology of : basins of (*continued*)—
       Petroliferous strata, classification of (*continued*)—
        Silurian [*continued*]—
          Crops out towards the north (*continued*)—
            Oil fields of—
              Bowling Green, 222.
              Corey, 203, 222.
              Findlay, 203, 222.
              Fremont, 203, 222.
              Lima, 203, 222-223.
              New York state, 203, 205.
              Northern Ohio, 203, 222.
              Sandusky, 222.
            Quartzose particles in rocks of, 207.
            Salt springs of, 207.
            Trenton limestone, 207.
        Tertiary strata, 204, 227, 276.
        Trias, 32, 227.
          Oil fields of—
            Colorado, 227.
            Connecticut, 32.
            Carolina, N., 32.
    History of petroleum—
      American Journal of Science quoted. *See* Hildreth, Dr., and Silliman, Benj.
      Bissel of New York, oil king, 192 *et seq.*
      Black Lake of Canada, 183.
      Brewer, Dr., 191-192.
      Brewer, Watson & Co., of Titusville, 191-192.
      Casing of oil wells introduced by Colonel Drâke, 194.
      Cherry Grove, its mushroom growth, 200.
      Cooper's heroes, petroleum in the age of, 186.
      Crosby, Dr., of Dartmouth, 192.
      Cyclopean tombs of mound builders, 182.
      Delaroche, the French missionary, quoted, 185.
      Drake, Colonel, 193-197.
      Fate of American petroleum decided by Drake's success, 196.
      Father Joseph quoted, 185.
      Fonvielle, W. de, quoted, 191-197.
      Fort Duquesne, 185, 201.
      France, king of, cared as little for petroleum as he did for Canada, 186.
      French settlers, early, and petroleum, 185.
      Hildreth, Dr., suggests petroleum as an illuminant, 187.
      Indians (Seneca) and petroleum, 188.
      Kier of Pittsburg pushes petroleum as a medicine, 189.
      Massachusetts Magazine quoted, 186.
      Medicinal sale of petroleum a failure, 190.
      Montcalm, Marquis de, reports to France on importance of petroleum, 185.
      Mound builders, the, their deep oil pits, 183.
      Mushroom towns, 199-200.
        Cherry Grove, 200.
        Oil City, 200.
        Pithole, 199, 200.
      Nile's Register quoted, 225.
      Oil Creek, price of land on, increased a thousand-fold, 198.
      Oil wells, the early spouting—
        " Burning Spring " (Virginia), 201.
        " David Wherloch," 199.
        " Densmore (3)," 199.
        " Empire," 198.
        " Eureka," 198.
        " Fountain," 198.
        " Grant " (Pithole), 199.
        " Llewellyn " (Virginia), 200.
        " Maple Shade," 199.
        " No. 54 " (Pithole), 199.
        " Noble and Delameter," 199.
        " Oil Creek Humbug," 198.
        " Phillips," 199.
        " Twin " (Pithole), 199.

America, United States of, oil fields of (*continued*)—
　　History of petroleum (*continued*)—
　　　Pennsylvania Rock Oil Co., 191 *et seq.*
　　　　Its early vicissitudes, 191.
　　　Peterson Lewis sells his petroleum to the apothecaries, 189.
　　　Pithole, its mushroom growth and equally rapid fall, 199-200.
　　　Saltmakers and petroleum, 187.
　　　Shale oil paves the way for introduction of petroleum as an illuminant, 190.
　　　Silliman, Benjamin—
　　　　Origin of petroleum, 189.
　　　　Recommends boring where natural fissures exist, 193.
　　　　This prophecy fulfilled, 195.
　　　Smith, Uncle Billy, drills the first oil well under Drake's management, 194.
　　　Tombs, cyclopean, of mound builders, 182.
　　　Trappers, 186.
　Hyde and Egbert's Maple Shade spouting well, 199.
　Indians (Seneca), their *rôle* in the history of petroleum, 185.
　Influx of *déclassés*, etc., to Pennsylvania during oil fever, 195.
　Instinct of deduction of Indians of, 183.
　Interest, land, in, 245.
　Iroquoise tribe of Indians, 185.
　" Jars," description and function of. *See* Oil-well Drilling.
　Kings, oil. *See* Oil Kings.
　King of France and America, petroleum, 186.
　Lake Erie, 203, 207.
　Lakes, great—
　　Lake Huron, 204 (map).
　　Lake St. Lawrence, 203.
　Land, purchase of, in, 5, 244.
　—— interest, 245.
　Mahoning sandstone, 205.
　Missionaries, their *rôle* in the history of petroleum of, 183, 185.
　Mississippi river, 202.
　Missouri river, 202.
　Mound builders, 181-183.
　Mushroom towns, 199, 200.
　Natural gas, 217, 225, 226.
　New Hampshire, Dr. Crosby of, 192.
　New York state, oil fields of, 207-208. *See* New York in General Index.
　Noble and Delameter spouting well, 199.
　Ohio, oil fields of, 202-204, 219, 222-224.
　Oil fever in, 6, 193-206.
　Oil-well drilling in—
　　Abandoned wells, law as to, 245.
　　Barren wells, ratio of, to productive, 244.
　　Cost of, compared with Galicia and Baku, 243.
　　Depth of wells, 244.
　　Duration of yield, 244.
　　Economical data of work, 245.
　　Land purchase, 244.
　　Minimum remunerative yield, 244.
　　Rope boring, system of, 374-378. *See* Oil-well Drilling in General portion of
　　　Index.
　Oil kings, 192.
　Oil producing centres of United States oil fields—
　　Alaska, 241.
　　Albany, New York, 203, 207, 279.
　　Alleghany, Pennsylvania, 189, 217, 279.
　　Allen, Ohio, 223.
　　Allentown, Pennsylvania, 501.
　　Auglaise, Lima, Ohio, 223.
　　Armstrong, Pennsylvania, 203, 213, 215, 279.
　　Baldbridge, Butler, Pennsylvania, 216.
　　Baltimore (Depôt), 481.
　　Baltown Forest Co., Pennsylvania, 213.
　　Beaver, Ohio, 203, 213, 215, 216, 222, 279.
　　Belden, Ohio, 223.
　　Bolivar, Pennsylvania, 501.
　　Bowling, Ohio, U.S.A., 223.
　　Bradford, U.S.A., 201, 204, 208, 501.

America, United States of, oil fields of (*continued*)—
  Oil producing centres (*continued*)—
    Brush Creek, Alleghany Pool, 217.
    Budlew, 206.
    Bullion Clarion Co., Pennsylvania, 216.
    Burkesville, Kentucky, 225.
    Eurning Spring, Western Virginia, 201, 220, 501.
    Butler Co., Pennsylvania, 203, 205, 213, 215, 216, 279.
    Butler Cross Belt, 216.
    California, 202, 204, 227-230, 501.
    Carolina, North, 33.
    Cattaraugus, Bradford, 208.
    Cherry Grove, Warren Co., Pennsylvania, 200, 211, 219, 240, 241.
    Clarendon, Warren Co., Pennsylvania, 210, 211.
    Clarion Co., Pennsylvania, 200, 203, 206, 213, 215, 219, 279, 501.
    Cleveland, Ohio (Depôt), 481.
    Clinton, Kentucky, 225.
    Colegrove, Bradford, 208.
    Colorado, 32, 202, 226, 278.
    Connecticut, 32.
    Cooper, Warren Co., Pennsylvania, 212.
    Cooperstown, Venango Co., Pennsylvania, 215.
    Coraopolis, Pennsylvania, 500.
    Corey, Ohio, 222, 279.
    Cranberry, Clarion Co., Pennsylvania, 216.
    Crawford, Pennsylvania, 206, 213, 501.
    Creek, French, Pennsylvania, 215.
    —— Oil, 215, 500, 501.
    —— Red Bank, Pennsylvania (lower field), 215.
    —— Sugar, 215.
    Cumberland, Kentucky, 225.
    Dakota, 202, 226.
    Elk Co., Pennsylvania, 216, 500.
    Evans City, Butler, Pennsylvania, 218, 500.
    Findlay, Ohio, 203, 222, 223, 279.
    Florida, 229.
    Fort Duquesne, Pittsburg, 185, 201, 216.
    Foster Creek, Pennsylvania, 501.
    Foxburgh, Clarion Co., Pennsylvania, 215, 500.
    Foxhill, 227.
    Franklin, Venango Co., Pennsylvania, 214, 215, 500, 501.
    Fremont, Ohio, 203, 222, 279.
    French Creek, Pennsylvania, 213, 214, 215.
    Gibsonbury, Ohio, 273.
    Glade Run, Butler Co., Pennsylvania, 216, 500.
    Green Co., Pennsylvania, 216, 218.
    Grove Town, Pennsylvania, 500.
    Harmony, Pennsylvania, 216, 500.
    Harrison, Ohio, 223.
    Henry's Bend, Venango, 215.
    Horse Neck, Virginia, 220.
    Illinois, 202, 226, 501.
    Indiana, 202, 224.
    Kane, Pennsylvania, 212.
    Kansas, 202, 227, 279.
    Kentucky, 32, 87, 202, 225.
    Kinzua, Pennsylvania, 208.
    Lawrence, 203, 213, 279.
    Lima, Ohio, 203, 222, 223, 279, 500.
    Los Angeles, California, 229, 230.
    Louisiana, 202, 226.
    MacCalmont Farm, Venango Co., Pennsylvania, 215.
    MacCurdy, Washington, Pennsylvania, 218.
    MacDonald, Pennsylvania, 218, 219, 500.
    MacKean, Pennsylvania, 203, 205, 208, 219, 279.
    Macksburg, Ohio, 222, 223, 279.
    Mahoning, Pennsylvania, 205, 213.
    Mannington, Pennsylvania, 500.
    Marietta, Ohio, 223.
    Mecca, Ohio, 223, 501.

America, United States of, oil fields of (*continued*)—
 Oil producing centres (*continued*)—
  Mercer, Ohio, 223.
  Minnesota, 202, 226.
  Mississippi River, 202.
  Missouri River, 202.
  —— State, 202, 227.
  Montpelier, Ohio, 224.
  Neff, Ohio, 222, 279.
  New York, 191, 193, 195, 200, 202, 203, 204, 207, 208, 209, 279, 501.
  Nobel, Macksburg, Ohio, 223, 279.
  Ohio, 202-204, 219, 222-224, 279, 500.
  Oil City, 200.
  Oil Creek, Pennsylvania, 5, 200, 214, 500, 501.
  Parker, 215.
  Pennsylvania, 184-201, 208-226, 279, 500, 501.
  Pierre, Colorado, 227.
  Pithole, Pennsylvania, 198, 215.
  Pittsburg, Pennsylvania, 187, 188, 200, 205, 206, 213, 217, 279.
  Potter Co., 208.
  Pottsville, 205, 213.
  Richburg, New York, 207, 208, 501.
  Rixford, Pennsylvania, 501.
  Roger's Gulch, Virginia, 501.
  Rumbaugh Farm, Butler Co., Pennsylvania, 216.
  Salt Lake City, Utah, 226.
  Sandusky, Ohio, 223.
  Santa Barbara, California, 228, 230.
  —— Clara, California, 230.
  —— Paula, California, 230.
  Scrubgas, Clarion Co., Pennsylvania, 216.
  Seattle, Alaska, 241.
  Sespé, California, 230.
  Shamburg, Venango Co., Pennsylvania, 215.
  Shannopin, Alleghany, 217.
  Sheffield, Warren Co., Pennsylvania, 212.
  Sisteville, Pennsylvania, 218.
  Smethport, Bradford, Pennsylvania, 208.
  Smith's Ferry, Beaver, Pennsylvania, 216, 501.
  Spencerville, Ohio, 223.
  St. Mary, Ohio, 223.
  Stillwell, Virginia, 220.
  Stoneham, Warren Co., Pennsylvania, 212, 500.
  Sugar Creek, Pennsylvania, 212.
  —— Run, Warren Co., Pennsylvania, 210.
  Syracuse, Richburg, 207.
  Tailorstown, Pennsylvania, 218, 500.
  Tarentum, Pennsylvania, 187.
  Tennessee, 32, 202, 203, 225.
  Terre Haute (High Land), Indiana, 225.
  Texas, 202, 205, 227.
  Tidioute, Warren Co., Pennsylvania, 212.
  Tiona, Pennsylvania, 212, 500.
  Titusville, Pennsylvania, 191-197, 214.
  Van Wert, 223.
  Venango, Pennsylvania, 214, 215, 279.
  Virginia, Western, 220, 221.
  Walnut Bend, Venango Co., Pennsylvania, 215.
  Wardwell, 210.
  Warren Co., Pennsylvania, 203, 204, 205 212, 219, 279, 501.
  Washington, Pennsylvania, 203, 205, 215, 216, 223, 279, 500.
  Wellsburg, Ohio, 222, 279.
  West Hickory, Pennsylvania, 213.
  Westmoreland, Alleghany, 217.
  Whisky Run, Butler, Pennsylvania, 216.
  White Oak, West Virginia, 220, 501.
  Whiting, Indiana, depôt, 481.
  Wildwood, Butler, Pennsylvania, 216.
  Windfall Run, Bradford, 208.
  Wirth, New York, 207.

America, United States of, oil fields of (*continued*)—
    Oil producing centres (*continued*)—
        Wisconsin, 202, 225.
        Wyandot, Lima, Ohio, 223.
        Wyoming, 220.
        Zelienople (pump), 216, 500.
    Oil, shale. *See* Shale Oil.
Ointment, petroleum as, in, 186.
Origin of petroleum, Silliman on, 188-190.
Pennsylvania oil field. *See* Pennsylvania in General Index.
Pension to Colonel Drake, 197.
Pharmacy and pharmacists and the sale of petroleum as a medicine, 189-190.
"Phillips" spouting well, 199.
Physic, petroleum as, 189-190.
Pipe lines of Pennsylvania, 218. *See* Transport of Petroleum.
Pipes, iron, first used as casing in, 194.
Pithole, 199, 200.
Pittsburg, 217, 220.
—— Elysée Reclus on, 217-218.
Pocono sandstones. *See* Geology.
Price of oil-well drilling machinery made in, runs dear in Japan, 165.
Progress or speed made daily in oil-well drilling in, 243.
Race defunct who occupied, prior to Indians, "mound builders," 182.
Railways of, network of, 195.
Rarity of game turns trappers into salt-makers and draws attention to petroleum, 187.
Ratio of barren to productive wells in Pennsylvania, 219.
Red Indians. *See* Indians.
Refineries, petroleum, of New York,
Report, Montcalm's to France, on petroleum of, 185.
Rig builder, 243.
Rocky Mountains. *See* Canada.
Romance of Pennsylvanian petroleum, 191 *et seq.*
Rope system of oil well drilling, 374-378.
Salary, Drake's, as a magistrate, 196.
Salt or brine wells associated with petroleum, 187.
—— industry abandoned on account of petroleum, the strong-smelling oil spoilt their brine, 187.
—— makers, trappers owing to scarcity of game turn, 187.
Saxons, Anglo-, in, 186.
Seneca Indians, 183.
Shale oil paves the way for petroleum as an illuminant, 190.
Shifting sands of Drake's first well, 194.
Silliman and origin of petroleum—the distillate from which anthracite is a residue, 188.
Silurian strata of. *See* Geology.
Simonin on Pithole, 200.
Soldiers, American, use crude petroleum as an embrocation, 188.
Speed made in oil-well drilling, 362.
Spouting wells. *See* History of American Petroleum.
Staff, enormous, required by American rope system of oil-well drilling, 300.
Statistics of production, 231.
Sunday, observance of, in, 215.
—— Well, 215.
Tarentum, working centre of salt industry, 187.
Tertiary strata of. *See* Geology of.
Testimony of Indians as to antiquity of oil pits of the mound builders, 182-183.
—— of trees as to antiquity of oil pits in which they stand, 183.
Thayer spouting well of Washington, 216.
Titusville, early operations at, 191-197.
Tombs, cyclopean, of the mound builders, 182.
"Torpedo," use of, in underground blasting of oil wells, 454-457.
Tramway or plank roads, 471.
Transport of petroleum in, 470-476.
Trappers, their *rôle* in the history of petroleum, 186.
Trees corroborate age ascribed by Indians to mound builders' oil pits, 182.
Trenches, deep, for collecting petroleum,
—— —— of Brewer, Watson & Co., at Titusville, 191. *See* Fettlöcher.
Trend of petroliferous strata of, 202.
Trias of, 32, 33.

America, United States of, oil fields of (*continued*)—
    Utah, *Zietriskisite* of, 226.
    Venango sands of. *See* Geology of.
    Zietriskisite of Utah, 226.
Ammi Moussa, Algeria, 261.
Ammonites (*A. Anceps; A. Bifrons; A. Biplex; A. Cordatus; A. Serpentinus; A. Spinatus*), 278.
Ammonium nitrate as an explosive blasting agent, 450-451.
Amoentaï, Borneo, 178.
Amotape Mountains, Peru, 251.
Amsterdam, ancient Jesuit MS. (*in re* Chinese system of rope boring) preserved at, 294.
Anaclie, Russia, discovery of petroleum, 80.
—— —— favourable situation of, 80.
Analysis of--
    Gas of burning fountain, 142.
    Mineral water encountered in Gabian boring, 133.
    White pyrites and marcassite of Gabian boring, 133.
        See Fractional Distillation Products.
Ancients and petroleum, 7.
Andaman Isles, 169.
Andidjane, Turkestan, 149.
Andrew, Samuel, one of the founders of the Standard Oil Co., 474.
Anglo-Saxons and the history of petroleum in America, 186.
Angier, Pennsylvania Rock Oil Co., 191.
Angoulême, France (petroleum depôt), 481.
Animal origin of petroleum, 12. *See* Petroleum, Origin of.
Animals (wild) drink petroleum of Black Lake, 183. *See* Camels, Cattle, Deer, Fish, Goat, Horse.
Ankavandra, Madagascar, 273.
Annam (French Indo-China), 170.
Annealing and tempering of blades of drill, 323.
—— —— cost of digging out a receptacle for, 336.
Antagonism. *See* Adversaries.
Anthracite, petroleum a distillate from, 11.
—— a sure sign of petroleum, 189.
—— of Pennsylvania, 12, 189, 205, 206.
Anticlines of petroliferous strata, Galicia, 30, 31.
—— —— Gaspé, Canada, 236, 237.
Antidotes for asphyxia from choke-damp poisoning, 284.
Antique Persian city, Baku an, 81.
—— Roman medals of Walsbronn, 117.
Antiquity of petroleum in—
    Algeria, 259-260.
    Alsace, 4, 6, 116-118.
    America, North (U.S.A.), 182-187.
    —— South (Peru), 252.
    Baku, 67-71, 97-98.
    Burmah, 153.
    China, 160.
    Europe, 5, 6, 7.
    France, 126, 129.
    Galicia, 6, 24, 27.
    Germany, 112, 113.
    Italy, 118.
    Japan, 163.
    Peru, 252.
Anvil, cost of, 289, 336.
Anzin, trials of explosives at, 451.
Aomori, Japan, 163.
Aoust, Virlet, d', origin of petroleum, 15.
Apathy of Chinese (commercial), 161.
Apostolic missionary (fire wells of Chinese Thibet), 163.
Apothecaries, American, and medicinal sale of petroleum, 189-190.
Apparatus of electric blasting outfit, 452-455.
Appearances (varied) of petroleum, 2.
Apennines, Italy, 120.
Appliances. *See* Tools.
Application to Galician Council of Mines for permission to drill oil wells, 52.
Appointment of oil well drilling manager to be noted to Galician Council of Mines, 52.

Appolonias of Rhodes, 68.
Apportionment of petroliferous land in Apsheron peninsula, 108-109.
Approximate depth of petroleum vein. *See* Depth.
Apshéron Peninsula. *See* Baku.
Apt, France, 136.
Aqueous rocks. *See* Sedimentary Strata.
Aquila, Italy, lignite mines of, 120.
Arab *raconteurs'* mythological origin of petroleum, 259-260.
Arakhan Isles, British Burmah, 153-155, 160, 277.
Ararat, No. 4 spouting well, 95.
Arboletes, Rio, Columbia, 250.
Arc of circle, Roumanian oil field comprised within, 58.
—— *See* Circle.
Arch or dome of strata and presence of oil in, 31.
Archipelago, Asiatic or Oceanic, 170.
Ardente, Fontaine, 139, 143.
Area of surfaces and solids, formulæ for calculating, 524-526.
Arenaceous of Galicia, 32.
Areometer, Baumé's : liquids heavier than water, 493.
—— —— liquids lighter than water, 494.
Argentine Republic, 258, 278.
Argillaceous shales. *See* Clay Shales.
Argilo-saline formation of Roumania, 59.
Argonauts, the, and mythology of petroleum, 68, 72.
Arguments as to origin of petroleum. *See* Origin of Petroleum.
—— as to different oil-well drilling systems. *See* Advocates.
Arianne, thread of (myth), 484.
Arid country, Baku situated in midst of, 81.
Aripero, Trinidad, 250.
Arithmetical progression formulæ, 524.
Armatures of Leyden jar electric blasting outfit, 452.
Armawir, No. 2 spouting well, Russia, 95.
Armenian caravans and early transport of petroleum, 71.
Armenia, Oriental, 81
Armenian Co., " Droojba " spouting well of, 92-93.
Armitage uses Canadian system at Aïn Zeft, Algeria, 266.
—— fractional distillation products of Aïn Zeft crude oil, 270.
Armstrong, Pennsylvania, 203, 213, 215, 279.
Aromatic hydrocarbides of Russian petroleum, 19.
Arrangement, geological, of strata. *See* Geological Structure.
Arrault succeeded Léon Dru, 406.
—— free-fall instrument of, 408.
Artesian wells, chronology of, 406.
—— brine wells, effect of striking oil in, 187.
Artichew pipe line, Baku, 477.
Artificial alumina, pure crystals of, as substitutes for diamonds in diamond rock
    boring, 394.
—— diamonds, 394.
—— petroleum from cod oil, 13-14.
—— respiration as a remedy for asphyxia by gas, 284.
Artois, France, first artesian well drilled at, 406.
Ascherleben, diamond boring results at, 400.
Ashwood. *See* Strength of Materials.
Asia, continuation of European line of strike through, 8.
—— oil fields of, 147-168.
Asiatic Archipelago, 169-181. *See* Oceania.
Asphaltum of—
        Barbadoes, 248.
        California, 228-230.
        Cuba, 247-248.
        Limagne, 136-139.
        Pechelbronn, 116-117.
        Peru, 251.
        Trinidad, 248-250.
—— cargo, shipmasters refuse, 250.
—— synthetical production of, 20.
Asphyxia, remedies for, 284.
Assam (British India), 152-153.
—— Oil syndicate, 152.
—— Railways and trading company, 152.

Asseretos, M., concession in Tunis, 273.
Association of petroleum with—
    Alkaline water, 277.
    Gypsum, 149, 263.
    Iodide springs, 41, 75.
    Salt, 24, 32, 58, 60, 75, 150, 187, 207, 225.
    Sulphur, 60, 78, 150, 263, 270.
Astchik spouting well, 95.
Astien marls, Algeria, 270.
Astrakhan, Russia, 110, 464, 468, 480.
Aswick Court, England, 125, 497.
Athabasca river, 239.
Atlantic, 127, 178.
Atlas, Mount, 261.
—— Tellien, Algeria, 260.
Atmosphere. *See* Air.
Auger, 392-395.
    Different modifications, 423.
    Illustrated, fig. 113, p. 392.
Auger band wheel, 354-357, 534.
—— stem, free-fall—
    Cost, 335.
    Described and *illustrated*, 321.
    Illustrated, 323.
    American system, 375.
    Canadian system—
        Cost of, 363.
        Described and *illustrated*, 359.
    Fauvelle system—
        Cost of, 389.
        Described, 382.
        Illustrated, 381.
—— stem guide—
    Described, 321.
    Function of, 321.
    Illustrated, 320.
Auglaise, Lima, Ohio, 223.
Augury of richness of French petroliferous strata, 127.
Austria—
    Consul of, on Baku petroleum industry, 101.
    Consumption of Galician petroleum by, 54.
    Emperor of, awards gold medal to Lukasiewicz, 40.
    Exports of Russian oil, 102.
    Government decrees against use of—
        Lamps in hand-excavated wells, 24.
        Locomobiles in drilling oil wells, 340.
    —— statistics of Galician production, 27.
    Imperial and Royal Geographical Institute of, 33.
    Oil-well, drilling machinery of, 405-406.
    Partially supplies Turkey with refined oil, 61.
    Petroleum prices compared with American, 35.
Austro-Belgian Co. of Galicia, 41.
Authentic statistics of Galician production awanting, 27.
Automatic free-fall systems. *See* Free-fall Systems and Instruments.
Autun shale, Selligues' attempts at distilling oil from, 4, 128, 190.
Auvergne—
    Bituminous shales of, 20, 127, 138.
    Eruptive phenomena of, and origin of petroleum, 16.
Available ground at Baku, 108.
Aveyron, France (Fauvelle drilling at Bourgnies Collieries), 381.
Avignon, France, 136.
Axes (hatchets), 285.
Axis of the bore hole, 347.

B.

Babadag, Roumania, 64.
Bacan, Roumania, 63.
Backward condition of petroleum industry in Roumania, causes of, 60-61.
Bad effects of protective tariffs on Roumanian petroleum, 61.

Bad-Iwonicz, Galicia, bathing establishments of, 41.
—— —— iodide springs of, 41.
Bag, seed. *See* Seed Bag and Well Cap.
Bahia, Brazil, 258.
Bailer. *See* Sand Pump.
Baille's official report on Aïn Zeft, Algeria, 265.
Ball spouting well, 235
Bakhtiari Mountains, Persia, 150.
Baku, including Apsheron Peninsula—
 Advanced post of civilisation, 81.
 Aligned streets of, 81.
 Analysis of crude oil of, 94.
 Antique Persian city, an, 81.
 Apportionment of petroliferous land, 108-109.
 Argonauts and mythological origin of petroleum of, 68.
 Arid country, situated in midst of, 81.
 Attraction of, for Guèbres, 68.
 Available, much ground in, still, 108.
 Average density of petroleum of, 3.
 Barbarism, a triumph over, 82.
 Black Sea, Duval says, communicates with Caspian, 69; ground between
  Caspian and, resounds when ridden across, 69.
 Black Town, 83.
 Blasting impracticable in oil field of, 88.
 Buildings, grandiose, of, 82.
 Buying land in, 105-109.
 Capital required to start, 107. *See* Schedule of Expenses.
 Caspian Sea, 8, 66, 67, 79, 109, 147, 464, 466, 468.
 City, the Holy, 67.
 Civil proprietorship, 108.
 Classification of land, 108, 109.
 Contractors' charges, 105.
 Cost of oil-well drilling, 104, 105.
 Crude oil, analysis of, 94.
 —— —— exports from, 109.
 Darius and petroleum mythology, 67.
 Day wages. See Monthly Wages.
 Deed, indemnity title, 108.
 Density of oil, average, 3, 496.
 —— —— of spouting wells, 95.
 Depôt of, 110.
 Depth at which oil is struck, 88, 95.
 Difference in oil level of adjacent wells, 84.
 Dupuis, Ch. Fr., 67.
 Duration of yield, 88.
 Duval, Father, on petroleum of, 7, 69.
 Electrically lighted quay of, 82.
 Engine fires extinguished during spouting, 93.
 Engines, locomobile, interdicted, 339.
 Eternal fire, temple of. 67.
 Factory, chimneys of, 81.
 Fire worship originated at, 67.
 Geology of, 85-87, 277.
 Grandiose buildings of, 82.
 Guèbres, Baku always attracted the, 68. *See* Parsees.
 Harbour, magnificent natural, of, 83.
 History of, 67, 71.
 Hotels of the first rank of, 82.
 Illuminating oil, percentage of, in crude oil compared with Galicia, 26, 94, 95.
 Indiscriminate sale of land stopped, 109.
 Intermittent action of spouting wells, 85-86.
 Leproux on petroleum statistics of Tiflis, 73.
 Locomobiles, interdiction of, 339.
 Loss and devastation by spouting, 90-94.
 Kerosene in crude oil of, 94-95.
 Kœchlin Schwartz on—
  Subterranean phenomena of Baku, 69.
  Temple of eternal fire, 97.
 Land, petroliferous—
  Buying, 105-109.
  State the proprietors of, 108-109.

Baku, including Apshéron Peninsula (*continued*)—
    Manufacturing town, a, 81.
    Marcus Polo and ancient knowledge of petroleum of, 6, 68.
    Monopoly, the Russian petroleum, 71-72, 108-109.
    —— effect on the industry, 108-109.
    —— its abolition, 108.
    —— the effect of abolition, 108.
    Mud volcanoes, 86-87.
    Mythology of petroleum of, 67, 68.
    Oil, crude, of, density of, 3, 95, 496.
    —— of, distillation products of—
        Nobel's refinery, 95.
        Topolnica refinery, 94.
    —— level, 84, 88.
    Parsees, 67, 68, 70.
    Petrolia, the Russian, 66.
    Petroliferous line of strike, 66, 77, 79, 83, 84.
    —— strata—
        Aix gypsum corresponds with level of, 85.
        Tertiary, oligocene, 85.
        Vertical sections of—
            Balakhany, 100.
            Binagadine, 101.
    Pipe lines of, 466, 476-477.
    Population rapidly increasing, 81.
    Price of petroliferous land, 105, 106, 107.
    Production, statistics of, 102.
    Proprietors of land, 108, 109.
    Quay, 81.
    Reserve, petroliferous land in, 108-109, 148.
    Reservoirs, natural, of, 90, 91, 461.
    Royalty, 107.
    Saline lakes of, 85.
    Schedule of expenses incidental to oil-well installation at Baku, 107.
    Sodium chloride, impregnations of, 85.
    Spouting wells—
        A list of (1889-1890), 95.
        Density of oil of, 95.
        Depth of each in metres, 95.
        Duration of spouting period, 95.
        Total production of, 95.
    Staff of an oil-well drilling installation at, 104, 105.
    Strikoff wells near village of, difference in oil level of, 84.
    Sultans, cruel, former seat of, 81.
    Temple of eternal fire, 68, 97.
    "Torpedo" unpracticable at, 88.
    Trans-Caucasian railway, 83-84.
    Turks near Caucasus respect fire, 68.
    Underground, hollow sound of, 69.
    Wages, 104, 105.
    Water, scarcity of, 83.
    "Watering" of streets with petroleum, 83.
    See Russia, oil fields of.
Bakhtiari Mountains, Persia, 150.
Baladchatni, Russia, 82.
Balakhany-Baku pipe line, 99, 464.
—— Russia, 87, 94, 95, 99, 102, 106, 110.
"Balancier." *See* "Walking Beam".
Balangan, Borneo, famous spouting well of, 175.
Balena, pliocene fossil, 276.
Baldbridge, Butler, Pennsylvania, 216.
Balkan, Little, 148.
Baltic, *Ludwig Nobel* tank steamer which plies on, 467.
Baltimore, U.S.A., petroleum depôt, 481.
—— pipe line, 474.
Baltown Forest Co., Pennsylvania, U.S.A., 213.
Bamboo, use of, in oil-well boring—
    China, 161.
    Japan, 164-165.

Band wheel. *See* Auger Band, etc.

Banes, Cuba, 248.

Bank cheque, pipe lines voucher equivalent to, 476.

Banking firm (French) and Anaclie, 80.

Bankruptcy of Colonel Drake, 196.

Barbadoes, West Indies, 248.

—— asphaltum of, 248.

Barbarism, petroleum conquers, at Baku, 82.

Bargain (fool's), allowing purchaser 25 per cent. tare on barrels, 57.

Barges, tank, 464, 467.

Barium carbide, synthetical production of, 20.

—— picrate of, 450.

Barmach, Mount, 69.

Barnes spouting well, 235.

Barnsdall spouting well, 209.

Barongah, Isle of, Burmah, 67, 154-155, 160, 170.

Barques, Persian, and transport of petroleum, 71.

Barrel, the, Gaspé, Canada, 237.

—— of petroleum, cost of delivering f.o.b. Port Sarnia before the days of pipe lines, 472.

—— —— bursting of, 50.

—— —— total cost of delivering at Liverpool from Port Sarnia before days of pipe lines, 472.

Barrels, American, Galician and British contents of, compared, 57, 462.

—— capacity of, tables for gauging, 515, 521.

—— —— formulæ for gauging, 526.

—— tare of, 57.

Barren beds, carboniferous strata of Pennsylvania, 205.

—— results of search for petroleum in France, 126.

—— wells, ratio of, to productive wells in Germany, 115.

—— —— in Pennsylvania, 219, 244.

—— —— increase of, in Pennsylvania, 219, 244.

Barrow, relays of, in wheeling up inclined plane, 529.

Barwineck, Galacia, 37.

Base, active, dynamite with an, 445.

—— inert, dynamite with an, 445.

Basin. *See* Oil Fields—

    Of the Danube, 59.

    Of the Dniester, 42.

    Rivers of America and their relation to the oil fields thereof, 202.

Batoum, 79-80, 467, 479.

Batoy, Jalomotei, Roumania, 64.

Batraki, Russian depôt, 481.

Battens, arrangement of, used in shoring up woodwork of hand-dug wells (*illustrated*), 286. *See* Derrick Estimates.

Battery, frictional electricity, of oil-well blasting outfit, 452.

Baumé's hydrometer, 493-494.

Bay of Gaspé, 236.

—— of Baku (a natural harbour), 88.

Beacon, Temple of Everlasting Fire formerly a magnificent, 97.

Beams. *See* Derrick Estimates and "Beam, Walking" under Oil-well Drilling Systems.

Beasts of the forest and petroleum, 183, 186.

Beaver, Ohio, U.S.A., 203, 213, 215, 216, 222, 279.

Becks of boards as improvised petroleum reservoirs, 233.

Bedden, Ohio, U.S.A., 223.

Bedding of vertical strata, planes of, liability to deviate in drilling parallel to, 427.

Beds, barren, of Pennsylvania, 205.

Beechwood as locomotive fuel, 340. *See also* Strength of Materials.

Beginners (prospectors) magnify importance of signs of petroleum, 483.

Bel, Antoine le, origin of petroleum, 13.

—— —— —— petroleum of Alsace, 6, 116.

Belden, Ohio, 223.

Belemnites mucromatus (fossil), 75.

Belgian-Austro Co. of Galicia, 46.

Belgium, exports of Russian oil to, 102.

Bellows (aerating) or air propeller in hand-dug wells—

    Galicia, 284.

    Japan, 164-165.

Belmont works of Price & Co. for treating Rangoon oil, 156.

Belo Telang, Sumatra, 170, 172, 177.
Beluchistan, European, petroliferous line of strike continued through, 8, 151.
Bème, British Burmah, 159, 277.
Benda's application of free fall to rope boring, 303, 419.
Beni Chougran, Algeria, 262.
Beni-Siar, Algeria, 272.
Beni Zenthis, Algeria, 264.
Ben Lakdar and the goat at the Black Spring, 259, 260.
Bennet Bros. spouting well, 235.
Benoit, origin of petroleum, 13.
Benzene (C_6H_6), absence of, from petroleum distillation products, 17.
—— dyers' and cleaners', 2.
—— fractional distillation product of petroleum, 3.
—— percentage in crude oil—
    Baku, 39.
    Galician, 37, 39.
    Peruvian, 255.
Bergheim and MacGarwey's refinery distillation products from Galician crude oil,
    38, 89.
    See Klobassa & M'Garwey.
Berlin depôt, 480.
Berthelot quoted, 16, 33, 447.
Besançon depôt, France, 420.
Bethlehem, near Liebau, boring results at, 400.
Beuthen, Silesia, record drilling speed at, 425.
Beziers, Bishop of, and Gabian oil, 6, 131.
Bhoema, 67.
Bibi Aibad, Russia, 100, 102, 110.
—— rope boring at, 100.
Biblisheim, Alsace, 277.
Big Fork, Gaspé, Canada, 237.
Bilin, Bohemia, speed in drilling at, 349.
Binagadine, Russia, 95, 100, 101.
Binitro-benzol as an explosive, 451.
Birkel, boring operations of, at Kertch, Russia, 74.
Bishop of Beziers and Gabian oil, 6, 131.
Bissel of New York (solicitor and oil king), 192 et seq.
Bit.   See Drill ; Oil-well Drilling.
Bitche, Alsace, 117.
Bitumen—
    Common origin of coal and, Virlet d'Aoust declaims against, 15.
    From synthetical carbides, 20.
    Of Jura, 10.
    Of which anthracite is residue could not have penetrated Alleghanies, 12.
    See Asphaltum.
Bituminous coal, 11, 12, 171.
—— deposits and eruptive phenomena, association of, 16.
—— shales of Autun, 4, 126, 190.
    Appear of organic origin, 20.
Bituminous shale, petroleum a product of natural destructive distillation of, 13.
—— products, synthetical, 20.
—— petroleum of Venezuela, 250.
Black and Mathewson spouting well, 235.
Black Lake, Canada, 183.
—— Sea, 8, 66, 69.
—— —— tank steamers on, 467.
—— Spring (Ain Zeft), Algeria, 259, 260.
—— Town.   See Baku.
Blade of Drill.   See Oil-well Drilling.
Blasting gelatine, 445 et seq.
Blasting impracticable at Baku, 88.   See Electric Blasting Outfit.   See also Ex-
    plosives and Torpedo.
—— levigation as substitute for, 457.
Blindfold operation, drilling a, 426.
Block fuel, 1.
Blunt tools, sharpening and tempering of, 323.
Bnito Co. of Balakhany, 99.
Bobrka, first Galician oil well, 39, 277, 304.
—— history of, 39, 40.
—— yield, duration of, 32.

Boge (France), recent discovery of petroleum indications, 130.
—— analogies with Pechelbronn, 159.
Boileau, Gauldrée. *See* Gauldrée Boileau.
Boiler (French), Hermann La Chapelle's recommended, 341.
—— furnaces should be put out when "spouting" occurs, 93, 442.
—— sheds, 351.
Bolivar, Pennsylvania, 501.
Bolivia, 258.
Bologna, Italy, 119-120.
Bologna-Padua, N. watershed of Apennines, 277.
Bond of security from Galician petroleum buyers, 56.
Bonfire, petroleum, of Senecas and French, 185.
Bonnefond, M. de, tank waggon builder, 462.
Boot blacking, early Galician use of petroleum as, 6, 24.
Bordeaux, France, depôt, 480.
Bore hole, axis of, importance of drilling in line of, 338.
Boring of oil wells. *See* Oil-well Drilling.
Borneo, 175-176.
   Bad climate, 175.
   Balangan spouting well, 175-176.
   Producing centres—
      Brunei, Kotei, Labuan, Malvoda, Martapoera, Seknati, Tandjony, 175.
Bort, use of, in diamond boring, 424.
Boryslaw—
   Accidents, high death rate from, 46.
   Distilling (Johan Mitis and Joseph Hecker's) in early days, 21.
   Early struggles, 24.
   Geological horizon, 276.
   Rush and subsequent failure, 25.
   Ozokerit, discovery of, saves situation, 25.
   —— annual production of, 26.
   Pipe line, 478.
   Pressure of gas (enormous), 46. *See* Galicia and Ozokerit.
Bosanquet, Ontario, 239.
Boston, U.S.A., depôt, 481.
Boulet's horizontal boiler, 341.
Bouncing post free fall, 320.
Boundary line of old continent, main petroliferous line of strike the, 8.
Bourgnie's collieries, Aveyron, France, result of use of hydraulic drilling, 381.
Bowling, Ohio, U.S.A., 222, 279.
Boxwood. *See* Strength of Materials.
Brachiopods fossil, 279.
Bradford, U.S.A., 201, 204, 205, 207, 208, 501.
Bradford, New Jersey, U.S.A., pipe line, 476.
Bradley spouting well, 235.
Braila, Roumania, 64.
Bravadoes flock to Pennsylvania after Drake's discovery, 195-196.
Brazil, 258.
Brea, Trinidad, 250.
Bredt of Ottynie, engines of, 342.
—— oil well, cap of (*illustrated*).
Bremen, Germany, 480.
Brewer, Dr., of Titusville, 191-192.
Brewer, Watson & Co., of Titusville, 191-192.
Breza, Roumania, 64.
Brickmaker, Dutch, discovers petroleum in his brickfield, 125.
Brick work or masonry of reservoirs, 462.
Brindisi, Italy, depôt, 480.
Brine. *See* Salt.
Briquette Fuel.
Britain, Great, 5, 125, 497.
British made engines recommended, Clayton and Shuttleworth's locomobiles, 342.
British made oil-well drilling machinery, Mather and Platts on Degoussée's system, 409.
—— Borneo, North, 175-176.
—— Columbia, 240.
—— East India, European line of strike continues through, 8, 15.
—— —— Oil fields of, 151-160, 277.
      Assam, 152, 277.
      Beluchistan, 151, 277.
      Burmah, 153-160, 277.

British East India, oil fields of (*continued*)—
    Punjaub, 152.
    General properties of oil, 160.
—— West Indies. *See* Barbadoes and Trinidad.
—— North America. *See* British Columbia, Canada, Newfoundland, Nova Scotia.
British *versus* French in—
    Canada, 185-186.
    China, 162.
Broken rods and drills. *See* Accidents.
Brooms as petroleum skimmers, 113.
Brunei, Borneo, 175.
Brunswick, Germany, 115.
Brush Creek, Alleghany Pool, 217.
Bucharest, Roumania (map), 59, 63, 64, 479.
Bucket lowering of men by, in hand-dug wells, 282.
—— raising detritus by, 283.
—— —— oil by, 165, 195.
Buda Pesth, Hungary, refineries of Galician petroleum, 55.
Budlew, U.S.A., 206.
Buffalo, U.S.A., pipe line, 474.
" Buffalo " motors, 342, 366.
Buja Dagh, Trans-Caspian territory, 277.
Builder, rig, 242.
Building of derricks, etc. *See* Derricks; Oil-well Drilling.
Bukowine (Bukovina), signs of petroleum lost in valleys of, 49, 63.
Bulgaria, exports of Russian petroleum to, 102.
Bulge of rods, 316.
Bulk transport of petroleum, 462-487.
Bullion, Clarion, Penn., U.S.A., 216.
Bulwheel, 375, 376, 533.
Bung, depth measurement of barrels at, 515-521.
Burgos, Spain, 124.
Burmah, 3, 8, 155-160, 170.
    Aeration of hand-dug wells, 156-157.
    Native methods, 153, 156-157.
    Oil, density of average, 8.
    —— general properties of, 155-160, 498.
    Arakhan Isles—
    Barongah, 154-155, 160, 498.
    Cheduba, 154, 160, 498.
    Flat, 154.
    Ramri, Letaung, Likmau, Kiank Phyn, 154, 160, 498.
    Coast or Lower Burmah oil field, 153.
    Dark, wells hand dug in, 156.
    Depth at which oil is struck, 155-156, 158-159.
    Geological structure, 153.
    Inter or Upper Burmah oil field—
    Bèma, 159, 277.
    Kodaung, 159.
    Mandalay, 156.
    Mimbain, 154.
    Mimbu, 498.
    Twingon, 160, 277, 498.
    Yenangyat, 160, 277, 498.
    Yenangyaung, 160, 277, 498.
Burning fountain, Gabian, France, 6, 126, 128-136.
—— Spring, Western Virginia, 201, 220, 501.
—— —— first operations in Virginia begun at, 201.
Bursting of barrels, 56.
Butler Co., Pennsylvania, U.S.A., 203, 205, 213, 215, 216, 279.
Butler cross belt, Pennsylvania, 216.
—— sandstone, 205.
Butter, nutty flavoured, adulterated with petroleum, 1.
Buyers of petroleum in Galicia, hints as to dealing with, 55, 56.
Buying, oil-leasing, etc., of petroliferous land in—
    America, 244, 245.
    Baku, 105-109.
    Canada, 5, 245-246.
    Galicia, 50-52.
    Madagascar, 274-275.

Buzau, Roumania, 63.
Bye-laws of continental railways as to conveyance of petroleum spirit, etc., 521-522.

C.

Cable, aloes, breaking strain of, 508.
—— —— used in Canadian system, cost of, 363.
—— —— used in Jobard's rope-boring system, 301.
—— chain, used in free-fall French plant, cost of, 404.
—— hemp, breaking strain, 509.
—— —— for raising and lowering man in hand-dug wells, 282.
    Cost of, 289.
—— —— round, load which it will carry, according to dimensions, tarred and un-
    tarred, 509.
—— iron, with hemp core, 301.
—— steel (hand-dug wells), 280.
    Cost of, 289.
—— —— (free-fall), cost of, 335.
—— —— (Fauck's system), cost of, 351.
—— —— round, load which it will carry according to dimensions, 510.
—— —— flat, load which it will carry according to dimensions, 510.
—— sand-pump (free-fall), cost of, 336.
—— —— (Fauck's system), cost of, 351.
—— —— French construction, 404.
Cadiz, Spain, 124, 481.
Cahours. *See* Pelouze and Cahours.
Cal of Sloboda, 2.
    Unworkable and unsaleable, 47.
Calamity, unforeseen spouting a, 439, 440, 441.
Calcareous strata. *See* Limestone, Marls, etc.
Calcium carbide, synthetical production of, 19, 20, 21.
—— carbonate. *See* Limestone.
—— sulphate. *See* Sulphate of Lime, Selenitic Marl and Gypsum.
Calculations, formulæ and tables to facilitate, 498-532.
California, 202, 204, 227-230, 276.
—— asphaltum of, 226-228.
—— cretaceous strata of, 204.
—— pliocene strata of, 209, 276.
Campbell spouting well, 235.
Campbell and Forsyth spouting well, 235.
Campina, Roumania, 65.
Calorific intensity of explosives, 447-451.
—— petroleum, 4, 340, 468.
Cambrian (upper), petroliferous strata of British Columbia, 240.
Camels, petroleum a cure for scab of, 7, 68.
Canadian wooden rod or pole system of oil-well drilling. *See* Oil-well Drilling,
    Canadian, etc., System.
—— and other North American oil fields, 230-241.
—— oil fields, 232-239, 276.
    Average density of oil, 3.
    Bad communication in early days, 470-472.
    Cost of sending barrel of oil to Port Sarnia and Liverpool, 471-472.
    French evacuation of, 186.
    History of, 232-235.
    John Shaw's famous spouting well, 232-235.
    Land, price of, during " oil fever," 3.
    Spouting wells, list of, 235.
      Gaspé oil field, divisions of, 236-238.
        Bay of Gaspé, 236.
        Boundaries, 236.
        Depth at which oil is struck, 238.
        Gas of, 238.
        Geology of, 236, 237. 238.
            Anticlines, 236, 237.
                Proximity of petroleum to, 236.
            Carboniferous strata, 236.
            Devonian strata, 236.
            Dip, 236.
            Dyke of eruptive trap, 237.

Canadian oil fields (*continued*)—
  Spouting wells, list of (*continued*)—
    Gaspé oil field, divisions of (*continued*)—
      Geology of (*continued*)—
        Enniskillen formation, 236.
        Eruptive trap dyke, 237.
        Gaspian formation, 236.
        Limestone, Silurian, 236.
        Logan on, 238.
        Osbalski on, 237-238.
        Sandstone, Devonian, 236.
        Trenton limestone, 236.
      Inter-collonia railway, 237.
      Intermittent working of deposits of, 237.
        Poor results of, 237.
      International Oil Co., 237.
      Minerals of Gaspé (Osbalski), 238.
      Oil, properties of, 237-238.
        Density, 237-238.
        Sulphur in, 237.
      Producing centres—
        Barrel, 237.
        Big Fork, 237.
        Douglastown, 237.
        Gall's Brook, 237.
        Haldimand, 236.
        Ile de la Traverse (Crossing Island), 238.
        Little Fork, 237.
        Pointe aux Trembles (Earthquake Point), 238.
        Rivière a la Rose (Rose River), 238.
        Sandy Beach, 236.
        Silver River, 237.
        Tar Cape (Cap au Goudron), 237.
  —— Ontario oil field—
    Geology of, 238-239.
      Anticline, petroliferous level extends in, 239.
      Chemung, Devonian beds, 238.
      Clay shale, petroliferous strata surmounted by, 239.
      Corniferous limestone (the petroliferous bed), 239.
      Depth (comparatively shallow) of petroliferous bed, 239.
      Dip of the beds, 239.
      Fossils, marine (animal, vegetable), 239.
      Limestone, corniferous, 239.
      —— Silurian, 238.
    Principal producing centres—
      Bosanquet, 239.
      Collingwood, 239.
      Enniskillen, 239, 279.
      Oil Spring, 239.
      Petrolia, 239.
      Sherkstown, 239.
    Minor producing centres—
      Cape Smith (Manitouline Island, Lake Huron), 239.
      Dereham, 239.
      Sainte Marie (St. Mary, Manitouline Island, Lake Huron), 239.
      Tilsonburg, 239.
  —— Western Canada, oil fields of—
    Sub-divisions—
      Alberta, 239.
      Gaseous manifestations, 239.
      Geographical situation—
        Eastern watershed of Rocky Mountains, 239.
      Geology—
        Cretaceous strata, 239.
        Rivers, Athabasca and Peace, 239.
        Sandstone, petroleum saturated outcrops of, 239.
  —— British Columbia, oilfields of—
    Geographical situation—
      Flathead River, 240.
      Kootenay Pass, 240.
      Western watershed of Rocky Mountains, 239.

Canadian oil fields (continued)—
    British Columbia, oil fields of (continued)—
        Geology—
            Cambrian upper formation, 240.
            Magnesian limestone, 240.
            Silicious sandstone, 240.
        Total production of Canada—
            Statistics of 1875-1894, 240.
            —— Newfoundland, 240.
            —— Nova Scotia, 240.
Candles, manufacture of, by Price & Co. at Belmont and Sherwood from Rangoon
    oil by Warren de la Rue's patent, 156.
Cannstadt, Wurtemburg, artesian well of, 406.
Canton, China, hostility of natives to strangers, 161.
Caoutchouc (vulcanised) insulators, blasting outfit, 453.
Capacity of barrels, 515-521.
—— tables for gauging, 526.
Cap au Goudron, 237.
Cape Negrais, Burmah, 169.
Capital, French diverted to Italian oil fields, 126.
—— required to start oil-well drilling at Baku, 107.
Capitalists deceived by speculators, 488.
"Capping" oil wells, 288, 442, 443.
Caprice of veins of oil, 91, 214.
Capsule, fulminating, blasting outfit, 454.
Caravans, Armenian, and early transport of Baku petroleum, 71.
Carbides (see hydro-carbides), metallic, 16, 17, 18, 19, 20, 21.
—— cyclic, 19.
Carbonic acid, formation of alkaline acetylides by contact with metals in interior of
    earth, 16.
Carboniferous strata of American oil fields, 204, 205, 206. See Coal Measures.
Carborundum, 425.
Carburetted hydrogen, antidote for toxic effects of, 284.
Cargoes of petroleum, ancient—
    Armenian caravans, 71.
    Burmese junks, 156.
    Persian barques, 71.
    —— —— modern. See Tank Steamers.
—— of pitch (Trinidad) refused by master mariners, 250.
Carmen, master, former transport of petroleum at mercy of, 472.
Carnen, J. Winter de, oil well drilling machines, 408.
Cars, tank. See Tank Waggons.
Carnival of French and Seneca Indians, 185.
Carolina, North, Trias oil fields of, 32.
Carpathian Mountains, boundaries and shape of, 28.
—— Coquand's geological study of, 16.
—— emeralds of, 28.
—— gold of flanks of, 28.
—·-· infiltration of petroleum across line of strike of petroliferous strata, 28.
Carriage of petroleum. See Conveyance, Pipe Lines, Transport.
—— —— cost of, to Liverpool, in early days, 471-472.
—— —— spirit, etc., by continental railways, 521-522.
Cart horses, 529, 531.
Cartage data and formulæ, 529, 531.
Carthagena (depôt), Spain, 481.
—— New Granada, mud volcanoes of, 87.
Cartridge, dynamite, of electric blasting outfit, 454.
Cashup, Venango, U.S.A., 215.
Casing of oil wells—
    Friction of sand brought up wears, 93.
    Hermetic, 180, 398, 436, 437.
    Introduced by Drake, 194. See Oil-well Drilling.
    Ratio of diameter of, to that of bore hole, 437.
Caspi pipe line, Baku, 477.
Caspian Sea—
    Daghestan oil field widens out towards shore of, 78.
    Mud volcanoes on shores, 87.
    Naphtha-impregnated sandstone of Caucasian oil fields projected beneath, 147.
    —— covered waters of, 67.
    Ocean, might well look like an infernal, 67.

Caspian Sea (*continued*)—
    Petroliferous line of strike crosses, 8.
    —— —— loses itself in, 8, 66.
    —— —— comes to a halt in, 66.
    Pipe lines, object of, to connect Balakhany with refineries on shores of, 99, 464.
    Ports of, 69.
    Route, the, and Volga, 467, 480.
    Southern Caucasian oil field ends on the shores of, at Baku, 79.
    Steamship companies of, refuse to support Nobel's tank steamer scheme, 466.
    Tank steamers cut across, with cargoes of petroleum for Russian markets, 109.
    —— fuel used by, 468.
    —— workshops for, 468.
    Underground of, the largest petroliferous basin of the world, 72.
Cast iron rods, etc. *See* Iron (Cast).
Castilian government and monopoly of Peruvian petroleum, 252.
Casual hands, derrick carpenters of America, 242.
Casualties to be reported to Galician council of mines, 54.
Catastrophe, unforeseen spouting a, 439-443.
Catch of free-fall instrument. *See* Free-fall Instrument.
Catholic Apostolic missionaries on Chinese gas (fire) wells, 163.
Catskill Mountains, U.S.A., 204.
Catskin rubber's electric blasting outfit, 452.
Cattaraugus, U.S.A., 208.
Cattle flock to salt and petroleum lakes, 183, 186.
—— disease of, petroleum as a remedy for, 39.
Caucase, Un touriste au, 69.
Caucasus, 66-110, 277, 278.
    Anaclie (Western Caucasus), recent researches, 80.
    Black Sea ports isolated from Europe by whole width of, 80.
    Communication, difficulties of, in other districts of, stopped attempts to develop, 73.
    Eocene formation, petroleum-bearing beds of, belong to, 8.
    Fires from natural petroleum gas numerous in ancient times, 67.
    Land, price of, in oil fields of, 104.
    Nobel promoter of industrial movement in, 73.
    Northern depôt of Wladekawkas, 77, 479.
    Oligocene, upper, petroliferous strata of, belong in greater part to, 85.
    Petroleum of mythological origin, 68.
    Production, only approximate estimate of, of all the districts of, can be made, 73.
    Prometheus enchained on, for having carried away the fire of heaven, 68.
    Tamansk, strata of, differ from remainder of, by consisting of hard rock, 75.
    Underground, inexhaustible stores of oil, 72.
    Watersheds of, petroliferous line of strike follows both northern and southern, 8.
    Western, recent researches in, 80.
        See Baku, Russia, etc.
Caulking ships with petroleum by—
    Algerian pirates, 260.
    Chinese (junks), 160.
Caumette, France, 129.
Causeway, ratio of traction to load on, 529-530.
Cave dwellers (Troglodytes), 294.
Caving in of bore hole. *See* Landslips.
Cavity. *See* Pockets.
Cébu, Isle of, Philippines, 178.
Ceiling of tank-car waggons, 463.
Celebes, Oceania, 170.
Celestial Empire. *See* China.
Celle, North Germany, 111, 115.
Cement, stone quarried for manufacture of, 141.
Cervus giganteus quaternary fossil, 276.
Chabat-Harmela, Algeria, 263.
Chaff as packing material for glass vessels containing petroleum spirit, 521-522.
Chaligny and Guyot-Sionnets' traction pumping engine, 341.
Chalk. *See* Cretaceous.
Chalon, French petroleum depôt, 480.
Chandler spouting well, 285.
Channel, Santa Cataline, 228.
—— (Thalweg) of the Danube, 58.
Chaotic geological structure of Yenangyaung oil field, 157.
Chapapote, Cuban, 248.
Chapelle, Hermann la, horizontal boiler of, 341.

Charcoal for smithy, cost of, 289, 290.
Chardin, Persia, natural oil wells of, 151, 498.
Charge of exploder in electric blasting outfit, 454.
—— of dynamite in torpedo, 457.
Charging tank steamers, 468.
Charleville, French petroleum depôt, 480.
Chatillon, petroliferous tertiary beds of, 128, 143-146.
—— natural gas of, 143-144.
—— domestic use of, 143.
—— petroleum store of, 143.
Chebka des Sellaoua, Algeria, 272.
Cheduba, Isle of, Burmah, 154, 160, 498.
Chee-Kong, China, 163.
Chekonspce, Russia, 77.
Cheliff River, Algeria, 262. See Map, 261.
Chemical composition of petroleum, 2.
—— theories as to origin of petroleum, 2, 3, 4, 10-23.
Chemists' *versus* geologists' theories as to origin of petroleum, 21.
Chemung, Devonian strata, 238. See America, Geology of.
Cheribon, Dutch East Indies, 499.
Cherry Grove, U.S.A., 200, 211-212, 219, 240, 241.
—— No. 646 spouting well, 212.
Chest affections, petroleum as a remedy for, 71.
Chief of the Senecas fêtes the French soldiers with a petroleum bonfire, 185.
Chieti, Italy, 120, 278.
Chili, 258.
Chimerical (*sic*) innovations of Nobel, 466.
Chimneys, factory, of Baku, 81.
China—
    Bamboo system of boring tools and pipes (conveyance and pumping), 161.
    Burmah, deposits of, correspond with those of, 161.
    Coal mines of, 160.
    Concessions, French, Government of, favourable to, 162.
    Copper mines of, 160.
    English scent rich oil fields, 162.
    Exports of Russian petroleum to, 102.
    Foo-Choo-Koo, petroleum of, 161.
    French engineers, Chinese Government to apply in first instance to, 162.
    Haï-tha oil field more prolific yield of, 161.
    Home consumption exceeds production, 161.
    Hostility, native, to strangers, 161.
    Houses, roofs of, coated with petroleum, 160.
    Huart's memoir on mining industry of, 160.
    India, oil fields correspond with those of, 162.
    Iron mines of, 160.
    Junks, petroleum used to caulk, 160.
    Kouang-Si, petroleum deposits of, 160, 162.
    Kouang-tonng, 162.
    Mathon-Jozet on Taï-li-Chen oil wells, 161.
    Mines and minerals, 160, 161.
    Pottery lamps, petroleum burned in, 160.
    Production, absence of data as to, 161.
    Sainte Claire Deville on petroleum of Foo-Choo-Koo, 161.
    Superstition, 161.
    Taï-li-Chen, greenish oil rich in paraffin of, 161.
    —— yield of, 161.
    —— tin mines of, 160.
    Transport, vile system of, between mining centres and waterways, 161.
    Yun-nan, 162.
Chinese Thibet, oil field of—
    Brine springs of, 162.
        Natural gas, heat from combustion of, used to crystallise salt of, 162.
    Coldre's, R. P., description of, 163.
    Continuation (supposed) of Fergane oil field, 162, 163.
    Derricks of, 163.
    Fire wells of (*tse-liou-tsin*), 162.
    Geology of—
        Tertiary sandstones of various description, 163.
            Song of the gas of a fire well, 163.

Chinese Thibet, oil field of (*continued*)—
    Geology of (*continued*)—
        Producing centres—
            Northern : Se-Tchouen, Tchouen-Pe, gas wells of—
                Chee-Kong, 163.
                Gan-Yo, 163.
                Ponk-Ki, 163.
                San-Tay-Le-Tche, 163.
                Su-Lin, 163.
                Tchong-Kiank, 163.
            Southern : Se-Tchouen and Se-Tchouen-Lan, gas wells of—
                Ho-Tsin, 163.
                Kong-Tsin, 163.
                Tse-Liou-Tsin, 163.
Chingalek, Russia, 74.
Chlorate of potash. *See* Potassic Chlorate.
Choke damp. *See* Asphyxia and Carburetted Hydrogen.
Choking. *See* Asphyxia.
Chorhot, India, 277.
Chorkowka, Galicia, 40.
Chouster, Persia, 150, 151, 498.
    Low density of oil, 150.
Chronology, French, of American petroleum, 185-186.
—— of inventions and improvements in rope boring of artesian wells, 303.
Chugorowa, Russia, 125.
Circle, geometrical formula connected with, 523 *et seq.*
Circuit, how completed in electric blasting of oil wells, 452, 453.
Circulation of water in hollow rods as an aid to oil-well drilling, sand pumping, etc., 379-389.
Circular iron rods, weight per running metre in air and water, according to diameter, 513, 514.
Circumference, geometrical formulæ, 523 *et seq.*
City, Fire, 217.
—— holy, 67.
—— iron, 217.
—— smoky, 217.
Civil proprietors of Baku, 108.
    Galicia, 50-52.
    Roumania, 62, 65.
Civilisation, Baku an advanced post of, 81.
Civilised people of America, ancient, 182.
Claims as to rights to concessions in Madagascar, how decided, 278.
Clairvoyance, Montcalm's, Marquis de, in reporting home to France on petroleum, 185.
Clarendon, Pennsylvania, U.S.A., 210, 211.
Clarification of crude Russian petroleum from sand by deposition, 461.
Clarion, U.S.A., 200, 203, 206, 213, 215, 219, 279.
Clays of Balakhany, grey, brown, etc., 100.
    Belowosa shales (red micaceous), 35.
    Binagadine, 101.
    Holowecko, 35.
    Magura sandstone, 35.
    Ropianka shales (micaceous and quartzose), 35.
Clay shales (argillaceous)—
    Barren, of America, 206.
    Belowosa, Galicia shales, 35.
    Crawford County, 206.
    Excavating oil-wells through, 290.
    Holowecko shales, green, black, brown, encountered in boring, 35.
    Pittsburg, 206.
    Ropianka shales, bluish-grey, 35.
    Southern Caucasian oil field, 80.
    Smilno shales, 35.
    Sumatra, 171, 172.
    Tamansk shales, eocene, 75.
Clayton and Shuttleworth's locomobile as pumping engine, 342.
Cleaners', dyers' and, benzene, 1.
Clearance of Cherry Grove Forest Co., 211.
Clercy, De, petroleum engineer, 256.
Clermont, France, 137.
—— Ferrand, 138.

Cleveland, Ohio, U.S.A., 481.
Clichy, French petroleum depôt, 480.
Clinton, Kentucky, 225.
Cloez's synthetical production of petroleum, 17.
Clouds, petroleum, of spouting wells, 94.
Coal—
    Alarmists, Nature replies to, with petroleum, 2.
    Bituminous, of Langkat, 171.
    British, liquid distillate from, 5.
    Calorific intensity compared with petroleum, 340.
    China, coal mines of, 160.
    Cost of, doubled by freight from station to works, 340.
    Differentiation between petroleum-stone and coal, 145.
    Liquid, petroleum a, 4.
    Measures of America, 205.
    Origin, common, of alleged, with petroleum, 3, 10, 11, 12, 13, 17, 21.
    Petroleum a liquid coal, 4.
    —— not mined like, 293.
    Smithy (free-fall system), cost of, 336.
    See Fuel.
Coast oil field of Burmah, 153, 155.
—— —— of San Angeles, 228-229.
Cobalescu, Roumania, 277.
Cod-liver, fractional distillation (under pressure) products analogous to or identical
    with petroleum, 13, 14.
Coefficient of contraction and expansion of Russian petroleum, 494.
Coinage of different countries reduced to French equivalents, 506.
Coin, Canadian borers propitiate genius of earth with, 490.
Coldre, R. P., on Chinese Thibet, 163.
Colegrove, Pennsylvania, U.S.A., 208.
Colibazi, Roumania, 64, 276.
Collapse of bore hole. See Landslips.
—— of casing. See Accidents.
Collecting trenches or pits, petroleum, of—
    Germans (Fettlöcher), 113.
    Mound builders, 182.
    Red Indians, 184.
    Titusville, U.S.A., 191.
Collingwood, Ontario, 239.
Collodion, 446.
Collonges, France, 139.
Cologne, German depôt, 480.
Colombes, French depôt, 480.
Colombia, South America, 230.
Colonel Drake. See Drake.
—— Roberts. See Roberts.
Colonies and dependencies, British, oil fields of. See Assam, Barbadoes, Beluchistan,
    Borneo, British Columbia, Burmah, Canada, Newfoundland, Nova Scotia,
    Punjaub, Trinidad.
—— Dutch. See Borneo, Java, Sumatra.
—— French. See Algeria, China (Tonquin), Madagascar, New Caledonia.
Colorado, 33, 202, 226, 278.
—— cretaceous oil fields of—
    Triumph Petroleum Co., 226.
    United Florence Rocky Mountain, 226.
Colours from petroleum, 1.
Columbia, British, 240. See Canada.
Comanesku, Roumania, 63.
Combat, petroleum rendered Indian warriors invulnerable in, 184.
Combe's improvements in rope boring, 303.
Combination of free-fall and Canadian oil-well drilling systems, 364.
Combinations of different oil-well drilling systems. See Oil-well Drilling Improve-
    ments, etc.
Combustible gases from cod oil distillation, 14.
—— petroleum a mineral, 21.
Commander-in-Chief of French-Canadian forces, 185.
Commerce, inertia of French, 127.
Commercial Geography, Paris Society of. See Dedication.
    Huart's memoir on China to, 160.
Comonesti, Roumania, 276.

Compagnie Française de Pétroles de l'Amérique du Sud, 256.
Comparative graphical chart of Russian and American petroleum statistics (facing),
   102.
Comparisons between—
   Calorific intensity of coal and petroleum, 4, 340, 465.
   Different systems of oil well drilling. *See* Oil-well Drilling.
   Geology of American, Baku, and Galician oil fields, 29-30, 32, 33, 74, 75, 87, 202.
   Space occupied by same weight of oil and coal respectively, 340.
   Wages in American, Baku, and Galician oil fields, 104, 242.
Compensation to workmen in Galicia, 54.
Competition between foremen borers in same oil field, 488.
—— between Canadian system and Fauck's free-fall system of oil-well drilling in
   Galicia, 405.
—— European, between American and Russian petroleum, 475.
Complicated strata of Yenangyaung oil field, 159.
Composition of petroleum not uniform, 2. *See* Fractional Distillation Products.
Concession, how obtained. *See* Conditions, etc.
—— natural gas, 142.
Concessionaire. *See* Drake.
Concussion, drilling by, Canadian system a system of, 371.
Condensation of gases of saline lagoons and origin of petroleum, 11.
—— products of marsh gas, 19.
Conditions under which petroliferous land may be bought or leased, and oil-well
     drilling commenced in—
   American oil fields, 244-245.
   Canadian oil fields, 245.
   Galician oil fields, 50-52.
   Madagascar oil fields, 274-275.
   Roumanian oil fields, 65.
Conduct. *See* Pipe Line.
Conductors of electric blasting outfit, 453.
Cones, formulæ for calculating area and solidity of, right, truncated and pyramidal, 525.
Conference of drilling engineers, 367-372, 415.
Conflagration, petroleum, brings ruin in train, 184.
Conformation of oil fields. *See* Geology of Respective Oil Fields.
Confused strata of Yenangyaung, 159.
Congealed dynamite, 446.
Congerian beds of Roumania, 59-60.
Conglomerate—
   Coarse grained, of Congerian beds of Roumania, 59-60.
   Galician, red hard, silicious of, 132, 134.
   Galician, 60.
   Quartzose of Holowecko, 35.
Congress of drilling engineers, Vienna, 370.
Coniferæ, fossil, characteristic of miocene, 276, 277.
Conil, Spain, oil field of, 124.
Connecting wires, 452-453.
Conquest, British, of Canada, and its effect on history of American petroleum, 186.
—— of barbarism by civilisation, 81.
—— of Gaul, Chinese rope boring contemporaneous with, 293.
—— Spanish, of Peru, and its effect on history of petroleum there, 252.
Consent of mining authorities before commencing to drill—
   In Galicia, 52.
   In Madagascar, 274.
Consignment note of petroleum spirit and other inflammable liquids, endorsement
   on, 521.
Consistency. *See* Viscosity.
Constantine, Algeria, Department of, 261, 272, 277.
Constitution of petroleum. *See* Composition, etc.
Construction of derricks, 306-313.
—— of reservoirs, 461-462.
—— of shed to cover hand-excavated wells, 288.
—— of smithy, 285-289, 336.
—— of tank waggons, 462-463.
—— of tools, 353.
Consul, Austrian, at Tiflis, 65.
   Report on the condition of Baku petroleum, 101.
Consumption, home, of Chinese petroleum, 161.
   Galician petroleum, 54.
   Japanese petroleum, 166.

Contact, electrical, how effected in electric blasting outfit, 452-454.
Contamination of dried Russian petroleum with sand. *See Kir.*
—— of Sloboda petroleum with sand. *See Cal.*
Contents of barrels, tables for gauging, 515-521.
—— —— formulæ for, 526.
Continent, old, boundary line of, passes through Asia, 8.
Continental railways, 521-522.
Continental Diamond Rock Boring Co., 400.
Continuity of European line of strike through Asia and Oceania, 8, 66, 148, 171.
Contorted strata of Galicia, 30.
Contour of the Carpathians, 29.
—— of the Alleghanies, 202, 208.
Contractors' prices for oil-well drilling—
    In American oil field, 243.
    In Baku oil field, 105.
    In Canadian oil field, 245.
    In Galician oil field, 105.
Contracts, selling petroleum, Galician, 56.
Contrast. *See* Comparison.
Contribution of master and servant in Galicia to workmen's accident fund, 54.
Controversies as to merits of oil-well drilling. *See* Advocates, etc.
Conveyance of petroleum. *See* Freight, Transport.
—— —— spirit, etc., 521-522.
Cooper, U.S.A., 212.
Cooperage, Nobel's mechanical, at Tzarytzin, 469.
Cooper's heroes, age of, 186.
Cooperstown, U.S.A., 215.
Copeck, equivalent in French money, 506.
Copper mines of China, 160.
Coquand's theory of origin of petroleum, 17.
—— geology of Caucasian strata, 85.
Corals, fossil, 276.
Coraopolis, Pennsylvania, U.S.A., 500.
Cordial, petroleum drunk as a, by Russians (Hanway), 71.
Corey, Ohio, 222, 279.
Cori-Derbent, 69.
Corpse wax. *See* Adipocere.
Correction of density of oil for each degree above and below 15° C., 494.
—— viscosity, 512.
Corroboration of antiquity of American petroleum, 188.
Corrosion of casing by sand brought up in train of petroleum of spouting wells, 93.
Corrupt people of Galicia, 50.
Cosmetic, petroleum as a, 2.
Cost of—
    Buildings. *See* Oil-well Drilling Buildings, Cost of.
    Derricks. *See* Oil-well Drilling Derricks, Cost of.
    Drilling per foot. *See* Oil-well Drilling, Respective Systems, Cost of Drilling
      per foot by.
    Engines. *See* Oil-well Drilling Engines, Cost of.
    Land in different oil fields. *See* Land, Cost of, in different Oil Fields.
    Oil well drilling tools. See Oil-well Drilling Tools, Cost of.
    Reservoirs (oil), iron and wooden, 461.
Cotton, gun. *See* Gun Cotton.
Council of mines, or equivalent of—
    Baku, 108.
    Galicia, 52.
    Madagascar, 274-275.
    Roumania, 65.
Counterbalancing " walking beam," 297, 314.
Counterfeit signs of petroleum, 483.
Counterpoise of "walking beam," 297, 314.
Coupling of " walking beam " to engine, 343.
Cowardice of oil-well drilling workmen under superstitious fears, 490.
Cow Creek, U.S.A., 220.
Cracow, Galicia, 40, 49.
Craft of tank lighters, Nobel's, 467.
Crane. *See* Oil-well Drilling Tools.
Craneberry, Pennsylvania, U.S.A., 206.
Crank coupling of " walking beam " to engine, 343.
Crawford, Pennsylvania, U.S.A., 206, 213, 501.

Credo, Le, France, 139, 278.

Creek, French, Oil, etc. *See* American Oil Field.

Credit, John Shaw's failure to obtain, the morning of day his famous well spouted, 233.

Cretaceous petroliferous strata of—
  America, U.S.A., 32, 204, 226, 278.
  Argentine Republic, 239, 278.
  Canada, 278.
  France, 278.
  Galicia, 32, 33, 48, 278.
  Germany, North, 112, 278.
  Italy, Southern, 120-123, 278.
  Palestine, 278.
  Portugal, 278.
  Russia, 278.
  Spain, 278.

Crevasse. *See* Pockets.

Crevice. *See* Pockets.

Crimea –
  Characteristic fossils—
    Belemnites mucromatus, 74.
    Nummulites Leymeriei, 74.
  Petroliferous strata confined to tertiary eocene (nummulitic beds), 8, 74, 277.
  Producing centres—
    Chingalek, 74.
    Kelechi, 74.
    Kertch, 74.
    Kop-Kut-Chigan, 74.
    Temseh, 74.
    Teschewli, 74.
    Zamoskaya, 74.

Crops, natural gas injures, 144.

Crosby, Dr., of Dartmouth, U.S.A., 192.

Crown. *See* Oil-well Drilling (Rotary).

Crude oil, American exports of, 246.
  Roumanian exports, effect of protective import tariffs on, 61.
  —— Russian exports of, 109.

Cruel sultans, Baku the former seat of, 81.

Crushing effect of Boryslaw gas pressure, 46.
  —— resistance to. *See* Strength of Materials.

Crustacea, fossil, 239.

Crystal, worship of image of Mars in, 67.

Crystalline alumina (pure) as diamond drilling substitute, 394.

Cuarazute, Bolivia, 258.

Cuba—
  Abandoned condition of mineral resources due to its disturbed history, 247.
  Asphaltum of, 247-248.
    Destructive distillation products of, 248.
    Lamp oil, its richness in, 248.
    Paraffin, its richness in, 248.
    Principal producing centre, Banes, 248.
    Soluble in spirits of turpentine, 248.
  Banes, producing centre, 248.
  Chapapote. *See* Asphaltum.
  Havana, asphaltum of, known to Spanish settlements of, 247.
  Political changes just supervened will lead to realisation of its deposits, 247.

Cube, lead measurement of force of explosives, 451.

Cubic contents of barrels, tables for gauging, 515-521.
  —— —— formulæ for gauging, 526.
  —— —— of timber, 526.

Cumberland, Kentucky, U.S.A., 225.

Cupolas of Pittsburg, 218.

Curative virtues of petroleum, 2, 7, 39, 68, 71, 117, 127, 129, 184, 186, 187, 189, 190.

Curler spouting well, 235.

Currency, Austrian, British, Galician, reduced to French, 506.

Current metre and current foot, cost of oil-well drilling per. *See* Oil-well Drilling, cost per current foot, etc.

Custody of samples brought up by drill, sand pump and core barrel, 486.

Customs (habits) of oil-well drillers in Galicia, 490.
  —— tariff in Japan and freight double price of American oil-well drilling machinery, 165.

Custom-house, Austrian, statistics unreliable, 27.
Cutting instruments, casing pipe—
    Drus' described and illustrated, 433.
    Fauck's described, 432.
    —— illustrated, 433.
Cyclic carbides, 19.
Cyclopean tombs of mound builders, 182.
Cylinder, area of, formula for calculating with parallel bases, 525.
—— truncated, 525.
—— volume of, formula for calculating, 525.
Cylindrical iron rods, solid, weight per current metre in air and water according to
    diameter, 513, 514.
Cypris fossil, 277.

D.

Daghestan, Russia, 78-79.
    Line of petroliferous strata, Tamansk-Derbent ends on, 79.
    Oil field widens out towards Caspian, 78.
    Producing centres—
        Derbent, 78.
        Petrowsk, 78.
        Temir Khan, 78.
    Bubbling sands of, 78.
Daily progress in oil-well drilling. See Oil-well Drilling.
—— wages of oil-well drilling staff. *See* respective systems of oil-well drilling. *See*
    also Monthly Wages.
Dalton's law, Mendéleef's application of, to origin of petroleum, 18.
Dakota, U.S.A., 202, 226.
Daliki, Persia, 151.
    Heavy oil of, 151.
Damage by unforeseen spouting, 88-94, 439-443.
Damp, choke, in hand-dug wells, 284-285.
—— fire, in hand-dug wells, 284-285.
Dandang-Ho, Dutch East Indies, 499.
Dangers incurred in excavated wells, 282-285.
Dangerous, nature of unforeseen spouting, 88-94, 439-443.
Danube, embouchure of, 58-59, 74.
Darfeldt, Germany, 278.
Darius and fire worship, 68, 72.
Dark, working in, whilst hand-digging oil wells, 24, 156.
Dartmoor River, Canada, 237.
Data, general, customary oil-well engineer's formulæ, 493-534.
David Wherlock spouting well, 199.
Davy lamp in hand-dug wells, 285.
Day book or journal, foreman borer's, 491-492, 488-489.
—— light, how Japanese reflect into bore hole, 164.
—— shift and night shift, rotation of, 489.
—— work *versus* piece work, 291.
Ddobesti, Roumania, 63.
Dead Sea, Palestine, 278.
Deadly nature of gas in Galician hand-dug wells, 283-284.
Dealing, difficulty of, with Galicians, 50.
Dearness of American oil-well drilling machinery. Price doubled in Japan by freight
    and customs tariff, 163.
Dearth of water and fuel at Puerto Grau, Peru, 256.
—— of water at Baku, 83.
Debarring locomobiles at Baku and Galicia, 340.
De Bur Brothers' No. 6 spouting well, 95.
Decadence of Pennsylvanian production, 218-220, 231.
Decaying organic matter, origin of petroleum from, 10-23.
Deceitful nature of Galicians, 50.
Deception of capitalists and prospectors by speculators, 482.
Decision as to speculators' rights to Madagascar mining concessions, 274.
—— —— mining departments, arbitrary, 274.
Declamations, Degoussée's, against Fauvelle system, 403.
Declaration on Continental railway, consignment note as to contents of packages
    containing petroleum spirit, 521-522.
—— before Galician Council of Mines, 52.
De Clercy, operations of, in Peru, 256-257.

Decomposition of organic matter (animal, vegetable and marine), origin of petroleum from, 10-23.
Decorations of temple of eternal fire, 98.
Decrease of Pennsylvanian production, 219, 220, 231, 244.
Decree interdicting use of—
    Lamps in hand-dug wells, 24.
    Locomobiles at Baku and in Galicia, 340.
—— French, classifying natural gas concessions with petroleum, 142.
Deductive reasoning, presence of petroleum determined by, 484.
Deed, advisable to purchase land by, in Galicia, 51.
Deep wells of Clarion Co., Pennsylvania, 215.
Deepening of "Barnsdall" causes it to spout, 209.
—— of Cherry Grove well leads to its resuming its ancient splendour, 212.
Definition of petroleum, 1-3.
Deformation of casing, 426-432.
Defunct race, the mound builders, 183.
Defrauding of capitalists by speculators, 482.
Degoussée hurls anathemas on Fauvelle, 403.
    Free-fall instrument of, 413.
    His method of keeping boring journals, 486.
    Oil-well drilling machinery of, 407.
    Petroleum of Schwabwiller, 116.
Dehulster free-fall system, 416.
Delaroche (missionary) and early history of American petroleum, 185.
De la Rue's, Warren, patent for treating Rangoon oil, 156.
Delegation (American) of French Institute of Civil Engineers, blasting an oil well before, 456.
Delesse on Chatillon gas, 144.
Deleterious nature of gas of hand-dug oil wells, 282-285.
Delicacy of French drilling machinery, 402, 403.
Deliquescence of explosives an undesirable feature, 448.
Deluge of petroleum. *See* Spouting Wells.
Demoralisation of workmen by joking at their superstition, 490.
Densmore oil wells, the three spouting, 199.
Denudation of beds above German petroliferous strata, 112.
Denunciation, Degoussée's, of the Fauvelle system, 403.
Denver, Colorado, U.S.A., 226.
Deposits. *See* Geology of respective Oil Fields.
Depôts, petroleum, 479-481.
    American, 481.
    French, 480.
    Galician, 479.
    German, 480.
    Roumanian, 479.
    Russian, 479-480.
    Spanish, 481.
Depreciation in price of Russian petroleum, 465.
Depression of strata in Bukovine, 49, 63.
Depth bored per day or month with different oil-well drilling systems. *See* Oil-well Drilling ; Speed.
Depth at which oil is struck in—
    Algeria, 261, 263, 264, 266-267, 268-269.
    America, U.S.A., 198, 199, 201, 209, 225, 226, 244.
    Baku, 88, 95.
    Balakhany, Baku, 100.
    Barbadoes, 248.
    Binagadine, Baku, 101.
    Burmah, 155, 156, 158, 159.
    Canada, Gaspé, 238.
    Crimea, 74.
    Ecuador, 250.
    Galicia, 30, 31, 35-36, 41, 42.
    Germany, North, 114, 115.
    Grosnaïa, 78.
    Japan, 164.
    Java, 174.
    Noworossirsk, 78.
    Peru, 251, 252, 253, 255, 257.
    Punjaub, 152.
    Sumatra, 172.
    Tchéléken, 147.

Depth at which oil varies in adjacent wells, 84.
Depth of oil well, formulæ for calculating by velocity of sound, 524.
Derbent, Russia, 79, 80.
Dereham, Canada, 279.
Derricks—
    American, described, 309, 310.
    —— illustrated, 309.
    Canadian, described, 311-312.
    —— illustrated, 311.
    Four-legged, old-fashioned, described and illustrated, 308.
    Instrumentation, described and illustrated, 312.
    Transportable, described and illustrated, 313.
    Tripod, described and illustrated, 308. See Oil-well Drilling under respective
        systems.
Derricks, installations without, 314.
Desolation, petroleum conflagrations spread, 184.
Despatches, French, Montcalm's, relative to petroleum, 185.
Despotism, enlightened, of Incas, 252.
Dessicated petroleum and sand kir, 80, 91.
Destruction caused by spouting wells, 88-94, 440.
Destructive distillation of cod oil under pressure yields artificial petroleum, 13-14.
Detonation, how produced, in the blasting of oil wells, 452-455.
Detractors of organic theory of origin of petroleum, 10.
Detritus from excavated wells raised in bucket, 283.
—— from drilled oil wells. See Sand Pumping.
Deutsch & Cie in Italy, 121.
—— depôts, 481.
—— tank steamers, 470.
Devastation, calamity and, petroleum conflagrations bring, 184.
Deviation in oil-well drilling—
    How it may be avoided, 427-428.
    Partial, causes of, 427.
    Remedies for partial and total, 428.
    Rope boring impracticable in Galicia owing to liability of, 296.
    Total, 428.
Devonian petroliferous strata of—
    America, U.S.A., 203, 205, 206, 207, 279.
    Canada, 236, 238, 279.
    France, 134, 135, 279. See Geology of Respective Countries.
    Gabian, 129.
Dhara, Algeria, 270.
Diagram, graphical, of fluctuation in American and Russian petroleum production,
    102.
Diamond replaces steel in rotary oil-well drilling, 394.
Diamonds, etc., artificial, in rotary oil-well drilling, 394.
Dicotyledons fossil, 277.
Dictionary definition of petroleum, 1.
Digboi, Assam, 152, 160, 277, 498.
Digging, hand-, of oil wells, 280-288.
Dijk, Van, free-fall instrument of, 417.
—— rods of, 369-370, 410.
Dijon, French depôt, 480.
Diluvial sand of North German oil field, 112.
Dimitresci, Roumania, 63.
Diodorus of Sicily, 118.
Dioscorides, 118.
Dip of petroliferous strata of—
    Algeria, sub-Apennine marls, 266.
    America, U.S.A., 202, 207.
    Canada, 236.
    Gabian, shown in section, 135.
    Galicia, shown in section, 43.
Diphtheria, petroleum a remedy for, 2.
Discipline in oil-well drilling, engine driver obeys foreman borer, 488.
Discharging tank steamers, 468.
Discourses. See Advocates of different Oil-well Drilling Systems.
Discussions. See Advocates of different Oil-well Drilling Systems.
Diseases of cattle, petroleum a remedy for, 39.
—— of camels, 7, 68.
Dishonesty of Galicians, 50.

Dislocation of strata—
   Galicia, 30.
   Yenangyaung, Burmah, 159.
Dislodging broken drill, 433.
—— —— rods, 434.
—— embedded drill, 430.
—— —— sand pump, 431.
—— unscrewed rods, 434.
—— —— drill, 434.
Dismissal of oil-well drilling staff in Galicia, 53, 54.
Distance of oil wells from habitations and fires generally, 52.
Distillation products, fractional, of crude oil of—
   Algeria, 270.
   Baku, 89, 95.
   Cuba, "Chapapote," 248.
   Galicia, 39.
   Lipinki (Galicia), 37.
   Nobel's refinery, 95.
   Peru, "London and Pacific," 255.
   Topolnica refinery, 94.
   Trinidad asphaltum, 249.
Distillers' petroleum, 85.
—— barrels, folly of using, 57.
—— early Galician, 24.
Distilleries, petroleum. *See* Refineries.
Distorted strata of Gabian, 129.
—— Yenangyaung, 159.
Distribution of petroleum over Russia and Europe, 464-470.
Diversion of French capital to Italian oil fields, 126.
Diversity of strata of Yenangyaung oil fields, 159.
Divide. *See* Watershed.
Djaba Kota, Dutch East Indies, 499.
Djebel Bou Kournine, 278.
Djebel el Dahr, Syria, 278.
Djebel Serra, Tunis, 273.
Djebel Zeit, Egypt, 276.
Djidjelli, Algeria, 272.
Dniester, valley of, 42.
Docile subjects of the Incas, 252.
Doctors, salt-makers had no traffic with, 187.
—— fees payable by master in case of accident, 54.
Documentary evidence as to title to land in Galicia, 50.
Doftaneti, Roumania, 65.
Dog fish, characteristic fossils of petroliferous pliocene, 276.
Doian, Roumania, 64.
Dolomite, 134.
Domestic uses of Chatillon gas, 143.
Domnino, Russia, depôt, 469.
Double dealing of Galicians, 50.
Douglastown, Gaspé, Canada, 237.
Doula, British India, 277.
Doyen, the, of the Pennsylvanian deposits, 213.
—— —— European oil fields, 118, 120.
Dragonese, Roumania, 60, 62, 276.
Dragoneza, Roumania, 64.
Drain to carry away water from surface level of hand-dug oil well, described and
   illustrated, 281.
Drake, Colonel, 193-197.
   Active man, a very, 193-194.
   Casing, first to use sheet-iron pipes to "case" bore hole, 193.
   "Colonel of industry," a, 193.
   First oil well, 193-195.
   Magistrate of Titusville, 196.
   New Yorker, a, but poorly educated, 193.
   Ousted by his neighbours, 196.
   Pensioned by a grateful country, 197.
   Ruined by speculation, 197.
   Success of, decided fate of American petroleum, 197-198.
Drastic purgative, petroleum a, 184.
Drawing wells. *See* Pumping Wells.

Drilling. *See* Oil-well Drilling.
Drinking of petroleum by Russians according to Hanway, 71.
Drohobycz, Galicia, 32, 49, 55.
—— crude oil refineries of, 55.
—— depôt, 479.
—— mining council of, 52.
—— salt deposits of, 32.
Droojba, spouting well of, Baku, 85, 93.
Drowning of John Shaw in his famous spouting well, 234.
Drug stores of United States vend petroleum as a medicine, 189, 190.
Dru, Leon, succeeded Mulot, 406, 407, 414.
—— —— free-fall instrument of, described, 413, 414.
—— —— illustrated, 414.
—— —— rods, improvements in connection of clasps, 410.
Drums of windlasses in American rope boring system, 376.
Dru, St. Just, free-fall instrument of, 415.
—— —— described, 416.
—— —— illustrated, 416.
—— —— expanding drill of, 433.
— — —— instrument of, for cutting casing pipes, 53.
Ducasse, signs of petroleum in Galicia, 30.
Dudeschi, Roumania, 8, 64, 66.
Dug, hand-, wells, 280-288.
Duhaigne, Albert, quoted, 233-234.
Duke, Grand, Michael, and No. 3. Schibaeff spouting well, 91.
Dunaborg, Russia, 110.
Dupuis, Ch., Fr., on Parsee religion, 67.
Duquesne Fort, 185, 201. *See* Pittsburg.
Duration of yield of oil wells of—
  Algeria, 267-268.
  America, U.S.A., 244.
  Galicia, 31-32.
  Roumania, 62.
Dutch East Indies, 169-176.
Duval, Paul, early writer (1688) on physical geography of Baku and on its crude oil, quoted, 69.
Dwellers, cave, 294.
Dwellings and fires, distance of, from oil wells, 52.
Dwiniaz, Galicia, 276.
Dyer's benzine, 1.
Dyke, eruptive, trap, of Gaspé, Canada, 237.
Dynamite, 444-457. *See* Explosives.
Dynasty of the oil kings, 192.

E.

Earth, The. *See* Soil.
—— internal heat of. *See* Origin of Petroleum.
—— reactions going on towards the centre of. *See* Origin of Petroleum.
—— robbed of its treasures would yield no more harvests, 161.
—— wooer of, the borer the, 490.
Earthenware, use of petroleum (asphaltum) in manufacture of—
  By ancient Californians, 229.
  By Incas, 252.
—— vessels, transport of Burmese petroleum in, 156.
East Indies, British, 158-160.
—— —— Dutch, 169-176.
—— —— Exports of Russian oil to, 102.
Eastern basin of America, oil fields in, 202.
—— Galician oil field, 46-49.
—— Roumanian oil field, 64.
Eccentric drill, 438.
Echery, Alsace, 277.
Echigo, Japan, 163-166, 168, 276, 499.
Economy, false, of sinking oil wells by hand digging, 293.
Ecuador, 251.
Edges, knife, of drill, transformations of, 331.
Effervescence, testing samples of detritus from bore hole for, 486.
Efficiency of oil-well drilling systems. *See* Oil-well Drilling.

Efflorescence of petroleum—
    Relizane, 271.
    Tiliouanet, 271.
Egypt, 276.
—— exports of Russian petroleum to, 102.
Ekaterinoslaw, Russia, 110, 480.
Elbe and Weser river, 111.
Election of Colonel Drake as magistrate of Titusville, 196.
Electric blasting outfit, 452-455.
" Electric " brand of American built engines, 342.
Elevation windlass for raising rods. *See* Oil-well Drilling Tools.
Elizabethpole, Russia, 73, 81, 479.
Elk Co., Pennsylvania, U.S.A., 218.
Ellipse, semi-, centre of gravity of, area of, 527.
Elmwood. *See* Strength of Materials.
Elongation of rods, windlass for. *See* Temper Screw, 348.
Emanation, gaseous. *See* Gas.
Emanations, petroleum. *See* Signs of Petroleum.
Embankments. *See* Reservoirs and Unforeseen Spouting.
Embedded tools, how raised. *See* Accidents, etc. ; Dislodging.
Embouchure of the Danube, 58, 59, 74.
—— of the Ingour, 80.
Embrocation, petroleum as an, 186.
Embryo prospectors cheated by speculators, 483.
Emeralds of the Carpathians, 28.
Emery replaced by carborundum, 425.
Emlenton, Clarion Co., U.S.A., 215.
Emoluments of Colonel Drake, 196.
" Empire " spouting well of Pennsylvania, 198.
Employer contributes to men's accident fund in Galicia, 54.
—— must give notice of dismissal in Galicia, 54.
Employés must have pass books in Galicia, 54.
Emporium, Baku an, 81-82.
Empties, petroleum spirit returned, etc., Continental rules as to, 521-522.
Enactment against use of locomobiles at Baku, 339.
—— —— in Galicia, 339.
Engineers, oil well drilling. *See* Armitage, De Clercy, Fairman, Lippmann,
    McIntosh, MacGarwey, Neuburger, Prentice, Zipperlen, etc., etc.
Engines, oil-well drilling—
    American, 341-342.
    Austrian, 341-342.
    " Buffalo " motor, 342.
    " Electric " motor, 342, 366.
    English, 341-342.
    French, 341-342.
    German, 341-342.
—— —— cost of, 351, 362, 363, 378, 389.
Engines, oil well drilling, makers of, *recommended*—
    Boulet, France (formerly Herman La Chapelle), horizontal boiler only, 341.
    Bredt, of Ottynia, Galicia, 342.
    Cail engine works (pumping), 341.
    Chaligny & Guyot-Sionnet's traction engines, 341.
    Clayton & Shuttleworths, England, locomobile as pumping engine, 342.
    Fauck, of Vienna, 342.
    Hammer, of Brunswick, 342.
    Seeger, of Czernowicz, 342.
England, 125, 497.
Engler's synthetical production of petroleum, 13-14.
Enlargement, safety bulge (shoulder), of rods, 316.
Enlightened despotism of Incas, 252.
Enniskillen, Canada, 279.
Enmity between foremen borers, 488.
Eocene strata of oil fields of—
    Baku, 8, 85.
    British India, 277.
    Burmah, 277.
    Carpathians, 127.
    Caucasus, 8, 127.
    Crimea, 8, 127.
    France, 7, 127, 277.

Eocene strata of oil fields of (*continued*)—
    Galicia, 7, 32, 34, 277.
    Germany, 277.
    Italy, 277.
    Persia, 150, 277.
    Roumania, 8, 59, 60, 277.
    Turkey in Asia, 277.
Epoch of the Incas, 252.
—— —— mound builders, 183.
—— —— Red Indians, 184.
—— —— trappers, 186.
—— —— troglodytes, 294.
Equalisation of sides of bore hole, 393.
Equations, algebraical, 524.
Equestrian, ground of Baku resounds under an, 69.
Equipment of tools and machinery incidental to sinking oil wells by—
    American rope boring system, 378.
    Canadian system, 363.
    Free fall, hand power, 335-336.
    —— steam power, 351-352.
    —— French make of machinery, 404.
    Hydraulic drilling, 389.
Equipoising " walking beam," 297.
Era. *See* Epoch.
Erie, Lake, 203, 204, 206, 207, 222.
Ermeling's free-fall system, 418.
Erratic career of Colonel Drake, 193-197.
Eruption, Isle of Tcheleken a gigantic, of ozokerit, etc., 146.
Eruptive force necessary to presence of petroleum, 133.
—— trap dyke, of Gaspé, Canada, 237.
Esche's free-fall instrument, 418.
Essentials, Zipperlen's, to presence of petroleum in any given bed, 133.
Estramadura, Portugal, 278.
Ether, inhalation of, as remedy for asphyxia, 284.
—— petroleum, 521-522.
Eureka spouting well, 190.
Europe, oil fields of, 6-9.
    Geology of—
        Carboniferous strata of, never petroliferous, 3.
        Line of strike—
            Main petroliferous, continued through Asia and Oceania, 8.
            Secondary or Mediterranean, 8. *See* Map of Oil Fields of the World facing 8.
        Vast petroliferous reservoir, 7.
    History of petroleum in, 4-7.
    Oil fields of—
        Alsace, 6, 116-118.
        Baku, 6, 7, 66-110.
        Britain, 125.
        Caucasus, 75-81.
        France, 126-146.
        Galicia, 6, 24-57.
        Germany, 6, 111-118.
        Holland, 125.
        Italy, 118-123.
        Russia, 124. *See* Baku, Caucasus.
        Spain, 124.
        Sweden, 125.
Evacuation of Canada by France terminates French chronology of American petroleum, 186.
Evans City, Butler Co., Pennsylvania, 216, 500.
Evaporated product of sand petroleum. *See* Kir.
Everlasting Fire, Temple of. *See* Eternal Fire, etc.
Evidence, documentary, as to title to petroliferous land in Galicia, 50-51.
Evoy Bros. spouting well, 235.
Evrard's reports on geology of Algerian oil fields, 265, 266, 268, 269, 270, 271.
Exaggeration of drilling, etc., results by foremen borers, 488.
Examination of tools, rods, drill, etc., 487.
Exasperation of foreman borer aggravates disaster, 429.
Excavation of oil wells, 280-288.

Exceptional ground of Perpignan boring, 379-380.
—— effort, tools to be examined after, 488.
Excessive cost of transport in early days, 471-473.
Exchange, petroleum—
    American, 200.
    German, 480.
Excitability of Chinese of Canton, 161.
Executive (Council of Mines, etc.). *See* Mining Administration.
Exertion, tools to be examined after special, 488.
Exfoliation of petroleum stone, 144.
Exhaustion, gradual, of Pennsylvanian oil field, 220.
Exhibition, Paris, congress of drilling engineers, 368, 372, 415.
Exorcisms of the foreman borer, 490.
Expanding drills (Reamers), 436-438.
Expenditure and reserve fund before starting oil-well drilling at Baku, 107.
Expense. *See* Cost.
Expensive, American oil-well drilling machinery, in Japan, 165.
Experience, oil prospector must have, 481.
Experiments on petroleum stone, 144-146.
—— Engler's on cod oil, 13-14.
Experimental boring, 485.
Exploration of Madagascar, 273-274.
Explosions during hand-digging of oil wells, danger of, 285.
Explosives, use of petroleum in, 2.
—— and the "torpedo" in oil-well drilling, 444-457.
Exports, American, of crude oil, 246.
Exportation of oil-well drilling machinery—
    America, 342, 404.
    Austria, 342, 405-406.
    France, 341, 403.
    Germany, 342, 405.
Exporters from Manila, nationality and methods of, 178-179.
Extinct race of America, 183.
Extinguish Davy lamp in hand-dug wells, how to, 285.
Extraneous substances in Baku petroleum, 39.
—— —— how removed, 461.
Extreme pumping, increased yield by, 459.
Exudations. *See* Signs of Petroleum.
Eye service of day labourers in Galicia, 291.

F.

Fabian free-fall system, 304-333, 410, 412. *See* Oil-well Drilling.
Fable of Prometheus, 68.
Factory, Nobel's, at Baku, refining results, 39.
Faculties of observation, Indians', 183.
Failure to strike oil at Gabian, cause of, 134-135.
Fainting. *See* Asphyxia.
Fairbank's spouting well, 235.
Fairman, E., prospection of Italian oil fields by, 118, 120.
Fairy tales. *See* Lore, Legendary.
Fall, free. *See* Free-fall system of Oil-well Drilling.
Fall of bodies, formula, 528.
Falling in of bore hole. *See* Landslips.
False economy of hand-dug wells, 293.
—— —— of casing *en colonne perdue*, 329-330.
Falsification of butter with petroleum, 1.
—— of ground with fictitious indications of petroleum, 483.
Famous Shaw well, 232-234.
Farms in American oil fields taken on long lease by speculators during oil fever, 195-196.
Fastening wires of casing to lowering cable, snapping of, 328.
Fat, stability of, 15.
"Fat holes" of Germany. *See Fettlöcher*.
Fatalities, etc., must be reported to state officials of Galicia, 54.
—— at Boryslaw, high percentage of, 46.
Fauck's oil-well drilling system, machinery and tools. See Oil-well Drilling, Free-fall System.

Fauvelle's hand power oil-well drilling system.
    Borings on, at Kertch, 74.
    —— —— at Perpignan, 379-380.
    —— —— at Rioni, 139.
    See Oil-well Drilling Hydraulic System.
Faults (geological), numerous, of Galician oil fields, 29.
Fauna, marine, of the Caspian, common to Black Sea, 69.
—— fossil, of Canada, 239.
Favour, Chinese recently disposed to favour the French *in re* concessions, 162.
Feats of speed in boring. *See* Record.
Features, characteristic, of oil-well drilling machinery of different countries, 402-406.
Fence, etc., land in Galicia to be taken possession of by, 50.
Fendul Saratu, Roumania, 64.
Ferdjionah, Algeria, 270.
Fergane (Turkestan) oil field, 148-150.
    Geological formation, 149.
        Alluvial deposits, 149.
        Altitude, 149.
        Gypsum of, 149.
        Jurassic rocks covered by tertiary beds, 149.
        Limestone, 149.
        Rock salt, 149.
        Undiscovered springs, 150.
            Producing centres—
                Aim-Kichlah, 149.
                Andidjane, 149.
                Isbakent, 149.
                Isfara, 149.
                Kara-Daria River, 149.
                Khokand, 149.
                Maila, 149.
                Makhram, 149.
                Namangane, 149.
Ferric carbide, 19.
Ferry, Smith's, Pennsylvania, 216.
Festal days of the planets, 67.
Festival. *See Fête.*
Festoons, calico printed, of Temple of Eternal Fire, 98.
Fête, the French and Senecas petroleum-illuminated, 185.
Fettlöcher, 6, 113.
Fever oil of America, 195-196.
Fibre, breaking point of strain parallel to, of timber and metals, 507.
Ficht, Samuel, and the Barnsdall spouting well, 209.
Fictitious signs of petroleum to deceive prospectors, 483.
Field. *See* Oil Field.
Fierce competition between Russian and American petroleum, 475.
Fiero spouting well, 235.
Fight, European, between the Nobels and the Rockefellers, 475.
Figuier, L., on explosives, 444.
Filipesci-Ilanica Railway Roumania, 64.
Filling barrels, advice as to, 56.
—— up abandoned oil wells in Pennsylvania under penalty, 245.
Findlay, Ohio, U.S.A., 203, 222, 223, 279.
Fir. *See* Pine.
—— or pine along with beech used as engine fuel, 340.
Fire, distance of, from oil wells, 52.
—— eternal, Temple of, 97-98.
—— wells, Caucasian, 67.
—— —— Chinese Thibet, 163.
—— worship, 7, 67.
Fiscal authorities of Galicia, 27.
—— measure, petroleum only serves Spaniards as a means for enacting a, 252.
Fish of Caspian common to Black Sea, 67.
—— fossil ganoid, 279.
—— —— placoid, 277.
Fissures in strata, richness in oil proportional to number of, 29, 133.
—— vertical, of Pennsylvania, Virginia and Ohio, presence of oil in, 29.
Fiume, Croatia, Austria, 55, 109, 480.
Flames of Châtillon gas, 144.
—— of Temple of Eternal Fire, 97-98.

Flanks of the Carpathians and petroliferous line of strike, 8, 58.
Flannel as filter for crude oil, 191.
Flat, Isle of, Burmah, 154.
—— iron, wire or steel cables. *See* Strength of Materials, Cables, etc.
Flathead River, British Columbia, 240.
Flatulence, petroleum cures, 184.
Fleece. the golden (mythology), 68.
Flesh of animals, easy decay of, 15.
Flooding of oil wells, 436-439.
    Use of refrigerators to prevent, 438-439.
Florida, U.S.A., 229.
Florisdorf, Vienna, 55.
Flow of a pipe, formulæ for calculating output of, 511-512.
Flowing wells, the first, 5.
Fluctuation in Russian and American production, comparative graphical chart, 66.
Fluorescence of cod oil distillate, 14.
—— of crude oil, 48 (a characteristic of mineral oils generally).
Fœtid smell of shale oil, 4, 5.
Folds, anticlinal, of Galicia, 30, 31, 34.
Foliaceous nature of petroleum stone, 144.
Fonvielle, W. de, quoted, 191-197, 214-215, 252.
Foo-choo-koo, China, 161.
Fool's bargain (allowing 25 per cent. tare on distiller's barrels), 56.
Foot (0·3048 metres), 504.
Foreigners, hostility of Chinese to, 161.
—— cannot own land in Roumania, 61.
Foreman borer, work and duties of—
    American, 376-377.
    Canadian, 360.
    Fabian, free-fall hand power, 305, 306, 318, 332, 337.
    Free-fall steam power, 352.
    Hydraulic drilling, 387.
    Rotary drilling, 399.
Forerunner in use of petroleum, 187.
Forest beasts drink at Black Lake, 183, 184, 186.
—— Co., Pennsylvania, 210, 213.
—— of Oulouchema, Daghestan, Russia, 79.
Forewarnings of presence of petroleum and unforeseen spouting, 439-440.
Forge. *See* Smithy.
Fork Scot, Kansas, 279.
Formation, petroliferous characteristic fossils of each, and oil fields in which they
    occur, 276-279.
Forms of petroleum, 1, 2.
Formulæ and memoranda, tables to facilitate calculations, 493-494, 503-535.
Fornova di Tarro, Italy, 121.
Fort Duquesne, 185, 201, 216.
—— —— Pittsburg built around, 201.
Fortnightly change of day gang to night gang, how effected, 489.
Fortune, Colonel Drake's, 5.
Fossils, characteristic, of each petroliferous formation, 276-279.
Foster Creek, Pennsylvania, 501.
Foster Oil Co., Bradford district, U.S.A., 209.
Foundations of petroleum reservoirs, 462.
Foundry, inspection of pipe line pipes at, 479.
Fountain, healing, of Gabian. *See* Gabian.
—— of bitumen (Delaroche), 185.
—— the, spouting well, 198.
Fountains, oil. *See* Spouting Wells.
Four-legged derrick, 308.
Foxburgh, Clarion Co., Pennsylvania, U.S.A., 215, 500.
Foxhill, U.S.A., 227.
Fractional distillation products of cod oil, 14.
—— —— of crude oil. *See* Distillation, Fractional.
Fragile. etc., labels for low flash petroleum products packed in glass vessels, 521-
    522.
Fragments, rock, brought up by drill, sand pump, and core barrel, sampling of, 486.
Framework. *See* Oil-well Drilling Derricks.
Française, la Société, de Boryslaw, 26.
France, exports of American crude oil to, 246.
—— —— —— Russian petroleum to, 162.

France, oil fields of, 6, 126-146, 277, 278, 497.
    Autun shale oil industry, 4, 128.
    Chatillon, 143-146, 497.
    Fontaine ardente, 139-143.
    Gabian, 128-136. 277, 279, 497.
    Le Credo, 139, 278, 497.
    Le Gua, 139, 497.
    Limagne, 137-139, 277.
    Vaucluse, 136.
—— petroliferous strata of, correlated geologically with Caucasian and Galician, 127.
François' free-fall instrument, 418.
Franklin, U.S.A., 214, 215, 500-501.
Fraudulent signs of petroleum, 482-483.
Free-fall. *See* Oil-well Drilling.
Freezing of water surrounding bore hole to prevent flooding, 438-439.
Freight trains of Galicia, 55.
Freminville's hydraulic borings in Isère, 380.
Fremont, Ohio, U.S.A., 203, 222, 279.
French boiler, 341.
—— alliance with Senecas, 185.
—— Canadian settler's chronology of petroleum, 185-186.
—— capital, 126, 178.
—— colonies. *See* Colonies, French.
—— consul. *See* Gauldree Boileau.
—— Creek, Pennsylvania, 213, 215.
—— engines, 341.
—— inertia, 126, 178, 470.
—— initiative, suggestion for, 162.
—— king regardless of both petroleum and Canada, 186.
—— missionaries, 7, 163, 185.
—— oil-well drilling machinery, 341.
—— —— delicacy of, 402-403.
—— petroleum depôts, 480.
—— renown, 185.
—— tank steamers, 470.
Friction of roads to load drawn, 529-531.
Fromman's improvements in rope boring, 303.
Frontage, extensive, of Baku Quay, 81.
Frost, effect of, upon barrels filled too full, 56.
Frustum, 526.
Fuchs and de Launay quoted, 204, 205.
Fucoids, 277.
—— of Canadian corniferous limestone, 239.
Fuel, natural gas as, 162, 217.
—— petroleum as, 2, 340, 468.
—— substitute for, compressed air as, 256.
—— wood (beech and fir), 340.
Fulgurite, Pictet's, 448, 449.
Fulminate of mercury, 449, 454.
Funk's "Fountain," the first American spouting oil well (Oil Creek Humbug), 198.
Furnace, the centre of the earth an incandescent, 17.
Fuse of electric blasting outfit, 452.

G.

Gabian, the healing fountain of—
    A. Hugo on, 6, 131.
    Bishop of Beziers on, 6, 131.
    Dr. Rivière on, 6, 130.
    Geology of—
        Devonian strata, 134, 279.
        Distorted strata, 129.
        Miocene strata, 7, 129, 277.
        Nummulitic strata, 129.
        Permian strata, 134.
        Tertiary strata, 135.
    Oil of—
        Average density, 3, 497.
        Universal panacea, 2.

Gabian, the healing fountain of (*continued*)—
    Oil-well drilling at, first boring—
        Canadian system rejected in favour of free fall by Dru, 132.
        Section of strata met with, depth, 132.
    Second boring—
        Canadian system, 134.
        Section of strata met with, depth, 134.
        The cause of failure of both borings, 134.
Gaiski, application of free-fall to rope boring, 303, 419.
Galician oil fields, 25-49, 276-278.
    Geology—
        Anticlines, presence of petroleum in arches of, described and illustrated, 30, 31.
        Arenaceous rocks (eocene sandstone, Ropianka *Schiefern*), 32.
        Argillaceous strata, 30.
        Carbon coal, 33.
        Cretaceous strata, 32, 33, 34, 48, 278.
        Eocene strata, 7, 32, 33, 34, 48, 277.
        Geological horizon of different producing centres, 48.
        —— —— compared with that of Canada and the United States, 32, 33.
        —— structure compared with that of Virginia, Pennsylvania and Ohio, 29.
            Classification of petroliferous strata, 35.
                Belowosa shales, 35.
                Magura sandstones, 35.
                Ropianka shales, 35.
                Smilno shales, 35.
        Highly inclined strata, 29.
        Lignite, 33.
        Line of strike, 28, 29.
        —— signs of petroleum lost in valley of Bukovina, 49, 63.
        Melenitic beds, 34.
        Miocene, 32, 33, 48.
        Neocomian, 43, 48, 278.
        Nummulitic beds, 7, 34.
        Oligocene beds, 34.
        Pliocene strata 276.
    History of, 24-27.
        Bobrka oil well, 32, 39, 40, 41, 48.
        Distillery of Joseph Hecker and Johann Mitis at Boryslaw, 24.
        Early struggles, 24.
        Final success, 25.
        Hand-dug wells, 24.
        Jews, rôle of, in, 25, 46.
        Ozokerit, discovery of, saves situation, 25.
        Statistics, 27.
        Subsequent failures, 25.
    Land-buying in—
        Possession, importance of taking, 50.
        Price of land, 51.
            At date of discovery of American petroleum, 5.
        Purchases, all, to be concluded before notary, 50.
        Leasing—
            Cost of concessions, 51.
            Possession, importance of taking, 51.
        Spouting wells increase price of neighbouring land, 51.
        Who may buy or lease—
            Restrictions on minors, 51.
            Women's rights, 51.
    Laws of master and servant—
        Accidents to be reported to mining authorities, 54.
        Accident fund, 54.
            Doctor's fees, 54.
            Masters' contribution, 54.
            Medicine, 54.
            Servants' contribution. 54.
        Engaging and dismissing labourers and professional staff, 53-54.
        Notice to workmen, 54.
        —— to professional staff, 54.
        Workmen's pass-book, 53.
    Mineralogy of—
        Carpathians, Jules-Radu on the gold, the gems, the metals and the salt of, 28.

Galician oil fields (*continued*)—
  Oil of—
    Bituminous or rich in paraffin of cretaceous strata, 33.
    Density of, 3, 47, 48, 57.
    Depth at which struck, 30-31, 35-36, 41-42.
    Distillation products of, 37, 39.
    Distilleries or refineries of, 55.
    Price of, 54, 55, 56.
      Causes which tend to raise, 54.
      —— —— —— to lower, 55.
      —— of variation in, 55.
      Influence of quality on, 55.
      —— of season on, 55.
    Prices ruling in different localities, 55.
    Selling of—
      American, Galician and English barrels, 56.
      Contracts, 55-56.
      Filling barrels, 56.
      Per fixed number of barrels, 56.
      Tare of barrels, 56.
      Whole production for fixed period, 56.
  Oil-well drilling and excavation—
    American rope boring impracticable, 296.
    Canadian system, 354, 370-373.
    Excavation (hand dug), 24, 25.
  —— —— aeration of, 282-285.
    Raky system, 423.
    Lamps, use of, interdicted in, 24.
  —— —— operators and owners of ozokerit mines past and present—
    Austro-Belgian Co., 46.
    Bergheim & MacGarwey, 38-39.
    Gartenberg, 26.
    Hecker, Joseph (1817), 24.
    La Société de la Banque Galicienne de credit, 45.
    La Société Française, 45.
    Liebermann, 26.
    Lindenbaum, 46.
    Lukasiewicz, 40.
    McIntosh, 46.
    Mitis, Johann (1817), 24.
    Neuburger (author), 51.
    Rice, 46.
    Schreiner, Abraham, 24.
    Wagmann, 26.
  Pipe lines, 477-478.
  Quotations and references bearing upon, from—
    Berthelot, 33.
    Ducasse, 30.
    Gauldrée-Boileau, 29.
    Gesner, 33.
    Heurteau, 50.
    Noth, 32.
    Petroleum depôts of, 479.
    Radu, 28
    Walter, 33-41.
  Subdivisions of oil field—
    Central, 41-46.
    Eastern, 46-49.
    Western, 33, 40, 41-42, 44, 74.
    Central oil field—
      Producing centres, quality and quantity of their oils, past history and present condition of the industry—
        Boryslaw, 24-26, 44, 46, 276, 478. *See* Ozokerit in General portion of Index.
        Galowka, 41.
        Holowecko, 42, 48, 51, 57, 495.
          Section of strata encountered in boring at, 35.
        Hoschow, 41.
        Leschzowate, 41.
        Lodyna, 41, 48.

Galician oil fields (*continued*)—
    Subdivisions of oil field (*continued*)—
        Central oil field (*continued*)—
            Producing centres, etc. (*continued*)—
                Olsanica, 41.
                Polana, 26, 42, 48.
                Potok, 41.
                Ropianka, 41.
                Schodnica, 32, 33, 46, 48, 478.
                Starzvwa, 41.
                Stopuciani, 42.
                Uherce, 41, 47.
                Ustianowa, 41, 48.
                Wankowa, 41, 48, 57, 477, 495.
        Eastern Galician oil field—
            Producing centres, quality and quantity of their oils, their past history and
            present condition of the industry—
                Nadworna, 46.
                Rungury, 46.
                Sloboda Rungurska, 46, 47, 48, 57.
                —— Kopalnia, 46, 47, 48, 57.
        Western Galician oil field—
            Producing centres, quality and quantity of their oils, their past history and
            present condition of the industry, and the geological horizon of the
            petroliferous strata—
                Bad-Iwonicz, 41.
                Barwineck, 32, 37.
                Bobrka, 39, 48.
                Iwonicz, 41, 48, 495.
                Kleczany, 36, 48, 57, 495.
                Klemkowka, 41, 48, 57.
                Kobilacka, 37.
                Krocienko, 37.
                Kryg, 37, 48, 57, 495.
                Libusza, 37, 48, 57, 495.
                Lipinki, 37, 48, 57.
                Mécina, 37, 48, 57, 495.
                Nowosielce-Gniewosz, 41.
                Pagorzin, 37, 48, 57, 495.
                Ropa, 36, 48, 57, 495.
                Ropianka, 37, 48.
                Ropica Polska, 37.
                —— Ruska, 37, 48, 57, 495.
                Rowna, 38.
                Rymanow, 41.
                Sekowa, 37, 57, 495.
                Siari, 37, 48, 57, 495.
                Stanislawow, 33, 36, 48.
                Starunia, 37, 48, 57, 495.
                Targowiska, 41.
                Vostkowa, 48.
                Wietrzno, 24, 32, 38-39, 48, 57, 477, 495.
                Wojtkowa, 37, 57, 495.
                Pipe lines, 477-478.
    Work—
        Application to council of mines for permission to, 52.
        Formalities to be observed before starting, 52.
        Registration by prefect of district, 52.
        —— of manager or overseer, 52.
Galleries. *See* Tunnels.
Gall's Brook, Canada, 237.
Galowka, Galicia, Walter's report on, 41.
Gang, day and night, how they change spells, 489.
Ganoid, fish, 279.
Gan-yo, Chinese Thibet, 163.
Gardagrelli, Italy, 120, 278.
Garfield, Warren Co., 210.
Garnetz, Russian dry measure (3·277 litre), 503.
Garnietz, Galician dry measure (1·59 litre), 503.
Gartenberg of Boryslaw, 26.

Garvan, Roumania, 64.

Gas, natural, of Châtillon and its uses, 143-144.

—— Chinese Thibet, 163.

—— Kansas, 279.

Gasoline an ingredient of crude oil, 3.

—— conveyance of, on continental railways, 521-522.

Gaspé, Canada, 232, 236, 237, 238. *See* Canada, oil field of, Gaspé, subdivision of.

Gas piping, failure of, in hydraulic oil-well drilling, 382.

Gau petroliferous tertiary (eocene) beds, 7.

Gauging barrels, tables for, 515-521.

—— —— formulæ for, 526.

Gault, free-fall instrument of, 412.

Gauldrée Boileau quoted, 29, 235, 471-472.

Gaymard's report on Freminville's boring, 380.

Gear of oil well drilling machinery. *See* Oil-well Drilling.

Gelatine, blasting, 446, 447, 448, 451.

Gelatinous animals and plants and origin of petroleum, 12.

Gems of the Carpathians, 29.

Generator of electricity of blasting outfit, 452-455.

Genesis of petroleum. *See* Origin of.

Genius, American, 475.

"Genius" of the well, propitiating, 490.

Genoa, Italy, 480.

Geological horizon of petroliferous strata of different localities with characteristic fossils, 276-279.

—— survey of Canada, origin of petroleum, 12.

Geologists' theory of origin of petroleum, 21-23.

Geology of the different oil fields, their structure and the horizon of the petroliferous strata. *See* respective Oil Fields.

Geometrical progression formulæ, 524.

—— formulæ (mensuration, etc.), 525.

Geoucena Saru, Java, 174.

George Graff spouting well, 235.

Gerace, Italy, 120, 278.

German oil field, North, bounded by lower courses of Weser and Elbe, 110.

    Geology of, 112, 276-279.

        Deluvial sands above petroliferous strata, 112.

        Denudation of beds above petroliferous, 112.

        Depth at which oil is struck, 114-115.

        Fettlöcher, 6, 113.

        Hydrostatics of, 112.

        Jurassic limestones and sandstones, 112, 278.

        Lithological modifications of petroliferous beds, 112.

        Marshes, a country of, 112.

        Petroliferous strata a transition between jurassic and lower cretaceous (tithonic bed), 112.

        Quaternary strata of, 278.

    History, remote—

        Fettlöcher, method of collecting oil from, 112.

    —— recent—

        Deutsch Pennsylvanien, 113.

        Oelheim, the German "Petrolia," 113.

        The petroleum swindle, 115.

    Line of strike, towns and districts passed by—

        Celle, 111, 116.

        Hanover, 111.

        Hartz Mountains (starts from), 111.

        Holstein terminates, 111.

        Plain of Luneburg, 111.

    Oil of, properties, 114, 116.

    Producing centres, quality and quantity of their oils, their past history and present condition of the industry—

        Brunswick, 115.

        Celle, 115.

        Hanigsen, 276.

        Hanover, 116.

        Koenigsen, 115.

        Linden, 276.

        Luneburger Heide, 115, 276.

        Mehldorf, 276.

German oil field, North (*continued*)—
    Producing centres, etc. (*continued*)—
        Neustadt, 116.
        Oberg, 114, 496.
        Oedesse, 113, 116, 496.
        Oelheim, 113-115, 116, 496.
        Sehnde, 115, 496.
        Soltau, 115.
        Steinforde, 114.
        Tegernsee, 276.
        Weetzen, 114, 116, 496.
        Wenigsen, 116, 496.
        Werden, 111, 113.
        Wietze, 276.
    Production, 116.
    See Alsace, Oil Field of.
Germany, statistics of American exports of crude oil to, 246.
—— petroleum depôts, 480.
—— Russian exports of petroleum to, 102.
Gernowicz, Roumanian forwarding station, 479.
Gesner, Abraham, origin of petroleum, 12, 33.
    Shale oil distilleries of, 190.
    Trinidad asphaltum, fractional distillation products of, 249.
    —— —— properties of, 250.
Gilbert well of Foster Oil Co., Bradford, U.S.A., 209.
Gilsonbury, Ohio, 223.
Gintl, Dr., statistics of Galician petroleum, 26.
Giorgi on lignite distillation products, 120.
Girgenti, Italy, 118, 120.
Glade Run, Butler Co., Pennsylvania, U.S.A., 216, 500.
Glass insulators used in electric blasting outfit, 453.
—— vessels filled with petroleum spirit, etc., continental railway regulations as to packing of, 521-522.
Glinoi Balka, Russia, 75, 110.
Glodeni, Roumania, 64.
Glucinum carbide, synthetical production of, 21.
—— —— yields methane with water, 21.
Glut of pipe line companies, former, in America, 473.
Gneiss, 60.
Goat, Ben Lakdar and the legend of the, 259-260.
Gods, petroleum the light of the, 185.
Goenoeng Sarie, Java, 174.
Gogor, natural gas well of, 174.
Gold of Peru, Roumanian petty landlords think they have found, when they strike a pocket of oil, 62.
Gold of Carpathians, 28.
Golden fleece (mythology), 68.
Gordon spouting well, Washington. U.S.A., 216.
Gorlice, Galicia, 40, 479.
Gouliani, South Caucasian oil field, 80.
Goullet Collet's rope boring improvements, 302.
Gouriamti, South Caucasian oil field, 80.
Gout, petroleum as a remedy for, 71.
Governments of different countries and regulations as to oil-well drilling, etc. *See* under respective countries.
Grabowski, origin of petroleum, 19.
Grabs, 480. *See* Tool wrenches, etc.
Grand Duke Michael and No. 3 Schibaeff spouting well, 91.
Grandeur of Baku, 82.
Granites, hard, and quartzose rocks and veins of sandstone, monthly progress in hand-digging oil wells through, 290.
Granitsa, Russian frontier petroleum depôt, 480.
" Grant," the, spouting well, 199.
Gravel, petroleum as a remedy for, 71.
—— time occupied in excavating cubic metre of, 506.
Gravitation, flow of oil through pipe lines by, 474.
Gravity, centre of, how to find—
    Lines, 527.
    Solids, 528.
    Surfaces, 527.

Gravity, specific, 493-494.
Grease waggons, one of the former uses of Galician petroleum, 24.
Great Oil Belt, Western Virginia, 222.
Green County, Pennsylvania, 217-218.
Greenish aspect of some natural petroleums, 2.
Greifenhagen, free-fall instrument of, 413.
Grenelle artesian well, 403, 406.
Grenoble, France, 139.
Griazi-Tzaritzin Railway Co., 466.
Grindstone for smithy, cost of, 336.
Grisoutine B, 451.
—— F, 451.
—— gelatine, 451.
Grosnaia, 73, 77.
Gross and nett weight of barrels and 25 per cent. tare thereon, 57.
Ground. *See* Land.
—— renting. *See* Land, Leasing of.
Groundwork. *See* Earthwork.
Grovetown, Pennsylvania, 500.
Growth, rapid, of—
   American mushroom towns, 199-201, 211.
   Baku, 81.
   Galician towns, 26.
Gua (Isere), 7.
Guano, petroleum of suggested similar origin to, 12.
Guard of honour to Baku, Elizabethpole a, 81.
Guasco, De, on small bitumen mines of Limagne, 138.
Guayaquil, Ecuador, 250.
Guèbres and worship of inflamed petroleum, 7, 68.
Guide, auger stem, described, 321.
   illustrated, 320.
Gun cotton, 451.
Gunda (Bengal), 277, 498.
Gypsum, 34, 85, 132, 149, 263, 267, 269, 270, 425.

H.

Habitations prescribed, distance from oil wells, 52.
Haf-Cheid, Persia, 151.
Hagenau, Germany, 116.
Haï-nan, Formosa, 167.
—— remunerative yield of trial borings, 167.
Haitha, China, 161.
Haldimand, Canada, 236.
Halifax, Nova Scotia, 190.
Hamburg, 480.
Hammer, G., Brunswick, oil-well drilling engines of, 342.
Hammers, cost of set of, for smithy. 289, 336.
Han and Kobolow, data of Russian spouting wells, 94-95.
Hand-digging of oil wells. *See* Oil Wells, Excavation of.
Hand-pumping of oil wells, 458-460.
Hanigsen, Germany, 276, 278.
Hanover, 110, 480. *See* German (North) Oil Field.
Hanway, Joseph, and early history of Caucasian petroleum, 69, 70, 71.
Harbour, Baku a magnificent natural, 82.
Hard ground, rope boring impracticable in, 377.
Harding and O'Connor, U.S.A., 208.
Hardy, "oilman," Roumania, 62.
Harklowa, Galicia, 32.
   Duration of yield of oil, 32.
Harmony, Butler Co., Pennsylvania, 216, 500.
Harness, horse in. *See* Horse.
Harrison, Ohio, 228.
Hartz Mountains, 111.
Hatchets. *See* Axes.
Hattan's application of steam to rope boring, 303.
Havre, France, 480.
Hay packing material for glass vessels filled with petroleum spirit, etc., 521-522.
Headache, petroleum a remedy for, 71.

Heat, internal, of earth, and synthetic origin of petroleum, 16-20.
Heavy oil. *See* Petroleum, Heavy.
Hecker, Joseph (pioneer, Galician), crude oil distiller, 24.
Hectare (2 acres, 1 rood, 35 poles), annual rent charge per, of land in Madagascar, 275.
Heide, Luneburger, Germany, 115, 116, 276, 278.
Hemp cable. *See* Cable, Hemp.
Henning, Germany, 278.
Henry's Bend, Venango, U.S.A., 215.
Herault, department of, France, 127-136.
—— River, 128.
Hermetic sealed casing, 180, 436.
Heurteau on Galician commercial morality, 50.
Hexane from cod oil, 14.
Hickory, West, Forest Co., Pennsylvania, U.S.A., 213.
Hidaka, Japan. 163.
Hieroglyphic markings of eocene strata of Galicia, 34.
Higgling, Drake's, over aliquot part of royalty, 193.
Highly inclined strata of Galicia, 29.
Highway, ratio of traction to load drawn on, 529-531.
Hihil, Algeria, 271.
Hirbach, Alsace, 277.
Hirsingen, Alsace, 277.
Hirsova, Roumania, 64.
Hissing noise of gas in oil wells, 163, 439.
Historical account of British trade over the Caucasus (Jonas Hanway), 69-71.
Historiographer of American petroleum (Gessner), 190.
History of petroleum in—
    Algeria, 259, 260, 265 *et seq*.
    Alsace, 4, 6, 114, 118.
    America, 182-201.
    Baku, 66-73.
    Burmah, 155-156.
    Canada, 232-235.
    China, 160.
    Europe, 4-7.
    France, 6, 126-145.
    Galicia, 24-27.
    Germany, 6, 112, 115.
    Italy, 6, 118.
    Japan, 163.
    Peru, 252.
Hit, Bagdad, Turkey in Asia, 277.
Hitchcock's theory of origin of petroleum, 11.
Hofen's theory of origin of petroleum, 15.
Hohne's spouting well, 235.
Holland, 125.
—— Russian petroleum, exports of, to, 102.
Holowecko, 57.
—— geographical situation of, 42.
—— oil field of, *illustrated*. Shared by French and German Co., 44.
—— section of strata encountered in boring at, 35.
—— Walter's report on, 42.
Holstein, North German line of strike terminates in, 111.
Holy Isle, 70.
—— City and the fire worshippers, 67.
Homologues of benzene. *See* Benzene.
Hong Kong, China, 162.
Horizon, geological, of petroliferous strata and characteristic fossils, 276-279.
Horn insulators of electric blasting outfit, 453.
Horsdorf, Germany, 278.
Horsfield on mud volcanoes, 176.
Horses, harness, effect of method, of on amount of work that can be got from, 529-531.
Horse power, 529-531.
—— Neck, Virginia, U.S.A., 220.
Hoschow, Galicia, 41.
Hose, india-rubber, use of, in steam jet levigation of oil wells, 457.
Hostility of Chinese, 160.
Hotels of Baku, 82.
Ho-Tsin, China gas well, 163.

House accommodation for workmen, 337.
Huber's German works at Montechino, 123.
Hue, Fernand, quoted, 64, 84-85, 156, 164-165, 179, 199-200, 211-212, 249, 258.
Hughes River, Western Virginia, 221.
Hugo, A., Gabian oil fountain, 6, 131.
    Roque-Salière bituminous shales, 136.
    The burning fountain, 139-140.
Huidobro, Spain, 124.
Human remains, fossil, 276.
Hungary, 124, 278.
Hunt, Professor Sterry, origin of petroleum, 11.
Huron, Lake, America, 204, 239.
Hutchinson's first pipe line, 473.
Hyde and Egbert's Maple Shade spouting well, 199.
Hydraulic drilling of oil wells. *See* Oil-well Drilling, Hydraulic System.
—— "jars," Fauck, 420.
Hydrogen of petroleum, 16.
Hydrocarbides, synthesis of and bearing thereof on origin of petroleum, 11, 16, 17, 18, 19, 20.
—— aromatic 19-20.
—— saturated, 18.
—— unsaturated, 20.
Hydrometer. *See* Baumé's Areometer.
Hypocritical population of Galicia, 50.
Hypotheses as to origin of petroleum. *See* Origin of Petroleum.

I.

Iburi, Japan, 168.
Ice of flood water, artificial drilling through, 438-439.
Idolatry of fire-worshippers, 67.
Ienikalé, Straits of, 66.
Igneous rocks, petroleum never found in, 22.
Ignition of fuse in electric blasting, 454.
Ilanica, Roumania, 64.
Illinois, 202, 226, 501.
    Abnormal density of its petroleum, 226.
Illuminating oil. *See* Petroleum, Illuminating.
—— —— natural, 75. *See* Petroleum, Natural Illuminating.
Illumination of French and Seneca fête by inflamed petroleum, 184.
Ilsanika, Roumania, 64.
Ilskaia, Valley of, Kuban, 75.
Imamzade, Persian convent of, 151.
Imbedded tools and rods. *See* Accidents.
Immersion, loss of weight of wooden rods on, 356.
—— effects of water pressure on wood, 369.
Impermeable superincumbent bed essential to presence of petroleum, 133.
Impetus given by Nobel to Russian petroleum industry, 464-470.
Impracticability of rope boring in European oil fields, 377.
Improvements in oil-well drilling machinery, systems and tools. *See* Oil-well Drilling, Improvements in.
Improvised reservoirs in case of unforeseen spouting, 90, 91.
Impurities of Sloboda (Galicia) petroleum (*Cal*), 247.
—— —— Baku petroleum, 39, 80, 91.
Inaccessability of Canadian oil fields in early days, 471.
—— —— Roumania, 61.
—— —— Tchéléken, 147.
—— —— Turkestan, 149.
Inaccuracy of Galician custom-house statistics, 27.
Inaction of French industry and commerce, 178.
Incandescent furnace in centre of earth, 17.
Incas, enlightened despotism of, 252.
Incendiary agent, petroleum as an, 184.
Incense in temples of fire-worshippers, 67.
Inches, table of reduction of, to millimetres, 505.
Inclined strata. *See* Deviation.
Incrustation of sand and petroleum (*Kir*), 91.
Indemnity title to petroliferous land—
    Russia, 108.
India, British East, 151-160.

India-rubber hose, use of, in steam jet levigation of oil wells, 457.
Indiana, U.S.A., 202, 204.
Indians, Seneca, 185.
Indies, Dutch East, 170-176.
Indigenous nature of petroleum, 21, 23.
Indurated sand and petroleum (*Kir*), 80, 91.
Industry, conquest of barbarism by, 82.
Inertia of French industry and commerce, 126, 178, 179.
Infiltrated water carries petroleum to collecting trenches, 113, 191.
Infiltration of petroleum into strata other than indigenous, 135.
—— —— across Carpathians into Hungary, 124.
Inflammability of petroleum stone, 144-146.
Inflammable, highly, label for vessels containing spirit, etc., on continental railways, 521-522.
Inflammation, petroleum a specific for, 184.
Inflations, surface of Baku, 69.
Influx of Déclassé's, etc., into Pennsylvania, during oil fever, 195.
Infusorial earth, kieselguhr as a packing material for vessels filled with petroleum spirit, etc., 521-522.
—— as inert base of explosives, 445 *et seq*.
Ingenuity, American, 475.
Ingour, embouchure of, 80.
Ingratitude of republics, 197.
Ingredients of crude oil, 3.
Inhalation of ether and strong vinegar as remedies for asphyxia, 284.
Initial expense of hand digging *versus* drilling oil wells, 292-293.
Initiative, French suggestion for, in China, 162.
Inland oil fields of Burmah, 156-160.
Innovations, Nobel's, in transport of Russian petroleum, 463-470.
Innovators. *See* Drake, Nobel, Roberts, etc., etc.
Inscriptions on temple of eternal fire at Baku, 97.
    At Walsbronn, 117.
    Roman, at Burning Fountain (*Fontaine Ardente*), 143.
Insolvency, Colonel Drake's, 196, 197.
Inspection of casing pipes at foundry, 479.
—— —— oil-well drilling tools and machinery. *See* Foreman Borer, Duties, etc., of.
Instinct of deduction, Indians', 183.
Instrument, free-fall. *See* Oil-well Drilling ; Free-fall Instruments.
Insulation of wires, etc., of electric blasting outfit, 452-455.
Intensity, calorific. *See* Calorific Intensity.
Intercourse between foremen borers, 488.
Interdiction of locomobiles in Baku and Galicia, 389.
Interest, land, in United States, 244-245.
Interference, foreman borer exempt from engine driver's, 488.
Interminable formalities and red tapeism of Russian officials, 109.
Intermittent action of spouting wells, theory of, 85-86.
Internal heat of the earth. *See* Origin of Petroleum.
Intoxicant-non (?), petroleum a (Hanway), 71.
Intumescence. *See* Mud Volcanoes.
Inundation. *See* Flooding.
Inured, Burmese, to bad atmosphere of dug wells, 157.
Invalidity, risk of, of title to Galician land, 50, 51.
Inventory of oil-well drilling tools. *See* Estimates under respective systems.
Invulnerable in combat, petroleum rendered Indians, 185.
Iodide springs—
    Bad-Iwonicz, Galicia, 41.
    Tamansk, Russia, 75.
Ionian Islands, 123.
Iquana River, Colombia, 252.
Iridescence of shower of spouting petroleum, 94.
—— of water due to petroleum, 188, 252.
Iridium, metallic, in rotary oil-well boring, 394.
Iron City, Pittsburg, 217.
Iron, bar and wrought, cost of, for smithy, 289, 336.
—— cables. *See* Cables, Iron.
—— casing pipes, cast iron, sheet iron, 300.
—— —— and cost of, per running foot, 437.
—— —— dimensions of, in rotary drilling, 398.
—— —— hermetically sealed, 436.

Iron, metallic, in meteorites and synthetical origin of petroleum, 19.
—— mines of China, 160.
—— percentage of, in Gabian Marcassite, 133.
—— resistance of (strength), cast, sheet, wrought, steel, 507.
—— rods, dimensions of, 383.
—— —— used in oil-well drilling, weight per running metre in air and water, 513-514.
—— tools, tempering of, 323.
—— weight of square metre of sheet, from thickness, 328.
Iroquoise Indians, 185.
Irrawaddy River and oil field, Burmah, 156, 159, 168-170.
Irreparable injury to bore hole by total deviation, 29, 428.
Irresolution of foreman borer aggravates accidental disasters, 429.
Isbakent, Turkestan, 149.
Isere, France, 7.
Isfara, Turkestan, 149.
Ishikari, Japan, 168.
Italy, exports of Russian petroleum to, 102.
—— oil fields of, 6, 118-123, 276, 277, 497.
    History of—
        Abbruzzi, the, 118, 119, 120.
        Ancient Italy and petroleum, 118.
        Agrigentum, the wells of, 118.
        Diodorus of Sicily on, 118.
        Dioscorides on, 118.
        French capital diverted to, 126, 143.
        Miano, use of petroleum from wells of, as an illuminant in Middle Ages, 118.
        Pliny on, 118.
        Sicilian oil, 118.
—— geology of, 7, 9, 276, 277, 278.
—— miocene strata of, 7, 119.
—— pliocene strata of, 276.
    Northern oil field, 119-123, 276-277 ; density of oil, 3.
        Producing centres—
            Bologna, 119, 120, 277.
            Faenza (Province of Emilie di san Colombo), 276.
            Fornova di Taro, 121.
            Medesano, 121.
            Miano, 120, 121, 123, 497.
            Milan, 120, 121, 123.
            Modena, 122.
            Montechino, 121.
            Neviano, 121.
            Ozzano de San Andréa del Taro, 121.
            Padua, 277.
            Parma, 120. 123, 497.
            Piedmont, 497.
            Pietra Mala, 121, 277.
            Plaisance (Placentia), 120, 121, 277.
            Salo, 123, 497.
            Salsomaggiore. 120.
            Salsominore, 120.
            Tocco, 119, 123, 497.
            Vergato, 277.
            Villeja, 121, 122.
        Sicilian oil field—
            Girgenti, doyen, the, of European oil fields, 118, 120.
            Pliny its godfather, 120.
        Southern oil field, 120, 278.
            Aquila, 120.
            Lignite mines of, 120
                Yield of crude oil distillate from, 120.
            Producing centres—
                Chieti, 120.
                Gardagrelli, 120.
                Gerace, 120.
                Manopello, 120.
                Rionero de Molise, 120.
                Tiriolo, 120.
                    Operators, etc., in Italian oil field—
                        Deutsch, 121.

Italy, oil fields of (*continued*)—
    Southern oil field (*continued*)—
        Producing centres (*continued*)—
            Operators, etc. in Italian oil field (*continued*) —
                Fairman, 120.
                Giorgi, 121.
                Huber, 121.
                Klobassa & MacGarwey, 122.
                Lippmann, 121.
                Marchand, 122.
                Zipperlen, 121-122.
    Statistics, 122, 497.
Itanescu, Roumania, 63.
Itch, petroleum cures, 7, 68.
Iwonicz, Galicia, 30, 41, 48, 495.

J.

Jacobi, Russia, 80.
Jakushkino, Russia, 125.
Japan, oil fields of, 163-168, 276, 499.
    American boring plant, cost of, in, 165.
    Ancient knowledge of petroleum in, 163.
    Excavation or hand-digging of oil wells in, 163-165.
        Air machine used in, 164-165.
        Depth of wells, 164.
        Earth and stones raised in a net, 164.
        Oil raised in buckets, 165.
        Reflector, primitive, used in, 164.
        Tunnels, horizontal, leading from bottom of well, 164.
        Ventilation, 164-165.
        Width of wells, 164.
    Geology of, 276.
    Hue, Fernand, quoted, 164-165.
    Oil of, density of, 166, 168.
    Producing centres—
        Akita-ken, 163.
        Aomori, 163.
        Echigo, 163, 166, 168, 276, 499.
        Hidaka, 163.
        Iburi, 163.
        Ishikari, 163.
        Kozodzu, 166.
        Mabana, 166.
        Machikata, 166.
        Miyôhôji, 166.
        Némuro, 163.
        Nïgata, 163.
        Orguni, 167-168, 499.
        Oshima, 163.
        Schinano, 163, 166.
        Tosan, 163, 167-168, 499.
        Toutomi, 163.
        Ugo, 163.
        Yamataga, 163.
        Yokoyama, 163.
    Production, total annual, 167, 499.
Japara, Java, 178.
"Jars," 359, 375, 418, 420. *See* Oil-well Drilling.
Jars, earthenware, and Burmese transport of petroleum in, 156.
Jason, the chief of the Argonauts, 68.
Java, oil fields of, 3, 169, 172-175, 177, 276, 499.
    Geology of, 172, 173, 276.
        Depth at which oil is struck, 174.
        Mud volcanoes of, 173.
        Tertiary strata, 172, 276.
    Horsfield quoted on Grobagan mud volcano, 173.

Java, oil fields of (*continued*)—
  Oil fields—
    South coast oil fields producing centres—
      Ngawi, 173.
      Poerwadi, 173. .
      Tjelatjap, 173.
    North coast oil field—
      Cheribon, 499.
      Dandang Ho, 499.
      Djaba Kota, 174, 499.
      Goenoeng Sarie, 174.
      Gogor, 174.
      Grobagan, 173.
      Japara, 173.
      Kotéï, 174.
      Lidah, 174.
      Madoera, 174.
      Matatœ, 174.
      Panolan, 173.
      Plœtœran, 173.
      Rembang, 173.
      Semarang, 173.
      Soerabaja, 174.
      Timor, 176-177, 499.
      Tinawen, 173.
      Toeban, 174.
      Wronoko, 174.
  Pipe lines of, 174, 478.
    See Timor, Isle of.
Jenery and Evoy spouting wells, 235.
Jesuit missionaries and rope boring, 294.
Jews in the history of Galician petroleum, 25, 46.
Jod, Hungary, 124.
Joiners. *See* Carpenters.
Joints, ball and socket, of pipe line pipes, 478.
Joists, etc. *See* Derricks.
Joke in *re* superstition of workmen, demoralising effect of, 489-490.
Jones, Rupert, on origin of petroleum, 12.
Jordan River, Palestine, 278.
Joseph, Father, quoted on history of American petroleum, 185.
Journal, oil-well drilling, 488, 489.
  Models of, 491, 492.
Jujuy, Argentine Republic, 258, 278.
Jumping on the " walking beam " to work it, 168.
Juncker, P., on crude oil from Puy de la Poix fluid bitumen, 139.
Junks, Burmese, transport of petroleum in, 156.
——— Chinese, caulking of, with petroleum, 160.
Jupiter, worship of, 67.
——— incense burned in honour of, 67.
Jura mountains, 11, 139.
Jurassic formation, petroliferous strata of, 7, 112, 227, 228, 278.

K.

Kahatan, Beluchistan, 151, 277, 498.
Kaitavotabassar, Daghestan, Russia, 79.
Kalantarow, No. 1 spouting well, 95.
Kama River, transport of petroleum to Perm, Russia, on the, 469.
Kamischki, Russia, 125, 279.
Kanawha River, Western Virginia, 221.
Kane, Pennsylvania, 212.
Kansas, U.S.A., 202, 227.
——— outcrop of carboniferous strata in, 203.
Kansas City, gas wells of, 279.
Karria-Daria, Turkestan, 149.
Kashavashirin, Persia, 150, 151 498.
Kastcheef spouting well, Baku, 95.
Kazan, Russian depôt, 125, 279, 480.

Kentucky, oil fields of, 32, 87, 202, 225, 279.
    American oil well, 225.
    Geology of, 32.
        Niles's Register, 225.
    History of, 225.
    Producing centres—
        Burkesville, 225.
        Clinton, 225.
        Cumberland, 225.
        Pulaski, 225.
        Rig Sandy, 225.
        Wayne Russel, 225.
    Production of, 225.
Kernel of centre of earth, influence of intense heat of, on synthetical origin of petroleum, 16-20.
Kerosene, 5. *See* Distillation Products (fractional) of Crude Oil.
Kertch, Russia, Birkel's boring at, 74.
—— Straits of, 75.
Key of pipe lines in safe custody of pipe line agent, 476.
Kharkow, Russia, 110, 480.
Khidirzind, Baku, 101.
Khokand, Turkestan, 149.
Kiauk Phyu, Isle of Ramri, Burmah, 154, 498.
Kier of Pittsburg and medicinal sale of crude oil, 189.
Kieri, Zante, 123, 124, 497.
Kieselguhr. *See* Infusorial Earth.
Kiew, Russia, 110, 480.
Kilogramme (2·205 lb.), 504.
Kilogrammetre defined, 528.
Kilometre (0·62 mile), 504.
Kimpolung, Galicia, 49.
Kind, free-fall instrument of, 410-412.
—— origin of wooden rod system of drilling, 368-369.
—— Plassy boring of, 343-345.
King of Burmah, 156.
—— of France as regardless of petroleum as of Canada, 186.
" Kings, Oil," 5, 192.
Kinzua, Pennsylvania, U.S.A., 208.
Kioi, Roumania, 64.
Kir (south Caucasian oil field), 80.
—— treated *in situ* for low-quality lamp oil, 80, 91.
Kism, Persia, 151, 498.
Kiurinsk, Daghestan, 79.
Klafter, Galician measure (1·93 metres), 503, 534.
Klecka's free-fall instrument, 412.
Kleczany, Galicia, 36, 495.
—— light oil of, 36.
—— oil-well drilling, free-fall system of, at, 370-371.
—— paraffin in, 36, 48, 57.
Klein, Oedesse, Germany, 278.
Klemkowka, Galicia, 41, 48, 57.
Kleritj's application of free-fall to rope boring, 303, 419.
Klobassa and Lukasiewicz, 38, 304.
—— and MacGarwey in Italy, 122.
—— —— in Galicia, 38, 304.
Kobilacka, Galicia, 37.
Kobrich's application of free-fall to hydraulic drilling, 303, 419.
Kœchlin Schwartz on Temple of Eternal Fire, 97.
—— geology of Apshéron underground, 69.
Koenigsen, Germany, 115.
Kokoreff, borings north of Petrowsk, 79.
Kolb, rope boring system of, 303.
Kolomea, Galicia, 55, 479.
Koller, Germany, oil-well drilling machines of, 408.
Konia, Hungary, 124.
Kong-Tsin, China, 163.
Kootenay Pass, British Columbia, 240.
Kop-Kut-Chigan, Crimea, 74.
Koretz, Galician measure (51·137 metres), 503.
Koronew, Russia, 110, 480.

Kosliakovsky, his Derbent borings, 79.
Koslow, Russia, 110, 480.
Kotei, Java, 174.
—— Borneo, 175.
Kouang-Tonng, China, 162.
Kouang-Tsi, China, 160, 162.
Kouban (Kuban), Russia, 73, 75, 76, 77.
—— ancient oil field, 75.
—— annual production, 78.
—— Ilskaia, valley of, 75.
Koudako, Russia, 77, 110, 496.
Kour River, 83, 84.
Koutais, Russia, 80, 479.
Kozlowa, Galicia, 278.
Kozodzu, Japan, 165.
Krasna, Galicia, 278.
Krementschong, Russia, 110, 480.
Krocienko, Galicia, 37.
Kronstadt, Austria, 64.
Krosno, Galicia, 477, 479.
Kryg, Galicia, 37, 48, 57, 495.
Kurds' method of working Persian petroleum, 150.

L.

Laboratory, artificial petroleum of the, 13, 21.
Labour, cost of, in America, 242-243.
    Galicia, 104.
    Russia, 104, 242.
—— See Oil-well Drilling, Cost of, by Different Systems.
Labourers. See Oil-well Drilling Systems, Staff of.
Labuan, Borneo, 175.
La Corunna, Spain, 481.
La Cruz, Peru, 254.
La Garita, Peru, 252.
Lagoon, primitive saline, and origin of petroleum, 11.
Laine, M., of Oued Ourizane, Algeria, gas well of, 271-272.
Lake Erie, 203, 204, 206, 222.
—— Huron, 204, 238-239.
—— Maracaibo, 250.
—— Ontario, 185, 239.
—— Pitch, 250.
—— Superior, 232.
Lakes, great, of America, 183.
Lambton Co., Ontario, 238.
Lamentations of Ben Lakdar and his sons on darkening of " Black Spring," 259.
Lamp oil. See Petroleum, Distillation Products (fractional).
Lanceolate gypsum absent from eocene shales of Galicia, 35.
Land buying and cost of in—
    America, U.S.A., 5, 244-245.
    Baku, 105-109.
    Canada, 5, 245.
    Galicia, 5, 50-51.
    Madagascar, 274-275.
—— Roumania, 65.
—— —— aliens cannot own land in, 61.
—— interest in U.S.A., 245.
—— leasing or renting in—
    America, 245.
    Galicia, 50-51.
—— selling in Galicia, 51.
Landing petroleum at Liverpool in early days, total cost of 1 barrel from Canada, 472.
Landlord, the states the, or owner of land in Baku, 108-109.
Landscape, anxiety to save beauty of, prevents drilling in proper place, 482.
Landslips. See Accidents.
Lang, " oilman," of Roumania, 62.
Langkat oil field, Sumatra, 170-171.
Languedoc, sick people of, attracted by Galician healing oil fountain, 129.

Lantern, miner's. *See* Davy Lamp.
La Place's hypothesis, Mendeleef's theory as to origin of petroleum starts from, 18.
Lapparent, origin of petroleum, theory of, 16.
Lathe, pipe line pipes must be capable of being turned on, 478.
Laurent's improvements on Degoussée's system, 407-408.
—— adopted by Schibaeff Co. at Baku, 409.
—— free-fall instrument, 415.
Lauristan, Persia, 150.
Lave's simultaneous sand pumping and drilling plant, 408.
Lawrence, Pennsylvania, U.S.A., 203, 213.
Lead cube measurement of force of explosives, 451.
Leakage reduced to minimum by tank waggons and tank steamers, 468.
Lease. *See* Land.
Leasing land. *See* Land, Leasing of.
Leather manufacture, use of petroleum by Dutch East Indians in, 169.
—— —— by Galicians, 39.
Leaves of a book rolled over each other, resemblance of Roque Salure shales to, 136.
Le Chatelier rope boring, 303.
Le Credo, France, 189.
Legal formalities in buying and leasing land in Galicia, 50-52.
Legality of title to land in Galicia, 50, 52.
Legends—Ben Lakdar and the goat, 259.
    The famous Shaw well, 232-234.
Le Gua, France, 128, 139, 143.
Leite de Bohol, Philippine Isles, 178.
Lemberg, Galicia, 58, 64.
Length, measures of, of different countries reduced to metric, 503-504.
Lengthening of rods, 307, 316.
Lenticular limestone of Colorado, 227.
Lepage's system of tank waggon construction, 462.
Lepan River of Sumatra, 171.
Leproux, Russian petroleum statistics, 101-102.
Leschowate, Galicia, 41.
Leschow's first use of diamond in rock boring, 394.
Lesley, origin of petroleum, 11.
Lesquereux, origin of petroleum, 11.
Lessee. *See* Land Leasing.
Lessor. *See* Land Leasing.
Letaung, Isle of Ramri, Burmah, 154, 160, 498.
Lethargic condition of French industry, 126, 178, 470.
Level (geological horizon) oil. *See* Oil Level.
Levigation by steam jet as a substitute for blasting, 457.
Leymerie on bitumen, 137.
—— on mud volcanoes, 86-87.
Leyden jar of electric blasting outfit, 452-454.
Lianosow No. 23 spouting well, 95.
—— pipe line, 477.
Lianworth spouting well, 235.
Libusza, Galicia, 37, 48, 57.
Lidah, Java, flowing wells of, 174.
Liebermann, Boryslaw, 26.
Life, fish, of Black Sea and Caspian Sea, 69.
Light oil, 75.
Lignite, Galicia, association of, with petroleum, 33.
    Crude oil distillate from that of Aquila mines, Italy, 120.
    Germany, petroleum of produced from adjacent (?), 13.
    Italy, Aquila mines of, 120.
Likmau, Isle of Ramri, Burmah, 154, 160
Lille (Du Nord), France, depôt, 480.
Lima, Ohio, oil field of, 203, 222, 223, 224, 479, 501. *See* Ohio, Oil Field of.
Limagne, petroliferous tertiary beds of, 7, 136-139, 227.
—— bitumen of, 136, 139.
Limestone, petroleum the result of action of volcanic gases on, 19.
—— of—
    Algeria (Constantine), 272.
    Canada, 205, 206, 238.
    —— corniferous, 239.
    Gabian boring (various), 182-134.
    —— dolomite, 134.

Limestone of—
 Galicia, shales of Holowecko boring, 35.
 —— menelitic (Bobrka Libusza), 48.
 Germany (neocomian and jurassic), 112.
 Turkestan, sandy, 149.
Lime, carbonate of, in Gabian boring mineral spring, 133.
—— sulphate. *See* Gypsum.
Limoges, France, depôt, 480.
Linares, Spain, 124.
Linden, Germany, 276, 278.
Line of strike of petroliferous strata—
 American, 202.
 European, main, 8.
 —— secondary, 8.
Lineal measures, European and their metric equivalents, 503-504.
Lining wells with timber, 299.
Liniment. *See* Embrocation.
Lion, tank steamer, 470.
Lippmann on oil-well drilling systems—
 Canadian, 368, 369.
 Casing *en colonne perdue*, 329.
 Free fall, 415-418.
 Hydraulic, Raky, rapid system, 422-423.
 Rope boring, 302.
 Sigmondi's free-fall instrument, 417.
 Van Dijk's wood enveloped iron rods, 369-370.
 —— free-fall instrument, 417.
—— Italian borings of, 121.
Lipinki, Galicia, 37, 48, 57, 495.
—— petroleum distillation products of, 37.
Liquid fuel, 4.
Lithium carbide, synthetical production of, 21.
Lithological modifications of upper jurassic petroliferous strata of Germany, 112.
Litigation caused by insecure titles to land in Galicia, 50-52.
Little Fork, Canada, 237.
—— Kabylie, Algeria, 272.
Liver oil, cod, destructive distillation of, yields petroleum, 13-14.
Liverpool, depôt, 480.
Llewellyn spouting well, Virginia, 201.
Load, ratio of traction to, on various roads, 529-531.
Lobsann, Alsace, 116, 277.
Locomobiles interdicted in Baku and Galicia, 339.
Lodynia, Galicia, 41, 48.
Logan, Sir W. E., on origin of petroleum, 12.
—— —— geology of Canadian petroliferous strata, 238.
Loire, river, France, 136.
Lomna, Galicia, 42, 44.
London & Pacific Petroleum Co., 254-256.
Lore (legendary) surpassed by famous Shaw well, 232-233.
Lorraine, 117.
Los Angeles, California, 228-230, 501.
Loss and devastation by spouting, 95, 439-441.
Loth (Russian weight, 12.797 grammes), 503.
Louisiana, U.S.A., 202, 226.
Low derricks involve short rods, etc., 307.
Lozere, France, 136.
Lubricating oil. *See* Petroleum, Lubricating.
Luçon, Philippine Isles, 178.
Ludwig Nobel, tank steamer plying on Baltic, 467.
Lukasiewicz of Galicia—
 Awarded medal by Austrian Emperor, 40.
 Deepening of his well by Fauck, 40.
 —— —— —— by Walter, 40.
 Establishes first petroleum refinery, 40.
 Identifies crude oil used to physic cattle as petroleum, 40.
 Introduces free-fall system into Galicia, 304.
Lundigar, India, 277.
Luneburger, 111.
—— Heide, 115.
Lyman on Japanese oil fields, 166.

Lymnæa palustris, pliocene fossil, 276.
—— *caudata*, eocene fossil, 277.
Lyons, France, petroleum depôt, 480.

M.

Mabana, China, 166.
MacCalmont's Farm, Venango, Pennsylvania, U.S.A., 215.
MacColl spouting well, 235.
MacCurdy, Washington, Pennsylvania, U.S.A., 218, 500.
Macdonald (Edward) well, U.S.A., "torpedoing" of, 457.
MacDonald Pool, Washington, Pennsylvania, U.S.A., 218.
MacGarwey introduces Canadian system into Galicia, 353.
- —— Galician "gushers" of, 463.
—— tank waggons of, 294.
  See Bergheim and MacGarwey, and Klobassa and MacGarwey.
Machikata, Japan, 166.
McIntosh, the American, succeeds at Sloboda where Rice, the Frenchman, failed, 46.
Mackean Co., 203, 205, 208, 219.
Macksburg, Ohio, U.S.A., 222, 223 279.
MacLane spouting well, 235.
Madagascar, 273-275.
    Explorers' accounts—
        Gautier, 273.
        Grosclaude, 274.
    Localities—
        Ambohitsaliko, 273.
        Ankavandra, 273.
        Manambolo, 273.
        Yankeli, 274.
        Mining laws, 275.
Madeleine, tank steamer, 470.
Madoera, Java, 174.
Madreporic strata, 276.
—— —— of Sumatra, 171.
Maestricht, Holland, 125.
Magalkot, Beluchistan, 151, 160, 498.
Magazine, The Massachusetts, 186.
Maghele, Russia, 80.
Magic formulæ of the foreman borer, 490.
Magicians, Indian, and virtues of petroleum, 184.
Magistrates, American, may trade, 196.
Magnesian limestone. *See* Dolomite.
Magnificence of Baku, 81.
Magnitude of Nobel's transport system, 468-470.
Magura mountains, 42, 43.
—— sandstone, 35.
Mahommedan priests' olive tree, Tarria, Algeria, 270.
Mahommedans, 68.
Mahoning River, valley of the, Washington, 217.
—— sandstone, 205.
Mailbag of Pithole, bulky, 199.
Maine, State of, 236.
Mainesci, Roumania, 63, 276.
Mainland of Burmah oil field, 155-156.
—— of Trans-Caspian oil field, 148-149.
Makhram, Turkestan, 149.
Makum, British India, 160, 498.
Maldener, free-fall instrument of, 413.
Malooda, Borneo, 175.
Mammoths, fossil, 276.
Manager, appointment of, in Galicia to be noted to state mining officials, 52.
—— registration of, by prefect, 52.
Manambola, Madagascar, 273.
Manambolatamy, Madagascar, 274.
Mancora, Peru, 254.
Mandalay, Burmah. 156.
Manganese carbide, synthetical, 17.
Manilla, Phillipine Isles, 170, 178.
Manitouline, Isle of, Lake Huron, 239, 279.

Manjak. *See* Asphaltum, Barbadoes.
Mannington, Pennsylvania, 500.
Manopello, Italy, 120, 278.
Manutahi, New Zealand, 180, 181, 499.
Maple Shade spouting well, 199.
Maquenne on origin of petroleum, 19.
Maracaibo, Lake, Venezuela, 250, 277.
Marcasite, 183.
Marchand (Frenchman) resumes working at Montechino, 122.
Marcus Polo, quoted, 7, 68.
Marietta, Ohio, U.S.A., 223.
Marine fossils of American silurian petroliferous strata, 203, 289.
—— limestone of Canada, 12.
Marl, weight of cubic metre of, 506.
Marls of Gabian, 132.
—— of Holowecko, 35.
—— selenitic, of Algeria, 267.
Mars, worship of, 67.
—— incense burned in temple of, 67.
Marseilles, France, 109, 480.
—— exports of Russian crude oil to, 480.
Marsh gas, petroleum a condensation product of, 19.
Martapoera River, Borneo, 175.
Marysville, Ohio, " torpedo " first used at, 455.
Mascara, Algeria, 261.
Masonry foundations of petroleum reservoirs, 462.
—— oil wells of mound builders lined with, 183.
Massis Co., Baku, No. 6 spouting well, 95.
Mastodontes, miocene fossil, 276.
Matatoe, Java, 174.
Materials, strength of, 507.
Mathematical formulæ, 508-512, 523-532.
Mather's rope boring system, 308.
Mather & Platt of Manchester, oil-well drilling machinery of, 409.
Mathon-Jozet on Tai-lï-Chen oil wells, China, 161.
Matitza, Roumania, 64, 65.
Mauch Chunk sandstone of U.S.A., 205.
Maximum daily progress in oil-well drilling, 387.
Mazouna, Algeria, 265.
Mazout, an ingredient of petroleum, 3.
—— as fuel for tank steamers, 468, 469.
—— —— oil-well drilling engines, 340.
—— —— Russian railway engines, 2.
Measures, European, etc., reduced to metric, 503-505.
Mecca, Ohio, U.S.A., 223, 501.
Mechanics, formulæ and data connected with problems in, 508-512, 523-532.
Mecina, Galicia, 37, 48, 57, 495.
Medals, antique, of Walsbronn, 117.
Medesano, Italy, 120.
Medical attendance, etc., on labourers in Galicia, employer bears cost, in accident
    cases, 54.
Medina, Columbia, 250, 502.
Mediterranean or secondary line of strike—
    Characterised more by signs than results, 9.
    Commences in Western Pyrenees, 9.
    Follows northern watershed, 9.
    Nummulite beds, parallel to, 9.
    Passes Vaucluse, 9.
    Italy, terminates in, 9.
Medjerda, Algeria, 272.
Megatherum cuvieri, pliocene fossil, 276.
Mehldorf, Germany, 276, 278.
Melinite, petroleum a constituent of, 2.
Menant, Rene, quoted, on French inertia in the Philippines, 178.
Mendeleef, origin of petroleum, 18.
Mendoza, Argentine Republic, 258.
Mercer, Ohio, U.S.A., 223.
Mercury, capsule of, fulminate of, of electric blasting outfit, 454.
Merits and defects of different oil-well drilling systems. *See* Advocates.
Merv, Russia, 66.

Meteorites, ferric carbide a constituent of, 19.
Meunier, Stanislas, origin of petroleum, 20.
Mexico, American exports of crude oil to, 246.
Miano, 120.
Mica in the petroliferous conglomerate of Roumania, 59-60.
Micaceous clay of Belowosa, Galicia, 35.
Michailoff-Quirilli pipe line, 467.
Michel Gabriel, Russia, 80.
Michailowka, Russia, 125, 279.
Middle field district of Pennsylvanian oil field, 210, 497.
—— Ages, Walsbronn petroleum fountain famous in, 117.
Mikhailowsk, Russia, 148, 497.
Milan, 120.
Mile (1·6093149 kilometres), 504. See Kilometre.
Milk antidote against carburetted gas, 284.
Millers ot Titusville (Brewer, Watson & Co.), 191.
Millet and Rozet, origin of petroleum, 11.
Mimbain, Ramri, 154.
Mimbu, Yenangyat, 159, 498.
Mindowsky, Mr., of Anaclie, 80.
Minerals of Canada, 238.
—— of Carpathians, 28.
—— of China, 160.
—— of Galicia, 28.
—— of Madagascar, 274-275.
Mineral industry, society of St. Etienne, 138, 381.
Minières de bitume, 138.
Minimum remunerative yield of an American oil well, 244.
Mingrelie, Prince of, 80.
Mining laws and customs of—
    Canada, 245.
    Galicia, 52.
    Madagascar, 274-275.
    Roumania, 65.
—— administration of Galicia, 52.
—— —— of Madagascar, 274-275.
Minnesota, 202, 226.
Minors, restrictions on, in Galicia, 51.
Minsk, Russian petroleum depot, 110.
Miocene petroliferous strata, 276-277.
    Alsace, 277.
    Algeria, 261, 277 (illustrated in section), 269.
    Gabian, 7, 129.
    Galicia (illustrated in section), 32, 43, 48, 277.
    Italy, 7, 277.
    Limagne, 7.
    New Zealand, 180.
    Roumania, 59, 277.
    Russia, 277.
    Trans-Caspian, 277.
    Trinidad, 277.
    Venezuela, 277.
Mirzocz pipe line, Baku, 477.
Misrepresentation of speculators leads capitalists astray, 482-483.
Missionaries, 163, 185.
—— Jesuit, 294.
Missouri River, 202.
—— State, 202, 227.
Mississippi River, 202.
Mitis, Johann, of Boryslaw, his primitive crude oil distillery, 24.
Miyôhôji, Japan, 166.
Mobility of petroleum. See Viscosity.
Model of Baku oil-well drilling plant, illustrated, 360.
Modena, Italy, 120, 122.
—— mud volcanoes of, 87.
Moinesti, Roumania, 276.
Moissan artificial diamonds, 424.
—— origin of petroleum, 20.
Mohlsheim, Alsace, 277.
Momentum formula, 529.

Monastery, Persian, at temple of eternal fire, 98.
Money (coins), European and equivalents in France, 506.
Monks, Persian, 98.
Monopoly, Russian, of petroleum, 71-72, 108-109.
Montechiuo, Italy, 121.
Montpelier, Ohio, gas wells of, 224.
Moorland, weight of a cubic metre of, 506.
Moscow, Russian, petroleum depot, 118.
Mostaganem, Algeria, 263.
Motors. *See* Engines.
Mouchekeloff on Turkestan (Fergane) Oil Fields, 148-150.
Mould, weight of a cubic metre of, 506.
Mound builders of America, 184.
Mountains—
    Alleghany, 202.
    Alps (French), 146.
    Apennines, 120.
    Atlas, 260.
    Carpathians, 28.
    Catskill, 205.
    Caucasus, 68.   *See* map, page 76.
    Rocky, 202, 239.
Mount Morris, Western Virginia, 221.
Mrasznica, Galicia, 278.
Mrzaniecz, Galicia, 53.
Muddy streets of Baku, 81.
Mud volcanoes. *See* Volcanoes, Mud.
Mulot, free-fall system of, 412.
—— Grenelle artesian well of, 406.
—— succeeded by Dru, and Dru by Arrault, 406.
Mushroom towns, 199, 200. *See* America, United States of.
Mutzig, Alsace, 277.
Mythological origin of Algerian petroleum, 259-260.
—— —— Caucasian petroleum, 68.

N.

Nadworna, Galicia, 46.
Nails, cost of, for nailing derricks, 309, 310, 312.
Namangane, Turkestan, 149, 278.
Nancy, France, depôt, 480.
Nantes, France, depot, 480.
Naphtha, 521-522. *See* Benzene.
Naphthnia Gora, Russia in Asia, 146, 277.
Naphthonia, Russia in Asia, 146.
Naroudja, Russia, 80.
Narratives, explorers, of travels in Madagascar, 273.
National Transit Co., 476.
Natives and pre-historical method of working and utilising petroleum—
    Algeria, pirates and Spaniards, 260.
    America, North, Indians, mound builders, salt makers, trappers, 182-187.
    —— South, Incas, 252.
    Burmah, 153.
    China, 160-161.
    Galicia, 24-25.
    Japan, 164-165.
    Roumania, 61-62.
Natural gas of—
    Algeria, 264, 268, 272.
    Boryslaw, great pressure of, 44.
    Burning fountain, Gua Isère, France, 139.
    Chatillon, France, 144.
    Chinese Thibet, 163.
    Indiana, 224.
    Pittsburg, U.S.A., 217.
    Wisconsin, 225.
Natural petroleum reservoirs of Baku, 461.
Nature replies to coal alarmists with petroleum, 4.
Navy, Italian, use of petroleum as fuel, 2.
—— merchant, use of petroleum as fuel, 2.

Necessaries to success in oil-well prospecting and development, 482-483.
Neff, Ohio, U.S.A., 222, 279.
Negara River, Borneo, 170, map, 176.
Negociability of pipe line vouchers or cheques, 476.
Negrais, Cape, 169.
Negritos, Peru, 254, 255.
Nemuro, Japan, 163.
Neocomian. petroliferous strata—
    Argentine Republic, 278.
    Colorado and Nevada, 227, 278.
    Galicia (shown in section, 43), 48, 278.
    Germany, 112, 278.
    Portugal, 278.
Nepitel, Russia, 77.
Neuburger, experience in buying land in Galicia, 51.
—— improvements in rods, 395.
—— —— in rope boring, 303.
Neu-Sandec, 29, 36.
Neu-Sandec-Zagorz-Chyrow-Stry-Kolomea Railway, 29.
Nevada, U.S.A., 278.
Nevers, France, 136.
Neviano, Italy, 121.
Newfoundland, 240.
New Brunswick, 236.
—— Caledonia, 180, 275.
—— Grenada, 87.
—— Hampshire, U.S.A., Dr. Crosby of, 192.
—— Jersey, Williamsport-Bradford pipe line, 476.
—— York, history of petroleum, 191, 193, 195.
    Oil fields of, 202, 203, 204, 207, 208, 209, 501.
—— —— pipe lines, 474.
—— Zealand oil field, 179, 181, 276.
    American contractors in, 179.
    Geographical situation, 179-180.
    Geology of—
        Eocene strata, 180.
        Fucoids, 180.
        Grey clays, 180.
        Marine fossils, 180.
        Miocene strata, 180.
        Non-volcanic nature of, 180.
        Oil bearing bed, brownish sandstone, 180.
        Presence of petroleum in New Zealand accidental, 180.
        Sandy shales, 180.
        Strata traversed, 180.
        Tertiary strata, 276.
        Volcano of Tongarino, 179.
        Volcanoes, mud of, 179.
    Principal producing centres, 179-180.
        Manutahi, 180, 499.
        Poverty Bay, 180, 499.
        Sugar Loaves, 180, 499.
    Water bearing beds involves hermetic casing, 180.
New Zealand petroleum and iron syndicate, 179-180.
Ngawi, Sumatra, map, 170.
Nicobar Isles, British Burmah, 169, 170.
Night gang, or night shift and day shift, routine of, 489.
Nijni Novgorod, Russia, 480.
Nitrate of ammonia. See Ammonium Nitrate.
Nitro-glycerine, 444-457.
Noble and Delamater spouting well, 199.
Nobel, Ohio, 203, 223, 279.
Nobel, Ludwig, 73, 466.
—— Messrs., improvements in transport of Russian petroleum, 464.
—— refinery of Baku, 95.
Nodules of pyrites, 133.
Non-conductors. See Blasting Outfit, 453-454.
Normal pentane fractional distillation product of ood oil, 14.
North-east to south-west, American line of strike, 202.
North-west to south-east, main European petroliferous line of strike, 8, 202.

Nossi-Bé, Madagascar, 273.
Noth, Julius application of free-fall to rope boring, 419.
Nova Scotia, 190, 241.
Noworossirsk, Russia, 75, 110, 129, 479, 496.
Nowosielce-Gniewosz, Galicia, 41.
Nowo-Semeckino, Russia, 125, 279.
Nugata, Japan, 163.
Nullaberg, Sweden, 125, 497.
Nummulites Leymeriei, fossil, 74, 276, 277.
—— *Scabra*, 277.
Nummulitic sandstones—
　　France, 9, 127, 129.
　　Galicia, 34, 48.
　　Russia (Crimea), 74.
Nutty-flavoured butter, 1.

O.

Oak. *See* Strength of Materials.
—— Creek, Wisconsin, 225.
—— rods of Canadian system, 352.
—— weight in air and in water, 358.
Obalski, J., quoted, geology of Gaspé, Canada, 237, 238.
Oberg, Germany, 114, 496.
Objections to Canadian system. *See* Lippmann.
—— to Fauvelle system. *See* Degoussée.
Objects, foreign, fall of, into bore hole. *See* Accidents.
Observation, alert faculties of, of Indians, 183.
Oceania, 169-181.
Octane, normal, from cod oil distillate, 14.
Octroi. *See* Custom House.
Odour of petroleum. *See* Petroleum, Smell of.
Oedesse, Germany, 113, 114, 115, 116, 278, 496.
Oelheim, Germany, 113-116, 278, 496.
Oenhausen inventor of "jars," 359, 418.
Officials, mining, of Galicia, 54.
Offing of Baku, 81.
Ohio, oil fields of, 202, 203, 204, 219, 222-224, 279, 501.
　　Geology of—
　　　　Devonian strata (Macksburg oil field), 222-223.
　　　　Silurian strata (Lima oil field), 222-223.
　　Oil properties of—
　　　　Lima, like Canadian, density of, 223.
　　　　—— sulphur in, 223.
　　　　Macksburg, like Great Oil Belt or Beaver, 223.
　　　　—— density heavier than average Pennsylvanian, 222.
　　Producing centres—
　　　　Lima oil field—
　　　　　　Allen, Auglaize, Findlay, Gibsonbury, Hancock, Mercer, New Baltimore, St. Mary, Seneca, Spencerville, Upper Sandusky, Van Wert, Wyandot, 223.
　　　　Macksburg oil field—
　　　　　　Belden, Duck Creek, Harrison, Marietta, Mecca, Nobel, Washington, 223.
　　　　Production (1875-1895), 224.
Oil Belt, Great, Virginia, 221.
—— Belt, Reno, 214.
—— City, 5, 200, 214.
—— Creek, 198, 214.
—— Creek humbug spouting well, 198.
—— crude. *See* Petroleum, Crude.
—— fever in America, 6, 193-206.
—— heavy. *See* Petroleum, Heavy.
—— illuminating. *See* Petroleum, Illuminating.
—— lamp. *See* Petroleum, Illuminating.
—— lubricating. *See* Petroleum, Lubricating.
—— reservoirs, 461-462.
—— shale, 4, 5, 128, 190-191.
—— Spring, Canada, 230.
Oil-well drilling, general ideas and history, 292-295.

Oil-well drilling, ancient rope system of—
  Advantages and disadvantages, 296, 374.
  Beam, walking, described and *illustrated*, 292.
  Bit. *See* Drill.
  Casing: Drake's discovery. Filling bore hole with water as a substitute for, 299.
  —— lining well with timber as a substitute for, 299.
  —— snapping the fastening, 300.
  —— lowering the, 300.
  —— method of, on Drake system, 300.
  Chinese origin of, 294, 296.
  Chronological nomenclature of different systems of, 303.
  Drill, described, 296-297.
      Illustrated, 297.
  Drilling plant, 300.
  Goullet Collet, described and *illustrated*, 301.
  Impracticable in European strata, 296.
  Improved systems—
      Jobard's, described, 301, 302.
      Illustrated, 301.
  Method of working, described and *illustrated*, 296.
  Sand pump described, 296-297.
  —— —— examination of detritus brought up by, 299.
      Illustrated, 297.
—— —— American rope system of—
  Advantages and disadvantages, 374.
  Beam, walking, described, 376.
      Illustrated, 375.
  Bit described and *illustrated*, 375.
  Buildings, cost of, 378.
  Cost of tools, etc., 378.
  Derrick, cost of, 378.
  Deviation, liability to, 375.
  Drilling machinery, surface, described, 375-376.
      Illustrated, 375.
  Drills, wide, heavy, impracticable with, 377.
  Engines, cost of, 378.
  European strata, impracticable in, 296, 374.
  Galicia, oil wells of, sunk by, 377.
      Speed, 378.
      Time occupied in the several operations, 377.
  Hard ground, slow progress in, 377.
  Harklowa, Galicia, results of, 377.
  Mencina, Galicia, speed, results by, 377.
  Rope socket described and *illustrated*, 394.
  Sand pumping, time occupied in, 377.
  Speed results got by, at Mencina, 377.
  Staff, 376.
  Temper screw, 375.
  Wages, 376.
Hand power, free-fall system of, Fabian—
  Advantages and disadvantages of, 304.
  Auger stem of, cost of, 335.
  —— —— described, 321.
      Illustrated, 321, 323.
  —— —— guide, its function described, 321.
      Illustrated, 320.
  Beam, walking, of, or bascule, cost of, 336.
  —— —— counterpoise of, described and *illustrated*, 314.
  —— —— double system of, described and *illustrated*, 315.
  —— —— fixed to woodwork of derrick, described, 314.
  —— —— how wrought, 314.
  —— —— how foreman borer escapes blows of end of, described and
      illustrated, 314.
  —— —— independent of derrick, 304-305.
      Illustrated, 305.
  —— —— point of support, 314.
  Bits. *See* Drill.
  Cable of hoisting winch (rods and drilling tools), described, 324.
  —— sand pump described, 306.
  —— —— —— —— lowering casing by, 328.

Oil-well drilling (*continued*)—
  Hand power, free fall system of, Fabian (*continued*)—
    Casing of the bore hole—
      Complete and partial, their advantages and disadvantages contrasted by Lippmann, 329-331.
        Illustrated, 329.
      Cost of, 336.
      En colonne perdue. *See* Complete Casing, etc.
      End of, provisional point-shaped, described and *illustrated*, 327.
      Holes or eyes for raising, 327.
      Lost tubes or pipes. *See* Complete and Partial, described, 329-331.
      Lowering the, by sand pump cables or rods, described and *illustrated*, 328.
      —— precautions in, 327-328.
      Method of, described, 326-332.
        Illustrated, 327, 328, 329, 331.
      Object of, prevention of landslips, 326.
      Partial. *See* Complete and Partial Casing.
      Resource, a last, 329.
      Riveting of, described and *illustrated*, 327.
      Rolling of, described, 327.
      Surface well, a, described, 331-332.
        Illustrated, 331.
      Thickness of, 327.
      Water, use of as a substitute for, weight of column of, 328.
      Cost of—
        Initial, greater than rope boring, 315.
        Total, drilling well 300 metres by, 337.
    Derricks—
      American, described, 309, 310.
        Illustrated, 309.
      —— cost, estimate and specifications of, 310.
      Canadian, described, 311-312.
        Illustrated, 311.
      —— cost, estimate and specifications, 312.
      Derivation of term, 306.
      Four-legged, described, 308, 309.
        Illustrated, 308.
      —— cost, estimate and specifications, 309.
      Height of, minimum, 307.
      —— lofty, of, enables longer rods to be used, 307.
      Installation without a, 313.
      Instrumentation, special kind of, described and *illustrated*, 312.
      Transportable, 313.
      Tripod, described and *illustrated*, 308.
    Drill of—
      Blades, small, on extremity of large blade of, function of, described, 322, 323.
        Illustrated, 322.
      Connections between rods and, screwed or collared, 323.
        Illustrated, 321, 322, 323.
      Neck, long, of, importance of, described, 323.
        Illustrated, 322.
      Scrutinising the, and *débris* simultaneously, 333-334.
      Shape, present, 322.
      Tempering of the, 323.
      Transformations and modifications described, 322.
        Illustrated, 321.
    Estimate and specification for derrick, rig and smithy (complete installation), according to, 335-336.
    Expenses, total monthly, of boring on, 337.
    Foreman borer—
      Drilling by, *illustrated*, 333.
      Drill, scrutinisation of, by, 333-334.
      Free-fall, how he produces the, 304, 305, 306, 320.
        Illustrated, 320.
      Wages of (Galicia), 336.
      —— of assistants, 336.
    Free-fall instrument, Fabian's, cost of, 335.
      Functions of, described, 318-320.
        Illustrated, 319, 320, 323.
    Fuel, monthly, bill, 337. *See* Smithy.

Oil-well drilling (*continued*)—
  Hand power, free-fall system of, Fabian (*continued*)—
    Guide, auger stem of. *See* Auger Stem, Guide of.
    Introduction of, into Galicia, first, 304.
    Ladder wheel (treadmill) windlass, described and *illustrated*, 324-325.
    Light illumination monthly bill, 337.
    Progress made in working by, 334.
    Rests or spells from working, 332.
    Rig, *illustrated*, 334.
    Rods, iron, of, described, 315-316.
      Illustrated, 316.
    —— collared or screwed connections with drill, 323.
    —— cost of, 335.
    —— dimensions of, 315, 316.
    —— lengthening the, system of, 316-318.
    —— —— adjustment pieces, 316.
    —— —— —— cost of, 335.
    —— —— the temper screw, described and *illustrated*, 317.
    —— long, save time in unscrewing, 307.
    —— retention key, 316.
    Sand-pump described, 325.
      Illustrated, 298.
    —— —— American piston described, 325-326.
        Illustrated, 326.
    —— —— cable, 306, 325.
    —— —— cost of, 335.
    —— —— windlass, 325.
    —— pumping, 306, 325, 326.
    —— —— signs of necessity for, 332.
    —— —— time occupied in, 333.
    Smithy, tools and stores, cost of, original, 336.
    —— —— —— —— monthly cost, 337.
    Staff, number of, 332.
    —— wages of, 336.
    Sundries, monthly bill of, 337.
    Surface well—
      Casing of, 332.
      Excavation of, described, 331-332.
        Illustrated, 331.
      Without a, method of commencing drilling, described, 337-338.
        Illustrated, 338.
      —— position of foreman borer, 337.
      —— provisional guiding casing, described and *illustrated*, 338.
    Time occupied in, by—
      Drilling, 333.
      Lowering the rods, 333.
      Raising the rods, 333.
      Sand pumping, 333.
    Temper screw of, described, 317.
      Illustrated, 317, 323, 338.
    —— —— cost of, 335.
    —— chain and windlass, substitute for, described, 317-318.
      Illustrated, 318.
    Tools, drilling of, assemblage of, *illustrated*, 323. *See* Estimates.
    Wages of staff, 336.
    Winch or windlass for hoisting rods or sand pump, described and *illustrated*, 324.
    —— treadmill, ladder wheel, substitute for, described and *illustrated*, 324.
    —— —— section and plan of installation, 325.
    —— —— *See* Derrick, Transportable, and Temper Screw.
    Working bar described and *illustrated*, 318.
    See Derricks, Cost of; Estimates; Fuel; Light; Smithy; Sundries.
  Pole (wooden) or Canadian system defined, 533.
    Auger band wheel described, 354-356.
      Illustrated, 354-357.
    —— stem, described and *illustrated*, 359.
    —— —— cost of, 363.
    Band wheel. *See* Auger Band Wheel.
    " Beam, walking," described, 353, 354.
      Illustrated, 354, 357, 358.
    —— —— bouncing post of, awanting, 354.

Oil-well drilling (*continued*)—
  Pole (wooden) or Canadian system (*continued*)—
    " Beam, walking," cost of, 362.
    Bit. *See* Drill.
    Blows of drill, number of, compared with Fauck's machine, 353.
    Boiler, cost of, 362-363.
    Bradford, U.S.A., Dennis No. 1 well, speed results, 362.
    Cable, aloes, cost of, 363.
    —— —— length of, 363.
    —— —— thickness, 363.
    Chain, retaining, of, cost of, 363.
    Cog-wheel gearing absent in, 353.
    Cost of drilling a well 300 metres deep by, 363.
    —— —— —— per metre and running foot, 363.
    Denis No. 1 well speed results, 362.
    Derrick of, cost of, 362, 363.
    Drill of, described and *illustrated*, 359.
    —— cost of, 363.
    —— size of, 363.
    Drilling tools, cost of, in Canada, 362.
    —— —— —— in Galicia, 363.
    Elongation of rods of, windlass for, described and *illustrated*, 354.
    Engine of, cost of, 362.
    —— horse power of, 353.
    Foreman borer at work, *illustrated*, 360.
    —— and carpenter can quickly erect rig, 353-354.
    —— —— duties of, described, 359-360.
    Galicia, Germany, Italy, use of, in, 373.
    Holowecko speed results, 361. *See* Time.
    " Jars " described, 353-359.
    —— cost of, 363.
    —— *illustrated*, 359.
    —— invented by Oenhausen (author), 359, 418.
    —— —— W. Morris (Redwood), 359.
    Lightness of all its parts, 353.
    Oak, comparative weight in air and in water, 358.
    —— poles or rods of, 354.
    Pine, comparative weight in air and in water, 358.
    Poplar, comparative weight in air and in water, 358.
    Rods, wooden, of—
      Lose weight in water, 356.
      Manipulation of screwing on and off, *illustrated*, 362.
      Quality of wood required to be used in making, 358.
      Screws of, male and female, 358.
      Weight of, per square centimetre in air and water, 358.
    Sand-pumps, cost of, 363.
    —— —— size of, 359.
    —— pumping, how done, 359.
    —— —— times occupied, 361.
    Smithy, repairs and renewal of tools, 363.
    Speed, 361-362, 371.
    —— Bradford, Denis No. 1, 362.
    —— Holowecko, Galicia, 361.
    —— Uherce, Galicia, 362.
    —— Więtrzno, 372.
    Staff, 359-360.
    Time occupied in—
      Drilling, 361.
      —— ratio of, to whole period of working compared with Fauck's, 362.
      Lowering the rods, 361-362.
      Raising the rods, 361-362.
      Sand-pumping, 361-362.
    Tools, drilling, cost of, in—
      Canada, 362.
      Galicia, 363.
    Uherce, Galicia, speed results, 362.
    United States, use of, in, 373.
    Wages less than Fauck's, 363.
    " Walking Beam." *See* Beam, Walking.
    Wooden rods, weight of, per square centimetre of section in air and water, 358.

Oil-well drilling (*continued*)—
  Steam power free-fall system—
    Adjustment pieces for connecting rods and drill, cost of, 351.
    Advance of drilling over excavation not realised until steam lent its aid, 339.
    Apostle of, Fauck the great, 346.
    Auger stem, cost of, 351.
    —— —— size of, 351.
    Beam, walking, drilling done by, as in hand power, 343.
    —— —— modifications described and *illustrated*, 347.
    Bit. *See* Drill.
    Boiler suitable for (French), 341.
    —— cost of, 351.
    —— crude oil may be used as fuel, 340.
    —— —— —— calorific intensity of, 340.
    —— furnace large and adapted to consume wood, 340.
    —— horse power of, 351.
    Cable, sand pumping, cost of, 351.
    —— steel, cost of, 351.
    Casing, cost of, 352.
    Costs, etc., of, 351-352.
    Drills, cost of, 351.
    —— size of, 351.
    Engines suitable for, 339, 340, 341, 342.
    —— American, 342.
    —— —— Buffalo and Electric motors, 342, 366.
    —— Austrian, 342.
    —— —— Fauck's, cost of, 351.
    —— British, 342.
    —— cost of, 351.
    —— French, 341.
    —— German, 342.
    —— horse power of, 345-351.
    —— locomobiles unsuitable, 339.
    —— pumping, 341.
    —— traction, interdicted, 339.
    —— uses which it has to serve, 341.
    Estimates, 351-352.
    Free-fall instruments, Fauck's, 308, cost of, 351.
    —— —— size of, 351.
    Installation on, first described, 343-346.
        Illustrated, 342-344.
    —— beam, walking of, 343.
    —— driving rod, 343.
    —— play of, 343.
    —— hoisting crane of, 343.
    —— Plassy boring, the first installation on, *illustrated*, 344.
    Kleczany results, 371.
    Leather belts and duplicates of parts for renewal, cost of, 352.
    Lodgings for men, cost of, 352.
    Machines on, first, combining the drilling and hoisting machinery in **one**—
        Described, 343-346.
        Illustrated, 345.
    —— drill effectively, 346.
    —— Fauck's improved, cost of, 351.
        Described, 346-349.
        Illustrated, 346, 347, 350.
        Average speed of, 349.
        Beam, walking, modifications, 347.
        Number of, at work, 347.
        Oscillations of—
            Number of, in drilling period, 349.
            Ratio of, to revolutions of fly-wheel, 349.
        Practical ideas on its working, 348.
        Temper screw, substituting, for, 348.
        Time occupied in drilling, raising the rods, sand pumping, lowering the rods, 349.
    —— horse power of, 345.
    —— objections to, 345.
    Plassy, boring by, *illustrated*, 344.
    Pulley for sand pump cable, cost of, 351.

·Oil-well drilling (*continued*)—
  Steam power, free fall system (*continued*)—
    Pulley for rod hoisting cable, cost of, 351.
    Rods, iron, cost of, 351.
    —— —— length of, total, 351.
    —— —— thickness of, 351.
    Sand-pumps, cost and size of, 351.
    Scientific instruments, cost of, 352.
    Shed for engines and machines, cost of, 351.
    —— for boilers, cost of, 351.
    Smithy, cost of, 352.
    Speed attained by, 349, 352.
    Staff, number of, 352.
    —— wages of, 352.
    Uherec speed results, 339.
    Windlasses, 843, 345.
  Drilling on the combined system—
    Adopted for pseudo-economical reasons, 364.
    Cost of drilling well of 300 metres by, 367.
    —— —— per running foot, 367.
    Definition of, 364.
    Fauck's machine for, described, 366.
      Illustrated, 365.
    Objections—
      Bending or buckling of rods, 365.
      Deficient unclutching or letting go of the free-fall, 364.
      Less effective blows, 365.
      Speed, the main advantage of Canadian system, lost, 365.
    Staff, same as Fauck's, 366.
    Time occupied in—
      Drilling, 366, 367.
      Lowering the rods, 366.
      Raising the rods, 366.
      Sand pumping, 366.
  Fauck and Canadian systems of, compared and contrasted, 368-373.
    Objections to Canadian—
      Fauck's, 370.
      Lippmann's, 368-70.
    Points in favour of Canadian—
      Author's, 373.
      Jurski, 371-2.
      Syroczinski, 372.
  Water Flush or Hydraulic System, Fauvelle's—
    American Rope System *versus*, 388.
    Auger stem of, described, 382.
      Illustrated, 381.
    —— cost of, 389.
    Beam, walking, of, *illustrated*, 382.
    Beart, an Englishman, first to put in practice at Perpignan, 379.
    Bit. *See* Drill.
    Bourgnies collieries, Aveyron, France, results, 381.
    Compared with American, 388.
    Cost per running foot of drilling by, 389.
    Degoussee's criticism of, 380.
    Derrick, *illustrated*, 382.
    Drill of, described, 381.
      Illustrated, 381.
    —— cost of, 389.
    —— diamond, for hard ground, 388.
    —— Fauvelle, for soft ground, 388.
    Drilling machine for drilling by combined Fauck's, fitted up for, *illustrated*, 385.
    —— hand power, *illustrated*, 384, 385, 386.
    —— —— in seeking for petroleum, few, 386.
    —— —— in seeking for water, Colonial use, 386.
    —— —— in trial borings, 386.
    —— —— or steam power, *illustrated*, 388.
    —— steam power, modified Canadian, *illustrated*, 382.
    —— —— cost of, 389.
    Drilling by, method of, rapid blows from low elevation, 379.
    Engine, cost of, 389.

Oil-well drilling (*continued*)—
  Water flush or hydraulic system, Fauvelle's (*continued*)—
    Fauck's machines for, described, 385.
        Illustrated, 382, 385, 388.
    —— successful employment of, 380.
    Foreman borer, duties of, in, 387.
    —— post of, in, 387.
    Freminville's boring in Isère, Gaymard's report on, 380.
    Gas piping for hollow rods, trial and failure of, 381.
    Horse power, calculating the force of the current to, 510-511.
    Isère, early borings by, ended in engulfment, 380.
    Joints of pipes, water conduit attenuated at, 382.
    Lippmann's adverse criticisms, 380.
    Nougarede on average daily speed, 389.
    —— results quoted, 381.
    Perpignan boring the famous and first well drilled, by, 379-380.
      Depth drilled in, 379.
      Time occupied in, 380.
    Principle of, 379.
    Przibilla of Cologne on the speed of the current of water with varying rocks, 386.
    Pumps, pressure drilling by, in, 388.
    Rods hollow or pipes used in, described, 381-383.
        Illustrated, 381.
    —— cost of, 389.
    —— adjustment system, *illustrated*, 385-387.
    —— —— cost of, 389.
    —— dimensions of, 383.
    —— gas piping, trial and failure of, as, 381.
    Sand pumping and drilling simultaneously, 379.
    —— hydraulic, by natural fall of water, 383.
    —— —— by pressure pumps, 384.
    —— —— by reversed current of water, 384.
    Speed of current of water used in, according to detritus, 386.
    —— of drilling by, daily (average), 389.
    Time occupied in drilling 300 metres by, 389.
  Rotary system of diamond boring—
    Accidents, sensitiveness of, renders it subject to, 401.
    —— time occupied in repairing, ratio of, to that spent in boring, 401.
    Advantages of, in hard ground, 390.
    America, extensively adopted in, 391.
    Artificial substitutes for diamonds in, 394.
    Augers of, described, 392-393.
        Illustrated, 392.
    —— corkscrew motion of, 392.
    —— metamorphoses of, 393.
    —— practicable only in soft ground and shallow wells, 393.
    —— replaced by " crown " in hard rocks, 393.
    Bore hole, equalising side of tool for (reamer), described and *illustrated*, 393.
    Bort, use of, in, 424.
    Casing iron, pipes of—
      Cast iron, cost of, and dimensions, 398.
      Screwing together, 398.
      Wrought iron, cost of, and dimensions, 398.
    Continental Diamond Rock Boring Co., their results, 400-401.
    Continuous process, 391.
    Core barrel (Carottier), 394, 395.
        Illustrated, 394.
    —— —— raising and emptying of, 399-400.
    Crown, described, 393.
        Illustrated, 394, 395, 399.
    —— alumina, crystals of, use of, in place of diamonds, 394.
    —— depths bored by, 401.
    —— diamond replaces steel in, 394.
        Illustrated, 395.
    —— duration of, 401.
    —— eccentric shape of, reason for, 398.
        Illustrated, 399.
    —— examination of, 400.
    —— iridium in place of diamond in, 394.
    —— metamorphoses of, 393-395.

Oil-well drilling (*continued*)—
  Rotary system of diamond boring (*continued*)—
    Crown, steel teeth of, fixed and movable, 394.
    —— *See* improvements, etc., in oil-well drilling.
    Diamond drill, no rock can resist, 390.
    —— —— cuts out circular section of ground and brings to surface intact in core barrel, 391.
    Foreman borer, duties of—
      Core barrel, raising of, 399-400.
      Crown, examining of, 400.
      Jerking, prevention of, 399.
    Hand systems of, described and *illustrated*, 396.
    Hard ground, advantages of, in, 390.
    Hollow rods, Neuburger's improvements, 395.
    Hydraulic cleaning out renders process continuous, 392.
    Lead, diamonds of crown embedded in layer of, 394.
    Leschow (Swiss engineer) first to use diamond in rock boring, 394.
    Machine for drilling by, described, 397-398.
      Illustrated, 397.
    Motive power, human, horse, steam, water, 397-398.
    Principles of, 390, 391.
    Reamer, 393.
    Rods of, 395-397.
    —— elongation of, 397.
    —— Neuburger's improvements in, 395-396.
    —— swivel, 396.
    Speed obtained in drilling by, 400-401.
    Staff, number of, 399.
    Time occupied in drilling a well of 300 metres by, 401.
    Trial borings used to ascertain nature of underground rocks, 391.
    Villefranche d'Allier boring, results obtained in, 400-401.
  Improvements and different systems of—
    Artesian wells, chronology of, 406.
    Estimate, French constructor's, for free-fall installation, 404.
    Improvements, etc., in machinery by makers of different nationalities and characteristic features of machinery of each country—
        American, 403.
        Austrian, 405-406.
        British, 405.
        French, 402-403, 404.
        German, 405.
    Improvements, etc., in machinery by different makers—
      Arrault, 406.
      Degoussée, 407.
      Dru, 406.
      Laurent, 407.
      —— modified (Baku), 409.
      Lavé, 408.
      Mulot, 406.
    Improvements, etc., in rods by different makers—
      Dru, Léon, 410.
      Dijk, Van, 410.
  Improvements in free-fall instruments—
    First free-fall instrument, described, 410-412.
      Illustrated, 411.
    Improvements of—
      Degoussée, 413-415.
        Illustrated, 415.
      Dehulster, 416-417.
        Illustrated, 416.
      Dru, Léon, 418.
      Dru, Saint-Just, 416.
      Ermeling, 413.
      Esché, 413.
      Fabian, 412.
      Fauck, 418.
      François, 418.
      Gault, 412.
      Greifenhagen, 413.
      Kind, 410-413.
        Illustrated, 411.

Oil-well drilling (*continued*)—
  Improvements in free-fall instruments (*continued*)—
    First free-fall instrument (*continued*)—
      Improvements of—
        Klecka, 412.
        Laurent, *illustrated*, 415.
        Lippmann, 418.
        Maldener, 413.
        Mulot, 412.
        Rost, 412.
        Seckendorff, 413.
        Sigmondi, 417.
        Van Dijk, 417.
        Werner, 412.
        Wilke's,
        Wlack, 412.
        Zobel, 412-413.
  Improvements and different systems of hydraulic drilling—
    Improvements of—
      Fauck's hand hydraulic system, described, 421-422.
        Illustrated, 421.
        Cost, 422.
        Speed, 422.
      —— jars, described and *illustrated*, 420.
      Kobrich, 419.
      Neuburger, 419.
      Przibilla, 419.
      Raky, described, 422-423.
      —— speed, 423.
      Schumacher, 420.
      Tecklenburg, de, 420.
    Improvements in—
      Rotary boring—diamond drilling—
        Augers, 423.
        Diamonds, substitutes for, 424.
        —— artificial, use of, 424.
        —— carborundum, 425.
        —— iridium (Przibilla), 425.
  Improvements in jars, 418.
  —— invented in 1834 by Oenhausen, 418.
  —— Fauck's improvements, 418.
  Improvements in rope drilling, 418-419.
  —— rope rods, 418-419.
  —— rope substitutes—
    Chains, 419.
    Hinged rods, 419.
  —— free-fall, application of, to rope drilling, 419.
  —— —— systems of Benda, 419.
  —— —— —— Gaiski, 419.
  —— —— —— Kleritj, 419.
  —— —— —— Noth, 419.
  —— —— —— Sisperlé, 419.
  —— —— —— Sontag, 419.
  —— —— —— Sparre, 419.
  —— —— —— Straka, 419.
Oil-well drilling, preventing and remedying of accidents—
  Casing pipes, deformed, cutting of, tool for, 432.
    Fauck's, described, 432-433.
      Illustrated, 433.
    Dru's, described and *illustrated*, 433.
    Expander, " Pear," described and *illustrated*, 432.
    Raising up of, by—
      Hooks attached to rods, 432.
      Tap borers, 432.
    Wrenching up of, 432.
  Drill, deviation of, 426-428.
    Partial, 427-428.
    —— cause of, inclined strata, 427.
    —— prevention of, 427-428.
    —— remedy, 428.

Oil-well drilling (*continued*)—
　Preventing and remedying of accidents (*continued*)—
　　Drill, deviation of (*continued*)—
　　　Total, 427, 428.
　　　—— cause of, 428.
　　　—— prevention of, 428.
　　　—— remedy, 428.
　　Drill, fracture of neck of, 426-428.
　　　Raising of, when resting—
　　　　Flat on bottom of bore hole, 434.
　　　　In vertical position, 434.
　　Drill, unscrewed, raising of, 434.
　　Foreign objects, fall of into bore hole, 434.
　　—— tools for " fishing " for, described, 434-436.
　　　Illustrated, 435-436.
　　Flooding, prevention of—
　　　Casing, hermetically sealed, the only preventive, 436.
　　　—— cost of, per running foot, 437.
　　　—— dimensions of, 437.
　　　—— inconvenience of, great thickness prevents telescopic casing, 436.
　　　Expanding drill used to work outside casing—
　　　　Dru's, described, 437.
　　　　　Illustrated, 436.
　　　　Fauck's, described, 437.
　　　　　Illustrated, 438.
　　　　Old-fashioned eccentric, 438.
　　Freezing mixtures to freeze water and then drill through resultant ice, 438-439.
　　Landslips—
　　　Cause of, 428-429.
　　　Exasperation aggravates disaster, 429.
　　　Measures to adopt, 429.
　　　Necessity for prompt action, 430.
　　　Preventive of, strong casing as, 429.
　　　—— water, filling bore hole with, as, 428.
　　　Raising embedded drill by temper screw, 430.
　　　—— —— —— by wrenching, 430.
　　　—— —— sand-pump by rods, 431.
　　　Sand-pumping, advisable to use rods for, in ground liable to slip, 431.
　　　Tool wrenches, described, 430-431.
　　　　Illustrated, 430.
　　　Wrenching, 430-431.
　　Rods, fractured, raising of, 434.
　　—— unscrewed, raising of, 434.
　　Spouting, unforeseen—
　　　Accident, regarded as, 439.
　　　Capping the well, 442-443.
　　　—— telescopic, 443.
　　　Potoki spouting well, 440.
　　　—— attendant phenomena, 440.
　　　—— damage caused by, 441.
　　　—— duration of spouting of, 441.
　　　Precautions and procedure to stop —
　　　　Boiler fires should be drowned out, 442.
　　　　Capping apparatus got ready, 442.
　　　Rare, 439.
　　　Signs, previous, 439.
　　　—— gas, evolution of, as, 440.
　　　—— interval between, and actual, 439.
　　　Well cap, Bredt's patent, 442-443.
　　　　Illustrated, 443.
Oil wells, abandoned, filling up of, 245.
—— levigation of, by steam jet, 457.
—— pumping of, 458-460.
—— submarine, 228.
—— underground blasting of, 452-457.
Ointment for scab of camels, 7, 68.
Ojai, California, 230.
Olhungen, Alsace, 116, 118, 497.
Oligocene, petroliferous strata of Baku, 85.
—— —— —— of Galicia, 34.

39

Olive tree, Mahommedan priests (*l'olivier du marabout*), Tarria, Algeria, 270.
Olsanica, Galicia, 41.
Omsk, Russian depôt, 467.
Ontario, Lake, 185, 230, 238.
—— Province of, 238-239. *See* Canada.
Oppel, geologist, quoted as to age of North German petroliferous strata, 112.
Oral tradition of Indians as to ancient knowledge of petroleum, 182-183.
Oran, Algeria, 261, 271, 272, 277.
—— *See* Algeria.
Orel, Russia, 110, 480.
Organic origin of petroleum. *See* Origin of Petroleum.
Organisms. *See* Fossils.
Orguni, Japan, 167-168, 499.
Oriental, Armenia, 81.
Origin of petroleum, 3, 4, 10, 23, 188, 239.
  Theories of, classification of, 10, 11.
    Chemical, 16-21.
    Mythological, 68, 259-260.
    Organic, 10-15.
    Volcanic, 15-16.
    Theories of—
      Abich, 18, 19.
      Aoust, Virlet d', 15.
      Bel, Le; 13.
      Benoit, 17.
      Berthelot, 16, 33.
      Buch, 10.
      Cahours, 17.
      Cloez, 17.
      Coquand, 17.
      Credner, 12.
      Daubrée, 16.
      Engler, 13-14.
      Gesner, 12-33.
      Grabowski, 19.
      Hitchcock, 11.
      Hofer, 15.
      Hunt, Sterry-, 11.
      Jones, Rupert, 12.
      Lapparent, 16.
      Lesley, 11.
      Lesquereux, 11.
      Logan, 12.
      Macquenne, 19.
      Mendéléef, 18.
      Meunier, 20.
      Millet, 10.
      Moissan, 20.
      Orton, 13.
      Peckham, 19.
      Pelouze, 17.
      Reichenbach, 10.
      Ross, 19.
      Rozet, 10.
      Silliman, 188.
      Smith, Watson, 12.
      Sokoloff, 19.
      Stromberg, 13.
      Turner, 10.
Orton quoted as to origin of petroleum, 13.
Oscillations of "walking beam," counting of, in intervals of working between spells in oil-well drilling by hand power, 332.
—— *See* Pendulum.
Oshima, Japan, 163.
Osmund, Mount, Sweden, 125.
Ostrea Deltoidea, upper jurassic fossil, 278.
Ouarsensis, Algeria, 262.
Oued, L', Crattouna, Tunis, 273.
—— —— el Ksob, Tunis, 273.
—— —— Houenet, Algeria, 261.

Oued, L', Ourizane, Algeria, 271.
—— —— Tarria, Algeria, 270.
Oulouchema, forest of, Daghestan, Russia, 79.
Outcrop defined, 29.
Outlay, sums to be disbursed in starting oil-well drilling at Baku, 105-107.
—— See Cost.
Overflow of oil. See Spouting Well Cap.
Overhauling of tools. See Foreman Borer.
Oxus river, valley of, now petroleum oil field, 148.
Ozera, Lake, Russia, 99.
Ozokerit mines of Boryslaw, 24, 25, 26, 44, 45, 46.
—— accidental deaths in, high rate of, 46.
—— annual production, 45.
—— gas pressure of, 46.
—— mining of, differentiated from oil-well drilling, 293.
—— of Tcheleken, 147.
Ozzano de San Andréa del Taro, Italy, 121.

P.

Pacific, The London and, Petroleum Co., 254. See London and Pacific Petroleum Co.
Packenham, Canada, 279.
Pacunetz, Roumania, 276.
Padua, Bologna, N. watershed of Apennines, 277.
Pagorzin, Galicia, 37, 48, 57, 495.
Pailfuls of petroleum, Billy Smith draws up, from first oil well, 195.
Pain, petroleum a cure for, 186.
Païta, Peru, 253, 254, 257.
Palembang, Sumatra, 170, 172, 173.
Palikao, Algeria, 261.
Paludina beds of Limagne, 7.
—— fossil (tertiary), 276, 277.
—— of Roumania, 59.
Panacea, petroleum a universal, 1, 2, 71, 186.
Panay archipelago, 178, 179.
Pane. See Window.
Panolan, Java, 173.
Pantin, Paris depôt, 480.
Paraffin wax, richness of crude oil in, 57, 156, 158, 160, 162, 177, 181, 495, 498, 499, 500, 502.
Parallelogram, centre of gravity of, 527.
Parisian tertiary strata, engulfment of Freminville's hydraulic boring in, 380.
Parker, Clarion Co., Pennsylvania, 215.
Parma, Italy, 120, 497.
Parsees, 68, 70.
Partial deviation. See Accidents.
Pass-book, workmen's, in Galicia, 53.
Passes, narrow, of Western Virginia, 221.
Patawagia Brook, Canada, 237, 502.
Patent for casing of oil wells, Drake did not take, 196.
—— Colonel Roberts' dynamite, in America, 455.
Pathar, 66.
Pax & Co., tank steamers of, 470.
Pay. See Oil-well Drilling, Wages and Staff.
Pays de la poix et de Cronelles, France, 137.
Peace, river, Canada, 239.
Pearl of the Apshéron, 97.
Pear pipe expander described and illustrated, 432.
Peasants of Galicia and history of Galician petroleum, 24, 25.
Peat, petroleum a decomposition product of, 23.
Pechelbronn oil field, Alsace, 116, 118, 277, 497.
—— asphaltum of, 117.
—— view of, illustrated, 117.
—— See Germany, Oil Fields of.
Peckham, origin of petroleum, 19.
Peine, North Germany, 111.
Pelisse on geology of Roumania, 60.
Pelouze on origin of petroleum, 17.
Pendulum, conical and simple, 528.

Peninsula, Apshéron. *See* Baku.
—— of Gaspé. *See* Gaspé.
Pennsylvania, 184-220. *See* America, U.S.A., Oil Fields of.
Pension to Colonel Drake, State of Pennsylvania awards, 197.
Pentane, normal, from cod oil, 14.
Penury, Colonel Drake reduced to, 196-197.
Percentage distillation products of crude oil. *See* Petroleum Fractional Distillation
    Products.
Perch (5·029 metres) defined, 504.
Percolation of petroleum through strata other than indigenous, 135.
Perdue, casing *en colonne*, Lippmann on disadvantages of, 329-331.
Perforation. *See* Oil Well Drilling.
Perfumed vaseline, 1.
Perimeter of Tarria, Algeria, 270.
Period of time occupied in oil-well drilling operations. *See* Time, etc.
Perm, Russia, 467.
Permian strata, characteristic fossils of, and oil fields in which they occur, 279.
Permit to work in Galicia, 52.
—— —— in Madagascar, 274-275.
Periqueux, France, depôt, 480.
Perpignan, France, hydraulic drilled well of, 379-380.
Persecutions caused by fire-worship, 68.
Persepolis, capital of Ancient Persia, 67.
—— planet and fire-worship at, 67.
Persians, grand *pyrée* of, at Seaté, 68.
Persia, 8, 67, 69, 150-151, 277, 498.
    Baku petroleum burned through all (A.D. 1688), 69.
    Barques of, convey petroleum from Baku, 71.
    European petroliferous line of strike passes through, 8, 67.
    Oil fields of, 150-151.
        Central section or Lauristan oil field, 150.
            Association of petroleum of, with salt and sulphur, 150.
            Geological formation eocene, 150.
            Producing centres—
                Chouster, 150, 151, 498.
                Chardin, 151.
                Haf-Cheid, 151, 498.
        Northern section or Kasharashirin oil field, 150, 498.
            Oil, properties of, 150-151, 498.
            Refined to certain extent on spot, 150.
            Rudimentary working by Kurds, 150.
            Yield, 150.
        Southern section producing centres—
            Daliki, 150, 498.
            Kism, 151, 498.
Peru, oil field of—
    Analogous to Amotape Mountains, 251.
    Boundaries and extent of, 251.
    California, petroliferous strata of, correlated with those of, 253.
    Depth at which oil is struck, 251-253, 255, 257.
    Distillation products, fractional, of London and Pacific (Négritos) crude oil, **255**.
    Incas, 252.
    Monopoly of former Castilian governments, 254.
    Reclus' delineation of, 251.
    Reventazon and La Garita mountains of "hardened tar," 252.
    Paita, 254-257.
    Subdivisions of oil field—
        La mina Bréa y Pariñas, 254.
        —— history of, 254.
        Producing centres—
            Mazout as fuel, 256.
            Négritos, 254-256.
            —— London and Pacific Company's operations at, 254.
            Pipe line connecting Négritos and Talara, 256.
            Sea water for boilers, 256.
            Spanish conquest, history of, goes back to, 254.
            Talara headquarters and refinery, 256.
            Tweddle, H., acquires concession from Peruvian government, **254**.
            —— cedes to London and Pacific, 255.
            Water and timber, want of, in, 256.

Peru, oil field of (*continued*)—
  Piura, 257.
    Dip of beds towards interior, 257.
    Least capable district of being developed, 257.
  Puerto Grau—
    Compagnie Française des pétroles de l'Amerique du Sud, operations at, 256.
    Concession, extent of, 256.
    De Clercy's operations, 256-257.
    Production, 257.
    Richness of, 257.
    State of the wells, 256.
    Tumbez, 253-254.
      Faustino Piaggio Co., 253.
      Prentice, A., 252, 253.
      Ruden, 253.
    Producing centres—
      La Cruz, 254.
      Mancora, 254.
      Quebrada, 253, 254.
      Tucillal, 254.
      Zorritos, 253-254.
    Production, total of, 257.
Peterson, Lewis, sells crude oil to apothecaries, 187.
Petroleum—
  Appearance of, 1.
  Artificial, of Engler, 13, 14.
  As—
    Adulterant, 1.
    Cattle medicine, 39.
    Chest affections, remedy for, 71.
    Cordial, a, 71.
    Cosmetic, 2.
    Diphtheria, a specific for, 71.
    Embrocation, 186.
    Explosive, 2.
    Flatulency, cure for, 71.
    Fuel, 2.
    Gout, remedy for, 71.
    Gravel, remedy for, 71.
    Grease eradicator, 71.
    Headache cure, 71.
    Illuminant, 71, 185, 187, 188, 190.
    Leather currying agent, 169.
    Liniment, 486.
    Lubricant, 24.
    Medicine, 71.
    Pain cure, 186.
    Pomatum, 1, 6.
    Purgative, drastic, 184.
    Reptiles, remedy for, bites of, 184.
    Rheumatism cure, 186.
    Scurvy cure, 71.
    Varnish, ingredient of, 71.
    Vermifuge, 184.
  Benzene, 2, 3, 17, 37, 39, 521, 522.
  Burning oil. *See* Illuminating Oil.
  Calorific intensity of, 4, 340.
  Coefficient of, expansion of, 494.
  Conveyance of, 458-481, 521-522.
  Density of, 495-502. *See under* Respective Oil Fields.
  Distillation, products of. *See* Fractional Distillation of.
  Essentials to presence of, in any given bed, 133.
  Ether. *See* Gasolene.
  Fractional distillation, products of crude oil of, 2, 3, 37, 39, 94, 95, 248, 255, 270.
  Freight trains of, 55.
  Gasolene, 3.
    Conveyance of, 521-522.
  Gravity of. *See* Density.
  Heavy oil of. *See* Fractional Distillation Products.
  Illuminating oil of. *See* Fractional Distillation Products.

Petroleum (*continued*)—
    Inflammability of, 185.
    Kerosene. *See* Illuminating Oil.
    Light Oil of. *See* Fractional Distillation Products.
    Lubricating Oil of. *See* Fractional Distillation Products.
    Mazout as fuel, 256, 468, 469. *See* Fractional Distillation Products.
    Naphtha. *See* Benzene and Fractional Distillation Products.
        Conveyance of, 521, 522.
    Origin of, 3, 4, 10, 23. *See* Origin of Petroleum.
    Philodermic preparation of, 1.
    Refineries, 39, 94, 95.
    Spirit, conveyance of, 521, 522. *See* Fractional Distillation Products.
    Theories as to origin of, 10-23.
Petroleum Oil Trust, 236.
Petrolia, Canada, 239.
—— Clarion Co., Pennsylvania, 215.
—— German, 113.
—— Russian, 66.
Petroliferous strata and characteristic fossils, 276-279. *See* Geology of Respective Oil
    Fields.
Petrowsk, Russia, 78, 79.
Pettit spouting well, 235.
Pharmacists, American, and petroleum, 189-190.
Pharmacy. *See* Petroleum as Medicine.
Phenomena, aerial, of spouting wells. *See* Spouting Wells.
—— underground, of Baku, 69. *See* Mud Volcanoes.
Philadelphia, U.S.A., depôt, 195, 481.
—— pipe line, 474.
—— —— reservoirs, 475.
Philippine Isles, 177-179.
    Producing centres—
      Isle of Cébu, 178, 500.
      Panay Archipelago, 178.
        Leite de Bohol, 178.
        Manilla, European population of, 179.
Phillips' spouting well, America, 199.
—— supports Engler's theory as to origin of petroleum, 14-15.
Philodermic preparations of petroleum, 1.
Physic, petroleum as. *See* Petroleum as Medicine.
Physico-chemical action of the soil of the *Minières de bitume*, 138.
Physics of the *fettlöcher*, 112.
Picks or pecks. *See* Excavation.
Picrate of baryta, 450.
—— of lead, 450.
—— of potash, 450.
—— of stroutia, 450.
Picric acid as an explosive, 450.
Piecework, advisable to excavate oil wells by, in Galicia, 291.
Pictet's fulgurite, 448-449.
—— requirements for a perfect explosive, 448.
Piedmont, Italy, 497.
Piero, J., and Grovier, A., inventors of seed bag, 234.
Pierpont & Havens, their *rôle* in history of American petroleum, 193.
Pierre, Colorado, 227.
Pietra Mala, Italy, 122, 277.
Piguerenda, Bolivia, 258.
Pine wood as fuel for boilers, 340. *See* Strength of Materials.
—— weight in air and water, 358.
Pint (0·57) litre, 504.
Pioneer of petroleum illumination, 187-188.
Pipe line companies, 473-476. *See* Standard Oil Company.
—— lines, 470-478.
—— —— American, 470-476.
—— —— —— former glut of, 473.
—— —— Baku, 477.
—— —— European, 476-477.
—— —— Galicia, 477-478.
—— —— Java, 478.
—— —— laying down a, 474.
—— —— specification for pipes of, 478.

Pipes, casing. *See* Casing.
Pirates, Algerian, caulk their ships with petroleum, 260.
Pissasphaltum, 187.
Piston, sand-pump. *See* Sand-pump.
Pitch. *See* Asphaltum and Bitumen.
—— Lake, Trinidad, 249.
Pithecus maritimus, 276.
Pithole, Pennsylvania, 199, 200.
Pits of mound builders, 182-183. *See Fettlöcher*.
Pittsburg, 187, 188, 200, 205, 206, 213, 217, 219.
—— carboniferous strata occupy irregular zone in environs of, 203.
—— coal, 205.
—— Elysée, Réclus, on, 217.
Piura, Peru, 253, 257.
Placoid fish, 279.
Plaisance (Placentia), Italy, 120, 122, 277.
Planes, inclined, relay of barrows, 529.
Planets, worship of, 6, 7 ; burning of incense in honour of, 67.
Plank road, 471.
Plants, fossil marine, 239.
Plassy boring, 344.
Plata, Bolivia, 258.
Plate, iron. *See* Iron, Wrought.
Plateau of Balakhany, 95.
—— of Sourakhany, 96.
—— central, of France, 127.
Platform of tank cars, 463.
Plausibility of speculators, 482-483.
Pliny on Agrigentum (Girgenti) petroleum, 118.
Ploesti, Roumania, 276, 495.
Ploeteran, Java, 173.
Plots of Russian petroliferous land, extent of, 108.
Pliocene strata of Algeria, 264, 276 ; *illustrated* in section, 269.
—— —— California, 229, 276.
—— —— Italy, 276.
—— —— Japan, 276.
—— —— Java, 276.
—— —— New Zealand, 276.
—— —— Roumania, 276.
Plutonic rocks, petroleum absent from, 22.
—— —— a sign when struck to stop boring, 22.
Po, river, Italy, and Mediterranean line of strike, 9, 120.
Pockets of oil (Fr., *Poche*), defined, 535.
—— —— in Roumania, 62.
Pocono sandstone. *See* America, U.S.A., geology of.
Poertwadi, Java, 173.
Pointed end of casing provisional. *See* Casing.
Pohar oil well, duration of yield, 32, 277.
Polana, Galicia, 26, 42, 48.
Poland, plain of richest oil fields, where it undulates, 8.
Politicians, American, try to tax petroleum, 195.
Polo, Marcus, on ancient knowledge of Baku petroleum, 7, 68.
Polotsk, depôt, Russia, 110.
Polygon formula for calculating area of, 487.
Pomatum, petroleum as, 1, 6.
Ponk-ki, China, 163.
Pont Euxin (Black Sea), 66.
Poplar, weight in air and water, 358. *See* Strength of Materials.
Porcelain, insulators of electric blasting outfit, 453.
Porphyritic sandstones of Galicia, 34.
Port-aux-poules, Algeria, 262.
Port of Spain, Trinidad, 248.
Port Sarnia, 471.
Possession, advisable to take, of petroliferous land when bought in Galicia, 51.
Post, bouncing, of free-fall system, 320.
—— —— absent from Canadian, 354.
Poti, Russia, 79, 84, 479.
Potters' earth, cubic metres of, time occupied in excavating, 506.
—— —— weight of, 506.

Pottery, lamps, Chinese, 160.
Pottsville Conglomerates. *See* America, U.S.A., geology of, 499.
Poverty Bay, New Zealand, 180.
Power, horse, 529-531. *See* Boilers and Engines.
Precision, oil-well drilling a work of, done blind-fold, 426.
Prêle boring, Châtillon, France, 144.
Premonitory symptoms of unforeseen spouting, 439-440.
Prentice, Pennsylvanian engineer, his *rôle* in history of Peruvian petroleum, 252.
Presence of petroleum indicated by characteristic fossils, 127.
     By exudations, 483.
     By petroleum stone, 144.
Pressure, gas, at Boryslaw, 46.
—— of wind, 507.
Price & Co., 156.
Price of Land. *See* Land Buying, and Land, Cost of.
—— of petroleum in Galicia, 54, 56. *See* Cost.
Priming of electric blasting outfit, 454.
Prism, area and volume of formulæ for calculating, 525.
—— centre of gravity of formulæ for finding, 528.
Production of petroleum, annual. *See* Graphical Chart and under Respective Oil
     Fields.
Proge river, Java, 173.
Progression, arithmetical, 524.
—— geometrical, 524.
Prohibition of locomobiles in Galician and Russian oil fields, 339.
Prometheus, fable of, 68.
Promotion of Pennsylvanian Rock Oil Co., 191-195.
Proofs of ancient working of American petroleum, 182-184.
Propeller, air. *See* Air Propeller.
Property in Galicia badly defined, 50.
—— —— possession of, a precaution, 50.
—— —— title to, 50.
—— —— transfer, mode of, uncertain, 50.
Proprietors of land. *See* Land, Proprietors of.
Prospectors deceived by speculators, 482-483.
—— points on which their examination of an oil field should bear, 484.
Protective tariffs, effect of, on Roumanian petroleum industry, 61.
Provisional point of casing. *See* Casing.
Prying into the bowels of the earth, 427.
Przbilla and use of artificial diamond in boring, 424.
—— hydraulic free-fall instrument of, 420.
Puente, California, U.S.A., 230.
Puerto Grau, Peru, 256, 502. *See* Peru.
Pulaski, Kentucky, 225.
Pulleys. *See* Oil-well Drilling Estimates.
Pumping of oil, 458-460.
—— daily accumulation of oil less than that removed by, 455.
—— hand pumps for, 460.
—— station, Galician, *illustrated*, 40.
—— —— distance between, 474.
—— steam pumps, 460.
Pumps, hydraulic, oil-well drilling by, 384.
Puncheon. *See* Barrel.
Punjaub, India, 66, 152.
—— Oil Prospecting Syndicate, 152.
Punoka, India, 277.
Punto d'Acaja, Venezuela, 277.
Purchase of land. *See* Land, Buying and Cost of.
Purdy spouting well, 235.
Putna, Galicia, 278.
Puy de Dome, France, 20, 136.
Puy de la Poix, France, bitumen of, 139.
Pyramid, centre of gravity of, formulæ for finding, 528.
Pyrée, grand, of the Persians, 68.
Pyrenees and Mediterranean line of strike, 9, 127, 129.
Pyrites, 133. *See* Marcasite.
Pyroxylin as an explosive, 450. *See* Gun Cotton.

Q.

Quadrilateral, centre of gravity of area of, 527.
Quarries, petroleum classed with in Roumania, 65.
Quaternary strata, 276.
Quebec, 232. *See* Canada.
Quebrada, Peru, 254.
Quirilli, 467.

R.

Race, defunct American, 182.
Radius of circle, ratio of, to circumference, etc., 523.
Radu, M. Jules, on minerals of Carpathians, 28.
Railways, American, network of, Bissel connects oil fields with, 195.
—— continental, regulations for conveyance of petroleum spirit, etc., 521-522.
—— Galician, Neu-Sandec-Zagorz-Chyrow-Stry-Kolomea, 29.
—— Russian Transcaucasian, 88.
—— —— —— effect of inauguration of, on Russian production, 72.
—— —— —— traffic served by, 467.
—— opposition to tank-waggons, tank-steamers, etc., 466.
—— petroleum as fuel for locomotives of, 2.
Ram, Jobard's rope boring, 301.
Ram Ormuz, Persia, 151.
Ramri, Isle of (Arakan Isles), Burmah, 151, 154.
Rapid growth of Baku, 83.
Rarity of game causes trappers to turn salt-makers, 187-188.
Ratio of barren wells to productive in Germany, 115.
Pennsylvania, 219, 244.
Rational work, necessity for, in oil-well prospecting and drilling, 482-484.
Ravines of Western Virginia, 221.
Reamer, 393.
—— expanding. *See* Expanding Drill.
Recent strata, fossils of, and oil fields in which they occur, 276.
Reclus, Elysée, on—
Peru, 251.
Pittsburg,
Sumatra, 169-170.
Record speeds in oil-well drilling. *See* Speeds.
Rectification of petroleum. *See* Refining of.
Red Indians and petroleum, 182-186.
Reeds, broom of, used to skim petroleum of *fettlöcher*, 113.
Refinery, petroleum. *See* Petroleum Refinery.
Reflector, primitive Japanese, 164.
Refrigeration of water flooding bore hole, 438-439.
Refusal of China to grant petroleum concessions to British, 162.
—— of Russian railway and steamship authorities to support Nobel, 466.
Reggio, Italy, 119.
Registration of manager or overseer in Galicia, 52.
—— by prefect of district of permission to drill oil wells in Galicia, 52.
Reichenbach's theory of origin of petroleum, 10.
Reitling, Germany, 278.
Relays of barrow in wheeling up an inclined plane, 529.
—— of pumping stations on pipe lines, 474.
Religion, Protestant, and observance of Sunday, 214.
—— Parsee, 67-68.
Relizane, Tarria, Algeria, 262, 271.
Rembang, Java, 173.
Remedy, petroleum as a general, 1, 2, 39, 68, 71, 126, 129, 131, 184-190.
Remedying accidents. *See* Accidents.
Remunerative yield, minimum, 244.
Renown of the French enhanced by being first to give to science exact ideas of Seneca oil, 185.
Repairs and renewals. *See* Oil-well Drilling Smithy.
Report, Montcalm's, to France on Seneca oil, 185.
Reptiles first occur in Devonian, 279.
—— petroleum wards off bites of, 184.
Repugnance, foreman borer's, to sacrifice ground he has drilled, 429.
Researches, Engler's, on origin of petroleum, 13.

Reserve fund, amount allocated to, in starting oil-well drilling,at Baku, 107.
—— Russian petroliferous land in, 108-109, 148.
Reservoirs, petroleum, 90, 461, 462, 474.
—— of Batoum, 467.
Resistance of derrick to wind, 507. *See* Strength of Materials.
Resumption of drilling at Barnsdall well, 209.
Reventazou, mountains of hardened tar of, Peru, 252.
Reversed strata of Galicia, 30.
Riazan, depôt, Russia, 110.
Richburg, New York, 207-208, 501.
Riches of oil kings, 5, 192.
—— mineral, of Carpathians, 28.
—— —— of China, 160.
Rig defined, 535.
Rig Sandy, Kentucky, 225.
Riga, depôt, Russia, 110.
Riom, France, 138.
Rionero di Molise, Italy, 120.
Rivière a la Rose, Canada, 238.
Rivière, Dr., on Gabian oil fountain, 6, 131.
Rixford, Pennsylvania, 501.
Roanne, France, depôt, 480.
Robustness of salt-makers, 187.
Roccamorice, Italy, 277.
Rocheron, Madagascar explorer, 278.
Rock. *See* Clay ; Conglomerate ; Granite ; Limestone ; Marl ; Sandstone ; Shale, etc.
Rockefeller, founder of the Standard Oil Co., 475.
Rocky Mountains, influence on Western U.S.A. oilfield, 204.
—— —— mud volcanoes of, 87.
—— —— watershed of—
   Eastern, 239, 278.
   Western, 239.
Roderen, Alsace, 277.
Rods. *See* Oil-well Drilling.
Roger's Gulch, Virginia, U.S.A., 501.
Roman inscription on burning fountain, 143.
—— —— at Walsbronn, 117.
—— conquest of Gaul, Chinese bored with the rope system for salt contemporaneously
   . with, 293.
Romance, an American, the history of petroleum, 191.
Romany, Russia, 102.
Romniculu Saratu, Roumania, 63, 479.
Roofs, petroleum imbued, of Baku, 88.
Ropa, Galicia, 36, 48, 57.
—— oil-well drilling speed results, etc., at, 366-367.
—— Fedorowitch, Galicia, 495.
Rope system of oil-well drilling. *See* Oil-well Drilling, Rope System.
—— rods, 418.
—— substitutes, 419.
—— *See* Cable.
Ropianka, Galicia, 32, 35, 37, 41, 48, 278.
Ropica Ruska, Galicia, 37, 48, 57, 495.
—— Polska, Galicia, 37.
Roque Salière, France, 136.
Ross, origin of petroleum, 19.
Rossdale, Kansas, gas wells of, 279.
Rost's free-fall instrument, 412.
Rotatory system of oil-well drilling. *See* Oil-well Drilling, rotary, etc.
Rouble (3½ francs), 506.
Rouen, France, depôt, 480.
Roumania, oil field of, 88, 58-65, 276.
   Aliens cannot own land in, 61.
   Analogy of, with Galician, 58.
   Arc of a circle of the Carpathians extends along, 58.
   Association of petroleum of, with—
      Common salt, 58, 60.
      Sulphur, 60.
   Bad roads of, 61.
   Bankruptcy, apparently perpetual, of, 60.
   Bukovina, signs of petroleum lost in valleys of, 49, 63.

Roumania, oil field of (*continued*)—
    Cantaguzène, petroleum well of, at Dragonèse, 60.
    Carpathians, petroliferous line of strike found in flanks of, 58.
    Cavities or pockets of oil frequent in, at shallow depths, 62.
    —— misleading effect on natives of, 62.
    Communication, means of, awanting, 61.
    Crude oil of, all exported, 60.
    —— density of, average, 3.
    Dragonese spouting well, 62-64.
    Duration of yield said not to last, 62.
    Geographical situation retards development of, 60.
    Geological formation of generally—
        Miocene, 59.
      Subdivisions of—
        Argilo saline formation, 59.
        Congerian beds, 59.
            Characteristics of, 59, 60.
            Chlorite of, 60.
            Conglomerates of, friable, analogous to Galician, 59, 60.
            Dragonèse, well developed at, 60.
            Gneiss of, 60.
            Mica of, 60.
            Quartzose, particles of, 60.
        Paludina beds, 59.
        Pelisse's classification of neogene beds—
            Congerian essentially petroliferous, 60.
            Sarmatic rich in salt, 60.
    Map of, 52.
    Mines of, as distinct from quarries, belong to State, 65. *See* Quarries and Royalties.
    Natives' futile attempts to work, 61.
    Oil fields of—
      Eastern oil field, 64.
        Producing centres—
            Babadag, 64.
            Batoy Jalomitei, 64.
            Doian, 64.
            Dudeschi, 64.
            Garvan, 64.
            Hirsova, 64.
            Kioi, 64.
            Tatar, 64.
      Western oil field, 63-64.
        Producing centres—
            Breza, 64.
            Buzau, 63, 65.
            Colibazi, 64.
            Comanesku, 63.
            Ddobesti, 63.
            Dimitresci, 63.
            Dragoneza, 64.
            Fendul Saratu, 64.
            Ilanica, 64.
            Itanescu, 63.
            Mainesci, 63, 276.
            Matitza, 64, 65.
            Romniculu Saratu, 63.
            Saveja, 63.
            Tetzaru, 63.
            Tirgu Okna, 63, 276.
            See for other centres, 276, 277.
    Pelisse on geology of, 60.
    Petroleum depôts of, 479.
    Pockets of oil, illusive influence of prevalence of, 62.
    Production, annual, of, 65.
    Protective tariffs of other countries retard development of petroleum industry of, 61.
    Quarries, including working of petroleum, belong to proprietor of ground, 65.
    Railways, list of, 60.
    Roads, bad, retard industry, 61.
    Royalty, 1 per cent. *ad valorem*, 65.

Routine of day and night shift in oil-well drilling, 489.
Rowna, Galicia, famous oil well of, 88.
Rows, John, Canadian petroleum pioneer, 232.
Royalty, Russian government on concessions, 107.
—— French, Madagascar, 275.
—— Roumanian government, 65.
Rozet and Millet, theory of, as to origin of petroleum, 10.
Rubbish. *See* Detritus.
Rugius, rope boring system of, 303.
Ruin of Colonel Drake, 196.
Rumbaugh Farm, Butler County, Pennsylvania, 216.
Rumsey spouting well, 235.
Rungury, Galicia, 47.
Rush to Pennsylvania during oil fever, 195.
—— to Boryslaw, 25.
Russia, oil fields of, 66-110, 277-279.
　　Description of, 74-110.
　　History of, 66-73.
　　Subdivisions of—
　　　　Anaclie, 80.
　　　　Baku, 81-95, 277.
　　　　Balakhany, 99.
　　　　Bibi Aibad, 100.
　　　　Binagadine, 100-101.
　　　　Central, 124-125, 279.
　　　　Crimea, 74.
　　　　Daghestan, 73, 78-79.
　　　　Elizabethpole, 73, 81.
　　　　Grosnaïa, 73, 78.
　　　　Kertch, 75.
　　　　Kouban, 73, 75.
　　　　Sourakhany, 96.
　　　　Tamansk, 73, 75, 277.
　　　　Tiflis, 73, 80, 81.
—— transport of petroleum through, 464-470.
Russian petroleum depôts, 110, 479, 480. *See* Baku; Caucasus.
Rustics, stupefied, Uncle Billy Smith draws up petroleum in bucketfuls before, 195.
Ruta Otar, India, 277.
Rymanow, Galicia, 41.

S.

Sabah, Borneo, 175.
Sabbath in Protestant countries, observance of, 215.
Sabountchany, Russia, 96, 110, 496.
—— pipe lines, 466.
Sacarite, Italy, 278.
Saczal, Hungary, 124.
Saddle back. *See* Anticlines.
Safety lamp,
—— explosives, trials of, at Anzin, 451. *See* Davy.
Sagene reduced to metric measurement (2·133 metres), 535.
Saghalien, Isle of, 167-168.
Sahalien strata of Algeria, 264, 266.
Sakatali, Tiflis, Russia, 81.
Salary, Colonel Drake's, as magistrate, 196.
Sale of land. *See* Land, Sale of.
Saline lagoons, primitive, and origin of petroleum, 11.
Saline lakes of Apshéron, 85.
Salo, Italy, 497.
Salses. *See* Mud Volcanoes.
Salsomaggiore, Italy, 120, 123.
Salsominore, Italy, 120.
Salt, association of, with petroleum, 24, 58, 75, 150, 187.
—— deposits of Galicia, 28.
—— —— of Roumania, 58.
—— industry of America, 187.
—— Lake City, Utah, 226.
Salta, Argentine Republic, 258, 278.

Saltmakers of America, grease that spoilt their brine, 187.
—— of Chinese Thibet use natural gas as fuel, 162.
—— petroleum an unknown factor to, 187.
Saltpetre as an ingredient of explosives, 450.
Salve, petroleum as a, 7, 68, 71.
Samara, Russia, 125.
Sambon and Shannon spouting well, 235.
Samennien strata. *See* Devonian.
Samkto, South Caucasus, Russia, 80.
San Carlos, Venezuela, 250.
San Diego, California, 276.
Sanctuaries or temples of the planets, 67.
San Raymondo, Ecuador, 251, 502.
San-tay-le-tche, China, 163.
Sand in Baku petroleum, 39.
Sandusky, Ohio, 223.
Sand brought up by spouting wells wears casing, 93.
—— shifting, 194.
—— time occupied in excavating a cubic metre, 506.
—— weight of a cubic metre of, fine dry, 506.
        See Cal and *Kir.*
Sand-pump of different oil-well drilling systems—
        Canadian, 363.
        Free-fall, 306, 325, 326, 351.
        Piston, American, described, 68, 69.
                Illustrated, 69.
        Cost of, 335.
—— —— cables. *See* Cables.
—— —— windlass (sand reel), 335, 404, 535.
Sand-pumping. *See* Oil-well Drilling.
Sandstone, Magura, Galicia, 35.
—— Mahoning, Pennsylvania, 205, 217.
—— M'Kean, Pennsylvania, 203, 205, 207-208.
—— soft granular Ropianka of Galicia, capacity for holding petroleum, 32.
—— Venango, Pennsylvania, 210-213.
Sandstone and sands, speed of hand-digging oil wells through, 200.
Sandusky, Ohio, 222, 223.
Sandy Beach, Canada, 236.
Santander, Spain, 481.
Santa Barbara, California, 228.
—— Clara, California, 230, 276.
—— Elena, Ecuador, 250.
—— Paula, California, 230.
—— —— Ecuador, 251.
Sarabilkowa, Russia, 125.
Saragossa, Spain, 110.
Saratoff, Russia, 467.
Saratow, Russia, 110.
Sarie, Java, 174.
Sassuolo, Italy, 122.
Saturated hydrocarbons. *See* Origin of Petroleum.
Saurians, fossil, of cretaceous system, 278.
Saveja, Roumania, 63.
Savoie, Haute (Upper Savoy), France, 127.
Saws, stock of, for use in hand-digging oil well rig, 285.
Sawdust packing material for glass vessels filled with petroleum spirit, 521-522.
Saxon, Anglo-, in America, 186.
Scab or itch of camels cured by petroleum, 7, 68.
Schedule of expenses in starting drilling at Baku, 107.
Schibaeff Co., No. 3 spouting well of, 91.
—— —— Geological section of their No. 8 well, 100.
—— —— oil-well drilling system of, at Baku, a modified Laurent, 409.
—— —— works at Khidirzind, 101.
Schinano, Japan, 163, 166, 168, 499.
Schmidt, German works of, at Montechino, Italy, 121.
Schodnica, Galicia, 32, 33, 46, 48, 277, 478.
Schonigsen, Sehnde, Germany, 278.
Schreiner, Abraham, of Boryslaw, Galician petroleum pioneer, 24.
Schugorowa, Russia, 279.
Schumacher's hydraulic free-fall instrument, 420.

Schutter on rope drilling speed results at Mencina, Galicia, 377.

Schwabwiller, Germany, 116, 118, 499.

Schwartz, Koechlin, quoted—
    Subterannean phenomena of Baku, 69.
    Temple of eternal fire, 97.

Scotland, 180.

Scrubgas, Clarion Co., Pennsylvania, 216.

Scrutinisation of drill and *débris* by foreman borer, 334, 486.
—— of tools, 487.

Scumming petroleum off the *fettlöcher*, 118.

Scurvy, petroleum a cure for, 71.

Sea. *See* Black Sea, Caspian Sea, etc.
—— water used for steam raising in Peru, 256.

Season of year, influence of, on prices on Galicia, 55, 56.

Seate, India, *pyrée* of, 68.

Seattle, Alaska, U.S.A., 241.

Sechura, Peru, 250.

Seckendorff's free-fall instrument, 413.

Sections, geological vertical, of petroliferous strata of—
    Balakhany, Baku, 100.
    Binagadine, Baku, 101.
    Canadian and Pennsylvanian oil fields, *illustrated*, 204.
    Gabian, France—
        First boring, 132.
        Second boring, 134, *illustrated*, 135.
    Galician, oil field, western, *illustrated*, 34.
    —— between the San and the Styria, *illustrated*, 43.
    Holowecko, Galicia, 35.
    Pennsylvanian, actual, 205-207.
    —— ideal, 205.

Sector, spherical, formulæ for calculating volume of, 526.

Sediment from Baku petroleum deposited in natural reservoirs, 461.

Sedimentary strata, affinity of petroleum for, 22.

Seed bag, capping oil well by, 234-235, 228.

Seeger, of Czernowicz, quality of his engines, 342.

Segment of a sphere, formulæ for calculating volume of, 526.

Sehnde, Germany, 115, 116, 496.

Seknati, Borneo, 175.

Sekoendar River, Sumatra, 171.

Selenitic marl of Algeria, 267.

Selligue rope boring, 303.

Selligues and shale oil, 4, 5, 128, 190.

Sello rope boring, 303.

Semarang, Java, 174.
—— River, 173.

Senecas (North American Indians), their *rôle* in the history of petroleum, 182-186.

Seneca oil a universal panacea, 2, 184, 186.
—— Lima, Ohio, 222.

Sereth River, boundary of W. Roumanian oil field, 63.

Series of Permian strata, American, 205.

Sespé, California, 230.

Se-Tchouen, 162, 163.

Settlers, French, and history of petroleum, 185.

Shaku, Persia, 150.

Shale (clay) (rock). *See* Clay Shales.
—— mineral, of Autun, 4, 128.
—— oil of—
    America, 190, 191.
    Autun, 4, 5, 128.
    Britain, 5.
—— industry swamped by petroleum—
    In America, 191.
    In France, 128.
—— —— works in America before oil fever, 191.

Shallow oil wells of Galicia, 25.
—— —— —— of Roumania, 61-62.

Sham signs of petroleum, 483.

Shamburg, Venango, Pennsylvania, 215.

Shannopin, Alleghany, 217.

Sharks, fossil, characteristic of pliocene strata, 276.

Sharpening of tools, 285-323.
Sheath of free-fall instrument, described and *illustrated*, 319-320.
Shed for covering smithy, 289, 336.
—— —— —— well, cost of, 288.
—— —— —— engines, cost of, 351.
—— —— —— boilers, cost of, 351.
Sheerness artesian well, 406.
Sheffield, Warren Co., Pennsylvania, U.S.A., 212.
Shells of Jura, volatile alkali in, 10.
Shepherd, the, and Ben Lakdar, 259, 260.
Sherkstone, Canada, 239.
Sherwood, Price & Co.'s works at, 156.
Shifting sands of Drake's first well, 194.
Shipmasters refuse asphaltum as cargo, 250.
Ships, Algerian pirates caulk, with petroleum, 260.
—— asphaltum, cargo of, strains, 250.
Shoemaker refuses boots on credit to John Shaw on morning his famous well spouted, 233.
Shoulder of rods, described and *illustrated*, 316.
Shovels, cost of, in oil-well excavating outfit, 289.
Shower of petroleum. *See* Spouting Well.
Shrinkage of wood under water pressure, 369.
Siberian-Russian petroleum transport route, 467.
Siam, 170.
Siari, Galicia, 37, 48, 57, 495.
Sichès, Peru, 254, 502.
Sicily, oil field of, 120.
—— *doyen*, the, of the world's oil fields, 120.
—— Pliny its godfather, 120.
Sidi Brahim, Algeria, 262, 264, 265.
Sifton, Gordon and Bennet spouting well, 235.
—— J. W., spouting well, 235.
Sig, Plain of, Algeria, 262.
Sighs of the earth, 168.
Sigmondi's free-fall instrument, 417.
Signs of petroleum in France, 126-146.
—— —— in Galicia, 30.
—— —— surface, too much importance not to be attached to, 483.
—— —— —— fictitious, 483.
Signals, code of, in hand-dug wells, 285.
Silicious conglomerates of Gabian, 134.
Silliman, Benjamin, theory of, as to origin of petroleum, 189.
—— recommends boring where natural crevices met with, 193.
—— prophecy of, fulfilled, 195.
Silver River, Canada, 237, 279.
Silurian petroliferous strata, 32, 502. *See* Canada, Geology of; America, U.S.A., Geology of.
Simonin quoted on Pithole, 200.
Sinking of oil wells. *See* Oil-well Drilling.
Sinking of strata in Bukovina, 63. *See* Dip.
Sisperle's application of free-fall, to rope boring, 303, 419.
Sisteville, Pennsylvania, 218.
Site of petroleum reservoirs, selection of, 462.
—— of trial borings, circumstances which decide selection of, 482.
—— negative results of badly selected, 134.
Situation, geographical, of Roumania against development of its oil fields, 60, 61.
Sivan (? Swan) spouting well, 235.
Skimming of *fettlöcher*, 113.
Skin diseases, petroleum as a curative for, 6, 68, 71.
Slipping of drill. *See* Deviation.
Sloboda, Galicia, 46-48, 57, 277, 495.
—— Rice, Frenchman, fails at, 46.
—— M'Intosh, American, succeeds, 46.
—— —— deep well of, at, 46.
—— one of most important places of Galicia, 46.
—— sinking of petroliferous strata from, to Bacau in Roumania, 49, 63.
Slope. *See* Watershed.
Slow decomposition of vegetable and animal matter. *See* Origin of Petroleum.
—— progress of rope-boring in hard ground, 377.
Smell, strong, of shale oil, 5.

Smethport, Bradford, U.S.A., 208.
Smilno, Hungary, 124.
Smith's Cape, Canada, 239.
Smith's Ferry, Beaver, Pennsylvania, 216, 501.
Smith's (black), wages in America and Baku, 248.
Smith, Uncle Billy, drilled first oil well, 194.
—— Watson, origin of petroleum, 12.
Smithy of oil-well drilling installation—
    Canadian system, 363.
    Excavation system, 285, 289.
    Free-fall system, hand-power, 336.
    —— —— steam-power, 352, 404.
Smolensk, Russia, 110, 480.
Smoothing or equalising sides of bore hole, 393.
Soda, sulphate of, 183.
Sodic acetylide, 16.
—— chloride, 85. See Salt.
Soerabaja, Java, 174.
Soft ground, Fauvelle drill adapted for, 388.
—— —— of Baku involves frequent casing, 87.
—— soap, petroleum sometimes resembles, 1.
Soil, time occupied in excavating a cubic metre of, 506.
—— weight of a cubic metre of, 506.
Sokoloff, origin of petroleum, 19.
Sola, Kansas, gas wells of, 279.
Solanti, Roumania, 276.
Soldiers, American, use petroleum as embrocation, 186.
Solis spouting well, 235.
Soltau, Germany, 115.
Somersetshire, England, 125.
Song of the earth (gas of fire well), 163.
Sontag's application of free-fall to rope boring, 419.
Soosmezo, Hungary, 124.
Sound, velocity of, finding depth of oil well by, formulæ for, 524.
Soundings. See Trial Borings.
Source. See Springs.
South America and West Indies, 247-258.
    Argentine Republic, 258.
    Barbadoes, 248.
    Bolivia, 258.
    Brazil, 258.
    Chili, 258.
    Colombo, 250.
    Cuba, 247-248.
    Ecuador, 250-251.
    Peru, 251-257.
    Trinidad, 248-250.
    Venezuela, 250.
Spain, oil fields of, 124, 497.
—— American exportation of crude oil to, 246.
—— petroleum depôts of, 481.
—— —— Russian exports of, to, 102.
—— South of Europe Exploration Co., 124.
Spaniards and petroleum of Black Spring, Algeria, 260.
Spanish domination of Peru and history of petroleum there, 252, 254.
Spark of electric blasting outfit, how obtained, 452.
Sparré's application of free-fall to rope boring, 303, 419.
Specific gravity of crude oil. See Areometer, Density.
Specification for pipes of pipe lines, 478-479.
—— See Estimates for Derricks, Smithies and Oil-well Drilling Installations, etc.
Spectacle of spouting oil wells. See Spouting Wells.
Speculators deceive capitalists, 482-483.
—— flock to Pennsylvania during oil fever, 195-196.
Speed obtained in oil-well drilling by—
    Canadian system, 361-363, 371, 372.
    Excavation, 290.
    Free-fall system, hand-power, 334, 335, 337.
    —— —— steam-power, 349, 352.
    —— —— combined with Canadian, 366-367.
    Hydraulic boring, 379-380, 388-389.

Speed obtained in oil-well drilling by—
    Rope boring, 302, 377, 378.
    Rotary drilling or diamond boring, 400-401.
Spencerville, Ohio, 223.
Sphere formulæ connected with area of zones, etc., and volumes of segments, etc.,
    525-526.
Spherical spindle, etc., area of, 525.
Spindle, spherical, 525.
Spirit, petroleum. *See* Naphtha.
Spirits of wine, Hanway's comparison of petroleum to, 71.
Spouting of oil wells, American, 198-201, 209, 211, 215-216, 220.
—— —— —— Russian, 87-95.
—— —— —— after underground blasting, 455-457.
Spring, the Black. *See* Algeria.
Springs, iodide, associated with petroleum of—
    Bad Iwonicz, 41.
    Tamansk, 75.
—— salt. *See* Salt Deposits.
Square iron rods, weight of, in air and in water, 513-514.
Squillaci, Italy, 278.
St. Andrew, Barbadoes, 248.
—— Bilt, Alsace, 277.
—— Claire Deville on properties of Foo-Choo-Koo petroleum, 161.
—— Croix, Alsace, 277.
—— Etienne, France, 138, 480.
—— —— Mineral Industry Society of, 138, 381.
—— —— trials of explosives at, 457.
—— John, Canada, 237.
—— Lawrence, Gulf of, 271.
—— —— river, 236.
—— Mary, Ohio, 223.
—— Petersburg, Russia, conveyance of petroleum to, 465, 480.
Staff of oil-well drilling systems, details of—
    Canadian system, 359, 363.
    Excavation system, 290.
    Free-fall (hand-power), 332, 336.
    —— (steam-power), 348, 352.
    Rope boring, 300, 302, 376-378.
    Rotary drilling, 399.
—— enormous, incidental to rope boring, 300.
—— professional, in Galicia, notice of dismissal or discharge to, 54.
Stagnant waters of the *fettlöcher*, 112.
Standard Oil Co., 473-476.
Stanislawow, Galicia, 33, 46, 48.
Starasol, Galicia, salt deposits of, 32.
Staremiasto, Galicia, 55, 67.
Stars, etc., Parsees still worship, 68.
Starunia, Galicia, 37, 48, 57, 276, 495.
Starzvwa, Galicia, 41.
Station, pumping, of Krosno, Galicia, *illustrated*, 40.
—— railway, of Uherce, Galicia, *illustrated*, 42.
Statistics, petroleum. *See* Individual Oil Fields.
Steam power drilling of oil wells, 339-352.
—— pumping, 458-460.
Steamers, merchant, use petroleum as fuel, 2.
—— tank, Russian, 464-470.
—— —— number of, flying French flag, 470.
—— —— of French South American Petroleum Co , 256.
Steinforde, Germany, 114, 276, 278.
Stillwell, Virginia, U.S.A., 200.
Stipulations in Galician petroleum contracts, 56.
Stirrup for drilling by in rope boring, 297.
Stock, overhauling, repairing and oiling of tools before putting back into, 487.
Stoneham, Warren Co., Pennsylvania, 212.
Stoneware lamps of Chinese, 160.
Stopuciani, Galicia, 42.
Storage and transport of petroleum, 458-481.
Store. *See* Smithy Estimates.
Straka's application of free-fall to rope boring, 303, 419.
Strands, rope, defined by number of, 509.

40

Strata, petroliferous, 276-279. *See* Geology of Respective Oil Fields. Plutonic,. Sedimentary, etc.

Straw as packing for glass vessels containing petroleum spirit, 521-522.

Streets of Baku aligned, 81.

—— —— muddy, 83.

Strength of materials, or breaking or crushing strain of—
    Aloes rope, 508-509.
    Ash, 507.
    Beech, 507.
    Boxwood, 507.
    Cast iron, 507.
    Copper, 507.
    Elm, 507.
    Hemp rope, 508-509.
    Iron wire cables, flat and round, 510.
    Lead, 507.
    Oak, 507.
    Pine, 507.
    —— red, 507.
    Poplar, 507.
    Sheet iron, 507.
    Steel, 507.
    —— wire cables, flat and round, 510.
    Wrought iron, 507.
    Zinc, 507.

—— relation of load to ratio of height to thickness of timber, 507.

Strike. *See* Petroliferous Line of.

Strikoff, Baku, variation in oil level of adjacent wells, 84.

Stromberg, von. *See* Origin of Petroleum.

Strontia, picrate of, 450.

Strontium carbide, 20.

Structure, geological. *See* Geology of Respective Oil Fields.

Strzelbycz, Galicia, 48, 55, 57.

Stupefying effect of spouting of oil wells, 90.

Submarine oil wells, 228, 229, 230.

Subterranean phenomena, 69.

Sudden spouting, damage caused by, regarded as accident, 426, 439.

Suffocation in hand-dug wells, preventing. *See* Aeration.

—— remedy for. *See* Asphyxia.

Sugar Creek, Pennsylvania, 212.

—— Loaves, New Zealand, 180, 181, 499.

—— Run, Warren Co., Pennsylvania, 210.

Sukkowo, Central Russia, 124.

Sulin, Chinese Thibet, 163.

Sulphate of lime and soda in mineral spring encountered in Gabian boring, 133.

—— of soda, 133.

Sulphur, result of volcanic gases on lime, 19.

—— association with petroleum, 60, 78, 150, 263, 270.

—— percentage of, in Gaspé crude oil, 297.

Sultan of Turkey's possessions in Georgia (Russia) (1688), 69.

Sumatra, Isle of, 169-172, 177, 499.

—— Langkat oil field, 170-172, 177, 499.
    Producing centres—
      Beloe Telang, 172, 177, 499.
      Palembang, 172.
      Soerabaja, 174.
      Sungie Penanti, 172, 177, 499.
      —— Rebah, 172, 177, 499.
      —— Sichino, 172, 177, 499.
      Telega Toengal, 171.
      —— Tiga, 171.
    Refinery, Balbalan, 170-172.

Summer, low price of Galician petroleum in, 55.

Sun, Daruis invokes, before going to battle, 67.

—— the festal day of, 67.

—— incense burned in honour of, 67.

Sunday, observance of, in Protestant countries, 215.

Sunday Well, 215.

Sungie, 170.

Sungie, Penanti, 172, 177, 499.
—— Rebah, 172, 177, 499.
—— Sichino, 172, 177, 499.
Superficial, etc., measures of different countries reduced to metric, 503, 504. *See* Area.
Superstition of oil-well drilling labourers, 490.
Supervision of day work in Galicia, 291.
Supsa, Russia, 80.
Surakhany, 96, 110, 496.
—— light petroleum of, 96.
Suram tunnel, 467.
Sure signs of petroleum. *See* Petroleum Stone.
Surety for fulfilling purchase contracts in Galicia, 56.
Surface well, free-fall system described, 331-332.
—— commencing to drill without a, 337-338.
—— excavation system, described and *illustrated*, 281.
Surveillance of day workmen in Galicia, 291.
Suspension, vertical, of drilling tools in axis of bore hole, 347.
Suszwycki, Zenon, Canadian drilled well of at Wietrzno, 372.
Sweatings. *See* Exudations.
Sweden, 125, 497.
—— diamond drilling in, 400.
Swindle, petroleum, of Hanover, 115, 116.
Synclines of Galicia, line of strike never follows, 30.
Synthetical origin of petroleum, 16-20.
Syracuse, Richburg, U.S.A., 207.
Syria, 278.
Syroczinski, Leon, on Canadian system, 372.
Szina, Hungary, 124.
Szczepanowski, Stanislas, extensive use of Canadian system, 372.

T.

Tagieff, Russia, 101.
Tagiew, Nos. 15 and 21 spouting wells, 95.
Taï-li-Chen, China, 161, 499.
Tailorstown, Pennyslvania, U.S.A., 218, 500.
Talara, Peru, 254, 255, 256, 502.
Tamansk, Russia, 75, 79, 110, 277, 496.
  Geology of—
    Eocene strata of (sandstones, clays and shales), 75.
    Hard rock, prevalence of, in petroliferous strata of, 75.
    Illuminating oil, high percentage of, in crude oil of, 75.
    Iodide springs, association of petroleum of, with, 75.
    Line of strike from, to Caucausus ends at Derbent, 79.
    Occurrence of petroliferous strata in narrow valleys, 75.
    Petroleum of, density (low) of, 75.
    Sands, shifting, rare, 75.
Tananarivo, Madagascar, 274.
Tandjony, Borneo, 175.
Tank steamers. *See* Steamer and Tank.
—— waggons. *See* Waggons, Tank.
Tap borer, pipe raising tool, 432.
Tar, Algerian pirates go to Black Spring for, 260.
—— *See* Mazout.
—— Cape, 502.
Taranaki, New Zealand, 179.
Tare, 25 per cent. allowance of, to buyers on casks, ridiculous, 56.
Tarentum, near Pittsburg, centre at one time of salt industry of America, 187.
Tariff, protective tariffs of foreign countries retard development of Roumanian petroleum, 61.
—— effect of, on American exports of crude oil, to France, etc., 426.
—— —— oil-well drilling machinery to Japan, 165.
Targowiska, Galicia, 41.
Tarria, Algeria, 262, 267, 270.
Taslau, Roumania, 276.
Tatar, Roumania, 64.
Tatra, Carpathians, 8.
Taylorstown, Pennsylvania, 500.
Tax, the possibility of imposing a, on Peruvian petroleum, the only light in which Spaniards regarded it, 252.
—— attempt to impose local, on petroleum in America, 195.

Tcheleken, Isle of, 147-148, 276.
    Borings, tentative, 147.
    Development retarded by situation, 147.
    Eruption of ozokerit mixed with *kir* and sand, island regarded as, 146.
    Nobel Brothers' concession, 147.
    Properties of the oil, 148.
    Russian government (sole proprietors), works of, 148.
    Yield, average, 148. *See* Mikhailowsk.
Tchernigow, Russia, 110, 480.
Tchong Kiank, China, 163.
Tchouen Pe, China, 163.
Tchouen-Lan, China, 163.
Team of horses, collective effective power when harnessed together compared with
    single harness, 530.
Tega, Roumania, 65.
Tegernsee, Bavaria, 277, 497.
Telega Campina, Roumania, 276.
Telega Toegal, Sumatra, 171.
—— Tiga, Sumatra, 171.
Telegraph, electric, in days of Drake, 195.
Telescopic casing impracticable with screwed cast-iron pipes, 436.
Tellien Atlas, Algeria, 260.
Telluric gases of Baku oil field increase spouting, 88.
Temir Khan, Daghestan, 68.
Temper screw. *See* Oil-well Drilling, Free-fall System.
Temperature, influence of, on gravity of oil, 494.
—— viscosity of oil, 512.
Temple of Eternal Fire, Baku, 97-98.
Tempseh, Crimea, 74.
Tenacious people, the Germans a, 113.
Tennessee, U.S.A., 32, 202, 203, 279.
Terek, Russia, 77, 78.
Terp, Olaf (Dane), 405.
—— use of artificial substitutes for diamond in rock boring, 424.
Terre Haute, Indiana, 225.
Tertiary, characteristic fossils and oil fields of. *See* Pliocene, Miocene, Eocene.
Teschewli, Crimea, 74.
Testimony of Indians as to antiquity of oil pits and mountain builders, 182-183.
—— of trees as to antiquity of oil pits, 182-183.
Tetzaru, Roumania, 63.
Texas, U.S.A., 127, 202, 225, 227.
Thayer spouting well, Washington, 216.
Theodoric (the hangman), derivation of derrick from, 306.
Theories as to origin of petroleum. *See* Origin of, etc.
Thomson, rope boring system of, 308.
Thorn Creek, phenomenal results of use of torpedo at, 455.
Thibet, China, 162-163.
Tide Water Pipe Line Co., 473.
Tidioute, Warren Co., Pennsylvania, U.S.A., 212.
Tiflis, Russia, 69, 479.
—— —— annual production of, 73.
Tiliouanet, Algeria, 262, 271.
Tilsonburg, Canada, 239.
Timber measuring, 526.
—— breaking strain of various, 507.
—— modified under heavy vertical water pressure, 369.
—— shrinkage of wooden poles from iron joints, 369.
Timor, Isle of, Java, 176, 177, 499.
Tin mines of China, 160.
Tinawen, Java, 173.
Tintea, Roumania, 276.
Tiona, U.S.A., 212, 500.
Tirgu Okna, Roumania, 63.
Tirioli, Italy, 120.
Tissue, soft, of animal, and origin of petroleum, 15.
Tithonic Beds, upper jurassic, of Hanover, 7, 112.
Title to land in Galicia, insecurity of, 50, 57,
—— indemnity, to land in Russia, 108.
Titusville, Pennsylvania, 191-197, 214.

Tjelatjap, Java, 178.
Tloki, Galicia, 38.
Tocco, Italy, 119, 120, 277, 278, 496.
Toeban, Java, 174.
Tombs, cyclopean, of the mound builders, 182.
Tongune. *See* Twingon.
Tools. *See* Oil-well Drilling.
Topolnica, Baku, petroleum refinery, results of fractional distillation of Baku crude, 94.
Torches of Seneca Indians, 185.
Toronto, Canada, 232.
Torpedo. *See* Explosives.
Torrey Cannon, California, 230.
Tosan, Japan, 163, 167-168, 499.
Total annual petroleum production. *See* Individual Oil Fields and Graphical Chart.
Tougue River, France, 128.
Toula, depôt, Russia, 110, 480.
Toulouse, depôt, France, 481.
Toutoma, Japan, 163.
Townshippe, Wisconsin, 225.
Toxic effects of carburetted gas, remedy for, 284.
Traces of petroleum. *See* Signs.
Traction, ratio of, to load drawn, 529-531.
—— engines. *See* Locomobiles.
Trains, running of winter freight, in Galicia, 55.
Tramway. *See* Plank Road.
Trans-Caspian oil field, 147-148, 276, 277, 497.
Trans-Caucasian railway, 83-84.
Transformations of petroleum, numerous, 2.
Transition strata. *See* Silurian and Devonian.
Transport of petroleum, 462-481, 521-522.
Transportable derricks, 313.
Trapezium, centre of gravity of area of, 527.
Trappers and petroleum, 186.
Travellers' reports on Madagascar, 273.
Travertine (calcareous tufa), Limagne, 137.
Trees corroborate age ascribed by Indians to mound builders' oil pits, 182-183.
Trenches, deep Indian, for collecting petroleum, 184.
—— —— of Brewer, Watson & Co., 191.
      See Fettlöcher.
Trend of petroliferous strata. *See* Line of Strike of.
Trial borings, hand-power hydraulic machine for drilling, 386.
—— selection of site for. *See* Site.
Triangle, centre of gravity of area of, 527.
Trias petroliferous strata of Carolina, N., 33.
—— —— —— of Connecticut, 33.
Trieste, Austria, 55.
Trilobites, 279.
Trinidad, Isle of, West Indies, 248, 250, 502.
      Producing centres—
            Aripero, 250.
            Brea, 250.
            Pitch Lake, 249-250.
Triolo, Italy, 278.
Tripod derricks, described and *illustrated*, 308.
Trodden ground, ratio of traction to load drawn on, 530.
Troops, American, use petroleum as embrocation, 186.
Troyes, depôt, France, 480.
Trunk of tree, stump of, cost of, for anvil bed, 289, 336.
Trzecieski, Titus, use of crude oil by, to doctor his cattle, leads to development of
      Galician petroleum industry, 39-40.
Tse-Liou-Tsin, China, 163.
Tuccilal, Peru, 254, 502.
Tumbez, Peru, 253.
Tunis, 273.
Tunnels. *See* Algeria, Gabian, etc.
Turco-Persian frontier, Persian oil field extends from, 150.
Turkestan, 278. *See* Fergane.
Turkey and Roumanian petroleum trade, 60-61.
—— exports of Russian petroleum to, 102.
—— receives all its refined petroleum from Russia and Austria, 60-61, 102.

Turkey in Asia, 277.
—— Foot, U.S.A., 221.
Turner, origin of petroleum, 17.
Tweddle cedes his Peruvian petroleum concession to London and Pacific, 254.
"Twin" spouting oil well of America, 199.
Twingon, Burmah, 156, 157, 160, 277, 498.
Tzaritzin, depôt, Russia, 109, 110.
—— cooperage, mechanical, of, 469.
—— docks of, 469.
Tzouloukidze, Prospector, Anaclie, 80.

U.

Ubiquity of British where petroleum is found, 69, 162.
Udwarhely, Hungary, 124.
Ugo, Japan, 163.
Uherce, Galicia, 41, 477, 479.
—— record speed in drilling at, 339, 362.
—— railway station of, *illustrated*, 42.
Underground disturbances at Baku, 85-87.
Undiscovered oil springs of Turkestan, 150.
Unforeseen spouting, 88-95.
United States oil fields. *See* America, United States of, Oil Fields of.
United Pipe Lines Co., 473.
Units of measure adopted in oil-well drilling, 504.
Universal panacea. *See* Curative Virtues of Petroleum.
Unscrewing of rods, time occupied in, 307.
Upheaved and distorted strata of Galicia, 30.
—— —— —— of Western Virginia, 221.
Upheaval of Tatra commencement of main European petroliferous line of strike, 8.
Urban, Hans, statistics of Galician petroleum, 26.
Ursus minutus, pliocene fossil, 276.
Ustianowa, Galicia, 48.
Ustrzki, oil fields of Galicia, 477.
Utah, U.S.A., 32, 202, 226.
—— —— cretaceous strata of, 204.
—— —— zietriskisite of, 226.
Utica, Venango Co., Pennsylvania, 215.
Utilisation of compressed air as motive power in oil-well drilling, 256.
—— of petroleum as fuel in oil-well drilling, 256.
—— of wood as fuel, 340.

V.

Vague ideas of petroleum, the French the first to give to science, 185.
Valence, France, depôt, 480.
Valencia, Spain, 481.
Valves of sand pump. *See* Sand Pumps.
Van der Made's Balangan spouting well, Borneo, 175.
Van Dijk, free-fall system of, 417.
—— Wood enveloped rods, 369-370.
Van Wert, Ohio, 223.
Variation of density of oil with temperature, 494.
—— of viscosity of oil with temperature, 512.
Varnish making, petroleum used in India in, 71.
Vaseline, perfumed, 1.
Vaucluse, France, 186.
Vedro, Russian measure of capacity (1·229 decilitres), 503.
Vegetable origin of petroleum. *See* Origin, Vegetable.
Vehicle, ratio of traction to load on different kinds of roads, 530-531.
Vein of petroleum. *See* Petroliferous Strata.
Velocity defined, 529.
Venango Co., 214-215, 279.
—— sands, 203, 205, 213, 214. *See* America, U.S.A., Geology of.
Vendres, France, signs of petroleum at, 129.
Venezuela, 250, 277, 502.
Ventilation. *See* Aeration.
Ventilator. *See* Air Propeller.
Ventura, California, 230.

Verbunnt's record results with combined Fauck and diamond drill, 425.
Verchok (0·044 metres), 503.
Verden, Germany, 277, 278.
Vergato, Italy, 277.
Verst (1067 metres), 503.
Vésine-Larue, 273.
Vian, M. (*Société Générale de dynamite*), 452.
Vibert's description of phenomena of spouting wells. 90.
Vice for smithy, cost of, 289, 336.
Vichau, M., Balakhany, spouting oil well of, 94.
Vieux Jardin, Algeria, 264.
Vieux plancher des vaches, 69.
Vif, France, 189.
Vigo, depôt, Spain, 481.
Vile means of communication in China, 161.
Villamina, Venezuela, 250.
Ville de Dieppe and *Ville de Douai*, tank steamers, 470.
Ville franche d'Allier, France, diamond rock boring at, 400-401.
Villeja, Italy, Zipperlen's borings at, 122.
    Illustrated, 121.
Vinegar, strong, inhalation of, for asphyxia, 284.
Virginia, Western, 3, 220, 221, 279, 501.
    Geology of—
        Carboniferous strata of Volcano, 221.
        Devonian strata, 221.
        Disturbed strata, 221.
        Ravines and narrow passes of, 221.
    Great oil belt—
        Producing centres, 220, 221. ·
        Production (annual, 1878-1894), 221.
        Standard Oil Co.'s policy, *in re* development of, 220.
Virlet d'Aoust, origin of petroleum, 15.
Virtues of petroleum. *See* Curative Virtues.
Viscosity of petroleum, 512.
Vitality or life of an oil well. *See* Duration, etc.
Volatile alkali in Jura shale, 10.
Volcanic upheavals necessary for presence of petroleum, 133.
Volcano, Virginia, 221.
Volcanoes, mud, of—
    Arakhan, 153.
    Asiatic Archipelago, 179.
    Baku, 86-87.
    Burmah, 179.
    Columbia, 250.
    Italy, 179.
    Java, 173.
    New Zealand, 179.
    Peru, 251.
    Sicily, 387.
    Trinidad, 173.
Volga River, Russia, petroleum transport routes, 465-476.
Voluta Lamberti, pliocene fossil, 276.
Von Veklé's works at Khidirzind, 101.
Voronège (depôt), Russia, 110, 480.
Vostkowa, Galicia, 48.
Voucher, pipe line, 476.
Vulcanite. *See* Ebonite.

W.

Wagap, New Caledonia, 275, 502
Wages
    America, 242-243.
    Baku, 104, 242-243.
    Galicia, 104.
    See Staff.
Waggons, tank, 462-470.
—— American, 463.
—— French state railway model, described and *illustrated*, 463.
—— grease, early use of petroleum as, 24.

Wagmann of Boryslaw, 26.
Waipava, New Zealand, 276.
Walking beam. *See* Oil-well Drilling.
Wallaf, Sweden, rotary boring results at, 400.
Walnut Bend, Venango, Pennsylvania, 215.
Walsbronn, Alsace, 117.
Walter, Henri, quoted, on geology of Galicia, 34-35.
——— deepens Bobrka oil well, 40.
——— Holowecko, report on, 42.
——— nummulitic beds of Carpathians, 74.
Walls of town of Baku stained with petroleum, 83.
——— of Temple of Eternal Fire, decoration, etc., of, 98.
Wankowa, Galicia, 41, 48, 57, 477, 495.
Wanless spouting well, Canada, 235.
Wardwell, U.S.A., 210.
Waren's Sumatra concession, 176.
Warren Co., 203, 204, 205, 212, 219, 500, 501.
——— Sands, 205, 210.
Warriors, Indian, shielded from bites of reptiles by petroleum, 185.
Warsaw, Russian Poland, 110, 469, 480.
Washington, Pennsylvania, U.S.A., 208, 205, 215, 216, 219, 223, 279, 500.
" Watering " the streets of Baku with petroleum, 83.
Water, table of the specific gravity of (a) liquids heavier than, and (b) lighter than, 493, 494.
——— flooding, freezing of, and drilling through, 438-439.
——— loss of weight of iron rods in, 513-514.
——— loss of weight of wooden rods in, 356.
——— preventing access of, to oil vein, by filling up abandoned wells, 245.
——— submarine spring of fresh, off coast of Florida, 229.
Watersheds of Caucasus, 8.
——— of Rocky Mountains, 239.
Wax. *See* Adipocere; Ozokerit; Zietriskisite.
Wayne Russell, Kentucky, 225.
Waynesbury series (geology) of America, 205.
Wealth of oil kings. *See* Oil Kings.
Webster and Shepley spouting well, Canada, 235.
Weerden, North German line of strike bends north near, 111.
Weetzen, North Germany, 114-116, 496.
Weenzen, Germany, 278.
Weinstein, density of Zorritos petroleum, 254.
Welding of pipe line pipes, 474.
Well cap, 442-443. *See* Seed Bag.
——— (oil), cap, Bredt's patent, 442-443.
Wells, artesian, chronology of, 406.
——— brine. *See* Salt.
——— oil. *See* Oil Wells.
Wellsburg, Ohio, U.S.A., 222, 279.
Wenigsen, 114, 116, 496.
Werner's free-fall instrument, 412.
Weser and Elbe, North German oil field comprised between, lower courses of, 111.
West Hickory, Pennsylvania, U.S.A., 218.
——— Indies, 246-250. *See* Barbadoes; Cuba; Trinidad.
     Bituminous deposits of, 246.
     Deep borings would strike oil, 248.
Western Virginia. *See* Virginia, Western.
Westmoreland, recent rapid development, 220.
Wettin, Saxony, 279.
Wharf, More's, California, 229.
Wheel band, 534.
——— bull, 534.
Whetstone. *See* Grindstone.
Whipple spouting well, 235.
Whisky Run, Butler, U.S.A., 216.
White Oak, West Virginia, U.S.A., 220, 501.
Whiting, depòt, Indiana, 481.
Wielycka, Galicia, salt deposits of, 32.
Wietrzno, Galicia, 24, 32, 38-39, 48, 57, 372, 477, 495.
Wietze, Germany, 277-278.
Wildwood, Butler, Pennsylvania, 216, 495.
Wilkes spouting well, 235.

Wilkes free-fall system, 418.
Williamsport pipe line, 476.
Wincquz, J. D., report on Aïn Zeft, Algeria, 265.
Wind, pressure of, 507. *See* Derrick, Resistance of.
—— of happiness (Chinese superstition), 161.
—— Fall Run, Bradford, U.S.A., 208.
Window, rudimentary, for lighting Japanese excavated wells, 164.
Wine, spirits of, petroleum an ancient substitute for, 71.
Winter prices of Galician petroleum, 55.
Wirbyca, Galicia, 82.
Wire cables, 510.
Wirth, New York State, 207.
Wischin, table of viscosity of petroleum, 512.
Wisconsin, U.S.A., 225.
" Witness " or sample (contents of core barrel), 400.
Wlach's free-fall instrument, 412.
Wladikawkas, Russia, 77, 479.
Woerth, Alsace, 116, 118, 497.
Wojtkowa, Galicia, 37, 57, 495.
Women's rights in Galicia, 51.
Wood. *See* Fuel, Strength of Materials, Timber, etc.
Wooden rod or pole system of drilling, 358, 368.
Woollen stuffs and Indian methods of petroleum collection, 184.
Workmen. *See* Labourers.
Workmanship of oil-well drilling machinery of different nationalities. *See* Oil-well Drilling Machinery Makers.
Workshops, Nobel's, at Tzaritzin, 468. *See* Smithy.
Worms, intestinal petroleum a remedy for, 184.
Worship. *See* Fire Worship.
Wounds, open, petroleum not to be applied to, 71.
Wrenches, tool, 480, 481.
Wyandot, Lima, Ohio, 223.
Wyoming, U.S.A., 202, 220.
—— railway station, 471.

Y.

Yamataga, Japan, 163.
Yankees and history of petroleum, 186.
Yanketi, Madagascar, 274.
Yard (lineal), 0·91438 metres, 504.
——— (square), 0·8360 metres, 504.
Yearly production. *See* Production, Annual.
Yenangyaung oil field, Burmah, 157-160.
Yenangyat, Burmah, 159-160.
Yen Tin, gas wells, Chinese Thibet, 163.
Yield, minimum remunerative, of an American oil well, 244.
Yokoyama, Japan, 163.
Yomadang Mountains, Burmah, 155.
Young & Tzouloukidzé, prospectors, Anaclie, 80.

Z.

Za Gleboka Galicia pumping results, 459.
Zagorz, Galicia, 41.
Zamoskaya, Crimea, 74.
Zanesville, 190.
Zante, 123-124, 497.
Zélienople, 216, 500.
Zendavesta, 67.
Zibo, Hungary, 124.
Zietriskisite of Utah, 226.
Zipperlen's borings at Gabian, 133.
—— on Canadian system in Italy, 373.
—— —— at Salsomaggiore, 120.
—— —— at Villega, 122.
—— —— views of, 121, 123.
—— use of Canadian system by, 373.

Zobel's, free-fall instrument of, 412, 413, 419.
Zole (0·044 grammes), 503.
Zolotnik (4·266 grammes), 503.
Zoroaster, founder of Parsee religion, 67.
Zorritos, Peru, 253, 254, 502.
    Faustino Piaggio Co. of, 253.
    Prentice's wells and monopoly, 252-253.
    Properties of the oil, 254.
    Refinery of, and its output, 254.
Zoubaloff spouting well, 496.
Zulia lagoons, 248.

THE ABERDEEN UNIVERSITY PRESS LIMITED.

AUGUST, 1904.

Catalogue

OF

Special Technical Works

FOR

Manufacturers, Students, and Technical Schools

BY EXPERT WRITERS

INDEX TO SUBJECTS.

	PAGE		PAGE		PAGE
Agricultural Chemistry	10	Dyers' Materials	21	Petroleum	6
Air, Industrial Use of	11	Dye-stuffs	23	Pigments, Chemistry of	2
Alum and its Sulphates	9	Enamelling Metal	18	Plumbers' Work	27
Ammonia	9	Enamels	18	Porcelain Painting	18
Aniline Colours	3	Engraving	31	Pottery Clays	16
Animal Fats	6	Essential Oils	7	Pottery Manufacture	14
Anti-corrosive Paints	4	Evaporating Apparatus	26	Power-loom Weaving	19
Architecture, Terms in	30	External Plumbing	27	Preserved Foods	30
Architectural Pottery	16	Fats	5, 6	Printers' Ready Reckoner	31
Artificial Perfumes	7	Faults in Woollen Goods	20	Printing Inks	3
Balsams	10	Gas Firing	26	Recipes for Oilmen, etc.	3
Bibliography	32	Glass-making Recipes	17	Resins	10
Bleaching	23	Glass Painting	17	Risks of Occupations	12
Bone Products	8	Glue Making and Testing	8	Rivetting China, etc.	16
Bookbinding	31	Greases	5	Sanitary Plumbing	28
Brick-making	15, 16	History of Staffs Potteries	16	Scheele's Essays	9
Burnishing Brass	28	Hops	28	Sealing Waxes	11
Carpet Yarn Printing	21	Hot-water Supply	28	Silk Dyeing	23
Ceramic Books	14, 15	How to make a Woollen Mill		Silk Throwing	19
Charcoal	8	Pay	21	Smoke Prevention	25
Chemical Essays	9	India-rubber	13	Soaps	7, 32
Chemistry of Pottery	17	Inks	3, 11	Spinning	20
Chemistry of Dye-stuffs	23	Iron-corrosion	4	Staining Marble, and Bone	31
Clay Analysis	16	Iron, Science of	26	Steam Drying	11
Coal-dust Firing	26	Japanning	28	Sugar Refining	32
Colour Matching	21	Lacquering	28	Steel Hardening	26
Colliery Recovery Work	25	Lake Pigments	3	Sweetmeats	30
Colour-mixing for Dyers	21	Lead and its Compounds	11	Terra-cotta	16
Colour Theory	22	Leather Industry	13	Testing Paint Materials	4
Combing Machines	24	Leather-working Materials	14	Testing Yarns	20
Compounding Oils	6	Lithography	31	Textile Fabrics	20
Condensing Apparatus	26	Lubricants	5, 6	Textile Materials	19, 20
Cosmetics	7	Manures	8, 10	Timber	29
Cotton Dyeing	22	Mineral Pigments	2	Varnishes	4
Cotton Spinning	24	Mine Ventilation	25	Vegetable Fats	7
Damask Weaving	20	Mine Haulage	25	Waste Utilisation	10
Dampness in Buildings	30	Oil and Colour Recipes	3	Water, Industrial Use	12
Decorators' Books	28	Oil Boiling	4	Waterproofing Fabrics	21
Decorative Textiles	20	Oil Merchants' Manual	7	Weaving Calculations	20
Dental Metallurgy	27	Oils	5	Wood Waste Utilisation	29
Dictionary of Paint Materials	3	Ozone, Industrial Use of	12	Wood Dyeing	31
Drying Oils	5	Paint Manufacture	2	Wool Dyeing	22
Drying with Air	11	Paint Materials	3	Writing Inks	11
Dyeing Marble	31	Paint-material Testing	4	X-Ray Work	13
Dyeing Woollen Fabrics	22	Paper-pulp Dyeing	18	Yarn Testing	20

PUBLISHED BY

SCOTT, GREENWOOD & CO.,

19 Ludgate Hill, London, E.C.

Tel. Address: "PRINTERIES, LONDON".   Tel. No. 5403, Bank.

Paints, Colours and Printing Inks.

THE CHEMISTRY OF PIGMENTS. By Ernest J. Parry, B.Sc. (Lond.), F.I.C., F.C.S., and J. H. Coste, F.I.C., F.C.S. Demy 8vo. Five Illustrations. 285 pp. 1902. Price 10s. 6d.; India and Colonies, 11s.; Other Countries, 12s.; strictly net.

Contents.

Introductory. Light—White Light—The Spectrum—The Invisible Spectrum—Normal Spectrum—Simple Nature of Pure Spectral Colour—The Recomposition of White Light—Primary and Complementary Colours—Coloured Bodies—Absorption Spectra—**The Application of Pigments.** Uses of Pigments: Artistic, Decorative, Protective—Methods of Application of Pigments: Pastels and Crayons, Water Colour, Tempera Painting, Fresco, Encaustic Painting, Oil-colour Painting, Keramic Art, Enamel, Stained and Painted Glass, Mosaic—**Inorganic Pigments.** White Lead—Zinc White—Enamel White—Whitening—Red Lead—Litharge—Vermilion—Royal Scarlet—The Chromium Greens—Chromates of Lead, Zinc, Silver and Mercury—Brunswick Green—The Ochres—Indian Red—Venetian Red—Siennas and Umbers—Light Red—Cappagh Brown—Red Oxides—Mars Colours—Terre Verte—Prussian Brown—Cobalt Colours—Cœruleum—Smalt—Copper Pigments—Malachite—Bremen Green—Scheele's Green—Emerald Green—Verdigris—Brunswick Green—Non-arsenical Greens—Copper Blues—Ultramarine—Carbon Pigments—Ivory Black—Lamp Black—Bistre—Naples Yellow—Arsenic Sulphides: Orpiment, Realgar—Cadmium Yellow—Vandyck Brown—**Organic Pigments.** Prussian Blue—Natural Lakes—Cochineal—Carmine—Crimson—Lac Dye—Scarlet—Madder—Alizarin—Campeachy—Quercitron—Rhamnus—Brazil Wood—Alkanet—Santal Wood—Archil—Coal-tar Lakes—Red Lakes—Alizarin Compounds—Orange and Yellow Lakes—Green and Blue Lakes—Indigo—Dragon's Blood—Gamboge—Sepia—Indian Yellow, Puree—Bitumen. Asphaltum, Mummy—**Index.**

THE MANUFACTURE OF PAINT. A Practical Handbook for Paint Manufacturers, Merchants and Painters. By J. Cruickshank Smith, B.Sc. Demy 8vo. 1901. 200 pp. Sixty Illustrations and One Large Diagram. Price 7s. 6d.; India and Colonies, 8s.; Other Countries, 8s. 6d.; strictly net.

Contents.

Preparation of Raw Material—Storing of Raw Material—Testing and Valuation of Raw Material—Paint Plant and Machinery—The Grinding of White Lead—Grinding of White Zinc—Grinding of other White Pigments—Grinding of Oxide Paints—Grinding of Staining Colours—Grinding of Black Paints—Grinding of Chemical Colours—Yellows—Grinding of Chemical Colours—Blues—Grinding Greens—Grinding Reds—Grinding Lakes—Grinding Colours in Water—Grinding Colours in Turpentine—The Uses of Paint—Testing and Matching Paints—Economic Considerations—Index.

THE MANUFACTURE OF MINERAL AND LAKE PIGMENTS. Containing Directions for the Manufacture of all Artificial, Artists and Painters' Colours, Enamel, Soot and Metallic Pigments. A Text-book for Manufacturers, Merchants, Artists and Painters. By Dr. Josef Bersch. Translated by A. C. Wright, M.A. (Oxon.), B.Sc. (Lond.). Forty-three Illustrations. 476 pp., demy 8vo. 1901. Price 12s. 6d.; India and Colonies 13s. 6d.; Other Countries, 15s.; strictly net.

Contents.

Introduction—Physico-chemical Behaviour of Pigments—Raw Materials Employed in the Manufacture of Pigments—Assistant Materials—Metallic Compounds—The Manufacture of Mineral Pigments—The Manufacture of White Lead—Enamel White—Washing Apparatus—Zinc White—Yellow Mineral Pigments—Chrome Yellow—Lead Oxide Pigments—Other Yellow Pigments—Mosaic Gold—Red Mineral Pigments—The Manufacture of Vermilion—Antimony Vermilion—Ferric Oxide Pigments—Other Red Mineral Pigments—Purple of Cassius—Blue Mineral Pigments—Ultramarine—Manufacture of Ultramarine—Blue Copper Pigments—Blue Cobalt Pigments—Smalts—Green Mineral Pigments—Emerald Green—Verdigris—Chromium Oxide—Other Chromium Pigments—Green Cobalt Pigments—Green Manganese Pigments—Compounded Green Pigments—Violet Mineral Pigments—Brown Mineral Pigments—Brown Decomposition Products—Black Pigments—Manufacture of Soot Pigments—Manufacture of Lamp Black—The Manufacture of Soot Black without Chambers—Indian Ink—Enamel Colours—Metallic Pigments—Bronze Pigments—Vegetable Bronze Pigments.

PIGMENTS OF ORGANIC ORIGIN—Lakes—Yellow Lakes—Red Lakes—Manufacture of Carmine—The Colouring Matter of Lac—Safflower or Carthamine Red—Madder and its Colouring Matters—Madder Lakes—Manjit (Indian Madder)—Lichen Colouring Matters—Red Wood Lakes—The Colouring Matters of Sandal Wood and Other Dye Woods—Blue Lakes—Indigo Carmine—The Colouring Matter of Log Wood—Green Lakes—Brown Organic Pigments—Sap Colours—Water Colours—Crayons—Confectionery Colours—The Preparation of Pigments for Painting—The Examination of Pigments—Examination of Lakes—The Testing of Dye-Woods—The Design of a Colour Works—Commercial Names of Pigments—Appendix : Conversion of Metric to English Weights and Measures—Centigrade and Fahrenheit Thermometer Scales—Index.

DICTIONARY OF CHEMICALS AND RAW PRODUCTS USED IN THE MANUFACTURE OF PAINTS, COLOURS, VARNISHES AND ALLIED PREPARATIONS. By GEORGE H. HURST, F.C.S. Demy 8vo. 380 pp. 1901. Price 7s. 6d.; India and Colonies, 8s. ; Other Countries, 8s. 6d. ; strictly net.

THE MANUFACTURE OF LAKE PIGMENTS FROM ARTIFICIAL COLOURS. By FRANCIS H. JENNISON, F.I.C., F.C.S. Sixteen Coloured Plates, showing Specimens of Eighty-nine Colours, specially prepared from the Recipes given in the Book. 136 pp. Demy 8vo. 1900. Price 7s. 6d.; India and Colonies, 8s.; Other Countries, 8s. 6d.; strictly net.

Contents.

The Groups of the Artificial Colouring Matters—The Nature and Manipulation of Artificial Colours—Lake-forming Bodies for Acid Colours—Lake-forming Bodies' Basic Colours—Lake Bases—The Principles of Lake Formation—Red Lakes—Orange, Yellow, Green, Blue, Violet and Black Lakes—The Production of Insoluble Azo Colours in the Form of Pigments—The General Properties of Lakes Produced from Artificial Colours—Washing, Filtering and Finishing—Matching and Testing Lake Pigments—Index.

RECIPES FOR THE COLOUR, PAINT, VARNISH, OIL, SOAP AND DRYSALTERY TRADES. Compiled by AN ANALYTICAL CHEMIST. 350 pp. 1902. Demy 8vo. Price 7s. 6d.; India and British Colonies, 8s.; Other Countries. 8s. 6d.; strictly net.

Contents.

Pigments or Colours for Paints, Lithographic and Letterpress Printing Inks, etc.—Mixed Paints and Preparations for Paint-making, Painting, Lime-washing, Paperhanging, etc.—Varnishes for Coach-builders. Cabinetmakers, Wood-workers, Metal-workers. Photographers, etc.—Soaps for Toilet, Cleansing, Polishing, etc.—Perfumes—Lubricating Greases, Oils, etc.—Cements, Pastes, Glues and Other Adhesive Preparations—Writing, Marking, Endorsing and Other Inks—Sealing-wax and Office Requisites—Preparations for the Laundry, Kitchen. Stable and General Household Uses—Disinfectant Preparations—Miscellaneous Preparations—Index.

OIL COLOURS AND PRINTING INKS. By LOUIS EDGAR ANDÉS. Translated from the German. 215 pp. Crown 8vo. 56 Illustrations. 1903. Price 5s.; India and British Colonies, 5s. 6d.; Other Countries, 6s.; strictly Net.

Contents.

Linseed Oil—Poppy Oil—Mechanical Purification of Linseed Oil—Chemical Purification of Linseed Oil—Bleaching Linseed Oil—Oxidizing Agents for Boiling Linseed Oil—Theory of Oil Boiling—Manufacture of Boiled Oil—Adulterations of Boiled Oil—Chinese Drying Oil and Other Specialities—Pigments for House and Artistic Painting and Inks—Pigment for Printers' Black Inks—Substitutes for Lampblack—Machinery for Colour Grinding and Rubbing—Machines for mixing Pigments with the Vehicle—Paint Mills—Manufacture of House Oil Paints—Ship Paints—Luminous Paint—Artists' Colours—Printers' Inks :—VEHICLES—Printers' Inks:—PIGMENTS and MANUFACTURE—Index.

(See also Writing Inks, p. 11.)

SIMPLE METHODS FOR TESTING PAINTERS' MATERIALS. By A. C. WRIGHT, M.A. (Oxon.), B.Sc. (Lond.). Crown 8vo. 160 pp. 1903. Price 5s.; India and British Colonies, 5s. 6d.; Other Countries, 6s.; strictly Net.

Contents.

Necessity for Testing—Standards—Arrangement—The Apparatus—The Reagents—Practical Tests—Dry Colours—Stiff Paints—Liquid and Enamel Paints—Oil Varnishes—Spirit Varnishes—Driers—Putty—Linseed Oil—Turpentine—Water Stains—The Chemical Examination—Dry Colours and Paints—White Pigments and Paints—Yellow Pigments and Paints—Blue Pigments and Paints—Green Pigments and Paints—Red Pigments and Paints—Brown Pigments and Paints—Black Pigments and Paints—Oil Varnishes—Linseed Oil—Turpentine.

IRON - CORROSION, ANTI - FOULING AND ANTI-CORROSIVE PAINTS. Translated from the German of LOUIS EDGAR ANDÉS. Sixty-two Illustrations. 275 pp. Demy 8vo. 1900. Price 10s. 6d.; India and Colonies, 11s.; Other Countries, 12s.; strictly net.

Contents.

Iron-rust and its Formation—Protection from Rusting by Paint—Grounding the Iron with Linseed Oil, etc.—Testing Paints—Use of Tar for Painting on Iron—Anti-corrosive Paints—Linseed Varnish—Chinese Wood Oil—Lead Pigments—Iron Pigments—Artificial Iron Oxides—Carbon—Preparation of Anti-corrosive Paints—Results of Examination of Several Anti-corrosive Paints—Paints for Ship's Bottoms—Anti-fouling Compositions—Various Anti-corrosive and Ship's Paints—Official Standard Specifications for Ironwork Paints—Index.

THE TESTING AND VALUATION OF RAW MATE-RIALS USED IN PAINT AND COLOUR MANU-FACTURE. By M. W. JONES, F.C.S. A Book for the Laboratories of Colour Works. 88 pp. Crown 8vo. 1900. Price 5s.; India and Colonies, 5s. 6d.; Other Countries, 6s.; strictly net.

Contents.

Aluminium Compounds—China Clay—Iron Compounds—Potassium Compounds—Sodium Compounds—Ammonium Hydrate—Acids—Chromium Compounds—Tin Compounds—Copper Compounds—Lead Compounds—Zinc Compounds—Manganese Compounds—Arsenic Compounds—Antimony Compounds—Calcium Compounds—Barium Compounds—Cadmium Compounds—Mercury Compounds—Ultramarine—Cobalt and Carbon Compounds—Oils—Index.

STUDENTS' MANUAL OF PAINTS, COLOURS, OILS AND VARNISHES. By JOHN FURNELL. Crown 8vo. 12 Illustrations. 96 pp. 1903. Price 2s. 6d.; Abroad, 3s.; strictly net.

Contents.

Plant—Chromes—Blues—Greens—Earth Colours—Blacks—Reds—Lakes—Whites—Painters' Oils—Turpentine—Oil Varnishes—Spirit Varnishes—Liquid Paints—Enamel Paints.

Varnishes and Drying Oils.

THE MANUFACTURE OF VARNISHES, OIL RE-FINING AND BOILING, AND KINDRED INDUS-TRIES. Translated from the French of ACH. LIVACHE, Ingénieur Civil des Mines. Greatly Extended and Adapted to English Practice, with numerous Original Recipes by JOHN GEDDES McINTOSH. 27 Illustrations. 400 pp. Demy 8vo. 1899. Price 12s. 6d.; India and Colonies, 13s. 6d.; Other Countries, 15s.; strictly net.

Contents.

Resins—Solvents: Natural, Artificial, Manufacture, Storage, Special Use—Colouring: Principles, Vegetable, Coal Tar, Coloured Resinates, Coloured Oleates and Linoleates—Gum Running: Melting Pots, Mixing Pans—Spirit Varnish Manufacture: Cold Solution Plant, Mechanical Agitators, Storage Plant—Manufacture, Characteristics and Uses of the Spirit Varnishes—Manufacture of Varnish Stains—Manufacture of Lacquers—Manufacture of Spirit Enamels—Analysis of Spirit Varnishes—Physical and Chemical Constants of Resins—Table of Solubility of Resins in different Menstrua—Systematic qualitative Analysis of Resins, Hirschop's tables—Drying Oils—Oil Refining: Processes—Oil Boiling—Driers—Liquid Driers — Solidified Boiled Oil — Manufacture of Linoleum — Manufacture of India Rubber Substitutes—Printing Ink Manufacture—Lithographic Ink Manufacture—Manufacture of Oil Varnishes—Running and Special Treatment of Amber, Copal, Kauri, Manilla—Addition of Oil to Resin—Addition of Resin to Oil—Mixed Processes—Solution in Cold of previously Fused Resin—Dissolving Resins in Oil, etc., under pressure—Filtration—Clarification—Storage—Ageing—Coachmakers' Varnishes and Japans—Oak Varnishes—Japanners' Stoving Varnishes—Japanners' Gold Size—Brunswick Black—Various Oil Varnishes—Oil-Varnish Stains—Varnishes for "Enamels"—India Rubber Varnishes—Varnishes Analysis: Processes, Matching—Faults in Varnishes: Cause, Prevention—Experiments and Exercises.

DRYING OILS, BOILED OIL AND SOLID AND LIQUID DRIERS. By L. E. ANDÉS. Expressly Written for this Series of Special Technical Books, and the Publishers hold the Copyright for English and Foreign Editions. Forty-two Illustrations. 342 pp. 1901. Demy 8vo. Price 12s. 6d.; India and Colonies, 13s. 6d.; Other Countries, 15s.; strictly net.

Contents.

Properties of the Drying Oils; Cause of the Drying Property; Absorption of Oxygen; Behaviour towards Metallic Oxides, etc.—The Properties of and Methods for obtaining the Drying Oils—Production of the Drying Oils by Expression and Extraction; Refining and Bleaching; Oil Cakes and Meal; The Refining and Bleaching of the Drying Oils; The Bleaching of Linseed Oil—The Manufacture of Boiled Oil; The Preparation of Drying Oils for Use in the Grinding of Paints and Artists' Colours and in the Manufacture of Boiled Oil and the Apparatus therefor; Livache's Process for Preparing a Good Drying Oil and its Practical Application—The Preparation of Varnishes for Letterpress, Lithographic and Copperplate Printing, for Oilcloth and Waterproof Fabrics; The Manufacture of Thickened Linseed Oil, Burnt Oil, Stand Oil by Fire Heat, Superheated Steam, and by a Current of Air—Behaviour of the Drying Oils and Boiled Oils towards Atmospheric Influences, Water, Acids and Alkalies—Boiled Oil Substitutes—The Manufacture of Solid and Liquid Driers from Linseed Oil and Rosin: Linolic Acid Compounds of the Driers—The Adulteration and Examination of the Drying Oils and Boiled Oil.

Oils, Fats, Soaps and Perfumes.

LUBRICATING OILS, FATS AND GREASES: Their Origin, Preparation, Properties, Uses and Analyses. A Handbook for Oil Manufacturers, Refiners and Merchants, and the Oil and Fat Industry in General. By GEORGE H. HURST, F.C.S. Second Revised and Enlarged Edition. Sixty-five Illustrations. 317 pp. Demy 8vo. 1902. Price 10s. 6d.; India and Colonies, 11s.; Other Countries. 12s strictly net.

Contents.

Introductory—Hydrocarbon Oils—Scotch Shale Oils—Petroleum—Vegetable and Animal Oils—Testing and Adulteration of Oils—Lubricating Greases—Lubrication—Appendices—Index.

TECHNOLOGY OF PETROLEUM : Oil Fields of the
World—Their History, Geography and Geology—Annual Production
and Development—Oil-well Drilling—Transport. By HENRY NEU-
BERGER and HENRY NOALHAT. Translated from the French by J. G.
MCINTOSH. 550 pp. 153 Illustrations. 26 Plates. Super Royal 8vo. 1901.
Price 21s. ; India and Colonies, 22s. ; Other Countries, 23s. 6d. ;
strictly net.

Contents.

Study of the Petroliferous Strata—Petroleum—Definition—The Genesis or Origin of
Petroleum—The Oil Fields of Galicia, their History—Physical Geography and Geology of
the Galician Oil Fields—Practical Notes on Galician Land Law—Economic Hints on Working,
etc.—Roumania—History, Geography, Geology—Petroleum in Russia—History—Russian
Petroleum (continued)—Geography and Geology of the Caucasian Oil Fields—Russian Petro-
leum (continued)—The Secondary Oil Fields of Europe, Northern Germany, Alsace, Italy, etc.—
Petroleum in France—Petroleum in Asia—Transcaspian and Turkestan Territory—Turkestan
—Persia—British India and Burmah—British Burmah or Lower Burmah—China—Chinese
Thibet—Japan, Formosa and Saghalien—Petroleum in Oceania—Sumatra, Java, Borneo—
Isle of Timor—Philippine Isles—New Zealand—The United States of America—History—
Physical Geology and Geography of the United States Oil Fields—Canadian and other North
American Oil Fields—Economic Data of Work in North America—Petroleum in the West
Indies and South America—Petroleum in the French Colonies.
   Excavations—Hand Excavation or Hand Digging of Oil Wells.
   Methods of Boring.
   Accidents—Boring Accidents—Methods of preventing them—Methods of remedying them
—Explosives and the use of the "Torpedo" Levigation—Storing and Transport of Petroleum
—General Advice—Prospecting, Management and carrying on of Petroleum Boring Operations.
   General Data—Customary Formulæ—Memento. Practical Part. General Data
bearing on Petroleum—Glossary of Technical Terms used in the Petroleum Industry—Copious
Index.

**THE PRACTICAL COMPOUNDING OF OILS, TAL-
LOW AND GREASE FOR LUBRICATION, ETC.**
By AN EXPERT OIL REFINER. 100 pp. 1898. Demy 8vo. Price 7s. 6d. ;
India and Colonies, 8s. ; Other Countries, 8s. 6d. ; strictly net.

Contents.

Introductory Remarks on the General Nomenclature of Oils, Tallow and Greases
suitable for Lubrication—Hydrocarbon Oils—Animal and Fish Oils—Compound
Oils—Vegetable Oils—Lamp Oils—Engine Tallow. Solidified Oils and Petroleum
Jelly—Machinery Greases: Loco and Anti-friction—Clarifying and Utilisation
of Waste Fats, Oils, Tank Bottoms, Drainings of Barrels and Drums, Pickings
Up, Dregs, etc.—The Fixing and Cleaning of Oil Tanks, etc.—Appendix and
General Information.

ANIMAL FATS AND OILS : Their Practical Production,
Purification and Uses for a great Variety of Purposes. Their Properties, Falsification and Examination. Translated from the German
of LOUIS EDGAR ANDÉS. Sixty-two Illustrations. 240 pp. 1898.
Demy 8vo. Price 10s. 6d. ; India and Colonies, 11s. ; Other Countries,
12s. ; strictly net.

Contents.

Introduction—Occurrence, Origin, Properties and Chemical Constitution of Animal Fats—
Preparation of Animal Fats and Oils—Machinery—Tallow-melting Plant—Extraction Plant
—Presses—Filtering Apparatus—Butter: Raw Material and Preparation, Properties, Adul-
terations, Beef Lard or Remelted Butter, Testing—Candle-fish Oil—Mutton-Tallow—Hare
Fat—Goose Fat—Neatsfoot Oil—Bone Fat: Bone Boiling, Steaming Bones, Extraction,
Refining—Bone Oil—Artificial Butter: Oleomargarine, Margarine Manufacture in France,
Grasso's Process, "Kaiser's Butter," Jahr & Münzberg's Method, Filbert's Process, Winter's
Method—Human Fat—Horse Fat—Beef Marrow—Turtle Oil—Hog's Lard: Raw Material—
Preparation, Properties, Adulterations, Examination—Lard Oil—Fish Oils—Liver Oils—
Artificial Train Oil—Wool Fat: Properties, Purified Wool Fat—Spermaceti: Examination
of Fats and Oils in General.

7

THE OIL MERCHANTS' MANUAL AND OIL TRADE READY RECKONER. Compiled by FRANK F. SHERIFF. Second Edition Revised and Enlarged. Demy 8vo. 214 pp. 1904. With Two Sheets of Tables. Price 7s. 6d.; India and Colonies, 8s.; Other Countries, 8s. 6d.; strictly net.

Contents.

Trade Terms and Customs—Tables to Ascertain Value of Oil sold per cwt. or ton—Specific Gravity Tables—Percentage Tare Tables—Petroleum Tables—Paraffine and Benzoline Calculations—Customary Drafts—Tables for Calculating Allowance for Dirt, Water, etc.—Capacity of Circular Tanks Tables, etc., etc.

VEGETABLE FATS AND OILS: Their Practical Preparation, Purification and Employment for Various Purposes, their Properties, Adulteration and Examination. Translated from the German of LOUIS EDGAR ANDÉS. Ninety-four Illustrations. 340 pp. Second Edition. 1902. Demy 8vo. Price 10s. 6d.; India and Colonies, 11s.; Other Countries, 12s.; strictly net.

Contents.

General Properties—Estimation of the Amount of Oil in Seeds—The Preparation of Vegetable Fats and Oils—Apparatus for Grinding Oil Seeds and Fruits—Installation of Oil and Fat Works—Extraction Method of Obtaining Oils and Fats—Oil Extraction Installations—Press Moulds—Non-drying Vegetable Oils—Vegetable drying Oils—Solid Vegetable Fats—Fruits Yielding Oils and Fats—Wool-softening Oils—Soluble Oils—Treatment of the Oil after Leaving the Press—Improved Methods of Refining—Bleaching Fats and Oils—Practical Experiments on the Treatment of Oils with regard to Refining and Bleaching—Testing Oils and Fats.

SOAPS. A Practical Manual of the Manufacture of Domestic, Toilet and other Soaps. By GEORGE H. HURST, F.C.S. 390 pp. 66 Illustrations. 1898. Price 12s. 6d.; India and Colonies, 13s. 6d.; Other Countries, 15s.; strictly net.

Contents.

Introductory—Soap-maker's Alkalies—Soap Fats and Oils—Perfumes—Water as a Soap Material—Soap Machinery—Technology of Soap-making—Glycerine in Soap Lyes—Laying out a Soap Factory—Soap Analysis—Appendices.

THE CHEMISTRY OF ESSENTIAL OILS AND ARTIFICIAL PERFUMES. By ERNEST J. PARRY, B.Sc. (Lond.), F.I.C., F.C.S. 411 pp. 20 Illustrations. 1899. Demy 8vo. Price 12s. 6d.; India and Colonies, 13s. 6d.; Other Countries. 15s.; strictly net.

Contents.

The General Properties of Essential Oils—Compounds occurring in Essential Oils—The Preparation of Essential Oils—The Analysis of Essential Oils—Systematic Study of the Essential Oils—Terpeneless Oils—The Chemistry of Artificial Perfumes—Appendix: Table of Constants—Index.

(For "Textile Soaps" see p. 32.)

Cosmetical Preparations.

COSMETICS: MANUFACTURE, EMPLOYMENT AND TESTING OF ALL COSMETIC MATERIALS AND COSMETIC SPECIALITIES. Translated from the German of Dr. THEODOR KOLLER. Crown 8vo. 262 pp. 1902. Price 5s.; India and Colonies, 5s. 6d.; Other Countries, 6s. net.

Contents.

Purposes and Uses of, and Ingredients used in the Preparation of Cosmetics—Preparation of Perfumes by Pressure, Distillation, Maceration, Absorption or Enfleurage, and Extraction Methods—Chemical and Animal Products used in the Preparation of Cosmetics—Oils and Fats used in the Preparation of Cosmetics—General Cosmetic Preparations—Mouth Washes and Tooth Pastes—Hair Dyes, Hair Restorers and Depilatories—Cosmetic Adjuncts and Specialities—Colouring Cosmetic Preparations—Antiseptic Washes and Soaps—Toilet and Hygienic Soaps—Secret Preparations for Skin, Complexion, Teeth, Mouth, etc.—Testing and Examining the Materials Employed in the Manufacture of Cosmetics—Index.

Glue, Bone Products and Manures.

GLUE AND GLUE TESTING. By Samuel Rideal, D.Sc. (Lond.), F.I.C. Fourteen Engravings. 144 pp. Demy 8vo. 1900. Price 10s. 6d.; India and Colonies, 11s.; Other Countries, 12s.; strictly net.

Contents.

Constitution and Properties: Definitions and Sources, Gelatine, Chondrin and Allied Bodies, Physical and Chemical Properties, Classification, Grades and Commercial Varieties—**Raw Materials and Manufacture:** Glue Stock, Lining, Extraction, Washing and Clarifying, Filter Presses, Water Supply, Use of Alkalies, Action of Bacteria and of Antiseptics, Various Processes, Cleansing, Forming, Drying, Crushing, etc., Secondary Products—**Uses of Glue:** Selection and Preparation for Use, Carpentry, Veneering, Paper-Making, Bookbinding, Printing Rollers, Hectographs, Match Manufacture, Sandpaper, etc., Substitutes for other Materials, Artificial Leather and Caoutchouc—**Gelatine:** General Characters, Liquid Gelatine, Photographic Uses, Size, Tanno-, Chrome and Formo-Gelatine, Artificial Silk, Cements, Pneumatic Tyres, Culinary, Meat Extracts, Isinglass, Medicinal and other Uses, Bacteriology—**Glue Testing:** Review of Processes, Chemical Examination, Adulteration, Physical Tests, Valuation of Raw Materials—**Commercial Aspects.**

BONE PRODUCTS AND MANURES: An Account of the most recent Improvements in the Manufacture of Fat, Glue, Animal Charcoal, Size, Gelatine and Manures. By Thomas Lambert, Technical and Consulting Chemist. Illustrated by Twenty-one Plans and Diagrams. 162 pp. Demy 8vo. 1901. Price 7s. 6d.; India and Colonies, 8s.; Other Countries, 8s. 6d.; strictly net.

Contents.

Chemical Composition of Bones—Arrangement of Factory—Crushing of Bones—Treatment with Benzene—Benzene in Crude Fat—Analyses of Clarified Fats—Mechanical Cleansing of Bones—Animal Charcoal—Tar and Ammoniacal Liquor, Char and Gases, from good quality Bones—Method of Retorting the Bones—Analyses of Chars—"Spent" Chars—Cooling of Tar and Ammoniacal Vapours—Value of Nitrogen for Cyanide of Potash—Bone Oil—Marrow Bones—Composition of Marrow Fat—Premier Juice—Buttons—Properties of Glue—Glutin and Chondrin—Skin Glue—Liming of Skins—Washing—Boiling of Skins—Clarification of Glue Liquors—Acid Steeping of Bones—Water System of Boiling Bones—Steam Method of Treating Bones—Nitrogen in the Treated Bones—Glue-Boiling and Clarifying-House—Plan showing Arrangement of Clarifying Vats—Plan showing Position of Evaporators—Description of Evaporators—Sulphurous Acid Generator—Clarification of Liquors—Section of Drying-House—Specification of a Glue—Size—Uses and Preparation and Composition of Size—Concentrated Size—Properties of Gelatine—Preparation of Skin Gelatine—Washing—Bleaching—Boiling—Clarification—Evaporation—Drying—Bone Gelatine—Selecting Bones—Crushing—Dissolving—Bleaching—Boiling—Properties of Glutin and Chondrin—Testing of Glues and Gelatines—The Uses of Glue, Gelatine and Size in Various Trades—Soluble and Liquid Glues—Steam and Waterproof Glues—**Manures**—Importation of Food Stuffs—Soils—Germination—Plant Life—**Natural Manures**—Water and Nitrogen in Farmyard Manure—Full Analysis of Farmyard Manure—Action on Crops—Water-Closet System—Sewage Manure—Green Manures—**Artificial Manures**—**Mineral Manures**—Nitrogenous Matters—Shoddy—Hoofs and Horns—Leather Waste—Dried Meat—Dried Blood—Superphosphates—Composition—Manufacture—Section of Manure-Shed—First and Ground Floor Plans of Manure-Shed—Quality of Acid Used—Mixings—Special Manures—Potato Manure—Dissolved Bones—Dissolved Bone Compound—Enriched Peruvian Guano—Special Manure for Garden Stuffs, etc.—Special Manures—Analyses of Raw and Finished Products—Common Raw Bones—Degreased Bones—Crude Fat—Refined Fat—Degelatinised Bones—Animal Charcoal—Bone Superphosphates—Guanos—Dried Animal Products—Potash Compounds—Sulphate of Ammonia—Extraction in Vacuo—French and British Gelatines compared—Index.

Chemicals, Waste Products and Agricultural Chemistry.

REISSUE OF **CHEMICAL ESSAYS OF C. W. SCHEELE.** First Published in English in 1786. Translated from the Academy of Sciences at Stockholm, with Additions. 300 pp. Demy 8vo. 1901. Price 5s.; India and Colonies, 5s. 6d.; Other Countries, 6s.; strictly net.

Contents.

Memoir: C. W. Scheele and his work (written for this edition by J. G. McIntosh)—On Fluor Mineral and its Acid—On Fluor Mineral—Chemical Investigation of Fluor Acid, with a View to the Earth which it Yields, by Mr. Wiegler—Additional Information Concerning Fluor Minerals—On Manganese, Magnesium, or Magnesia Vitrariorum—On Arsenic and its Acid—Remarks upon Salts of Benzoin—On Silex, Clay and Alum—Analysis of the Calculus Vesical—Method of Preparing Mercurius Dulcis Via Humida—Cheaper and more Convenient Method of Preparing Pulvis Algarothi—Experiments upon Molybdæna —Experiments on Plumbago—Method of Preparing a New Green Colour—Of the Decomposition of Neutral Salts by Unslaked Lime and Iron—On the Quantity of Pure Air which is Daily Present in our Atmosphere—On Milk and its Acid—On the Acid of Saccharum Lactis —On the Constituent Parts of Lapis Ponderosus or Tungsten—Experiments and Observations on Ether—Index.

THE MANUFACTURE OF ALUM AND THE SULPHATES AND OTHER SALTS OF ALUMINA AND IRON. Their Uses and Applications as Mordants in Dyeing and Calico Printing, and their other Applications in the Arts, Manufactures, Sanitary Engineering, Agriculture and Horticulture. Translated from the French of LUCIEN GESCHWIND. 195 Illustrations. 400 pp. Royal 8vo. 1901. Price 12s. 6d.; India and Colonies, 13s. 6d.; Other Countries, 15s.; strictly net.

Contents.

Theoretical Study of Aluminium, Iron, and Compounds of these Metals— Aluminium and its Compounds—Iron and Iron Compounds.

Manufacture of Aluminium Sulphates and Sulphates of Iron—Manufacture of Aluminium Sulphate and the Alums—Manufacture of Sulphates of Iron.

Uses of the Sulphates of Aluminium and Iron—Uses of Aluminium Sulphate and Alums—Application to Wool and Silk—Preparing and using Aluminium Acetates—Employment of Aluminium Sulphate in Carbonising Wool—The Manufacture of Lake Pigments—Manufacture of Prussian Blue—Hide and Leather Industry—Paper Making—Hardening Plaster—Lime Washes—Preparation of Non-inflammable Wood, etc.—Purification of Waste Waters —**Uses and Applications of Ferrous Sulphate and Ferric Sulphates**—Dyeing—Manufacture of Pigments—Writing Inks—Purification of Lighting Gas —Agriculture—Cotton Dyeing —Disinfectant—Purifying Waste Liquors—Manufacture of Nordhausen Sulphuric Acid—Fertilising.

Chemical Characteristics of Iron and Aluminium—Analysis of Various Aluminous or Ferruginous Products—Aluminium —**Analysing Aluminium Products**—Alunite Alumina—Sodium Aluminate—Aluminium Sulphate—Iron—Analytical Characteristics of Iron Salts—Analysis of Pyritic Lignite—Ferrous and Ferric Sulphates—Rouil Mordant—Index.

AMMONIA AND ITS COMPOUNDS: Their Manufacture and Uses. By CAMILLE VINCENT, Professor at the Central School of Arts and Manufactures, Paris. Translated from the French by M. J. SALTER. Royal 8vo. 114 pp. 1901. Thirty-two Illustrations. Price 5s.; India and Colonies, 5s. 6d.; Other Countries, 6s.; strictly net.

Contents.

General Considerations: Various Sources of Ammoniacal Products; Human Urine as a Source of Ammonia—**Extraction of Ammoniacal Products from Sewage**— Extraction of Ammonia from Gas Liquor—**Manufacture of Ammoniacal Compounds from Bones, Nitrogenous Waste, Beetroot Wash and Peat**—Manufacture of Caustic Ammonia, and Ammonium Chloride, Phosphate and Carbonate—Recovery of Ammonia from the Ammonia-Soda Mother Liquors—Index.

ANALYSIS OF RESINS AND BALSAMS. Translated from the German of Dr. KARL DIETERICH. Demy 8vo. 340 pp. 1901. Price 7s. 6d.; India and Colonies, 8s.; Other Countries, 8s. 6d.; strictly net.

Contents.

Definition of Resins in General—Definition of Balsams, and especially the Gum Resins—External and Superficial Characteristics of Resinous Bodies—Distinction between Resinous Bodies and Fats and Oils—Origin, Occurrence and Collection of Resinous Substances—Classification—Chemical Constituents of Resinous Substances—Resinols—Resinot Annols—Behaviour of Resin Constituents towards the Cholesterine Reactions—Uses and Identification of Resins—Melting-point—Solvents—Acid Value—Saponification Value—Resin Value—Ester and Ether Values—Acetyl and Corbonyl Value—Methyl Value—Resin Acid—Systematic Résumé of the Performance of the Acid and Saponification Value Tests.
Balsams— Introduction — Definitions—Canada Balsam—Copaiba Balsam — Angostura Copaiba Balsam — Babia Copaiba Balsam — Carthagena Copaiba Balsam — Maracaibo Copaiba Balsam—Maturin Copaiba Balsam—Gurjum Copaiba Balsam—Para Copaiba Balsam —Surinam Copaiba Balsam—West African Copaiba Balsam—Mecca Balsam—Peruvian Balsam—Tolu Balsam—Acaroid Resin—Amine—Amber—African and West Indian Kino—Bengal Kino—Labdanum—Mastic—Pine Resin—Sandarach—Scammonium—Shellac—Storax —Adulteration of Styrax Liquidus Crudus—Purified Storax—Styrax Crudus Colatus—Tacamahac—Thapsia Resin — Turpentine—Chios Turpentine — Strassburg Turpentine—Turpeth Turpentine. **Gum Resins**—Ammoniacum—Bdellium—Euphorbium — Galbanum—Gamboge —Lactucarium—Myrrh—Opopanax—Sagapenum—Olibanum or Incense—Acaroid Resin—Amber—Thapsia Resin—Index.

MANUAL OF AGRICULTURAL CHEMISTRY. By HERBERT INGLE, F.I.C., Lecturer on Agricultural Chemistry, the Yorkshire College; Lecturer in the Victoria University. 388 pp. 11 Illustrations. 1902. Demy 8vo. Price 7s. 6d.; India and Colonies, 8s.; Other Countries, 8s. 6d. net.

Contents.

Introduction—The Atmosphere—The Soil—The Reactions occurring in Soils—The Analysis of Soils—Manures, Natural—Manures (continued)—The Analysis of Manures—The Constituents of Plants—The Plant—Crops —The Animal—Foods and Feeding—Milk and Milk Products—The Analysis of Milk and Milk Products—Miscellaneous Products used in Agriculture—Appendix—Index.

THE UTILISATION OF WASTE PRODUCTS. A Treatise on the Rational Utilisation, Recovery and Treatment of Waste Products of all kinds. By Dr. THEODOR KOLLER. Translated from the Second Revised German Edition. Twenty-two Illustrations. Demy 8vo. 280 pp. 1902. Price 7s. 6d.; India and Colonies, 8s.; Other Countries, 8s. 6d.; strictly net.

Contents.

The Waste of Towns—**Ammonia and Sal-Ammoniac**— Rational Processes for Obtaining these Substances by Treating Residues and Waste—Residues in the Manufacture of Aniline Dyes—Amber Waste—Brewers' Waste— Blood and Slaughter-House Refuse—Manufactured Fuels—Waste Paper and Bookbinders' Waste—Iron Slags—Excrement—Colouring Matters from Waste—Dyers' Waste Waters—Fat from Waste—Fish Waste—Calamine Sludge—Tannery Waste—Gold and Silver Waste—India-rubber and Caoutchouc Waste—Residues in the Manufacture of Rosin Oil—Wood Waste—Horn Waste—Infusorial Earth—Iridium from Goldsmiths' Sweepings—Jute Waste—Cork Waste—Leather Waste—Glue Makers' Waste —Illuminating Gas from Waste and the By-Products of the Manufacture of Coal Gas—Meerschum—Molasses—Metal Waste—By-Products in the Manufacture of Mineral Waters —Fruit—The By-Products of Paper and Paper Pulp Works—By-Products in the Treatment of Coal Tar Oils—Fur Waste—The Waste Matter in the Manufacture of Parchment Paper —Mother of Pearl Waste—Petroleum Residues—Platinum Residues—Broken Porcelain. Earthenware and Glass—Salt Waste—Slate Waste—Sulphur—Burnt Pyrites—Silk Waste—Soap Makers' Waste—Alkali Waste and the Recovery of Soda—Waste Produced in Grinding Mirrors—Waste Products in the Manufacture of Starch—Stearic Acid—Vegetable Ivory Waste—Turf—Waste Waters of Cloth Factories—Wine Residues—Tinplate Waste—Wool Waste—Wool Sweat—The Waste Liquids from Sugar Works—Index.

Writing Inks and Sealing Waxes.

INK MANUFACTURE : Including Writing, Copying, Litho-
graphic, Marking, Stamping, and Laundry Inks. By SIGMUND LEHNER.
Three Illustrations. Crown 8vo. 162 pp. 1902. Translated from the
German of the Fifth Edition. Price 5s.; India and Colonies, 5s. 6d.;
Other Countries, 6s.; net.

Contents.
Varieties of Ink—Writing Inks—Raw Materials of Tannin Inks—The Chemical Constitution
of the Tannin Inks—Recipes for Tannin Inks—Logwood Tannin Inks—Ferric Inks—Alizarine
Inks—Extract Inks—Logwood Inks—Copying Inks—Hektographs—Hektograph Inks—Safety
Inks—Ink Extracts and Powders—Preserving Inks—Changes in Ink and the Restoration of
Faded Writing—Coloured Inks—Red Inks—Blue Inks—Violet Inks—Yellow Inks—Green
Inks—Metallic Inks—Indian Ink—Lithographic Inks and Pencils—Ink Pencils—Marking Inks
—Ink Specialities—Sympathetic Inks—Stamping Inks—Laundry or Washing Blue—Index

**SEALING-WAXES, WAFERS AND OTHER ADHES-
IVES FOR THE HOUSEHOLD, OFFICE, WORK-
SHOP AND FACTORY.** By H. C. STANDAGE. Crown
8vo. 96 pp. 1902. Price 5s.; India and Colonies, 5s. 6d.; Other
Countries, 6s.; strictly net.

Contents.
Materials Used for Making Sealing-Waxes—The Manufacture of Sealing-Waxes—
Wafers—Notes on the Nature of the Materials Used in Making Adhesive Compounds—Cements
for Use in the Household—Office Gums, Pastes and Mucilages—Adhesive Compounds for
Factory and Workshop Use.

Lead Ores and Compounds.

LEAD AND ITS COMPOUNDS. By THOS. LAMBERT,
Technical and Consulting Chemist. Demy 8vo. 226 pp. Forty Illustrations. 1902. Price 7s. 6d.; India and Colonies, 8s.; Other Countries,
8s. 6d.; net. Plans and Diagrams.

Contents.
History—Ores of Lead—Geographical Distribution of the Lead Industry—Chemical and
Physical Properties of Lead—Alloys of Lead—Compounds of Lead—Dressing of Lead Ores
—Smelting of Lead Ores—Smelting in the Scotch or American Ore-hearth—Smelting in the
Shaft or Blast Furnace—Condensation of Lead Fume—Desilverisation, or the Separation
of Silver from Argentiferous Lead—Cupellation—The Manufacture of Lead Pipes and
Sheets—Protoxide of Lead—Litharge and Massicot—Red Lead or Minium—Lead Poisoning
—Lead Substitutes—Zinc and its Compounds—Pumice Stone—Drying Oils and Siccatives
—Oil of Turpentine Resin—Classification of Mineral Pigments—Analysis of Raw and Finished
Products—Tables—Index.

NOTES ON LEAD ORES : Their Distribution and Properties.
By JAS. FAIRIE, F.G.S. Crown 8vo. 1901. 64 pages. Price 2s. 6d.;
Abroad, 3s.; strictly net.

Industrial Uses of Air, Steam and Water.

DRYING BY MEANS OF AIR AND STEAM. Explana-
tions, Formulæ, and Tables for Use in Practice. Translated from the
German of E. HAUSBRAND. Two folding Diagrams and Thirteen Tables.
Crown 8vo. 1901. 72 pp. Price 5s.; India and Colonies, 5s. 6d.;
Other Countries, 6s.; strictly net.

(For Contents see next page.)

Contents.

British and Metric Systems Compared—Centigrade and Fahr. Thermometers—Estimation of the Maximum Weight of Saturated Aqueous Vapour which can be contained in 1 kilo. of Air at Different Pressure and Temperatures—Calculation of the Necessary Weight and Volume of Air, and of the Least Expenditure of Heat, per Drying Apparatus with Heated Air, at the Atmospheric Pressure: *A*, With the Assumption that the Air is *Completely Saturated* with Vapour both before Entry and after Exit from the Apparatus—*B*, When the Atmospheric Air is Completely Saturated *before entry*, but at its *exit* is only ⅔, ½ or ¼ Saturated —*C*, When the Atmospheric Air is *not* Saturated with Moisture before Entering the Drying Apparatus—Drying Apparatus, in which, in the Drying Chamber, a Pressure is Artificially Created, Higher or Lower than that of the Atmosphere—Drying by Means of Superheated Steam, without Air—Heating Surface, Velocity of the Air Current, Dimensions of the Drying Room, Surface of the Drying Material. Losses of Heat—Index.

(*See also* "*Evaporating, Condensing and Cooling Apparatus,*" *p.* 26.)

PURE AIR, OZONE AND WATER. A Practical Treatise
of their Utilisation and Value in Oil, Grease, Soap, Paint, Glue and other Industries. By W. B. COWELL. Twelve Illustrations. Crown 8vo. 85 pp. 1900. Price 5s. ; India and Colonies, 5s. 6d. ; Other Countries, 6s. ; strictly net.

Contents.

Atmospheric Air; Lifting of Liquids; Suction Process; Preparing Blown Oils; Preparing Siccative Drying Oils—Compressed Air; Whitewash—Liquid Air; Retrocession—Purification of Water; Water Hardness—Fleshings and Bones—Ozonised Air in the Bleaching and Deodorising of Fats, Glues, etc.; Bleaching Textile Fibres—Appendix: Air and Gases; Pressure of Air at Various Temperatures; Fuel; Table of Combustibles; Saving of Fuel by Heating Feed Water; Table of Solubilities of Scale Making Minerals; British Thermal Units Tables; Volume of the Flow of Steam into the Atmosphere; Temperature of Steam—Index.

THE INDUSTRIAL USES OF WATER. COMPOSI-TION — EFFECTS—TROUBLES — REMEDIES—RE-SIDUARY WATERS—PURIFICATION—ANALYSIS.
By H. DE LA COUX. Royal 8vo. Translated from the French and Revised by ARTHUR MORRIS. 364 pp. 135 Illustrations. 1903. Price 10s. 6d. ; Colonies, 11s. ; Other Countries, 12s. ; strictly net.

Contents.

Chemical Action of Water in Nature and in Industrial Use—Composition of Waters—Solubility of Certain Salts in Water Considered from the Industrial Point of View—Effects on the Boiling of Water—Effects of Water in the Industries—Difficulties with Water—Feed Water for Boilers—Water in Dyeworks, Print Works, and Bleach Works—Water in the Textile Industries and in Conditioning—Water in Soap Works—Water in Laundries and Washhouses—Water in Tanning—Water in Preparing Tannin and Dyewood Extracts—Water in Papermaking—Water in Photography—Water in Sugar Refining—Water in Making Ices and Beverages—Water in Cider Making—Water in Brewing—Water in Distilling—Preliminary Treatment and Apparatus—Substances Used for Preliminary Chemical Purification—Commercial Specialities and their Employment—Precipitation of Matters in Suspension in Water —Apparatus for the Preliminary Chemical Purification of Water—Industrial Filters—Industrial Sterilisation of Water—Residuary Waters and their Purification—Soil Filtration—Purification by Chemical Processes—Analyses—Index.

(*See Books on Smoke Prevention, Engineering and Metallurgy, p.* 26, *etc.*)

Industrial Hygiene.

THE RISKS AND DANGERS TO HEALTH OF VARI-OUS OCCUPATIONS AND THEIR PREVENTION.
By LEONARD A. PARRY, M.D., B.S. (Lond.). 196 pp. Demy 8vo. 1900. Price 7s. 6d. ; India and Colonies, 8s. ; Other Countries, 8s. 6d. ; strictly net.

Contents.

Occupations which are Accompanied by the Generation and Scattering of Abnormal Quantities of Dust—Trades in which there is Danger of Metallic Poisoning—Certain Chemical Trades—Some Miscellaneous Occupations—Trades in which Various Poisonous Vapours are Inhaled—General Hygienic Considerations—Index.

X Rays.

PRACTICAL X RAY WORK. By FRANK T. ADDYMAN, B.Sc. (Lond.), F.I.C., Member of the Roentgen Society of London; Radiographer to St. George's Hospital; Demonstrator of Physics and Chemistry, and Teacher of Radiography in St. George's Hospital Medical School. Demy 8vo. Twelve Plates from Photographs of X Ray Work. Fifty-two Illustrations. 200 pp. 1901. Price 10s. 6d.; India and Colonies, 11s.; Other Countries, 12s.; strictly net.

Contents.

Historical—Work leading up to the Discovery of the X Rays—The Discovery—**Apparatus and its Management**—Electrical Terms—Sources of Electricity—Induction Coils—Electrostatic Machines—Tubes—Air Pumps—Tube Holders and Stereoscopic Apparatus—Fluorescent Screens—**Practical X Ray Work**—Installations—Radioscopy—Radiography—X Rays in Dentistry—X Rays in Chemistry—X Rays in War—Index.

List of Plates.

Frontispiece—Congenital Dislocation of Hip-Joint.—I., Needle in Finger.—II., Needle in Foot.—III., Revolver Bullet in Calf and Leg.—IV., A Method of Localisation.—V., Stellate Fracture of Patella showing shadow of "Strapping".—VI., Sarcoma.—VII., Six-weeks-old Injury to Elbow showing new Growth of Bone.—VIII., Old Fracture of Tibia and Fibula badly set.—IX., Heart Shadow.—X., Fractured Femur showing Grain of Splint.—XI., Barrell's Method of Localisation.

India-Rubber and Gutta Percha.

INDIA-RUBBER AND GUTTA PERCHA. Translated from the French of T. SEELIGMANN, G. LAMY TORVILHON and H. FALCONNET by JOHN GEDDES MCINTOSH. Royal 8vo. Eighty-six Illustrations. Three Plates. 412 pages. 1903. Price 12s. 6d.; India and Colonies, 13s. 6d.; Other Countries, 15s.; strictly net.

Contents.

India-Rubber—Botanical Origin—Climatology—Soil—Rational Culture and Acclimation of the Different Species of India-Rubber Plants—Methods of Obtaining the Latex—Methods of Preparing Raw or Crude India-Rubber—Classification of the Commercial Species of Raw Rubber—Physical and Chemical Properties of the Latex and of India-Rubber—Mechanical Transformation of Natural Caoutchouc into Washed or Normal Caoutchouc (Purification) and Normal Rubber into Masticated Rubber—Softening, Cutting, Washing, Drying—Preliminary Observations—Vulcanisation of Normal Rubber—Chemical and Physical Properties of Vulcanised Rubber—General Considerations—Hardened Rubber or Ebonite—Considerations on Mineralisation and other Mixtures—Coloration and Dyeing—Analysis of Natural or Normal Rubber and Vulcanised Rubber—Rubber Substitutes—Imitation Rubber.

Gutta Percha—Botanical Origin—Climatology—Soil—Rational Culture—Methods of Collection—Classification of the Different Species of Commercial Gutta Percha—Physical and Chemical Properties—Mechanical Transformation—Methods of Analysing—Gutta Percha Substitutes—Index.

Leather Trades.

PRACTICAL TREATISE ON THE LEATHER INDUSTRY. By A. M. VILLON. Translated by FRANK T. ADDYMAN, B.Sc. (Lond.), F.I.C., F.C.S.; and Corrected by an Eminent Member of the Trade. 500 pp., royal 8vo. 1901. 123 Illustrations. Price 21s.; India and Colonies, 22s.; Other Countries, 23s. 6d.; strictly net.

Contents.

Preface—Translator's Preface—List of Illustrations.

Part I., **Materials used in Tanning**—Skins: Skin and its Structure; Skins used in Tanning; Various Skins and their Uses—Tannin and Tanning Substances: Tannin; Barks (Oak); Barks other than Oak; Tanning Woods; Tannin-bearing Leaves; Excrescences; Tan-bearing Fruits; Tan-bearing Roots and Bulbs; Tanning Juices; Tanning Substances used in Various Countries; Tannin Extracts: Estimation of Tannin and Tannin Principles.

Part II., **Tanning**—The Installation of a Tannery: Tan Furnaces; Chimneys, Boilers, etc.; Steam Engines—Grinding and Trituration of Tanning Substances: Cutting up Bark; Grinding Bark; The Grinding of Tan Woods; Powdering Fruit, Galls and Grains; Notes on

ARCHITECTURAL POTTERY. Bricks, Tiles, Pipes, Ena-
melled Terra-cottas, Ordinary and Incrusted Quarries, Stoneware
Mosaics, Faïences and Architectural Stoneware. By LEON LEFÈVRE.
With Five Plates. 950 Illustrations in the Text, and numerous estimates.
500 pp., royal 8vo. 1900. Translated from the French by K. H. BIRD,
M.A., and W. MOORE BINNS. Price 15s.; India and Colonies, 16s.;
Other Countries, 17s. 6d.; strictly net.

Contents.

Part I. Plain Undecorated Pottery.—Clays, Bricks, Tiles, Pipes, Chimney Flues,
Terra-cotta.
Part II. Made-up or Decorated Pottery.

**THE ART OF RIVETING GLASS, CHINA AND
EARTHENWARE.** By J. HOWARTH. Second Edition.
1900. Paper Cover. Price 1s. net; by post, home or abroad, 1s. 1d.

HOW TO ANALYSE CLAY. Practical Methods for Practical Men. By HOLDEN M. ASHBY, Professor of Organic Chemistry,
Harvey Medical College, U.S.A. 74 pp. Twenty Illus. 1901. Price
2s. 6d.; Abroad, 3s.; strictly net.

NOTES ON POTTERY CLAYS. Their Distribution, Properties, Uses and Analyses of Ball Clays, China Clays and China
Stone. By JAS. FAIRIE, F.G.S. 1901. 132 pp. Crown 8vo. Price
3s. 6d.; India and Colonies, 4s.; Other Countries, 4s. 6d.; strictly net.

A Reissue of

**THE HISTORY OF THE STAFFORDSHIRE POTTER-
IES; AND THE RISE AND PROGRESS OF THE
MANUFACTURE OF POTTERY AND PORCELAIN.**
With References to Genuine Specimens, and Notices of Eminent Pot-
ters. By SIMEON SHAW. (Originally Published in 1829.) 265 pp.
1900. Demy 8vo. Price 7s. 6d.; India and Colonies, 8s.; Other
Countries, 8s. 6d.; strictly net.

Contents.

Introductory Chapter showing the position of the Pottery Trade at the present time
1899)—Preliminary Remarks—The Potteries, comprising Tunstall, Brownhills, Green-
field and New Field, Golden Hill, Latebrook, Green Lane, Burslem, Longport and Dale Hall,
Hot Lane and Cobridge, Hanley and Shelton, Etruria, Stoke, Penkhull, Fenton, Lane Delph,
Foley, Lane End—On the Origin of the Art, and its Practice among the early Nations—
Manufacture of Pottery, prior to 1700—The Introduction of Red Porcelain by Messrs
Elers, of Bradwell, 1690—Progress of the Manufacture from 1700 to Mr. Wedgwood's
commencement in 1760—Introduction of Fluid Glaze—Extension of the Manufacture of
Cream Colour—Mr. Wedgwood's Queen's Ware—Jasper, and Appointment of Potter to Her
Majesty—Black Printing—Introduction of Porcelain. Mr. W. Littler's Porcelain—Mr
Cookworthy's Discovery of Kaolin and Petuntse, and Patent—Sold to Mr. Champion—re-
sold to the New Hall Com.—Extension of Term—Blue Printed Pottery. Mr. Turner, Mr
Spode (1), Mr. Baddeley, Mr. Spode (2), Messrs. Turner, Mr. Wood, Mr. Wilson, Mr. Minton—
Great Change in Patterns of Blue Printed—Introduction of Lustre Pottery. Improve-
ments in Pottery and Porcelain subsequent to 1800.

17

A Reissue of

THE CHEMISTRY OF THE SEVERAL NATURAL
AND ARTIFICIAL HETEROGENEOUS COM-
POUNDS USED IN MANUFACTURING POR-
CELAIN, GLASS AND POTTERY. By SIMEON SHAW.
(Originally published in 1837.) 750 pp. 1900. Royal 8vo. Price 14s. ;
India and Colonies, 15s. ; Other Countries, 16s. 6d. ; strictly net.

Contents.

PART I., ANALYSIS AND MATERIALS.—Introduction : Laboratory and Apparatus
Elements—Temperature—Acids and Alkalies—The Earths—Metals.
PART II., SYNTHESIS AND COMPOUNDS.—Science of Mixing—Bodies : Porcelain
—Hard, Porcelain—Fritted Bodies, Porcelain—Raw Bodies, Porcelain—Soft, Fritted Bodies,
Raw Bodies, Stone Bodies, Ironstone, Dry Bodies, Chemical Utensils, Fritted Jasper, Fritted
Pearl, Fritted Drab, Raw Chemical Utensils, Raw Stone, Raw Jasper, Raw Pearl, Raw Mortar,
Raw Drab, Raw Brown, Raw Fawn, Raw Cane, Raw Red Porous, Raw Egyptian, Earthenware,
Queen's Ware, Cream Colour, Blue and Fancy Printed, Dipped and Mocha, Chalky, Rings,
Stilts, etc.—Glazes : Porcelain—Hard Fritted Porcelain—Soft Fritted Porcelain — Soft
Raw, Cream Colour Porcelain, Blue Printed Porcelain, Fritted Glazes, Analysis of Fritt,
Analysis of Glaze, Coloured Glazes, Dips, Smears and Washes ; Glasses : Flint Glass,
Coloured Glasses, Artificial Garnet, Artificial Emerald, Artificial Amethyst, Artificial Sap-
phire, Artificial Opal, Plate Glass, Crown Glass, Broad Glass, Bottle Glass, Phosphoric Glass,
British Steel Glass, Glass-Staining and Painting, Engraving on Glass, Dr. Faraday's Experi-
ments—Colours : Colour Making, Fluxes or Solvents, Components of the Colours : Reds,
etc., from Gold, Carmine or Rose Colour, Purple, Reds, etc., from Iron, Blues, Yellows,
Greens, Blacks, White, Silver for Burnishing, Gold for Burnishing, Printer's Oil, Lustres.
TABLES OF THE CHARACTERISTICS OF CHEMICAL SUBSTANCES.

Glassware, Glass Staining and Painting.

RECIPES FOR FLINT GLASS MAKING. By a British
Glass Master and Mixer. Sixty Recipes. Being Leaves from the
Mixing Book of several experts in the Flint Glass Trade, containing
up-to-date recipes and valuable information as to Crystal, Demi-crystal
and Coloured Glass in its many varieties. It contains the recipes for
cheap metal suited to pressing, blowing, etc., as well as the most costly
crystal and ruby. Crown 8vo. 1900. Price for United Kingdom,
10s. 6d. ; Abroad, 15s. ; United States, $4; strictly net.

Contents.

Ruby—Ruby from Copper—Flint for using with the Ruby for Coating—A German Metal—
Cornelian, or Alabaster—Sapphire Blue—Crysophis—Opal—Turquoise Blue—Gold Colour—
Dark Green—Green (common)—Green for Malachite—Blue for Malachite—Black for Mala-
chite—Black—Common Canary Batch—Canary—White Opaque Glass—Sealing-wax Red—
Flint—Flint Glass (Crystal and Demi)—Achromatic Glass—Paste Glass—White Enamel—
Firestone—Dead White (for moons)—White Agate—Canary—Canary Enamel—Index.

A TREATISE ON THE ART OF GLASS PAINTING.
Prefaced with a Review of Ancient Glass. By ERNEST R. SUFFLING.
With One Coloured Plate and Thirty-seven Illustrations. Demy 8vo.
140 pp. 1902. Price 7s. 6d. ; India and Colonies, 8s. ; Other Countries,
8s. 6d. net.

Contents.

A Short History of Stained Glass—Designing Scale Drawings—Cartoons and the Cut Line
—Various Kinds of Glass Cutting for Windows—The Colours and Brushes used in Glass
Painting—Painting on Glass, Dispersed Patterns—Diapered Patterns—Aciding—Firing—
Fret Lead Glazing—Index.

THE TECHNICAL TESTING OF YARNS AND TEX-TILE FABRICS.
With Reference to Official Specifica-tions. Translated from the German of Dr. J. HERZFELD. Second Edition. Sixty-nine Illustrations. 200 pp. Demy 8vo. 1902. Price 10s. 6d.; India and Colonies, 11s.; Other Countries, 12s.; strictly net.

Contents.

Yarn Testing. Determining the Yarn Number—Testing the Length of Yarns—Examination of the External Appearance of Yarn—Determining the Twist of Yarn and Twist—Determination of Tensile Strength and Elasticity—Estimating the Percentage of Fat in Yarn—Determination of Moisture (Conditioning)—Appendix.

DECORATIVE AND FANCY TEXTILE FABRICS.
By R. T. LORD. Manufacturers and Designers of Carpets, Damask, Dress and all Textile Fabrics. 200 pp. 1898. Demy 8vo. 132 Designs and Illustrations. Price 7s. 6d.; India and Colonies, 8s.; Other Countries, 8s. 6d.; strictly net.

Contents.

A Few Hints on Designing Ornamental Textile Fabrics—A Few Hints on Designing Orna-mental Textile Fabrics (continued)—A Few Hints on Designing Ornamental Textile Fabrics (continued)—A Few Hints on Designing Ornamental Textile Fabrics (continued)—Hints for Ruled-paper Draughtsmen—The Jacquard Machine—Brussels and Wilton Carpets—Tapestry Carpets—Ingrain Carpets—Axminster Carpets—Damask and Tapestry Fabrics—Scarf Silks and Ribbons—Silk Handkerchiefs—Dress Fabrics—Mantle Cloths—Figured Plush—Bed Quilts—Calico Printing.

THEORY AND PRACTICE OF DAMASK WEAVING.
By H. KINZER and K. WALTER. Royal 8vo. Eighteen Folding Plates. Six Illustrations. Translated from the German. 110 pp. 1903. Price 8s. 6d.; Colonies, 9s.; Other Countries, 9s. 6d.; strictly net.

Contents.

The Various Sorts of Damask Fabrics—Drill (Ticking, Handloom-made)—Whole Damask for Tablecloths—Damask with Ground- and Connecting-warp Threads—Furniture Damask—Lampas or Hangings—Church Damasks—The Manufacture of Whole Damask —Damask Arrangement with and without Cross-Shedding—The Altered Cone-arrangement—The Principle of the Corner Lifting Cord—The Roller Principle—The Combination of the Jacquard with the so-called Damask Machine—The Special Damask Machine—The Combina-tion of Two Tyings.

FAULTS IN THE MANUFACTURE OF WOOLLEN GOODS AND THEIR PREVENTION.
By NICOLAS REISER. Translated from the Second German Edition. Crown 8vo. Sixty-three Illustrations. 170 pp 1903. Price 5s.; Colonies, 5s. 6d.; Other Countries, 6s.; strictly net.

Contents.

Improperly Chosen Raw Material or Improper Mixtures—Wrong Treatment of the Material in Washing, Carbonisation, Drying, Dyeing and Spinning—Improper Spacing of the Goods in the Loom—Wrong Placing of Colours—Wrong Weight or Width of the Goods —Breaking of Warp and Weft Threads—Presence of Doubles, Singles, Thick, Loose, and too Hard Twisted Threads as well as Tangles, Thick Knots and the Like—Errors in Cross-weaving—Inequalities, i.e., Bands and Stripes—Dirty Borders—Defective Selvedges—Holes and Buttons—Rubbed Places—Creases—Spots—Loose and Bad Colours—Badly Dyed Selvedges—Hard Goods—Brittle Goods—Uneven Goods—Removal of Bands, Stripes, Creases and Spots.

SPINNING AND WEAVING CALCULATIONS, especially relating to Woollens.
From the German of N. REISER. Thirty-four Illustrations. Tables. 170 pp Demy 8vo. 1904. Price 10s. 6d.; India and Colonies, 11s.; Other Countries, 12s.; strictly net.

Contents.

Calculating the Raw Material—Proportion of Different Grades of Wool to Furnish a Mixture at a Given Price—Quantity to Produce a Given Length—Yarn Calculations—Yarn Number—Working Calculations—Calculating the Reed Count—Cost of Weaving, etc.

WATERPROOFING OF FABRICS. By Dr. S. MIERZINSKI.
Crown 8vo. 104 pp. 29 Illus. 1903. Price 5s.; Colonies, 5s. 6d.;
Other Countries, 6s.; strictly net.

Contents.
Introduction—Preliminary Treatment of the Fabric—Waterproofing with Acetate of Alumina—Impregnation of the Fabric—Drying—Waterproofing with Paraffin—Waterproofing with Ammonium Cuprate—Waterproofing with Metallic Oxides—Coloured Waterproof Fabrics—Waterproofing with Gelatine, Tannin, Caseinate of Lime and other Bodies—Manufacture of Tarpaulin—British Waterproofing Patents—Index.

HOW TO MAKE A WOOLLEN MILL PAY. By JOHN
MACKIE. Crown 8vo. 76 pp. 1904. Price 3s. 6d.; Colonies, 4s.;
Other Countries, 4s. 6d.; net.

Contents.
Blends, Pil·s, or Mixtures of Clean Scoured Wools—Dyed Wool Book—The Order Book—Pattern Duplicate Books—Management and Oversight—Constant Inspection of Mill Departments—Importance of Delivering Goods to Time, Shade, Strength, etc.—Plums.
(For "Textile Soaps" see p. 32.)

Dyeing, Colour Printing, Matching and Dye-stuffs.

THE COLOUR PRINTING OF CARPET YARNS. Manual
for Colour Chemists and Textile Printers. By DAVID PATERSON,
F.C.S. Seventeen Illustrations. 136 pp. Demy 8vo. 1900. Price
7s. 6d.; India and Colonies, 8s.; Other Countries, 8s. 6d.; strictly net.

Contents.
Structure and Constitution of Wool Fibre—Yarn Scouring—Scouring Materials—Water for Scouring—Bleaching Carpet Yarns—Colour Making for Yarn Printing—Colour Printing Pastes—Colour Recipes for Yarn Printing—Science of Colour Mixing—Matching of Colours—"Hank" Printing—Printing Tapestry Carpet Yarns—Yarn Printing—Steaming Printed Yarns—Washing of Steamed Yarns—Aniline Colours Suitable for Yarn Printing—Glossary of Dyes and Dye-wares used in Wood Yarn Printing—Appendix.

THE SCIENCE OF COLOUR MIXING. A Manual intended for the use of Dyers, Calico Printers and Colour Chemists. By
DAVID PATERSON, F.C.S. Forty-one Illustrations, **Five Coloured Plates,
and Four Plates showing Eleven Dyed Specimens of Fabrics.** 132
pp. Demy 8vo. 1900. Price 7s. 6d.; India and Colonies, 8s.; Other
Countries, 8s. 6d.; strictly net.

Contents.
Colour a Sensation; Colours of Illuminated Bodies; Colours of Opaque and Transparent Bodies; Surface Colour—Analysis of Light; Spectrum; Homogeneous Colours; Ready Method of Obtaining a Spectrum—Examination of Solar Spectrum; The Spectroscope and Its Construction; Colourists' Use of the Spectroscope—Colour by Absorption; Solutions and Dyed Fabrics; Dichroic Coloured Fabrics in Gaslight—Colour Primaries of the Scientist versus the Dyer and Artist; Colour Mixing by Rotation and Lye Dyeing; Hue, Purity, Brightness; Tints; Shades, Scales, Tones, Sad and Sombre Colours—Colour Mixing; Pure and Impure Greens, Orange and Violets; Large Variety of Shades from few Colours; Consideration of the Practical Primaries; Red, Yellow and Blue—Secondary Colours; Nomenclature of Violet and Purple Group; Tints and Shades of Violet; Changes in Artificial Light—Tertiary Shades; Broken Hues; Absorption Spectra of Tertiary Shades—Appendix: Four Plates with Dyed Specimens Illustrating Text—Index.

DYERS' MATERIALS: An Introduction to the Examination,
Evaluation and Application of the most important Substances used in
Dyeing, Printing, Bleaching and Finishing. By PAUL HEERMAN, Ph.D.
Translated from the German by. A C. WRIGHT, M.A. (Oxon.), B.Sc.
(Lond.). Twenty-four Illustrations. Crown 8vo. 150 pp. 1901. Price
5s.; India and Colonies, 5s. 6d.; Other Countries, 6s.; strictly net.

and Carbonated Alkali (Soda)—Chlorometry—Titration—Wagner's Chlorometric Method—
Preparation of Standard Solutions—Apparatus for Chlorine Valuation—Alkali in Excess in
Decolourising Chlorides—Chlorine and Decolourising Chlorides—Synopsis—Chlorine—
Chloride of Lime—Hypochlorite of Soda—Brochoki's Chlorozone—Various Decolourising
Hypochlorites—Comparison of Chloride of Lime and Hypochlorite of Soda—Water—
Qualities of Water—Hardness—Dervaux's Purifier—Testing the Purified Water—Different
Plant for Purification—Filters—Bleaching of Yarn—Weight of Yarn—Lye Boiling—
Chemicking—Washing—Bleaching of Cotton Yarn—The Installation of a Bleach Works—
Water Supply—Steam Boilers—Steam Distribution Pipes—Engines—Keirs—Washing—
Machines—Stocks—Wash Wheels—Chemicking and Souring Cisterns—Various—Buildings—
Addenda—Energy of Decolourising Chlorides and Bleaching by Electricity and Ozone—
Energy of Decolourising Chlorides—Chlorides—Production of Chlorine and Hypochlorites
by Electrolysis—Lunge's Process for increasing the intensity of the Bleaching Power of
Chloride of Lime—Trilfer's Process for Removing the Excess of Lime or Soda from De-
colourising Chlorides—Bleaching by Ozone.

Cotton Spinning and Combing.

COTTON SPINNING (First Year). By THOMAS THORNLEY,
Spinning Master, Bolton Technical School. 160 pp. Eighty-four Illustrations. Crown 8vo. 1901. Price 3s.; Abroad, 3s. 6d.; strictly net.

Contents.

Syllabus and Examination Papers of the City and Guilds of London Institute—Cultivation, Classification, Ginning, Baling and Mixing of the Raw Cotton—Bale-Breakers, Mixing
Lattices and Hopper Feeders—Opening and Scutching—Carding—Indexes.

COTTON SPINNING (Intermediate, or Second Year). By
THOMAS THORNLEY. 180 pp. Seventy Illustrations. Crown 8vo. 1901.
Price 5s.; India and British Colonies, 5s. 6d.; Other Countries, 6s.;
strictly net.

Contents.

Syllabuses and Examination Papers of the City and Guilds of London Institute—The
Combing Process—The Drawing Frame—Bobbin and Fly Frames—Mule Spinning—Ring
Spinning—General Indexes.

COTTON SPINNING (Honours, or Third Year). By THOMAS
THORNLEY. 216 pp. Seventy-four Illustrations. Crown 8vo. 1901.
Price 5s.; India and British Colonies, 5s. 6d.; Other Countries, 6s.;
strictly net.

Contents.

Syllabuses and Examination Papers of the City and Guilds of London Institute—Cotton—
The Practical Manipulation of Cotton Spinning Machinery—Doubling and Winding—Reeling
—Warping—Production and Costs—Main Driving—Arrangement of Machinery and Mill
Planning—Waste and Waste Spinning—Indexes.

COTTON COMBING MACHINES. By THOS. THORNLEY,
Spinning Master, Technical School, Bolton. Demy 8vo. 117 Illustra-
tions. 300 pp. 1902. Price 7s. 6d.; India and Colonies, 8s.; Other
Countries, 8s. 6d. net.

Contents.

The Sliver Lap Machine and the Ribbon Cap Machine—General Description of the Heilmann
Comber—The Cam Shaft—On the Detaching and Attaching Mechanism of the Comber—
Resetting of Combers—The Erection of a Heilmann Comber—Stop Motions: Various Calcu-
lations—Various Notes and Discussions—Cotton Combing Machines of Continental Make—
Index.

Collieries and Mines.

RECOVERY WORK AFTER PIT FIRES. A Description of the Principal Methods Pursued, especially in Fiery Mines, and of the Various Appliances Employed, such as Respiratory and Rescue Apparatus, Dams, etc. By ROBERT LAMPRECHT, Mining Engineer and Manager. Translated from the German. Illustrated by Six large Plates, containing Seventy-six Illustrations. 175 pp., demy 8vo. 1901. Price 10s. 6d. ; India and Colonies, 11s. ; Other Countries, 12s. ; strictly net.

Contents.

Causes of Pit Fires—Preventive Regulations : (1) The Outbreak and Rapid Extension of a Shaft Fire can be most reliably prevented by Employing little or no Combustible Material in the Construction of the Shaft : (2) Precautions for Rapidly Localising an Outbreak of Fire in the Shaft ; (3) Precautions to be Adopted in case those under 1 and 2 Fail or Prove Inefficient. Precautions g inst Spontaneous Ignition of Coal. Precautions for Preventing Explosions of Fire-damp and Coal Dust. Employment of Electricity in Mining, particularly in Fiery Pits. Experiments on the ignition of Fire-damp Mixtures and Clouds of Coal Dust by Electricity—Indications of an Existing or Incipient Fire—Appliances for Working in Irrespirable Gases : Respiratory Apparatus; Apparatus with Air Supply Pipes; Reservoir Apparatus; Oxygen Apparatus—Extinguishing Pit Fires : (a) Chemical Means ; (b) Extinction with Water. Dragging down the Burning Masses and Packing with Clay ; (c) Insulating the Seat of the Fire by Dams. Dam Building. Analyses of Fire Gases. Isolating the Seat of a Fire with Dams : Working in Irrespirable Gases ("Gas-diving") ; Air-Lock Work. Complete Isolation of the Pit. Flooding a Burning Section isolated by means of Dams. Wooden Dams : Masonry Dams. Examples of Cylindrical and Dome-shaped Dams. Dam Doors : Flooding the Whole Pit—Rescue Stations : (a) Stations above Ground ; (b) Underground Rescue Stations—Spontaneous Ignition of Coal in Bulk—Index.

VENTILATION IN MINES. By ROBERT WABNER, Mining Engineer. Translated from the German. Royal 8vo. Thirty Plates and Twenty-two Illustrations. 240 pp. 1903. Price 10s. 6d. ; India and Colonies, 11s. ; Other Countries. 12s. ; strictly net.

Contents.

The Causes of the Contamination of Pit Air—The Means of Preventing the Dangers resulting from the Contamination of Pit Air—Calculating the Volume of Ventilating Current necessary to free Pit Air from Contamination—Determination of the Resistance Opposed to the Passage of Air through the Pit—Laws of Resistance and Formulæ therefor—Fluctuations in the Temperament or Specific Resistance of a Pit—Means for Providing a Ventilating Current in the Pit—Mechanical Ventilation—Ventilators and Fans—Determining the Theoretical, Initial, and True (Effective) Depression of the Centrifugal Fan—New Types of Centrifugal Fan of Small Diameter and High Working Speed—Utilising the Ventilating Current to the utmost Advantage and distributing the same through the Workings—Artificially retarding the Ventilating Current—Ventilating Preliminary Workings—Blind Headings—Separate Ventilation—Supervision of Ventilation—INDEX.

HAULAGE AND WINDING APPLIANCES USED IN MINES. By CARL VOLK. Translated from the German. Royal 8vo. With Six Plates and 148 Illustrations. 150 pp. 1903. Price 8s. 6d. ; Colonies, 9s. ; Other Countries, 9s. 6d. ; strictly net.

Contents.

Haulage Appliances—Ropes—Haulage Tubs and Tracks—Cages and Winding Appliances—Winding Engines for Vertical Shafts—Winding without Ropes—Haulage in Levels and Inclines—The Working of Underground Engine—Machinery for Downhill Haulage.

Engineering, Smoke Prevention and Metallurgy.

THE PREVENTION OF SMOKE. Combined with the Economical Combustion of Fuel. By W. C. POPPLEWELL, M.Sc., A.M.Inst., C E., Consulting Engineer. Forty-six Illustrations. 190 pp. 1901. Demy 8vo. Price 7s. 6d. : India and Colonies, 8s. ; Other Countries, 8s. 6d. , strictly net.

(For Contents see next page.)

Contents.

Fuel and Combustion— Hand Firing in Boiler Furnaces—Stoking by Mechanical Means—
Powdered Fuel—Gaseous Fuel—Efficiency and Smoke Tests of Boilers—Some Standard
Smoke Trials—The Legal Aspect of the Smoke Question—The Best Means to be adopted for
the Prevention of Smoke—Index.

GAS AND COAL DUST FIRING. A Critical Review of
the Various Appliances Patented in Germany for this purpose since
1885. By ALBERT PÜTSCH. 130 pp. Demy 8vo. 1901. Translated
from the German. With 103 Illustrations. Price 7s. 6d.; India and
Colonies, 8s.; Other Countries, 8s. 6d.; strictly net.

Contents.

Generators—Generators Employing Steam—Stirring and Feed Regulating Appliances—
Direct Generators—Burners—Regenerators and Recuperators—Glass Smelting Furnaces—
Metallurgical Furnaces—Pottery Furnace—Coal Dust Firing—Index.

THE HARDENING AND TEMPERING OF STEEL
IN THEORY AND PRACTICE. By FRIDOLIN REISER.
Translated from the German of the Third Edition. Crown 8vo.
120 pp. 1903. Price 5s.: India and British Colonies, 5s. 6d.; Other
Countries, 6s.; strictly net.

Contents.

Steel—Chemical and Physical Properties of Steel, and their Casual Connection—
Classification of Steel according to Use—Testing the Quality of Steel — Steel-
Hardening—Investigation of the Causes of Failure in Hardening—Regeneration of
Steel Spoilt in the Furnace—Welding Steel—Index.

SIDEROLOGY: THE SCIENCE OF IRON (The Con-
stitution of Iron Alloys and Slags). Translated from German of
HANNS FREIHERR V. JÜPTNER. 350 pp. Demy 8vo. Eleven Plates
and Ten Illustrations. 1902. Price 10s. 6d.; India and Colonies, 11s.;
Other Countries, 12s.; net.

Contents.

The Theory of Solution.—Solutions—Molten Alloys—Varieties of Solutions—Osmotic
Pressure—Relation between Osmotic Pressure and other Properties of Solutions—Osmotic
Pressure and Molecular Weight of the Dissolved Substance—Solutions of Gases—Solid Solu-
tions—Solubility—Diffusion—Electrical Conductivity—Constitution of Electrolytes and Metals
—Thermal Expansion. Micrography.—Microstructure—The Micrographic Constituents of
Iron—Relation between Micrographical Composition, Carbon-Content, and Thermal Treat-
ment of Iron Alloys—The Microstructure of Slags. Chemical Composition of the Alloys
of Iron.—Constituents of Iron Alloys—Carbon—Constituents of the Iron Alloys, Carbon—
Opinions and Researches on Combined Carbon—Opinions and Researches on Combined
Carbon—Applying the Curves of Solution deduced from the Curves of Recalescence to the De-
termination of the Chemical Composition of the Carbon present in Iron Alloys—The Constitu-
ents of Iron—Iron—The Constituents of Iron Alloys—Manganese—Remaining Constituents of
Iron Alloys—A Silicon—Gases. The Chemical Composition of Slag.—Silicate Slags—
Calculating the Composition of Silicate Slags—Phosphate Slags—Oxide Slags—Appendix—
Index.

EVAPORATING, CONDENSING AND COOLING AP-
PARATUS. Explanations, Formulæ and Tables for Use
in Practice. By E. HAUSBRAND, Engineer. Translated by A. C.
WRIGHT, M.A. (Oxon.), B.Sc. (Lond.). With Twenty-one Illustrations and Seventy-six Tables. 400 pp. Demy 8vo. 1903. Price
10s. 6d.; India and Colonies 11s.; Other Countries, 12s.; net.

Contents.

k-Coefficient of Transmission of Heat, $k/$, and the Mean Temperature Difference, θ/m—
Parallel and Opposite Currents—Apparatus for Heating with Direct Fire—The Injection of
Saturated Steam—Superheated Steam—Evaporation by Means of Hot Liquids—The Transference of Heat in General, and Transference by means of Saturated Steam in Particular
—The Transference of Heat from Saturated Steam in Pipes (Coils) and Double Bottoms
—Evaporation in a Vacuum—The Multiple-effect Evaporator—Multiple-effect Evaporators
from which Extra Steam is Taken—The Weight of Water which must be Evaporated from
100 Kilos. of Liquor in order its Original Percentage of Dry Materials from 1-25 per cent.
up to 20-70 per cent.—The Relative Proportion of the Heating Surfaces in the Elements
of the Multiple Evaporator and their Actual Dimensions—The Pressure Exerted by Currents
of Steam and Gas upon Floating Drops of Water—The Motion of Floating Drops of Water

upon which Press Currents of Steam—The Splashing of Evaporating Liquids—The Diameter of Pipes for Steam, Alcohol, Vapour and Air—The Diameter of Water Pipes—The Loss of Heat from Apparatus and Pipes to the Surrounding Air, and Means for Preventing the Loss—Condensers—Heating Liquids by Means of Steam—The Cooling of Liquids—The Volumes to be Exhausted from Condensers by the Air-pumps—A Few Remarks on Air-pumps and the Vacua they Produce—The Volumetric Efficiency of Air-pumps—The Volumes of Air which must be Exhausted from a Vessel in order to Reduce its Original Pressure to a Certain Lower Pressure—Index.

Dental Metallurgy.

DENTAL METALLURGY: MANUAL FOR STUDENTS AND DENTISTS. By A. B. GRIFFITHS, Ph.D. Demy 8vo. Thirty-six Illustrations. 190£. 200 pp. Price 7s. 6d.; India and Colonies, 8s.; Other Countries, 8s. 6d.; strictly net.

Contents.

Introduction—Physical Properties of the Metals—Action of Certain Agents on Metals—Alloys—Action of Oral Bacteria on Alloys—Theory and Varieties of Blowpipes—Fluxes—Furnaces and Appliances—Heat and Temperature—Gold—Mercury—Silver—Iron—Copper—Zinc—Magnesium—Cadmium—Tin—Lead—Aluminium—Antimony—Bismuth—Palladium—Platinum—Iridium—Nickel—Practical Work—Weights and Measures.

Plumbing, Decorating, Metal Work, etc., etc.

EXTERNAL PLUMBING WORK. A Treatise on Lead Work for Roofs. By JOHN W. HART, R.P.C. 180 Illustrations. 272 pp. Demy 8vo. Second Edition Revised. 1902. Price 7s. 6d.; India and Colonies, 8s.; Other Countries, 8s. 6d.; strictly net.

Contents.

Cast Sheet Lead—Milled Sheet Lead—Roof Cesspools—Socket Pipes—Drips—Gutters—Gutters (continued)—Breaks—Circular Breaks—Flats—Flats (continued)—Rolls on Flats—Roll Ends—Roll Intersections—Seam Rolls—Seam Rolls (continued)—Tack Fixings—Step Flashings—Step Flashings (continued)—Secret Gutters—Soakers—Hip and Valley Soakers—Dormer Windows—Dormer Windows (continued)—Dormer Tops—Internal Dormers—Skylights—Hips and Ridging—Hips and Ridging (continued)—Fixings for Hips and Ridging—Ornamental Ridging—Ornamental Curb Rolls—Curb Rolls—Cornices—Towers and Finials—Towers and Finials (continued)—Towers and Finials (continued)—Domes—Domes (continued)—Ornamental Lead Work—Rain Water Heads—Rain Water Heads (continued)—Rain Water Heads (continued).

HINTS TO PLUMBERS ON JOINT WIPING, PIPE BENDING AND LEAD BURNING. Third Edition, Revised and Corrected. By JOHN W. HART, R.P.C. 184 Illustrations. 313 pp. Demy 8vo. 1901. Price 7s. 6d.; India and Colonies, 8s.; Other Countries, 8s. 6d.; strictly net.

Contents.

Pipe Bending — Pipe Bending (continued) — Pipe Bending (continued) — Square Pipe Bendings—Half-circular Elbows—Curved Bends on Square Pipe—Bossed Bends—Curved Plinth Bends—Rain-water Shoes on Square Pipe—Curved and Angle Bends—Square Pipe Fixings—Joint-wiping—Substitutes for Wiped Joints—Preparing Wiped Joints—Joint Fixings—Plumbing Irons—Joint Fixings—Use of "Touch" in Soldering—Underhand Joints—Blown and Copper Bit Joints—Branch Joints—Branch Joints (continued)—Block Joints—Block Joints (continued)—Block Fixings—Astragal Joints—Pipe Fixings—Large Branch Joints—Large Underhand Joints—Solders—Autogenous Soldering or Lead Burning—Index.

WORKSHOP WRINKLES for Decorators, Painters, Paper-hangers and Others. By W. N. BROWN. Crown 8vo. 128 pp. 1901. Price 2s. 6d.; Abroad, 3s.; strictly net.

SANITARY PLUMBING AND DRAINAGE. By JOHN
W. HART. Demy 8vo. With 208 Illustrations. 250 pp. 1904. Price
7s. 6d.; India and Colonies, 8s.; Other Countries, 8s. 6d.; strictly net.

Contents.

Sanitary Surveys—Drain Testing—Drain Testing with Smoke—Testing Drains with Water
—Drain Plugs for Testing—Sanitary Defects—Closets—Baths and Lavatories—House Drains
—Manholes—Iron Soil Pipes—Lead Soil Pipes—Ventilating Pipes—Water-closets—Flushing
Cisterns—Baths—Bath Fittings—Lavatories—Lavatory Fittings—Sinks—Waste Pipes—
Water Supply—Ball Valves—Town House Sanitary Arrangements—Drainage—Jointing
Pipes—Accessible Drains—Iron Drains—Iron Junctions—Index.

**THE PRINCIPLES AND PRACTICE OF DIPPING,
BURNISHING, LACQUERING AND BRONZING
BRASS WARE.** By W. NORMAN BROWN. 35 pp. Crown
8vo. 1900. Price 2s.; Abroad, 2s. 6d.; strictly net.

HOUSE DECORATING AND PAINTING. By W.
NORMAN BROWN. Eighty-eight Illustrations. 150 pp. Crown 8vo.
1900. Price 3s. 6d.; India and Colonies, 4s.; Other Countries, 4s. 6d.;
strictly net.

A HISTORY OF DECORATIVE ART. By W. NORMAN
BROWN. Thirty-nine Illustrations. 96 pp. Crown 8vo. 1900. Price
2s. 6d.; Abroad, 3s.; strictly net.

**A HANDBOOK ON JAPANNING AND ENAMELLING
FOR CYCLES, BEDSTEADS, TINWARE, ETC.** By
WILLIAM NORMAN BROWN. 52 pp. and Illustrations. Crown 8vo.
1901. Price 2s.; Abroad, 2s. 6d.; net.

THE PRINCIPLES OF HOT WATER SUPPLY. By
JOHN W. HART, R.P.C. With 129 Illustrations. 1900. 177 pp., demy
8vo. Price 7s. 6d.; India and Colonies, 8s.; Other Countries, 8s. 6d.;
strictly net.

Contents.

Water Circulation—The Tank System—Pipes and Joints—The Cylinder System—Boilers
for the Cylinder System—The Cylinder System—The Combined Tank and Cylinder System
—Combined Independent and Kitchen Boiler—Combined Cylinder and Tank System with
Duplicate Boilers—Indirect Heating and Boiler Explosions—Pipe Boilers—Safety Valves—
Safety Valves—The American System—Heating Water by Steam—Steam Kettles and Jets
—Heating Power of Steam—Covering for Hot Water Pipes—Index.

Brewing and Botanical.

**HOPS IN THEIR BOTANICAL, AGRICULTURAL
AND TECHNICAL ASPECT, AND AS AN ARTICLE
OF COMMERCE.** By EMMANUEL GROSS, Professor at
the Higher Agricultural College, Tetschen-Liebwerd. Translated
from the German. Seventy-eight Illustrations. 1900. 340 pp. Demy
8vo. Price 12s. 6d.; India and Colonies, 13s. 6d.; Other Countries,
15s.; strictly net.

Contents.

HISTORY OF THE HOP—THE HOP PLANT—Introductory—The Roots—The Stem—
and Leaves—Inflorescence and Flower: Inflorescence and Flower of the Male Hop; Inflorescence and Flower of the Female Hop—The Fruit and its Glandular Structure: The
Fruit and Seed—Propagation and Selection of the Hop—Varieties of the Hop: (a) Red Hops;
(b) Green Hops; (c) Pale Green Hops—Classification according to the Period of Ripening:
Early August Hops; Medium Early Hops: Late Hops—Injuries to Growth—Leaves Turning
Yellow, Summer or Sunbrand, Cones Dropping Off, Honey Dew, Damage from Wind, Hail

and Rain ; Vegetable Enemies of the Hop: Animal Enemies of the Hop—Beneficial Insects on Hops—CULTIVATION—The Requirements of the Hop in Respect of Climate, Soil and Situation: Climate: Soil; Situation—Selection of Variety and Cuttings—Planting a Hop Garden: Drainage; Preparing the Ground; Marking-out for Planting; Planting; Cultivation and Cropping of the Hop Garden in the First Year—Work to be Performed Annually in the Hop Garden: Working the Ground; Cutting; The Non-cutting System; The Proper Performance of the Operation of Cutting: Method of Cutting; Close Cutting, Ordinary Cutting, The Long Cut, The Topping Cut; Proper Season for Cutting: Autumn Cutting, Spring Cutting; Manuring; Training the Hop Plant: Poled Gardens, Frame Training: Principal Types of Frames; Pruning, Cropping, Topping, and Leaf Stripping the Hop Plant; Picking, Drying and Bagging—Principal and Subsidiary Utilisation of Hops and Hop Gardens—Life of a Hop Garden; Subsequent Cropping—Cost of Production, Yield and Selling Prices.

Preservation and Storage—Physical and Chemical Structure of the Hop Cone —Judging the Value of Hops.

Statistics of Production—The Hop Trade—Index.

Timber and Wood Waste.

TIMBER : A Comprehensive Study of Wood in all its Aspects (Commercial and Botanical), showing the Different Applications and Uses of Timber in Various Trades, etc. Translated from the French of PAUL CHARPENTIER. Royal 8vo. 437 pp. 178 Illustrations. 1902. Price 12s. 6d. ; India and Colonies, 13s. 6d. ; Other Countries, 15s. ; net.

Contents.

Physical and Chemical Properties of Timber—Composition of the Vegetable Bodies—Chief Elements—M. Fremy's Researches—Elementary Organs of Plants and especially of Forests—Different Parts of Wood Anatomically and Chemically Considered—General Properties of Wood—**Description of the Different Kinds of Wood**—Principal Essences with Caducous Leaves—Coniferous Resinous Trees—**Division of the Useful Varieties of Timber in the Different Countries of the Globe**—European Timber—Asiatic Timber—American Timber—Timber of Oceania—**Forests**—General Notes as to Forests ; their Influence—Opinions as to Sylviculture—Improvement of Forests—Unwooding and Rewooding—Preservation of Forests—Exploitation of Forests—Damage caused to Forests—Different Alterations—**The Preservation of Timber**—Generalities—Causes and Progress of Deterioration—History of Different Proposed Processes—Dessication—Superficial Carbonisation of Timber—Processes by Immersion—Generalities as to Antiseptics Employed—Injection Processes in Closed Vessels—The Boucherie System, Based upon the Displacement of the Sap—Processes for Making Timber Uninflammable—**Applications of Timber**—Generalities—Working Timber—Paving—Timber for Mines—Railway Traverses—Accessory Products—Gums—Works of M. Fremy—Resins—Barks—Tan—Application of Cork—The Application of Wood to Art and Dyeing—Different Applications of Wood—Hard Wood—Distillation of Wood—Pyroligneous Acid—Oil of Wood—Distillation of Resins—Index.

THE UTILISATION OF WOOD WASTE. Translated from the German of ERNST HUBBARD. Crown 8vo. 192 pp. 1902. Fifty Illustrations. Price 5s. ; India and Colonies, 5s. 6d. ; Other Countries, 6s. ; net.

Contents.

General Remarks on the Utilisation of Sawdust—Employment of Sawdust as Fuel, with and without Simultaneous Recovery of Charcoal and the Products of Distillation—Manufacture of Oxalic Acid from Sawdust—Process with Soda Lye ; Thorn's Process ; Bohlig's Process—Manufacture of Spirit (Ethyl Alcohol) from Wood Waste—Patent Dyes (Organic Sulphides, Sulphur Dyes, or Mercapto Dyes)—Artificial Wood and Plastic Compositions from Sawdust—Production of Artificial Wood Compositions for Moulded Decorations—Employment of Sawdust for Blasting Powders and Gunpowders—Employment of Sawdust for Briquettes—Employment of Sawdust in the Ceramic Industry and as an Addition to Mortar—Manufacture of Paper Pulp from Wood—Casks—Various Applications of Sawdust and Wood Refuse—Calcium Carbide—Manure—Wood Mosaic Plaques—Bottle Stoppers—Parquetry—Fire-lighters—Carborundum—The Production of Wood Wool—Bark—Index.

Building and Architecture.

THE PREVENTION OF DAMPNESS IN BUILDINGS;
with Remarks on the Causes, Nature and Effects of Saline, Efflorescences and Dry-rot, for Architects, Builders, Overseers, Plasterers, Painters and House Owners. By ADOLF WILHELM KEIM. Translated from the German of the second revised Edition by M. J. SALTER, F.I.C., F.C.S. Eight Coloured Plates and Thirteen Illustrations. Crown 8vo. 115 pp. 1902. Price 5s.; India and Colonies, 5s. 6d.; Other Countries, 6s.; net.

Contents.

The Various Causes of Dampness and Decay of the Masonry of Buildings, and the Structural and Hygienic Evils of the Same—Precautionary Measures during Building against Dampness and Efflorescence—Methods of Remedying Dampness and Efflorescences in the Walls of Old Buildings—The Artificial Drying of New Houses, as well as Old Damp Dwellings, and the Theory of the Hardening of Mortar—New, Certain and Permanently Efficient Methods for Drying Old Damp Walls and Dwellings—The Cause and Origin of Dry-rot: its Injurious Effect on Health, its Destructive Action on Buildings, and its Successful Repression—Methods of Preventing Dry-rot to be Adopted During Construction—Old Methods of Preventing Dry-rot—Recent and More Efficient Remedies for Dry-rot—Index.

HANDBOOK OF TECHNICAL TERMS USED IN ARCHITECTURE AND BUILDING, AND THEIR ALLIED TRADES AND SUBJECTS. By AUGUSTINE C. PASSMORE. Demy 8vo. 380 pp. 1904. Price 7s. 6d.; India and Colonies, 8s.; Other Countries, 8s. 6d.; strictly net.

Foods and Sweetmeats.

THE MANUFACTURE OF PRESERVED FOODS AND SWEETMEATS. By A. HAUSNER. With Twenty-eight Illustrations. Translated from the German of the third enlarged Edition. Crown 8vo. 225 pp. 1902. Price 7s. 6d.; India and Colonies, 8s.; Other Countries, 8s. 6d.; net.

Contents.

The Manufacture of Conserves—Introduction—The Causes of the Putrefaction of Food—The Chemical Composition of Foods—The Products of Decomposition—The Causes of Fermentation and Putrefaction—Preservative Bodies—The Various Methods of Preserving Food—The Preservation of Animal Food—Preserving Meat by Means of Ice—The Preservation of Meat by Charcoal—Preservation of Meat by Drying—The Preservation of Meat by the Exclusion of Air—The Appert Method—Preserving Flesh by Smoking—Quick Smoking—Preserving Meat with Salt—Quick Salting by Air Pressure—Quick Salting by Liquid Pressure—Gamgee's Method of Preserving Meat—The Preservation of Eggs—Preservation of White and Yolk of Egg—Milk Preservation—Condensed Milk—The Preservation of Fat—Manufacture of Soup Tablets—Meat Biscuits—Extract of Beef—The Preservation of Vegetable Foods in General—Compressing Vegetables—Preservation of Vegetables by Appert's Method—The Preservation of Fruit—Preservation of Fruit by Storage—The Preservation of Fruit by Drying—Drying Fruit by Artificial Heat—Roasting Fruit—The Preservation of Fruit with Sugar—Boiled Preserved Fruit—The Preservation of Fruit in Spirit, Acetic Acid or Glycerine—Preservation of Fruit without Boiling—Jam Manufacture—The Manufacture of Fruit Jellies—The Making of Gelatine Jellies—The Manufacture of "Sulzen"—The Preservation of Fermented Beverages—**The Manufacture of Candies**—Introduction—The Manufacture of Candied Fruit—The Manufacture of Boiled Sugar and Caramel—The Candying of Fruit—Caramelised Fruit—The Manufacture of Sugar Sticks, or Barley Sugar—Bonbon Making—Fruit Drops—The Manufacture of Dragées—The Machinery and Appliances used in Candy Manufacture—Dyeing Candies and Bonbons—Essential Oils used in Candy Making—Fruit Essences—The Manufacture of Filled Bonbons, Liqueur Bonbons and Stamped Lozenges—Recipes for Jams and Jellies—Recipes for Bonbon Making—Dragées—Appendix—Index.

Dyeing Fancy Goods.

THE ART OF DYEING AND STAINING MARBLE, ARTIFICIAL STONE, BONE, HORN, IVORY AND WOOD, AND OF IMITATING ALL SORTS OF WOOD. A Practical Handbook for the Use of Joiners, Turners, Manufacturers of Fancy Goods, Stick and Umbrella Makers, Comb Makers, etc. Translated from the German of D. H. SOXHLET, Technical Chemist. Crown 8vo. 168 pp. 1902. Price 5s.; India and Colonies, 5s. 6d.; Other Countries, 6s.; net.

Contents.
Mordants and Stains — Natural Dyes—Artificial Pigments—Coal Tar Dyes — Staining Marble and Artificial Stone—Dyeing, Bleaching and Imitation of Bone, Horn and Ivory—Imitation of Tortoiseshell for Combs: Yellows, Dyeing Nuts—Ivory—Wood Dyeing—Imitation of Mahogany: Dark Walnut, Oak, Birch-Bark, Elder-Marquetry, Walnut, Walnut-Marquetry, Mahogany, Spanish Mahogany, Palisander and Rose Wood, Tortoiseshell, Oak, Ebony, Pear Tree—Black Dyeing Processes with Penetrating Colours—Varnishes and Polishes: English Furniture Polish, Vienna Furniture Polish, Amber Varnish, Copal Varnish, Composition for Preserving Furniture—Index.

Lithography, Printing and Engraving.

PRACTICAL LITHOGRAPHY. By ALFRED SEYMOUR. Demy 8vo. With Frontispiece and 33 Illus. 120 pp. 1903. Price 5s.; Colonies, 5s. 6d.; Other Countries, 6s.; net.

Contents.
Stones—Transfer Inks—Transfer Papers—Transfer Printing—Litho Press—Press Work—Machine Printing—Colour Printing—Substitutes for Lithographic Stones—Tin Plate Printing and Decoration—Photo-Lithography.

PRINTERS' AND STATIONERS' READY RECKONER AND COMPENDIUM. Compiled by VICTOR GRAHAM. Crown 8vo. 1904. [*In the press.*

Contents.
Price of Paper per Sheet, Quire, Ream and Lb.—Cost of 100 to 1000 Sheets at various Sizes and Prices per Ream—Cost of Cards—Quantity Table—Sizes and Weights of Paper, Cards, etc.—Notes on Account Books—Discount Tables—Sizes of spaces — Leads to a lb.—Dictionary—Measure for Bookwork—Correcting Proofs, etc.

ENGRAVING FOR ILLUSTRATION. HISTORICAL AND PRACTICAL NOTES. By J. KIRKBRIDE. 72 pp. Two Plates and 6 Illustrations. Crown 8vo. 1903. Price 2s. 6d.; Abroad, 3s.; strictly net.

Contents.
Its Inception—Wood Engraving—Metal Engraving—Engraving in England—Etching—Mezzotint —Photo-Process Engraving – The Engraver's Task—Appreciative Criticism —Index.

Bookbinding.

PRACTICAL BOOKBINDING. By PAUL ADAM. Translated from the German. Crown 8vo. 180 pp. 127 Illustrations. 1903. Price 5s.; Colonies, 5s. 6d.; Other Countries, 6s.; net.

Contents.
Materials for Sewing and Pasting—Materials for Covering the Book—Materials for Decorating and Finishing — Tools—General Preparatory Work — Sewing — Forwarding, Cutting, Rounding and Backing—Forwarding, Decoration of Edges and Headbanding—Boarding—Preparing the Cover—Work with the Blocking Press—Treatment of Sewn Books, Pastening in Covers, and Finishing Off—Handtooling and Other Decoration—Account Books—School Books, Mounting Maps, Drawings, etc.—Index.

Sugar Refining.

THE TECHNOLOGY OF SUGAR: Practical Treatise on the Modern Methods of Manufacture of Sugar from the Sugar Cane and Sugar Beet. By JOHN GEDDES McINTOSH. Demy 8vo. 83 Illustrations. 420 pp. Seventy-six Tables. 1903. Price 10s. 6d.; Colonies, 11s.; Other Countries, 12s.; net.
(*See "Evaporating, Condensing, etc., Apparatus," p. 26.*)

Contents.

Chemistry of Sucrose, Lactose, Maltose, Glucose, Invert Sugar, etc.—Purchase and Analysis of Beets—Treatment of Beets—Diffusion—Filtration—Concentration—Evaporation—Sugar Cane: Cultivation—Milling—Diffusion—Sugar Refining—Analysis of Raw Sugars—Chemistry of Molasses, etc.

Bibliography.

CLASSIFIED GUIDE TO TECHNICAL AND COMMERCIAL BOOKS. Compiled by EDGAR GREENWOOD. Demy 8vo. 250 pp. 1904. Being a Subject-list of the Principal British and American Books in print; giving Title, Author, Size, Date, Publisher and Price. [*In the press.*]

Contents.

Agriculture — Architecture — Art — Book Production—Building—Chemicals—Commercial—Electricity—Engineering—Farming—Gardening—Glass—Hygiene—Legal (not pure Law)—Metallurgy — Mining — Military — Music — Naval — Oils — Paints — Photography — Physical Training—Plumbing — Pottery — Printing — Public Health — Railways — Roads — Soaps — Surveying—Teaching—Textile—Veterinary—Water—Index, etc., etc.

Textile Soaps.

TEXTILE SOAPS AND OILS. Handbook on the Preparation, Properties and Analysis of the Soaps and Oils used in Textile Manufacturing, Dyeing and Printing. By GEORGE H. HURST, F.C.S. Crown 8vo. 195 pp. 1904. Price 5s.; India and Colonies, 5s. 6d.; Other Countries, 6s.; strictly net.

Contents.

Methods of Making Soaps—Hard Soap—Soft Soap. **Special Textile Soaps**—Wool Soaps—Calico Printers' Soaps—Dyers' Soaps. **Relation of Soap to Water for Industrial Purposes**—Treating Waste Soap Liquors—Boiled Off Liquor—Calico Printers and Dyers' Soap Liquors—**Soap Analysis—Fat in Soap.**
ANIMAL AND VEGETABLE OILS AND FATS—Tallow—Lard—Bone Grease—Tallow Oil. **Vegetable Soap, Oils and Fats**—Palm Oil—Coco-nut Oil—Olive Oil—Cotton-seed Oil—Linseed Oil—Castor Oil—Corn Oil—Whale Oil or Train Oil—Repe Oil.
GLYCERINE.
TEXTILE OILS—Oleic Acid—Blended Wool Oils—Oils for Cotton Dyeing, Printing and Finishing—Turkey Red Oil—Alizarine Oil—Oleine—Oxy Turkey Red Oils—Soluble Oil—Analysis of Turkey Red Oil—Finisher's Soluble Oil—Finisher's Soap Softening—Testing and Adulteration of Oils—Index.

SCOTT, GREENWOOD & Co. will forward any of the above Books, *post free*, upon receipt of remittance at the published price, or they can be obtained through all Booksellers.

Full List of Contents of any of the books will be sent on application, and particulars of books in the press will be sent when ready to persons sending name and address.

SCOTT, GREENWOOD & ·CO.,
Technical Book Publishers,
19 LUDGATE HILL, LONDON, E.C.

Ingram Content Group UK Ltd.
Milton Keynes UK
UKHW021103210423
420563UK00005B/254